ANDERSONIAN LIBRARY
WITHDRAWN
FROM
LIBRARY
STOCK
UNIVERSITY OF STRATHCLYDE

D1556741

Array Signal Processing

PRENTICE-HALL SIGNAL PROCESSING SERIES

Alan V. Oppenheim, Editor

ANDREWS and HUNT *Digital Image Restoration*
BRIGHAM *The Fast Fourier Transform*
BURDIC *Underwater Acoustic System Analysis*
CASTLEMAN *Digital Image Processing*
CROCHIERE and RABINER *Multirate Digital Signal Processing*
DUDGEON and MERSEREAU *Multidimensional Digital Signal Processing*
HAMMING *Digital Filters, 2e*
HAYKIN, ED. *Array Signal Processing*
LEA, ED. *Trends in Speech Recognition*
LIM, ED. *Speech Enhancement*
MCCLELLAN and RADER *Number Theory in Digital Signal Processing*
OPPENHEIM, ED. *Applications of Digital Signal Processing*
OPPENHEIM, WILLSKY, with YOUNG *Signals and Systems*
OPPENHEIM and SCHAFER *Digital Signal Processing*
PHILLIPS and NAGLE *Digital Control System Analysis*
RABINER and GOLD *Theory and Applications of Digital Signal Processing*
RABINER and SCHAFER *Digital Processing of Speech Signals*
ROBINSON and TREITEL *Geophysical Signal Analysis*
TRIBOLET *Seismic Applications of Homomorphic Signal Processing*
WIDROW and STEARNS *Adaptive Signal Processing*

ARRAY SIGNAL PROCESSING

Simon Haykin, Editor
Director, Communications Research Laboratory
McMaster University, Hamilton, Ontario

James H. Justice
Chair in Exploration Geophysics
University of Calgary, Alberta

Norman L. Owsley
U.S. Naval Underwater Systems Center
Newport, Rhode Island

J. L. Yen
Department of Electrical Engineering
University of Toronto, Ontario

Avi C. Kak
School of Electrical Engineering
Purdue University, W. Lafayette, Indiana

PRENTICE-HALL, INC., Englewood Cliffs, New Jersey 07632

Library of Congress Cataloging in Publication Data

Main entry under title:
Array signal processing.
(Prentice-Hall signal processing series)

Includes bibliographical references.
1. Signal processing. I. Haykin, Simon S., 1931– . II. Justice, James H., 1941– .
III. Series.
TK5102.5.A73 1985 621.38'043 84-6974
ISBN 0-13-046482-1

Editorial/production supervision and
interior design: Tom Aloisi
Cover design: George Cornell
Manufacturing buyer: Tony Caruso

Printed in the United States of America

10 9 8 7 6 5 4 3 2 1

ISBN 0-13-046482-1

PRENTICE-HALL INTERNATIONAL, INC., *London*
PRENTICE-HALL OF AUSTRALIA PTY. LIMITED, *Sydney*
EDITORA PRENTICE-HALL DO BRASIL, LTDA., *Rio de Janeiro*
PRENTICE-HALL CANADA INC., *Toronto*
PRENTICE-HALL OF INDIA PRIVATE LIMITED, *New Delhi*
PRENTICE-HALL OF JAPAN, INC., *Tokyo*
PRENTICE-HALL OF SOUTHEAST ASIA PTE. LTD., *Singapore*
WHITEHALL BOOKS LIMITED, *Wellington, New Zealand*

Contents

Preface ix

1 INTRODUCTION: S. HAYKIN

1.1 Array Processing 1
1.2 Wave Propagation 3
References 5

2 ARRAY PROCESSING IN EXPLORATION SEISMOLOGY: J. H. JUSTICE 6

2.1 Introduction 6
2.2 The Seismic Experiment 8
2.3 Wave Propagation and Models for the Seismic Experiment 14
2.4 Seismic Exploration Objectives 25
2.5 Identifying Events on Seismic Data 29
2.6 Seismic Data Acquisition 31
2.7 Data Processing 43
2.8 Spectral Estimation 46
2.9 Multidimensional and Multichannel Filtering 59

2.10 Velocity Filtering 63
2.11 Polarization and Dispersion Analysis 68
2.12 Deconvolution 73
2.13 Deghosting Filters 75
2.14 Velocity Analysis 77
2.15 Stacking Filters and Multiple Suppression 80
2.16 Migration 84
2.17 Additional Topics in Seismic Array Processing 91
2.18 Summary 97
References and Bibliography 99

3 SONAR ARRAY PROCESSING: N. L. OWSLEY 115

3.1 Introduction 115
3.2 Sonar Optimum Beamforming: Is It Necessary? 119
3.3 A Sonar Array Optimum Beamformer 123
3.4 Minimum Complexity Implementation of an Optimum Beamformer 132
3.5 Signal Models and Noise Cancellation Performance 137
3.6 Sonar Array Processor Realization 145
3.7 Implementation Considerations 155
3.8 Sonar Array Processing Examples 157
3.9 High-Wavenumber Noise 161
3.10 High-Resolution Array Processing 164
3.11 Resolution Performance 173
3.12 Two-Dimensional, Space–Time Fourier Transform 185
3.13 Summary 187
References 188

4 RADAR ARRAY PROCESSING FOR ANGLE OF ARRIVAL ESTIMATION: S. HAYKIN 194

4.1 Introduction 194
4.2 Basic Signal and Noise Models 195
4.3 Multipath Model 200
4.4 The Fourier Method 205
4.5 Maximum-Likelihood Estimation 208
4.6 The Maximum-Likelihood Receiver for Symmetric Multipath 208
4.7 The Maximum-Likelihood Receiver: General Case 225
4.8 An Adaptive Antenna with a Calibration Curve: Symmetric Multipath 234
4.9 Experimental Results 239
4.10 Parametric Spectral Estimation Methods 242

4.11 Linear Prediction 243
4.12 Autoregressive/Maximum-Entropy Spectral Analysis 249
4.13 The Burg Technique 252
4.14 The Forward-Backward Linear Prediction Method 262
4.15 Limitations of the FBLP Method 267
4.16 Eigenvector Representation of Prediction Filter 267
4.17 Special Case: Noiseless Data 270
4.18 The Modified FBLP Method 275
4.19 Computer Simulation Results 278
4.20 Operation in a Diffuse Multipath Environment 286
4.21 Other Methods 287
4.22 Summary 288
Acknowledgments 289
References 289

5 IMAGE RECONSTRUCTION IN SYNTHESIS RADIO TELESCOPE ARRAYS: J. L. YEN 293

5.1 Introduction 293
5.2 Wave Fields of Radio Sources 296
5.3 Measurement of Mutual Coherence and Mutual Spectral Density 300
5.4 Array Geometry and Rotational Sampling of Baseline Space 308
5.5 Image Reconstruction by Fourier Inversion 320
5.6 Image Restoration by the Method "CLEAN" 330
5.7 Phase and Amplitude Errors: Self-Calibration of Visibilities in Image Restoration 338
5.8 Maximum-Entropy Image Restoration 342
5.9 Summary 345
References 346

6 TOMOGRAPHIC IMAGING WITH DIFFRACTING AND NONDIFFRACTING SOURCES: A. C. KAK 351

6.1 Introduction 351
6.2 Some Applications of Tomographic Imaging 354
6.3 Definitions and Theoretical Preliminaries 357
6.4 Two Fundamental Theorems for Tomographic Imaging 364
6.5 Interpolation and A Filtered-Backpropagation Algorithm for Diffracting Sources 374
6.6 Filtered-Backprojection Algorithms for Nondiffracting Sources 390
6.7 Algebraic Reconstruction Algorithms 408

6.8 Bibliographical Notes 419
6.9 Summary 422
Acknowledgments 422
Bibliography and References 423

Index 429

Preface

This is the first book to be devoted completely to array signal processing, a subject that has become increasingly important in recent years. The book consists of six chapters. Chapter 1, which is introductory, reviews some basic concepts in wave propagation. The remaining five chapters deal with the theory and applications of array signal processing in (a) exploration seismology, (b) passive sonar, (c) radar, (d) radio astronomy, and (e) tomographic imaging. The various chapters of the book are self-contained.

The book is written by a team of five active researchers, who are specialists in the individual fields covered by the pertinent chapters.

Much of the material covered in the book has not appeared in book form before. It is hoped that it will be found useful by researchers in array signal processing, its theory and applications, and to newcomers to the subject. Also, by bringing the various applications of array signal processing under one cover, it is hoped that the book will lead to further cross-fertilization among the various disciplines of interest.

Simon Haykin

Array Signal Processing

1

SIMON HAYKIN
Communications Research Laboratory
McMaster University, Hamilton, Ontario

Introduction

1.1 ARRAY PROCESSING

Array processing deals with the processing of signals carried by propagating wave phenomena [1]. The received signal is obtained by means of an *array* of *sensors* located at different points in space in the field of interest. The aim of array processing is to extract useful characteristics of the received signal field (e.g., its signature, direction, speed of propagation).

The sources of energy responsible for illuminating the array may assume a variety of different forms. They may be *noncoherent* (i.e., independent of each other) or *coherently* related to each other. Equally, as seen from the location of the array, the radiation may be from diffused media and therefore distributed in nature, or it may be from isolated sources of finite angular extent.

The array itself takes on a variety of different geometries depending on the application of interest [2, 3]. The most commonly used configuration is the *linear array*, in which the sensors (all of a common type) are uniformly spaced along a straight line. Another common configuration is a *planar array*, in which the sensors form a rectangular grid or lie on concentric circles.

In this book we will study theoretical aspects of array signal processing and its application in exploration seismology, sonar, radar, radio astronomy, and tomography.

In *exploration seismology*, array processing is used to unravel the physical characteristics of a limited region of the interior of the earth which may have potential for trapping commercial quantities of hydrocarbons. In particular, we use it to image the interior of the object from some finite aperture on the surface, with the measurements being made at a finite number of points within the limited aperture. The earth tends to act like an elastic medium for the propagation of acoustic energy. Accordingly, for the source of acoustic energy we use a *shot* (e.g., a stick of dynamite) that applies an impulse to the earth, and to record the received signal we use a *geophone*. Typically, a number of shots are laid out at equally spaced intervals along a surveyed line, and geophones are located at equally spaced intervals on one or both sides of each shot. The resultant geophone outputs are due to signals reflected, diffracted, or refracted back to the earth surface from the original source of disturbance.

In *passive, listening-only sonar*, the received signal is externally generated, and the primary requirement of array processing is to estimate both the temporal and spatial structure of the received signal field. The array sensors consist of sound pressure-sensing electromechanical transducers known as *hydrophones*, which are immersed in the underwater medium. Basically, the processing applied to the sensor outputs involves some form of spectral and/or wavenumber analysis so as to determine a directional map of the background sound power, with emphasis on the detection of signals that are characterized by low signal-to-noise ratios and extended duration.

In *radar* array processing, a transmitting antenna is used to floodlight the environment surrounding the radar site, and a receiving array of *antenna* elements is used to listen to the radar returns caused by reflections from targets located in the path of the propagating electromagnetic wave. Here again, we may use array processing to estimate the wavenumber power spectrum of the received signal, with emphasis on spatial resolution.

In *radio astronomy*, the interest is in radio emission from celestial sources. The emission, depending on the radiation mechanism and the state of the emitting region, shows broad continuum spectral features, narrow band, or absorption line structures. The arrays used here consist of tens of *antenna* elements that extend from hundreds of meters to nearly the diameter of the earth. The requirement is to use array processing for *image reconstruction* of unpolarized or partially polarized continuum and spectral line radio sources, with emphasis on resolution, ambiguity, and dynamic range of the reconstructed maps.

In *tomography*, array processing is used to obtain *cross-sectional images* of objects from either transmission or reflection data. In most cases, the object is illuminated from many different directions either sequentially or simultaneously, and the image is reconstructed from data collected either in transmission or reflection. The most spectacular success of tomography has thus far been in medical imaging with x-rays. There is also active interest today in extending tomographic imaging to ultrasound and microwaves for use in medical imaging, seismic exploration, and nondestructive testing.

Before going on to a detailed discussion of these topics, one by one, we will complete this introductory chapter by reviewing some concepts in wave propagation [4] which are basic to the development of array signal processing theory.

1.2 WAVE PROPAGATION

Whenever a driving force is coupled to an open medium, we find that *traveling waves* are generated. They are so-called because they travel away from the source of the disturbance. Traveling waves have the important property that they transport energy, the form of which depends on the physical nature of the driving force. Also, in the *far field* or a large distance away from the source, the waves become essentially plane. As a matter of fact, the *plane wave* is probably the most common of all the different forms of wave propagation.

Suppose we have a harmonic traveling wave that propagates through a homogeneous dispersive medium in the direction of the unit vector $\hat{\mathbf{z}}$, along the z-axis of the Cartesian coordinate system. At the plane defined by a fixed value of z, the wave function has the time dependence

$$g(t, z) = A \cos [2\pi(ft - \nu z)] \tag{1.1}$$

where A is the amplitude, f the frequency, and t the time. The parameter ν is called the *wavenumber*. The physical meaning of the wavenumber ν is that in a distance z, measured along the propagation direction $\hat{\mathbf{z}}$, the phase accumulates by $2\pi\nu z$ radians. (Frequently, the wavenumber is defined as $k = 2\pi\nu$, so that the phase accumulated in a distance z along the propagation direction $\hat{\mathbf{z}}$ equals kz radians [4].)

Let $\mathbf{r}$ denote a point in space as measured from the origin of the Cartesian coordinate system. The plane $z =$ constant is described by the plane $z = \hat{\mathbf{z}} \cdot \mathbf{r} =$ constant, where $\hat{\mathbf{z}} \cdot \mathbf{r}$ is the *dot product* of the unit vector $\hat{\mathbf{z}}$ along the propagation direction and the vector $\mathbf{r}$. Thus we may express the quantity $-\nu z$ in Eq. (1.1) as

$$\begin{aligned} -\nu z &= -\nu(\hat{\mathbf{z}} \cdot \mathbf{r}) \\ &= \boldsymbol{\nu} \cdot \mathbf{r} \end{aligned} \tag{1.2}$$

where

$$\boldsymbol{\nu} = -\nu\hat{\mathbf{z}} \tag{1.3}$$

We refer to $\boldsymbol{\nu}$ as the *wavenumber vector*. Note that this vector points in the opposite direction to the direction of propagation $\hat{\mathbf{z}}$.

Using this new notation, we may now express the traveling wave of Eq. (1.1) in the equivalent form

$$g(t, \mathbf{r}) = A \cos [2\pi(ft + \boldsymbol{\nu} \cdot \mathbf{r})] \tag{1.4}$$

The argument of the sinusoidal wave function is called the *phase* $\phi(t, z)$; that is,

$$\begin{aligned} \phi(t, z) &= 2\pi(ft - \nu z) \\ &= 2\pi(ft + \boldsymbol{\nu} \cdot \mathbf{r}) \end{aligned} \tag{1.5}$$

At fixed time t, the points with equal phase ϕ define a plane called a *wavefront*, which is described by

$$\begin{aligned} d\phi &= 2\pi(f\,dt + \mathbf{v}\cdot d\mathbf{r}) \\ &= 2\pi(0 + \mathbf{v}\cdot d\mathbf{r}) \qquad \text{at fixed time} \\ &= 0 \qquad \text{only if } d\mathbf{r} \text{ is perpendicular to } \mathbf{v} \end{aligned} \tag{1.6}$$

Accordingly, at fixed time t the phase will have the same value at all points reached by adding up vectors $d\mathbf{r}$ that are perpendicular to the vector wavenumber $\mathbf{v}$. That is, the phase increment $d\phi$ equals zero in moving from one such point to another, which means moving about in a plane. It is for this reason that such a wave is referred to as a plane wave.

Another quantity used in describing wave propagation is the *phase velocity*, which is defined as dz/dt for a fixed phase $\phi(t, z)$. Using Eq. (1.5), we may write

$$d\phi = 2\pi(f\,dt - v\,dz)$$

Putting $d\phi = 0$, and solving for dz/dt, we get the following formula for the phase velocity:

$$v_\phi = \frac{dz}{dt} = \frac{f}{v} \tag{1.7}$$

Thus a plane wave propagates with respect to a fixed point in space in a direction defined by the vector wavenumber $-\mathbf{v}$ and with a phase velocity or speed equal to f/v, where $v = |\mathbf{v}|$ is the magnitude of $\mathbf{v}$.

In the case of electromagnetic plane waves the electric and magnetic fields are transverse to the direction of propagation [4]. Assuming that the direction of propagation is along $\hat{\mathbf{z}}$, there are two transverse directions, $\hat{\mathbf{x}}$ and $\hat{\mathbf{y}}$, and the fields with one orientation with respect to $\hat{\mathbf{x}}$ and $\hat{\mathbf{y}}$ are independent of those with an orientation differing by 90°. Accordingly, we may have various amplitudes of the fields in each of the two transverse directions and various possible relative phases. We refer to a specific relation of the amplitudes and phases of the two independent transverse fields as a state of *polarization* [4].

All the waves that we study consist of some physical quantity whose displacement from its equilibrium value varies with both position and time. Thus for plane waves propagating along $\hat{\mathbf{z}}$, we may express the displacement as the vector

$$\mathbf{g}(z, t) = \hat{\mathbf{x}}g_x(z, t) + \hat{\mathbf{y}}g_y(z, t) + \hat{\mathbf{z}}g_z(z, t) \tag{1.8}$$

where $\hat{\mathbf{x}}$, $\hat{\mathbf{y}}$, and $\hat{\mathbf{z}}$ are unit vectors along the x-, y-, and z-axes, respectively. In the case of waves with transverse polarization, the vector $\mathbf{g}(z, t)$ has only x- and y-components.

It is important to realize, however, that the concept of polarization applies only to waves that have at least two independent "polarization directions." For sound waves in air, for example, the displacement is along the propagation direction. These waves are called *longitudinal waves*; we do not ordinarily say that these waves are longitudinally polarized. Rather, we reserve the term "polarization" to describe waves for which there are at least two alternative polarization directions. In the case of sound waves in solids, there are one longitudinal and two transverse polarization directions available. Accordingly, in this case we may have longitudinally polarized waves or two different transversely polarized waves or a superposition of all three polarizations.

REFERENCES

1 D. E. Dudgeon, "Fundamentals of Digital Array Processing," Proc. IEEE, Vol. 65, pp. 898–904, June 1977.

2 R. E. Collins and F. J. Zucker, *Antenna Theory*, Part I, McGraw-Hill, New York, 1969.

3 R. S. Elliott, *Antenna Theory and Design*, Prentice-Hall, Englewood Cliffs, N. J., 1981.

4 F. S. Crawford, Jr., *Waves*, Berkeley Physics Course, Vol. 3, McGraw-Hill, New York, 1968.

2

JAMES H. JUSTICE
Chair in Exploration Geophysics
University of Calgary, Alberta

Array Processing in Exploration Seismology

2.1 INTRODUCTION

Seismic signal processing represents an extremely important area of application in array processing, and it is an area in which development of array processing procedures has been underway for some time. Seismic data acquisition as practiced today is inherently multidimensional in character; that is, the seismic experiment attempts to record the multidimensional character of acoustic wave fields which have both temporal and spatial characteristics. As a result, much of seismic signal processing is concerned with utilizing the full multidimensional character of the recorded wave field, and so, in one way or another, may be classified as array processing. While there are important exceptions to this rule, it is generally true, and therefore many of the procedures which have been developed for seismic data acquisition and processing will be discussed here.

Because seismic exploration holds a great fascination, but is not widely understood, we shall endeavor to provide necessary background information to create a framework and a context for these discussions. In particular, we shall examine wave propagation in a nonhomogeneous medium, the foundation for the seismic method, and from this we shall consider the origin of some of the models which form the basis for many of our array processing procedures.

The ultimate goal of the exploration process is to reconstruct as accurately as possible an image of the subsurface in which both structure and physical properties are delineated. We shall therefore enter into a brief discussion of the ultimate objectives of seismic exploration, that is, a discussion of hydrocarbon traps and trapping mechanisms which we hope to identify when our processed data is interpreted. The drill bit becomes the final arbiter in the determination of our success or failure in this process.

The exploration process may be broken down into three large categories once a prospective area has been identified. These are: data acquisition, data processing, and interpretation. We shall devote most of our discussion to the first two, the latter belonging to that gray area between art and science where only understanding, past experience, and logical inference can make the difference between success and failure. It is quite possible that some day this process will be well-explained or enhanced by techniques of artificial intelligence, unknown or in development today, but for now we leave this realm to the skill of the interpreter and shall only try to suggest what he looks for.

Finally, we shall consider many of the techniques which have been developed or are being developed to provide the pieces of the puzzle which must be fitted together by the interpreter.

The field of exploration seismology offers especially fertile ground to the signal processor since most areas of inquiry in signal processing seem to have some application to the problems in exploration seismology. These may range from the popular areas of spectral estimation to parameter estimation or multidimensional filtering, to image processing or antenna design. These would include the areas of adaptive signal processing, noise cancellation, time-delay estimation, and many other active areas of inquiry in signal processing. Indeed, the concept of holography as applied to acoustic wave fields, and even tomography, have been considered in exploration seismology. Thus it seems safe to predict that the areas of pattern recognition and artificial intelligence will find more and more application in exploration seismology as time goes on.

The signal processing procedures that have been developed for exploration seismology have one goal—to solve the geophysical inverse problem in which an acoustic signal is generated at the surface of the earth and is propagated through the earth, undergoing reflection where changes in acoustic impedance are encountered. The reflected signal is propagated back to the surface of the earth, suffering transmission losses, frequency-selective attenuation, and dispersion as it propagates. It is further contaminated by various types of noise both random and coherent, then modified by the response of the recording system, and mixed with near-field effects of the source, including surface and near-surface modes of propagation as well as a variety of other signal contaminants. From this multidimensional recorded data set, obtained by successive reflections within the medium of interest, we must somehow, in the final analysis, correctly infer the structure of that medium. It is interesting to note that today work is underway to provide a direct solution to this inverse imaging process. Although we still have a long way to go, encouraging results are being reported. It is quite possible that within the next few decades we will see significant changes in the ways we process seismic exploration data.

Looking beyond what has been accomplished in the past and what is being done at present, it seems safe to speculate that whereas we have been content before to attempt a reconstruction of subsurface structure of the earth (which is more or less equivalent to saying that we have been content to infer the velocity profiles of the earth), the future seems to hold a greater potential and a greater challenge. Where past processing methods have been largely based on attempting to determine times of arrival of events from which velocity estimates, and thereby structure in depth of the earth can be inferred, the future tells us that the recorded wave field is carrying much more information than simple time of arrival, and that contained within these additional parameters, relating to the amplitude and phase characteristics of the recorded signal, is possibly information sufficient to make at least some judgment concerning the actual lithologic environment within the earth (rock type, porosity, saturation, etc.). Although it is true today that we can only poorly understand the mechanisms of attenuation and dispersion that operate within the earth, we are seeing increased activity in a relatively new field of petrophysics or rock properties intended to answer the question of how the various parameters of lithology express themselves in the seismic signature. With success, we may eventually be able to read the full seismic message that is sent to us by the propagating medium and infer with some sense of completeness a full knowledge of its structure. When we add these new parameters and variables to the seismic problem, then that which was already multidimensional becomes even more so and emphasizes the need to extend many of our mathematical tools and processes to higher-dimensional settings, so that they will be available as we increase our demand for these tools in order to deal with the increasing complexities of the seismic inverse problem.

In summary, exploration seismology as practiced today is indeed a multidimensional problem employing multidimensional processes in its (approximate) solution. Looking ahead to the future, we can predict that more sophisticated multidimensional processing, based on a large number of variables or parameters, will become increasingly important and will result in much more information eventually being derived from the seismic experiment. Understanding that our discussions can never be complete, we now go on to consider the concepts and methods which are a part of array processing in exploration seismology.

2.2 THE SEISMIC EXPERIMENT

Before proceeding with a detailed discussion of the various array processing procedures that are applied to seismic data in any of the many forms in which it may be displayed, it may be of value to review the procedures that are in use today to generate multidimensional data sets and to give some justification for these procedures. Let us recall that our primary objective is to unravel the complexities of structure and even the physical characteristics of that structure where possible, contained within limited regions of the interior of the earth which may have the potential for trapping commercial quantities of hydrocarbons. Mathematical physics tells us that if we wish to determine the structure of the interior of a solid by making measurements on its surface, then if we know the *Green's function* for the medium, and if we can conduct measurement at all points on the surface obtained by impulsing the object with an infinitely broad frequency source located outside the body, we may in fact completely and accurately infer the interior structure of the body [210]. The seismic problem is very much like this; we have a region in the earth which we suspect to contain commercial quantities of hydrocarbon and our problem is to determine its structure and probable physical properties while being restricted to make our measurements on the surface of the body. In the seismic problem, however, we have many restrictions that render the mathematical physics solution inapplicable. In particular, we cannot make measurements on an entire surface surrounding the region of interest. This means in fact that we must image the interior of the object from some finite aperture on the surface that does not include an entire surface surrounding the region of interest, and further, we are able to make measurements at only a finite number of points within this limited aperture. In addition, we do not know the appropriate Green's function, although we may approximate it in various ways, and finally, in the seismic experiment the source is actually interior to the region of interest, which means that we cannot neglect the volume integral whose data are unavailable to us. As a result, there is no simple direct solution to the seismic inversion problem, to date, which parallels the elegant solution to the mathematical physics problem alluded to, although progress is being made in this direction. Instead, we must resort to a wide variety of methods, each designed to fill in one piece of a very complex puzzle, with the hope that taken together they may yield sufficient information to at least justify or preclude the possibility of drilling. Indeed, we are only partially successful in solving this inverse problem, as is evident from the large number of dry holes drilled every year in the search for commercial quantities of hydrocarbons (over 33,000 dry holes were drilled in the United States in 1981 [14]).

Let us now consider the *seismic imaging problem.* To begin with, we must have a source of illumination for the subsurface of the earth, that is, some signal which can be propagated into the earth and will return to us carrying information about the subsurface structure. Since we are constrained to remain on the surface of the earth and since our targets typically occupy very limited regions in the near surface ("near surface" may mean the first few tens of thousands feet of the earth's crust), it follows that we will be able to record only signals that have been reflected, diffracted, or refracted to us from our surface source of illumination. That is, in the absence of a bore hole penetrating deeply into the earth, we will not receive transmitted signals which have passed through the medium directly without reflection. To be successful, then, it is important that the earth will, in fact, return a signal to us which is propagated into it from the surface.

The earth tends to act very much like an elastic medium for the propagation of acoustic energy. Since the velocity of acoustic wave propagation tends to increase with depth in the earth due to increasing compaction and overburden pressure and since acoustic impedance within this elastic medium is a function of density and velocity, it follows that we may expect to encounter changes in acoustic impedance as we propagate an acoustic wave through this medium. In particular, when we encounter changes in the lithologic sequence within the earth, we may expect corresponding changes in density and velocity, which means that we may expect changes in the acoustic impedance. At these boundaries where the acoustic impedance may be expected to undergo an abrupt or discontinuous change, we may expect to obtain a reflection of the acoustic energy propagating downward and this energy will, ultimately, return to us at least in part, at the surface of the earth.

The first question that would be natural to ask then is: What kind of acoustic energy is most likely to propagate deeply into the earth, at least in the order of tens of thousands of feet, and return to us with sufficient energy to be detectable at the surface? The answer is that acoustic energy in the range of a few hertz to several hundred hertz may be expected to fulfill these requirements, with the majority of reflected seismic energy being concentrated at around 30 Hz.

Let us now conceive of the simplest seismic experiment that we may carry out in the interest of achieving our goal. We shall need a source of energy which spans the range that is most likely to propagate through the earth with sufficient energy to give us a recordable signal, and we shall need a sensor to obtain the record of the energy returning to us from the subsurface. These two devices are typically referred to as the *shot* and the *geophone* (or "phone"), respectively, and our problem now is to impulse the earth with a source of acoustic energy from the shot, which might be, for example, a stick of dynamite, and then to record the signal that comes back to us at the phone. Generally, we record several seconds of signal at the phone in order to observe the energy reflected from the portion of the near surface that is of interest to us.

If this experiment were carried out as described, the question is: Could we conceivably reconstruct the subsurface structure of the earth from this information? In a very limited sense the answer might be yes. If the energy had traveled more or less vertically through the earth before being successively reflected due to changes in acoustic impedance that occurred between the idealized lithologic boundaries in the earth, our ideal seismic trace, the recording obtained by our phone, would consist of a repetition of our source pulse displaced in time and diminished in amplitude as we progress down the trace. The positions of these repetitions of the source pulse on the seismic trace should correspond in some way to the positions of the reflected boundaries within the earth.

Here we encounter our first problem in seismic exploration; that is, to reconstruct the subsurface structure of the earth in terms of depth, we must be able to convert our time record to a depth record. Simple physics tells us that to accomplish this feat we must have a knowledge of the velocity structure of the earth in the region through which our signal propagated. Our simple experiment, however, leaves us little hope of determining this velocity structure since we have only the information of time of arrival of our pulses and we have no knowledge of either the depth or the velocity structure that we seek. The situation is not as simple as this. It would be nice if our signal represented only a sequence of pulses reflected from each of the reflecting boundaries within the earth; indeed, it would be nice if all these reflections were actually reflections. In general, this is not so. Although we may in fact be recording reflections, there is no reason to believe that these reflections are what we call *primary reflections*, that is, those that have traversed vertically downward until encountering some reflective boundary and then have propagated vertically upward until they reach our phone.

Each time a reflecting boundary is encountered either while the wave is propagating downward or upward, both a reflection and a transmission, or equivalently, a partitioning of energy, into an upgoing and a downgoing wave will be generated. This process multiplies as more and more interfaces are encountered by more and more reflected waves from the original source and much of this energy returns to us ultimately at the surface to be recorded by our phone. Events that have undergone more than one reflection in passes through the earth are referred to as *multiple events*, and it is much more difficult to determine the structure of reflecting horizons within the earth when our recorded signal may contain many multiple reflections (see Fig. 2.1).

In addition to the multiple reflections, we may expect that energy has propagated not downward, into the earth and back to us, but indeed may have propagated directly from the shot to the phone over the surface of the earth and this energy is known as a *surface wave*. Surface waves typically carry large amounts of energy and propagate with a very low apparent velocity. They contaminate our recordings of the much weaker reflections returning to us from depth and tend to obscure these reflections so that they become impossible to see.

In addition to these effects, the phone is sensitive to all acoustic energy, and while we thought it was recording reflected energy from the interior of the earth, it may very well have been listening to the song of the June bug perched on top of the case, the tractor plowing in the nearby field, or the pleasantness of the wind rustling the leaves in the nearby tree. Our signal is contaminated by a potentially wide

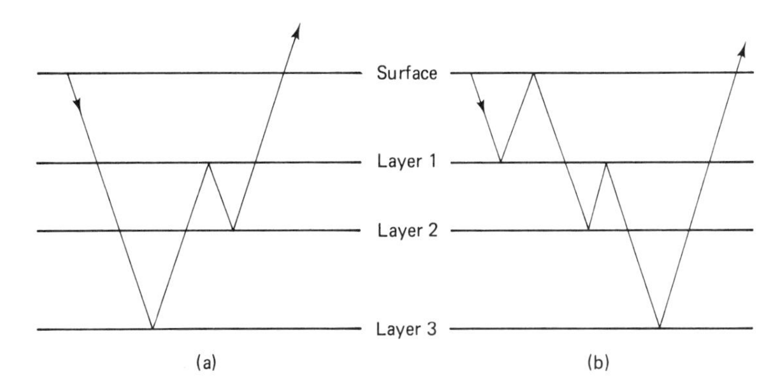

Figure 2.1 Examples of multiple reflections.

variety of coherent and incoherent noise sources which also tend to obscure the relatively weak signal returning to us from the interior of the earth. The primary signal desired may be further obscured by edge diffractions from buried faults, for example, and waveguide phenomena which may scatter energy back to us from what appear to be secondary sources.

Finally, although the effect is often not considered to be significant, the earth behaves more or less as an elastic medium. This means that *compressional or shear waves* striking a boundary at other than normal incidence will generate reflected and transmitted compressional and shear waves which travel at different velocities, and these waves generate further conversions and multiples at each interface encountered.

The question remains: Given this jumble of information contained in our recorded seismogram, can we infer the structure of the earth in the region of interest? At this point the problem seems to become a bit bewildering and we may begin to appreciate why the seismic experiment is carried out as a multidimensional experiment employing many geophones and many shot-point locations with the possibility of using an array of either or both of the shot points and geophones. That is, we must add dimensionality to have any hope of solving it. Since the latter part of this presentation will be devoted to the methods that utilize the multidimensionality of our data set in order to extract the information that we seek (in spite of the many complications that are going to be recorded with our data), at this point, let us simply consider the way the seismic experiment is actually carried out today so that we may understand the nature of our data sets.

The seismic experiment as it is carried out today may take one of many forms, but basically all have the characteristic that a number of shot points will be laid out, typically along a *surveyed line, usually at equally spaced intervals.* Geophone locations will be laid out on one or both sides of each shot point with equally spaced intervals, the geophone groups extending to some prescribed distance from the shot. Each geophone group may, in fact, itself consist of a large number of geophones connected in parallel or series, forming an array for the purpose of discriminating against certain types of wave motion.

If geophones are laid out on both sides of the shot point, we refer to this as a *split-spread configuration.* If, on the other hand, the geophones are laid out on only one side of the shot point, we refer to this as an *end-on configuration.* In a typical seismic exploration experiment, shot points and receivers will be moved along surveyed lines at equally spaced intervals which may be on the order of a few hundred feet for distances of several miles or even tens of miles.

While we have described *linear arrays*, that is, those which are laid out on a long surveyed line, they are indeed not the only arrays used in seismic exploration, although they are probably the most commonly used arrays today. It is also possible to lay out recording geometries in both spatial dimensions, resulting in what is called a *three-dimensional survey* or in more limited cases a *wide-line survey.* Perhaps the most common type of recording geometry in use today is the *common depth point* (CDP) or *roll-along geometry.* In this method of seismic surveying, an idealized

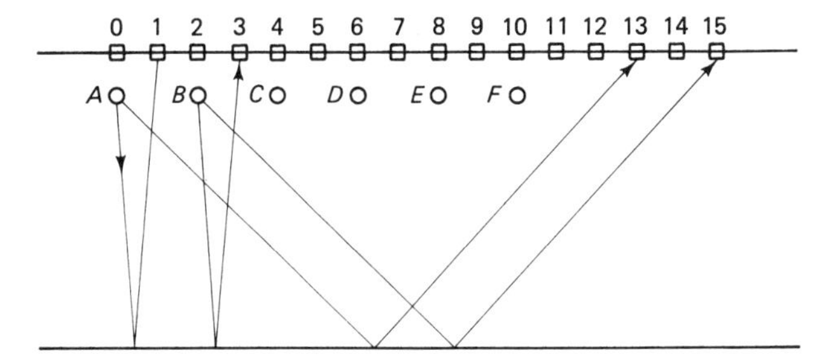

Figure 2.2 CDP method for seismic data acquisition. Shot point *A* is recorded on phones 1 to 24; shot point *B* is recorded on phones 3 to 26, for example.

reflecting point half way between the shot and the receiver, known as the *common depth point*, *common reflection point*, or *common midpoint*, is imaged many times between many combinations of shots and receivers. In fact, each common midpoint may be imaged as many as 24 times, depending on the manner in which the survey is carried out. The reader is referred to Fig. 2.2 for a diagram showing the CDP method of seismic surveying. Once a CDP survey has been carried out there are a number of ways in which the traces can be "gathered" or displayed as groups. Some of the more common ways of displaying the data are the common shot gather, the common offset gather, the common geophone gather, and the common depth point gather. These types of gathers are illustrated in Fig. 2.3, and each has its own particular uses in unraveling the complex structure of the subsurface from the recorded information we have obtained.

At this point we should have a reasonably clear idea of how seismic field data are generated and how these multidimensional data sets may be displayed. Rather than launch into a further detailed discussion of data acquisition at this point, let us defer that discussion until later and go on to consider some of the mathematical models on which we base many of our processing procedures.

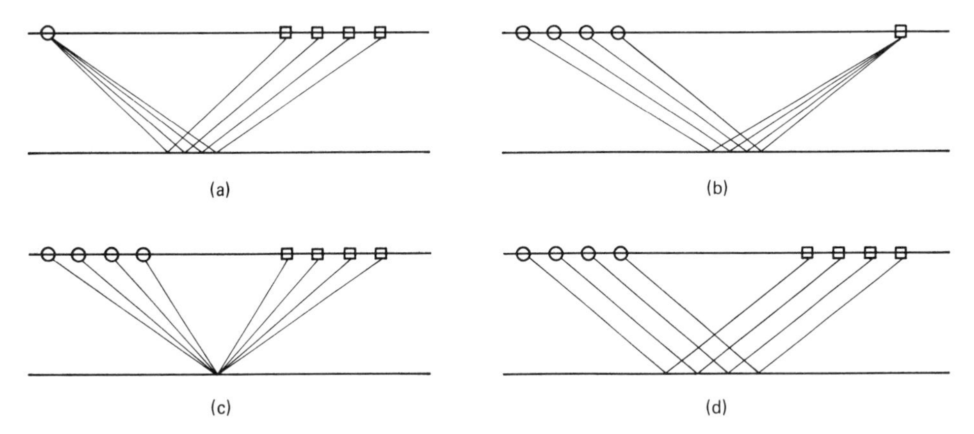

Figure 2.3 (a) Common shot point gather; (b) common receiver gather; (c) common midpoint gather (common depth point gather); (d) common offset gather.

2.3 WAVE PROPAGATION AND MODELS FOR THE SEISMIC EXPERIMENT

To gain a better understanding of the true nature and complexity of seismic data, it is beneficial at this point to consider some of the characteristics of acoustic wave propagation in an inhomogeneous elastic medium. Although it would be improper to claim that we have gained a complete understanding of wave propagation in the earth or that we can accurately write the equations describing the propagation process, it does seem to be true that in some gross sense the earth behaves, or at least can be modeled, as an elastic medium which, for small displacements, may be assumed to satisfy the condition for linear elasticity.

Although it would be nice to treat the problem of wave propagation in an elastic medium in its full generality, it is uncommon in seismology to deal with the problem at this level. In particular, rather than assuming that the earth is an inhomogeneous elastic medium, we generally make the assumption that the earth is partitioned into regions or layers within which the elastic parameters remain constant but may differ as we progress from one layer to another. In practice, we may frequently treat the problem as if it were two-dimensional rather than three-dimensional and we frequently base our processing methods on the plane-wave solution, and indeed some of our seismic processing techniques are designed to reduce the problem to a form in which this assumption becomes approximately valid. In the limiting case, we may go so far as to assume that the earth consists of a series of flat layers stacked one on top of the other, the elastic parameters being constant within each layer, and we assume plane-wave propagation in a direction normal to the boundaries between the layers.

In summary, this says that geophysical data processing procedures may be based on a variety of assumptions, from the very simplest to those which are quite complex. Indeed, we may even attempt to include the effects of attenuation and to associate these effects with various lithologic properties, such as density, porosity, saturation, overburden pressure, temperature, and so on, with the understanding that many of these relationships are, at best, poorly understood.

Let us begin our analysis of wave propagation by assuming that the medium of interest satisfies the conditions of linear elasticity (which is approximately true for the small displacements encountered in seismic wave propagation) and that the medium is homogeneous, that is, the elastic parameters are assumed to be constant throughout the medium. To derive the equation for wave propagation in this medium it is necessary to understand the nature of the forces acting on a small element of volume and to relate these to the resulting strains which represent the response of the elastic medium to these stresses. The wave equation is obtained by first applying *Newton's* law. We then substitute for the stress term, which defines the force in Newton's equation, the corresponding elastic relationship, which relates stresses to strains in the medium to obtain the elastic wave equation. An outline of this process follows.

The Elastic Wave Equation

The equations of motion for a small volume element of an elastic solid under stress are given by

$$\rho \frac{\partial^2 u_\kappa}{\partial t^2} = \sum_{l=1}^{3} \frac{\partial \sigma_{\kappa l}}{\partial \chi_l} \tag{2.1}$$

where ρ is the density of the volume element, and $\sigma_{\kappa l}$ is the stress on a face perpendicular to the χ_κ-axis in the direction χ_l (normal or shear stress). The stresses on the remaining faces may be calculated from the Taylor expansion. Now the condition of moment equilibrium implies that $\sigma_{ij} = \sigma_{ji}$, so the tensor

$$\begin{pmatrix} \sigma_{11} & \sigma_{12} & \sigma_{13} \\ \sigma_{21} & \sigma_{22} & \sigma_{23} \\ \sigma_{31} & \sigma_{32} & \sigma_{33} \end{pmatrix}$$

is a symmetric tensor of second order. Equation (2.1) is obtained by summing all the forces (stress on each face multiplied by the area of the face) and equating the result to

$$\rho \prod_{\kappa=1}^{3} \Delta\chi_\kappa \frac{\partial^2 \mathbf{u}}{\partial t^2}$$

(mass times acceleration), where $\mathbf{u}$ is the *displacement vector* of a point within the (infinitesimal) volume element. Note that we have neglected external force fields such as gravity. This is common practice in seismology.

To complete the derivation, it only remains to relate stress to strain. Following Hooke (Hooke's law is valid for small deformations) we may assume that strain is linearly related to stress. If we align our stress with the three principal axes (shear stresses do not occur), the strains also align with these axes and we arrive at

$$\sigma_{\kappa\kappa} = 2\mu\epsilon_{\kappa\kappa} + \lambda \sum_{\kappa=1}^{3} \epsilon_{\kappa\kappa}$$

where we have assumed elastic isotropy and λ and μ are the *Lamé constants* for the medium. Finally, applying the appropriate tensor transformation rule for changing coordinates in our tensors gives (see [274])

$$\sigma_{i\kappa} = 2\mu\epsilon_{i\kappa} + \lambda\delta_{i\kappa}\theta \tag{2.2}$$

where $\delta_{i\kappa}$ is the *Kronecker delta* equal to 1 for $i = k$ and zero otherwise, and

$$\theta = \sum_{\kappa=1}^{3} \epsilon_{\kappa\kappa} = \operatorname{div} \mathbf{u}$$

since

$$\epsilon_{ij} = \frac{1}{2}\left(\frac{\partial u_j}{\partial \chi_i} + \frac{\partial u_i}{\partial \chi_j}\right)$$

If we now substitute Eq. (2.2) for the strains into Eq. (2.1), we obtain

$$\rho \frac{\partial^2 \mathbf{u}}{\partial t^2} = (\mu + \lambda) \text{ grad } \theta + \mu \nabla^2 \mathbf{u} \tag{2.3}$$

where $\mathbf{u}$ is the displacement vector. Equation (2.3) is sometimes called *Navier's equation* and is useful for seismology since it is written entirely in terms of the displacement vector $\mathbf{u}$.

It will be worthwhile, now, to examine some of the properties of Eq. (2.3). If we now take the divergence of both sides of Eq. (2.3), we obtain

$$\rho \frac{\partial^2 \theta}{\partial t^2} = (2\mu + \lambda)\nabla^2 \theta \tag{2.4}$$

which is called the *dilatational wave equation*. Its solution, θ, represents a (compressional) wave traveling with velocity

$$V_c = \left(\frac{\lambda + 2\mu}{\rho}\right)^{1/2}$$

and whose displacement is in the direction of propagation. If we take the curl of both sides of Eq. (2.3), we obtain

$$\rho \frac{\partial^2 \boldsymbol{\tau}}{\partial t^2} = \mu \nabla^2 \boldsymbol{\tau} \tag{2.5}$$

where $\boldsymbol{\tau} = \nabla \times \mathbf{u}$. The solution, $\boldsymbol{\tau}$, represents a vector (shear) wave field perpendicular to the direction of propagation with velocity of propagation given by

$$V_s = \left(\frac{\mu}{\rho}\right)^{1/2} \tag{2.6}$$

Since it would be difficult to obtain the components of the displacement, $\mathbf{u}$, from the solutions above, it is generally possible to rewrite these equations in a more general form. When $\mathbf{u}$ can be written in the form

$$\mathbf{u} = \nabla \phi - \nabla \times \boldsymbol{\psi} \tag{2.7}$$

(*Helmholtz theorem* [1]) where ϕ is a scalar field and $\boldsymbol{\psi}$ is a vector field, we find by substitution that

$$\frac{\partial^2 \phi}{\partial t^2} = \frac{2\mu + \lambda}{\rho} \nabla^2 \phi \tag{2.8}$$

$$\frac{\partial^2 \boldsymbol{\psi}}{\partial t^2} = \frac{\mu}{\rho} \nabla^2 \boldsymbol{\psi} \tag{2.9}$$

and the components of **u** can be determined directly from derivatives of these solutions by invoking Eq. (2.7).

Reviewing our derivation above, we find that we can separate the elastic wave equation into two separate modes of propagation: *compressional*, or *p*, and *shear*, or *s*, waves. Further, as long as the *Lamé parameters* λ and μ (the elastic parameters of the medium) are constant, the steps invoked in the separation are valid.

Partitioning of Energy at a Boundary

If the earth is modeled as a collection of layers separated by their boundaries, with different Lamé constants in each layer, the wave equation cannot be uncoupled across a boundary and we must expect that exchange of energy between the compressional and shear modes may occur. In addition, since the velocity of propagation (a function of the Lamé constants) and/or the density may change as we cross a boundary, we may expect a partitioning of energy between upgoing and downgoing waves. That is, energy will be divided between reflected and transmitted waves.

To simplify the discussion, we point out that *Snell's law* holds governing changes in direction of propagation at a boundary, and since this direction is dependent on velocity as well as angle of incidence, it follows that *p* and *s* waves will refract differently upon passing from one layer to another. Further, our earlier observation on uncoupling the vector (elastic) wave equation tells us that we should expect *p* waves to give rise to *s* waves, and vice versa, upon encountering a boundary.

Let us now consider in more detail the partitioning of energy that occurs between reflection and transmission at a boundary. To solve the problem, we must specify the boundary conditions which determine how the solutions on both sides of the interface match at the boundary.

We assume that a boundary between layers constitutes a "welded" contact in which both displacements and stresses are continuous across the interface. If we denote the two layers by the subscripts *A* and *B*, we require that on the boundary

$$\mathbf{u}_A = \mathbf{u}_B \tag{2.10}$$

and

$$\sigma_{k3_A} = \sigma_{k3_B}, \qquad k = 1, 2, 3 \tag{2.11}$$

where the χ_3 axis is assumed to be normal to the boundary (continuity of normal and tangential stresses on the boundary).

Now, for convenience, consider two homogeneous half-spaces welded along $\chi_3 = 0$. Let us assume that the scalar and vector potentials ϕ and ψ are represented by

$$\phi = \exp\,(\pm j\beta_p \chi_3)\,\exp\,[j(\alpha\chi_1 - \omega t)]$$

$$\psi = \exp\,(\pm j\beta_s \chi_3)\,\exp\,[j(\alpha\chi_1 - \omega t)]$$

Note that we are confining attention to the χ_1–χ_3 plane and the $\pm$ sign indicates that the wave may be upgoing (+) or downgoing (−). β_p and β_s are the p and s wave vertical wavenumbers, respectively.

If we now consider an incident p wave, it will give rise to reflected and transmitted p and s waves as in Figs. 2.4 and 2.5. If we denote the reflection and transmission coefficients for p and s waves by R_{pp}, R_{ps}, T_{pp}, and T_{ps}, where R_{pp} is an amplitude factor multiplied by the reflected p wave generated by the incident p wave, and so on, the ϕ and ψ potentials in media A and B will be [291a]

$$\phi_A = \exp(j\beta_{p_A}\chi_3) + R_{pp}\exp(-j\beta_{p_A}\chi_3)$$

$$\psi_A = R_{ps}\exp(-j\beta_{s_A}\chi_3)$$

$$\phi_B = T_{pp}\exp(j\beta_{p_B}\chi_3)$$

$$\psi_B = T_{ps}\exp(j\beta_{s_B}\chi_3)$$

where we have dropped the factor $\exp[j(\alpha\chi_1 - \omega t)]$ which is common to all waves.

The boundary conditions applied to these fields give four equations (*Zoeppritz equations*) in four unknowns which can be solved for the reflection and transmission coefficients [160, 292, 321a]. Further, we may derive expressions for reflection and transmission in terms of pressure, particle velocity, or displacement from the appropriate relations for the potentials. In exploration seismology, it is uncommon to deal with the complex problem of multiple-wave-type conversions at many interfaces which are traversed in propagation through the earth and the problem is simplified significantly.

At normal incidence, p-to-s conversions do not occur and even though the seismic field experiment may involve large offsets, it is generally assumed that we are working at near-normal incidence and wave-type conversions may be ignored as a significant factor in energy partitioning, and indeed, we generally neglect the depen-

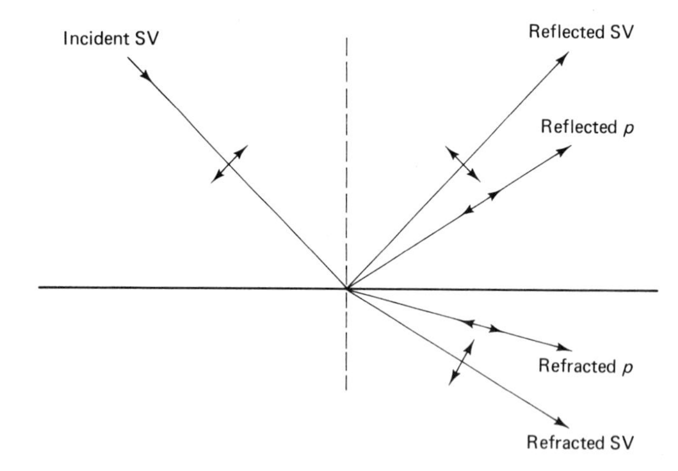

Figure 2.4 Waves generated by an SV wave incident on a plane boundary.

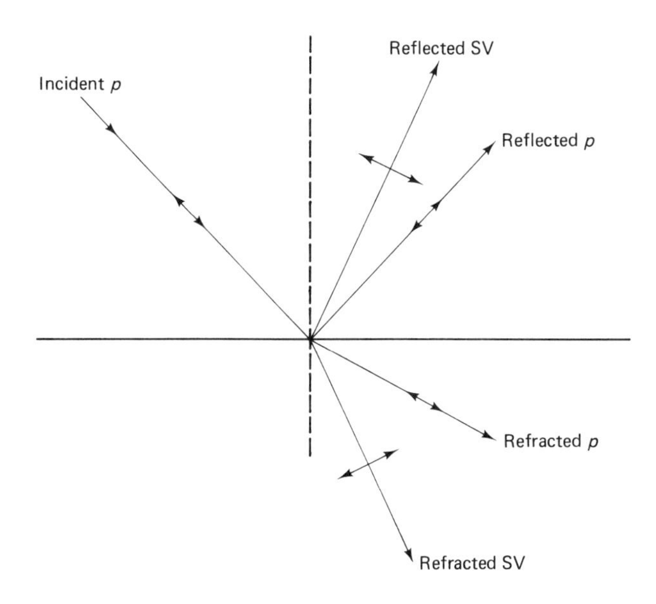

Figure 2.5 Waves generated by a p wave incident on a plane boundary.

dence on angle of the reflection coefficients and use those coefficients corresponding to vertical incidence.

In this case we obtain for the pressure reflection and transmission coefficients at normal incidence without conversion of wave type at the boundary the equations

$$R = \frac{\rho_B V_B - \rho_A V_A}{\rho_A V_A + \rho_B V_B} \tag{2.12}$$

where ρV is the *acoustic impedance* of the medium and R is the *reflection coefficient.* The *transmission coefficient* is given by

$$T = \frac{2\rho_B V_B}{\rho_A V_A + \rho_B V_B} \equiv 1 + R \tag{2.13}$$

We have given the equations for reflection and transmission coefficients in terms of pressure to be consistent with standard use of these symbols. However, we can derive similar formulations for the displacement and particle velocity and these are obtained by simply replacing the reflection coefficient, R, in Eqs. (2.12) and (2.13) by $-R$. Henceforth in our discussions, we shall assume that the measurement of interest is acoustic pressure. We recall that these derivations are strictly valid only for compressional plane waves at normal incidence to the (welded) contact between two homogeneous layers. While the assumptions associated with these derivations may seem unrealistic in practice, some of our data processing serves the purpose of transforming our data into a form for which these are not such bad assumptions in many cases. In any event, we are able now to build some simple models for seismograms recorded in the field.

Models for the Seismic Experiment

Imagine a plane impulse traveling vertically downward through a stack of (plane) layers spaced at some fixed Δt unit in time apart. As the wave encounters each boundary (with associated reflection and transmission coefficients), part of its energy is reflected back to the surface and part continues downward. In reality, the part reflected upward would continue to undergo further reflection and transmission at each interface encountered, but let us ignore all downgoing reflections generated by an upgoing wave as if they did not occur. In fact, let us assume that there is no change in wavelet energy upon transmission through an interface. Given these seemingly unrealistic assumptions, we may view our model as shown in Fig. 2.6, where the pressure response at the surface is indicated for each reflected arrival.

It is obvious that the seismogram recorded at the surface consists of the convolution of the input wavelet with a sequence of δ's appropriately spaced in time and weighted by the reflection coefficient sequence $R_1, \ldots, R_n$. In the limiting case of diminishing (equal) travel time between layers, it may be shown that the reflectivity sequence is replaced by a (continuous) reflectivity function given by

$$r(t) = \frac{1}{2} \frac{d \log \rho v}{dt} \tag{2.14}$$

where v is velocity as a function of two-way travel time, and the response of the earth to an input signal is modeled by the equation $S(t) = r(t) * f(t)$, where f is the *source signature* or *seismic wavelet* and $*$ denotes convolution. This approach is very practical for generating simple synthetic seismograms in areas where drilling has occurred, since we commonly have a sonic log available from the borehole which gives the function $v(z)$, that is, velocity as a function of depth. We may easily convert depth to two-way travel time and differentiate the logarithm of the result to obtain

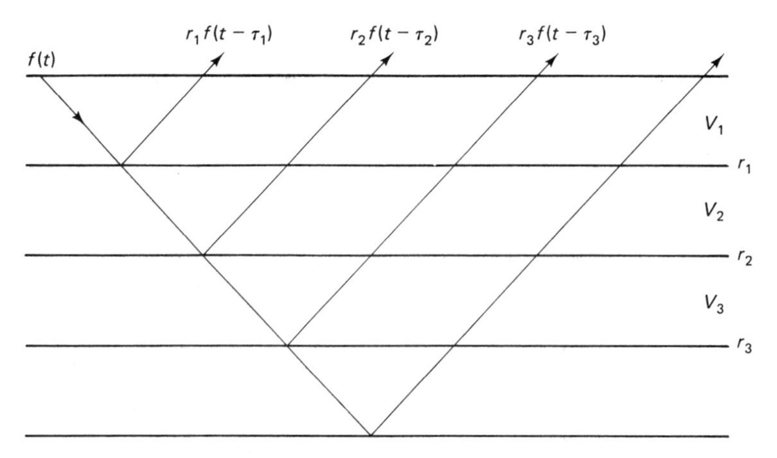

Figure 2.6 Simplified normal incidence model showing primary reflections.

the required (continuous) reflectivity function. We generally add noise, denoted by $n(t)$, to the model, as shown by

$$S(t) = r(t) * f(t) + n(t) \tag{2.15}$$

since some form of noise is always present on seismic recordings. We may also attempt to account for many other effects on the recorded signal due to spherical divergence, and the filtering response of the earth and our recording system; for example, we may write

$$S(t) = r(t) * f(t) * e(t) + n(t) \tag{2.16}$$

While models such as these are commonly employed as guides to developing signal processing techniques for seismic data, let us consider the problem of generating models once more, from a slightly more realistic point of view. Again, we begin with a (plane) layered model whose interfaces are separated by equal travel-time intervals (Fig. 2.7). This is not really a loss in generality, since setting a reflection coefficient to zero effectively removes any (artificial) boundary from the model. Denoting the consecutive reflection coefficients by $r_0, r_1, \ldots$, we find that successive arrivals generate the (time) sequence

$$r_0, (1 - r_0^2)r_1, (1 - r_0^2)(1 - r_1^2)r_2 - (1 - r_0^2)r_1^2 r_0, \ldots$$

Recall that a reflection coefficient changes sign for an upgoing wave. If we take the *z-transform* of the time series recorded at the surface (upgoing waves at the zeroth

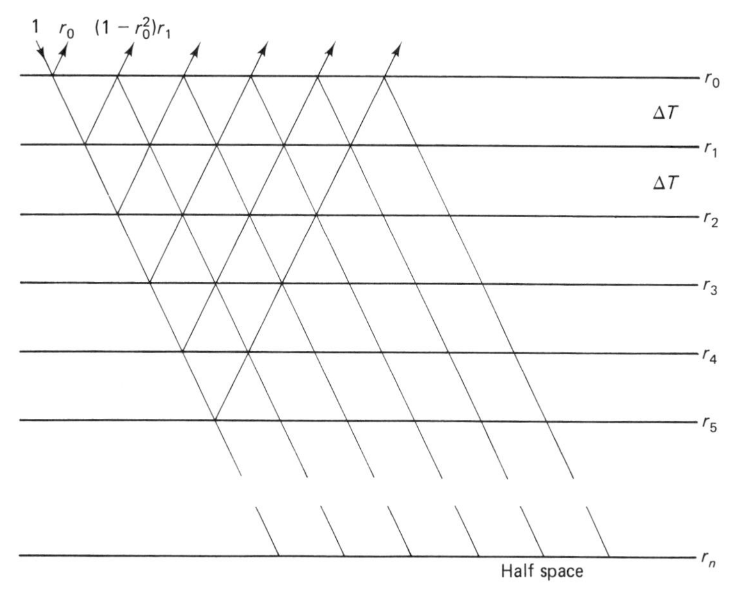

Figure 2.7 Normal incidence seismic model with equal travel time in layers.

interface), we obtain

$$U(s) = \frac{r_0 P_n(z) - Q_n(z)}{P_n(z) - r_0 Q_n(z)} D_0(s), \qquad z^{-1} = \exp(-s) \tag{2.17}$$

where $U(s)$ is the Laplace transform of the response at the zeroth interface and $D(s)$ is the Laplace transform of the input wavelet at the zeroth interface (see [241]), and the polynomials $P_n(z)$ and $Q_n(z)$ are generated by the recursion formulas

$$\begin{aligned} P_0(z) &= 1 \\ Q(z) &= 0 \\ P(z) &= P(z) - r_K z^{-1} Q_{K-1}^R(z) \\ Q(z) &= Q(z) - r_K z^{-1} P_{K-1}^R(z) \end{aligned} \tag{2.18}$$

where the superscript R denotes that the polynomial coefficients are taken in reverse order. The variable z^{-1} represents a time-domain delay.

Now we expect that this formula for the response of a layered system (which is not difficult to compute) should be more accurate than the simpler models derived earlier. One obvious difference is that the new response contains multiple reflections of all orders, whereas the former model contained only "primary" reflections, that is, those which have undergone only one reflection before returning to the surface.

A glance at the first few values generated, however, suggests that

$$\begin{aligned} r_0 &= r_0 \\ (1 - r_0^2) r_1 &\simeq r_1 \\ (1 - r_0^2)(1 - r_1^2) r_2 - (1 - r_0^2) r_1^2 r_0 &\simeq r_2 \end{aligned}$$

for very small values of the reflection coefficients. In general, the response at time $2n\,\Delta t$ equals

$$r_n \prod_{K=1}^{n-1} (1 - r_K)^2 + (\text{higher-order terms in } r_K)$$

which may be considered to be an approximation to r_n if the reflection coefficients are sufficiently small, and n is not large. When these assumptions are not valid with sufficient accuracy, the correct equations for the surface response should be used. However, the first (simple) model may not be in great error provided that the reflection coefficients (which are bounded by 1 in magnitude) are sufficiently small, which is frequently true in practice. When this is not true, we should replace this simple approximation by the more accurate one.

The logical next step is to consider the extension of this model to the case of plane waves encountering plane layers at nonnormal incidence, and accounting for the compressional and shear conversions that occur at each interface. (Only one component of the shear wave couples with the p wave in that case and it is usually termed the *vertical shear component* (SV). The other, horizontal shear (SH) component is orthogonal to the page on which we draw the model and does not couple with the p wave.) This program has also been carried out (see [89]), but we shall not comment on it further here except that it is important to understand the context of the models with which we deal. Although we are improving the potential accuracy of our models with each step we take, arguments like the one given above, and the argument that our experiment often involves near-normal incidence, are invoked to justify the use of the simpler models.

Given the simple (convolution) models that we have arrived at, it is natural to suspect that some attempt might be made to build back into these models various effects which may be of interest and which they ignore. While the debate continues concerning the importance of *frequency-selective attenuation* (and therefore, *dispersion*) in the earth, a reasonable approach to include these effects based on our simple models was made by Trorey [296], who assumed a linear relationship between frequency and attenuation, the exact relationship being determined for each layer in the Goupillaud model independently (see Fig. 2.8; see also [102]).

Trorey implemented the model by introducing a frequency-selective filter determined by each individual layer into each transmission path through that layer. The time-domain response of this filter for the nth layer is given by

$$F_n(t) = \frac{2/D_n \tau \pi}{1 + 4(t - \tau)^2/(D_n \tau)^2}$$

where D_n is the *dissipation factor* (reciprocal of Q) for the nth layer and τ is the *one-way travel time* (constant) for a layer.

To become much more sophisticated with these simple models would probably justify a move to ray tracing in nonuniformly layered media. There is some interest today in designing processing procedures based on more sophisticated approaches; however, the simple convolution models are still the basis for most procedures used in seismic data processing.

A point of clarification may be in order at this point. The models we have been referring to may be used in very different ways. However, if they are actually used to

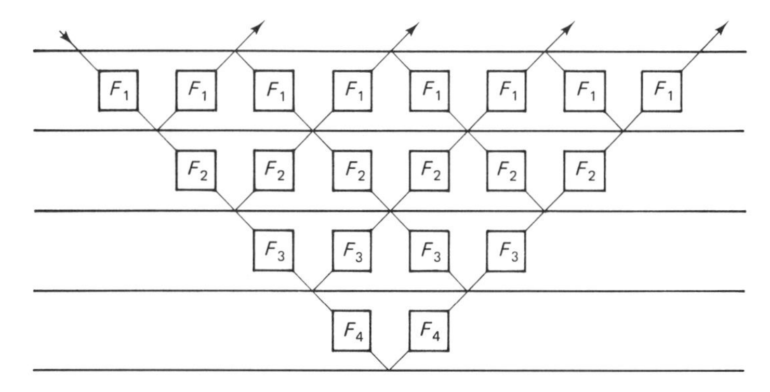

Figure 2.8 Trorey model for normal incidence seismogram with absorption.

model the response of the earth subject to various choices of the elastic parameters, we may go far beyond these simple models and approach reality via solutions of the elastic or scalar wave equations with considerable accuracy using methods which may be quite different from those described here. It may be anticipated that these more accurate methods will become an integral part of seismic data processing in the future as a result of increased computing power which will make this possible.

Additional Modes of Propagation

Although we should not dwell too long on such a sketchy approach to wave propagation, there are a few more important wave types which figure prominently in exploration seismology and which must be mentioned.

If we consider wave motion in a homogeneous half-space bounded by a free surface (normal stresses all vanish), a new wave type is possible which propagates along the surface with a velocity less than the shear-wave velocity and with an amplitude that decays with depth. This is a wave type studied by Lord Rayleigh and named for him, and is an element of what is referred to in geophysics as *ground roll.* Rayleigh studied the problem of wave motion along a free surface over an elastic half-space. He found that indeed such waves exist and that they decay exponentially with depth. He further derived an important property of these waves—that the particle motion for the fundamental mode is retrograde elliptic in a vertical plane orthogonal to the direction of propagation. In theory, for the ideal case considered, these waves are nondispersive. However, in practice, propagation does not occur in a homogeneous half-space but occurs typically in a low-velocity surface layer overlying higher-velocity material. In this case the velocity of propagation becomes frequency dependent and the wave is said to be *dispersive.* For short wavelengths the velocity approximates nine-tenths of the near-surface shear velocity and increases with increasing wavelength to about nine-tenths of the shear velocity of the higher-velocity material beneath. In any event, velocity of propagation is bounded above by the appropriate shear velocity, so surface waves typically exhibit low velocities as far as the seismic experiment is concerned.

In exploration seismology, with near-surface sources, ground roll may carry significant amounts of energy, swamping the reflections we would like to see and severely reducing the usable dynamic range of our recording instruments. Ground roll is frequently accompanied by an even lower-velocity wave which travels through the air, coupling to the ground surface and known as the *air wave,* which again may carry significant quantities of energy (Fig. 2.9). In addition to these wave types, another relatively strong wave is generated by refraction (at the critical angle) along high-velocity near-surface interfaces and is generally the first arrival on seismic records with sufficient offset, and is known as the *head wave.* This wave is not generally a problem in exploration seismology, as it typically arrives before the reflections of interest. Finally, we may mention various scattered waves generated by waveguides and reflectors, diffractors, and so on, within the earth which may tend to obscure the correct interpretation of our data.

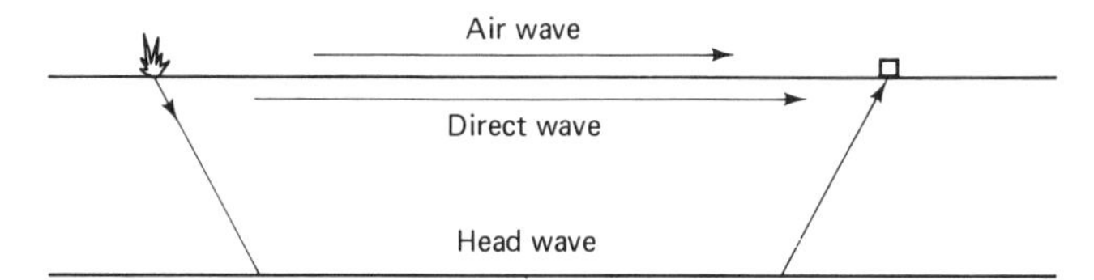

Figure 2.9 Near surface events on seismic data.

2.4 SEISMIC EXPLORATION OBJECTIVES

We have now defined the nature of a seismic experiment and have considered some of the properties of wave propagation in an elastic stratified earth medium. Before continuing to discuss the array processing procedures which are used to produce a final estimate of the subsurface structure of the earth, we would do well, in the interest of completeness, to consider those features of the earth's subsurface structure which are of interest to us and which we hope to identify as the end product of our data acquisition, processing, and interpretation efforts.

Petroleum hydrocarbons tend to accumulate in naturally formed reservoirs or *traps* within the earth. It is the locations and characteristics of these reservoirs which are of interest to us in our search for commercial quantities of petroleum hydrocarbons. We begin by considering the origin of petroleum and its subsequent migration and accumulation as well as the nature of some common types of reservoirs or traps into which the petroleum migrates.

It is fairly well established that petroleum hydrocarbons are the results of natural processes working on organic plant and animal material over a significant period of geological time, within certain narrow temperature ranges. Indeed, it is thought that the most important factor in the formation of petroleum hydrocarbons is temperature, with time playing a secondary role in the process. Because organic materials, which represent residues of plant and animal life, tend to accumulate in sedimentary environments, it follows that, with very few exceptions, the world's petroleum supply is associated with sedimentary rock and so it is to the areas of the world where sedimentary basins have formed that we primarily direct our attention in the search for oil and gas. The largest accumulations of plant and animal material will occur in near-shore and offshore marine environments with near-shore features, such as deltas, providing a major site for the accumulation of land-derived materials, and continental margins, providing sites for major accumulations of aquatic or marine sources of plant and animal material. Although hydrocarbons are widespread throughout the sedimentary basins of the world, nevertheless, favorable conditions for the conversion of organic material to petroleum hydrocarbons, combined with subsequent trapping and accumulation in commercial quantities, are very localized and present a challenge to the seismic explorationists trying to locate them.

To begin the process of oil generation, the depositional environment, whether it be clastic or carbonate, must contain a certain minimal amount of organic matter, which may be as high as 0.5 percent in shales and less in carbonates in order that they may become what is known as a *source rock*. Given that suitable quantities of organic matter are available, we may imagine that as the sedimentary deposits are buried, they undergo a variety of changes. Initial decomposition of the material by aerobic bacteria begins to break down the organic matter, releasing carbon dioxide in the process. With increasing depth of burial, anaerobic bacteria take over and begin the generation of methane. As the temperature of the organic matter reaches 50°C, we begin to see the onset of oil generation from the decomposed organic materials that remain. Oil generation continues with increasing temperature, reaching a peak somewhere in the range 50 to 150°C. If temperatures continue to increase above this level, all organic matter will eventually be converted to methane or graphite.

Since temperatures in the earth increase on the order of 2 to 5°C per 100 m, we may in some sense associate increasing temperatures with depth of burial. However, other factors, such as intrusion of igneous material into sedimentary basins, may affect the temperature and depth relationship and may very well be important factors in the formation of hydrocarbons. In addition, we have mentioned that time is an important factor in petroleum generation, and there is reason to believe that petroleum hydrocarbons undergo a maturation process with the result that the oldest and deepest reservoirs, in general, contain the lightest oils. We begin to see that there is a "window" which is influenced most by thermal history of the sedimentary basin, but which may also be correlated thereby to depth of burial and geologic age, which is auspicious for the generation of petroleum hydrocarbons. Outside this window the probability of the production and preservation of hydrocarbons decreases rapidly. Studies have tended to indicate that this petroleum hydrocarbon window probably does not extend to more than about 30,000 ft in depth in the earth, and whereas, in chronological terms, it extends from the Precambrian to the present, most of the oil reserves in the world today are associated with sediments of the Mesozoic–Tertiary periods. The age of the sediment in which the oil is trapped may have very little to do with the actual age of the oil that is found there, because of the migration process.

As petroleum hydrocarbons are generated in the source rock, we may in general assume that this rock is undergoing successive burial and compaction. A very fine water layer attaches itself firmly to the individual grains of the matrix and the petroleum hydrocarbons are held in suspension in the interstitial water or connate water within the formation. As the grains of the rock matrix undergo compaction, the free water and petroleum hydrocarbons are squeezed out of the rock matrix and will move or migrate into any free pore space that is available. This migration will continue as long as available pore space exists, that is, as long as the porosity is high enough and as long as the ability to flow through the associated rock matrix exists, or as long as permeability is high enough. If there were no barriers in the path of this migration process, the oil and water would eventually find their way to the surface. Indeed, there are numerous examples of this around the world where oil seeps occur, with petroleum hydrocarbons emerging at the surface. However, in most cases we may assume that a barrier will eventually be reached which will prevent the further migration of the petroleum hydrocarbons and they will come to rest in some *reservoir rock*. This rock will be sealed in some way by a rock matrix which is impermeable to the flow of water and petroleum hydrocarbons. The combination of a reservoir rock and a seal is called a *trap* and there are many kinds of traps that have been identified for hydrocarbons, including *structural traps*, generally formed by folding or faulting of sediment; and *stratigraphic traps*, created by various types of stratigraphic effects, such as a change in porosity and permeability within a rock matrix or the termination of dipping beds by unconformities. It is also possible to have a trap generated by the flow of surface water into subsurface strata, creating pressure sufficient to stop the flow of hydrocarbons, and these are called *hydrologic traps*. It is also possible to find oil closely associated with the source rock, which may still be in the process of generation, or to find oil which is in the process of migration and which has not entered its final confinement in a trap.

In any event, our search for commercial deposits of petroleum hydrocarbons is restricted to perhaps the first 10 km or so of the earth's surface and is restricted primarily to the younger rocks of major sedimentary basins. Of course, in some cases, these may be overlain by significantly older rocks, as is the case in the overthrust belt of the Rockies, for example.

Several types of traps are shown in Fig. 2.10. Part (a) of the figure shows an *anticline trap*, which is the type of trap associated with the very first oil discovery and for some time was thought to be the only trapping mechanism for petroleum hydrocarbons. The existence of free gas in this type of trap depends on the pressures present in the reservoir. At high pressures, the gas will go into solution in the oil and at lower pressures will appear as shown. This type of trap is an example of a structural trap. Figure 2.10(b) shows a salt dome formed by the plastic flow of buried salts, which may be the result of evaporation of an ancient seabed. As the salt is buried and subjected to increasing pressure, it is capable of plastic flow and can force its way up through the sedimentary overburden, pulling the beds up alongside its flanks. This upward dip, terminating against the impermeable salt, produces a potential trapping mechanism and, indeed, this type of trap is very common in the Gulf Coast region of the United States. Many early traps that were found in this region were of this type. In Fig. 2.10(c) a stratigraphic trap is shown consisting of a dipping bed in which a change of porosity occurs due to diagenesis of the layer to the left, creating pore space. Stratigraphic traps can be extremely subtle and difficult to detect using seismic methods, but they are becoming more and more important as the more easily detected structural traps which have not been found become fewer in number. In Fig. 2.10(d) a *reef trap* is shown, which is important in many pools found in North America, including Texas and New Mexico, Michigan, and western Canada, for example. These reefs frequently still contain porous material and often are associated with source rocks, such as bituminous shales, surrounding them. In Fig. 2.10(e) another common type of trap is shown, known as the *fault trap*, which is produced as a result of faulting in the earth, which may bring an impermeable sediment into contact with a porous and permeable one, as shown in the

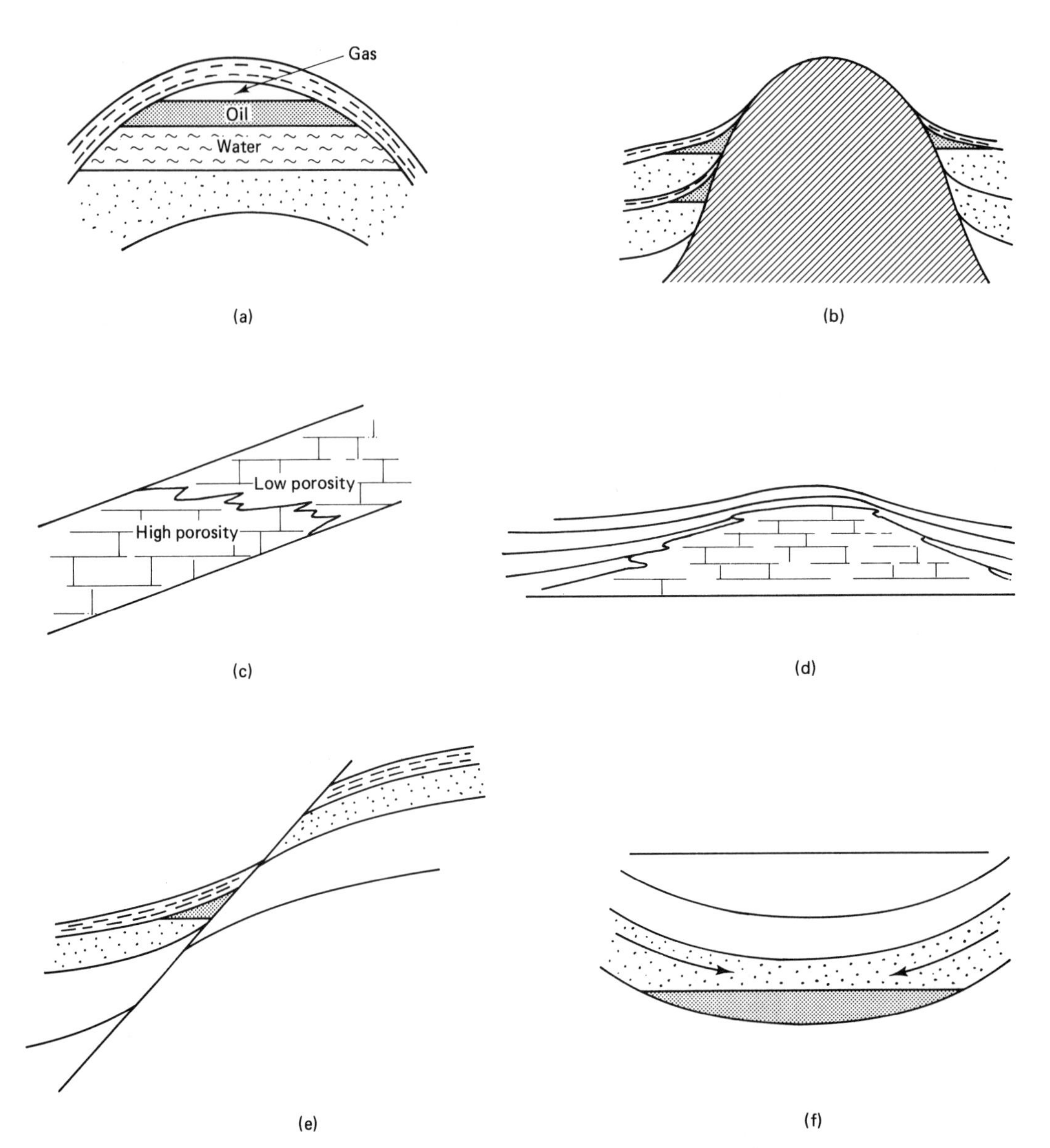

Figure 2.10 (a) Anticline trap; (b) salt dome; (c) stratigraphic trap; (d) reef trap; (e) fault trap; (f) hydrologic trap.

illustration. The oil migrates through the permeable layer until it encounters the impermeable material into which it cannot flow, and so a trap is formed. Finally, in Fig. 2.10(f) we see an example of a *hydrologic trap,* in which water pressure due to surface waters entering folded strata at the surface holds the oil in place. In some cases, this type of trap might be referred to as a hydrodynamic trap since pressures generated by the actual movement of waters through the reservoir appear to provide the trapping mechanism. An example where this type of trap occurs would be the San Juan Basin in New Mexico, or possibly the Alberta Deep Basin [189a].

Hopefully, this brief introduction has given the reader some feeling for what we are after in petroleum exploration. Further information on these matters can be found in [177], for example. Recalling now that we are interested in probing sedimentary basins to significant depths using surface-source-generated acoustic energy, let us go on to consider the methods used in acquiring and processing seismic exploration data, in an attempt to identify the existence of potential hydrocarbon traps, and the presence or absence of hydrocarbons, if possible.

2.5 IDENTIFYING EVENTS ON SEISMIC DATA

We have previously considered the seismic experiment and have discussed the usual recording geometries which are used in carrying out that experiment. Further, we have discussed the variety of ways in which the recorded data may be sorted and gathered for various data processing applications. In this section we consider briefly the identification of various types of events of interest on seismic data, and for this purpose we restrict ourselves to the *common depth point (CDP) gather*. The simplest events to detect, and those which are usually the first to arrive on seismic data, would be the surface waves, which exhibit a low but constant apparent velocity across the recording spread and are illustrated in Fig. 2.9. The *head wave* is a refracted wave which diffracts at the critical angle and travels along the interface between the near-surface material and the first high-velocity layer beneath, and which, because of its higher velocity, will arrive at far offset phones before the surface waves arrive. Both of these events appear to have nearly constant velocity across the array, and so appear with a linear alignment on the recorded data, as shown in Fig. 2.11. The *air wave* is, of course, a type of surface wave, moving with a very low velocity and so exhibits a very large *moveout* in time as it progresses from one recording group to the next. Finally, we come to the events which are really of interest to us, and which are broadly classified as *body waves*. This class would include the primary reflection, which is what we would really like to see on our records, as well as all multiple reflections and mode conversions that occur within the earth.

We begin the analysis of the character of reflected body waves as expressed on the CDP gather with a simple example. Given a shot-to-receiver distance of X units and a single horizontal, constant velocity (V) reflecting layer over a half-space (see Fig. 2.12), we may readily verify that

$$T_x^2 = \frac{4D^2}{V^2} + \frac{X^2}{V^2}$$

This can be written in the form

$$T_x^2 = T_0^2 + \frac{X^2}{V^2}$$

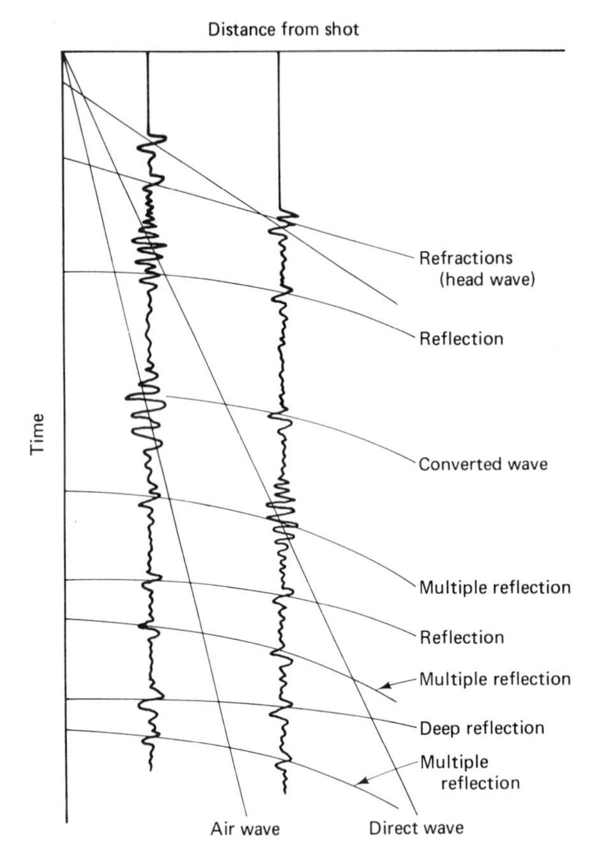

Figure 2.11 Interpreting events on seismic data. Two simulated traces at different offsets are shown.

where T_0 is the two-way normal incidence (source and receiver coincident) travel time. We see that the (two-way) travel time (T_x) versus offset (x) curve is a hyperbola.

If we now consider a stack of horizontal layers with varying thicknesses, d_1, ..., d_n, and velocities, $V_1, \ldots, V_n$, and invoke *Fermat's principle*, which says that a ray path from source to receiver follows the minimum time path, it follows that the two-way travel time for the nth layer is given by [284]

$$T_{x,n}^2 = C_1 + C_2 X^2 + C_4 X^4 + C_6 X^6 + \cdots$$

Taner and Koehler [284] indicate that in current practice, truncating this series

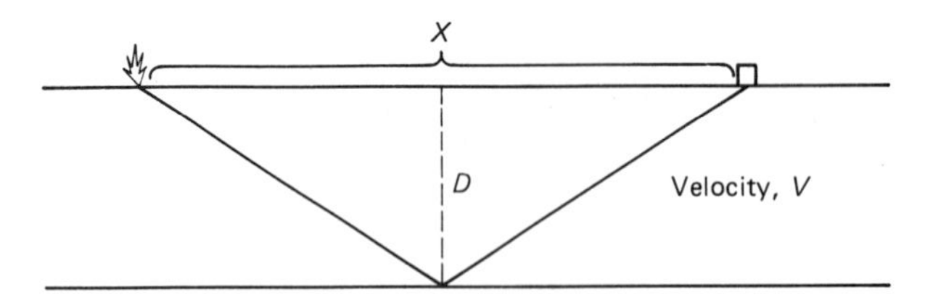

Figure 2.12 Diagram for travel-time versus offset in a flat-layered constant velocity medium.

after two terms corresponds to velocity errors of several percent and is, perhaps, not a bad approximation. If we make this approximation, it can be shown [284] that

$$C_1 = \left(\sum_{k=1}^{n} t_k \right)^2 \equiv T_{0,n}^2$$

and

$$C_2 = \frac{\sum_{k=1}^{n} t_k}{\sum_{k=1}^{n} t_k V_k^2} \equiv \frac{1}{\bar{V}_n^2}$$

Thus we may write

$$T_{x,n}^2 = T_{0,n}^2 + \frac{X^2}{\bar{V}_n^2}$$

and $\bar{V}_n$ is commonly called the root-mean-square (rms) velocity.

In any event, a hyperbolic fit using the appropriate rms velocity is not a bad approximation to the shape of the moveout curve for reflected body waves. Readers interested in higher-order approximations may see [192], for example. However, this approximation is good enough for our purposes now and we may assume that reflection events will be approximately hyperbolic in character. The extension of these procedures to arbitrary dipping plane layers in three dimensions has been carried out [81, 173], but the result is that we may still assume that body reflections exhibit a basically hyperbolic moveout curve. Similar analysis of diffraction patterns yields the same result [74] and, of course, multiple reflections will show this character. Figure 2.11 has been provided (primarily to aid our visualization of the data at this point) to summarize these remarks.

2.6 SEISMIC DATA ACQUISITION

We referred earlier to the seismic experiment, which may serve as an introduction to seismic data acquisition. We now consider the process of data acquisition in more detail, as an application of array processing procedures in its own right as well as to provide necessary preliminary material to the array processing procedures which are applied in seismic data processing which will follow. Earlier we justified treating the seismic data acquisition problem as an array problem and pointed out that there are basically two types of arrays that are important in seismic data acquisition. These arrays are termed *linear*, which indicates that the shooting and receiving geometry is laid out along a line in one spatial dimension, and the *areal* array in which the sources and/or receivers may occupy some geometric configuration involving two spatial variables. In today's exploration practice, the linear arrays are by far the most commonly employed, primarily for economic reasons. We are, however, seeing increasing interest in the use of areal surveying techniques, particularly those related to the three-dimensional method of seismic exploration. Although three-dimensional seismic exploration methods are much more expensive than linear array methods, they can often be justified in the field, particularly when a discovery well has been made and the economic potential of the region has proved itself. At this point it may become very economic to carry out a thorough three-dimensional survey to delineate accurately the reservoir and its characteristics, so that we may optimally plan the location and production strategies of further development wells in the area. Indeed, it may be anticipated that a major area of activity for the seismic method, in the future, will be directed toward the problem of field development to which we have just referred. We anticipate that this will attach further importance to the development of three-dimensional field and seismic processing techniques and will lead to the development of many new procedures as well.

In discussing the use of arrays for seismic data acquisition, we confine our attention primarily to the design of arrays for linear data acquisition. The first question that we may reasonably ask is: Why apply array procedures in seismic data acquisition at all since we may envision that essentially all array processing activities could be simulated after the fact by applying these procedures to our recorded data using digital computers to simulate a wide variety of array processing tasks, thereby preserving many of our options which may not be available if these procedures are applied before the data are recorded? In fact, this is a valid question and there are pros and cons to applying array processing procedures in the field. To gain insight into this question, we would do well to consider the purpose of array techniques in data acquisition.

Our principal concern in data acquisition is to obtain records that contain as much usable information as possible and which are as free as possible from noise. This noise may contribute little to the ultimate objective of our efforts, which is to gain a representation that is as accurate as possible of the structure and conditions present in the subsurface. Having stated these objectives, we may immediately identify several problems which will be encountered in data acquisition from the point of view of reflection seismology. In particular, most of our processing and interpretation will be based on the attempt to identify primary reflections generated at the interfaces between layers of differing lithology which may be at considerable depth within the earth and which we may record at relatively small offset distances. This means that the surface waves and air waves, for example, as well as other near-surface phenomena, such as surface-generated multiples, while carrying useful information in their own right concerning the near surface, can tell us very little about conditions at the depth in which we may be interested. Because surface waves, in particular, and near-surface multiples can carry significant amounts of energy compared to the energy present in reflections from stratigraphic sections deep within the earth, the first problem that we encounter in data acquisition is a dynamic range problem. That is, low-velocity high-energy near-surface-wave motion may arrive simultaneously with much deeper body wave reflections which have significantly lower energy, giving, at best, recordings with an extremely low signal-to-noise ratio, or at worst may actually swamp the dynamic range of the recording system. In this case we completely lose all information concerning the reflections at depths that are of interest to us. As a result, there is some motivation to control the amount of energy that we record due to the near-surface-wave motion at the data acquisition stage to avoid obtaining data from which we have little chance of ever recovering the information of interest. Most array techniques, which are applied in the field at the data acquisition stage, are designed to deal with the problem of reducing significantly the amount of recorded energy due to near-surface disturbances, including air waves, ground roll, surface-generated multiples, 60-Hz pickup in our cables, land and noise sources such as wind, or cultural noise such as traffic on nearby highways, and even coherent noise sources such as machinery that may be operating in the recording environment. Given that it is advisable to consider the possibility of designing arrays to discriminate against these various noise sources, we may now ask on what basis we can perform this discrimination. Certainly, temporal frequency is one candidate that is valid for several of these noise types, and *spatial frequency* or *wavenumber* is another candidate. If we combine these two discrimination factors, we find that they are related through apparent velocity: that is, the velocity at which any wave appears to cross the receiving array. In many seismic applications apparent velocity becomes a very valuable discriminator for separating various wave components, including those that are desirable and those that are not desirable on our recordings. In particular, the ideal seismic recording system in the field should pass very high apparent velocities and reject very low apparent velocities or, at least, should represent a low-pass filter for spatial frequencies. A reference to Fig. 2.13 helps to clarify this comment. In this figure we have shown the approximate locations of some of the important wave types that we have made reference to, in the temporal frequency versus spatial frequency (F–K) domain. The diagram illustrates

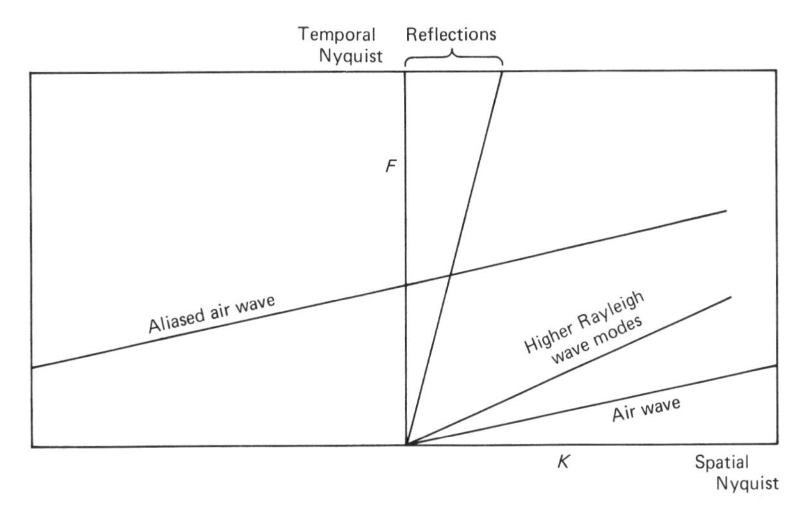

Figure 2.13 Interpretation of events on F–K plot.

the justification for separating signals consisting of body- or surface-wave components on the basis of selective pass and reject characteristics determined by apparent velocity. It also justifies the statement that the seismic array should respond more or less as a low-pass filter for spatial frequencies. In essence we may think of the ideal seismic array as a beamformer with maximum sensitivity in the vertical downward direction and minimal sensitivity or nulls in the horizontal direction. This is easily verified by imagining a plane wave approaching the array at various angles of incidence: if approaching the array vertically from below, the apparent velocity at the array would be infinite, whereas this apparent velocity would reach a minimum if approaching the array from a horizontal direction in line with the receiver elements.

It is generally not possible to achieve the theoretical ideal response for a finite discrete array in the field. Indeed, we can only hope to approximate such a response and we would undoubtedly be forced to pass several components which we would prefer to reject. The most we can hope for is to attenuate these components as much as possible. This goal has led to a more or less standard pattern for laying out linear arrays in the field. A reference to Fig. 2.14 reveals the most common field layout used today, consisting of a cable laid along a line surveyed through a shot point, consisting of a number of receiver locations (which may be several hundred feet apart). At each receiver location we find not one single receiver but indeed a small array of receivers that may be laid out in a linear or an areal pattern, depending on the field conditions. For purposes of this discussion we assume that the small subarrays are laid out in a linear pattern. The spacing of these elements may be quite close, and in particular we must be mindful of the *Nyquist criterion*, [148] which implies that they must be laid out closely enough together that we sample the shortest wavelength that we are likely to encounter at least several times per cycle. The elements in each subarray, that is, the group of receivers for each receiver location, are generally weighted and summed to produce a single output for that receiver location. That is, the output from all these receivers at each receiver location generates a single output trace representing a recording at the approximate offset of the group.

In many cases the output from each group of receivers is simply summed to produce the output trace for a given offset location. A moment's reflection reveals that this results in a somewhat suboptimal beamformer of the type which we wish to achieve, in that a plane wave arriving at the array at vertical incidence from below will be passed, whereas plane waves arriving at other angles of incidence may be attenuated due to the phase difference of the arrival as it passes each of the geophones. In addition, any coherent noise that exhibits a random phase distribution among the different sensors which are summed would then tend to be canceled in the summing process. Although this is not a very sophisticated approach to array

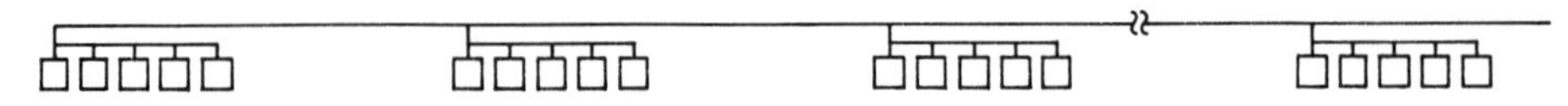

Figure 2.14 Standard field array with geophone subarrays for noise attenuation.

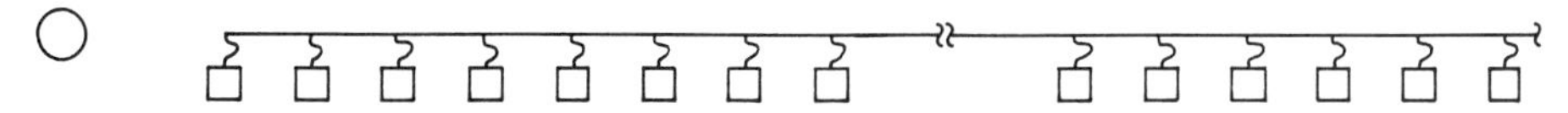

Figure 2.15 Arrangement of shot and phones for a noise test.

design, it obviously does have some desirable characteristics and is employed quite often in the field.

To be more sophisticated in our approach to array design, using a pattern shown in Fig. 2.14, it would be most helpful to have some means of determining the actual apparent velocities of various identifiable noise sources which we would like to attenuate with our array design. To obtain this information we frequently carry out what is called a *noise test.* The layout for a noise test is shown in Fig. 2.15, where an array of receivers is again laid out along a surveyed line to the shot point, but in this case the receivers are individually recorded and are typically spaced relatively close together, perhaps on the order of only a few feet apart. A shot is fired and as there is no summing of receivers, the full noise wave field is recorded and displayed. Major components of this noise field, that is, those that are carrying significant amounts of energy, are identified with their apparent velocity across the array. We may then take this information and use it as a basis for the design of a field array which will have nulls, ideally, at the apparent velocities corresponding to these major components of the noise wave field. A diagram showing some of the types of components that might be identified on a standard noise test is shown in Fig. 2.16.

To consider the possibilities for designing an array, based on the information that might be obtained from a noise test, let us now consider the set of parameters that we may use to carry out this design. In field applications, we are typically

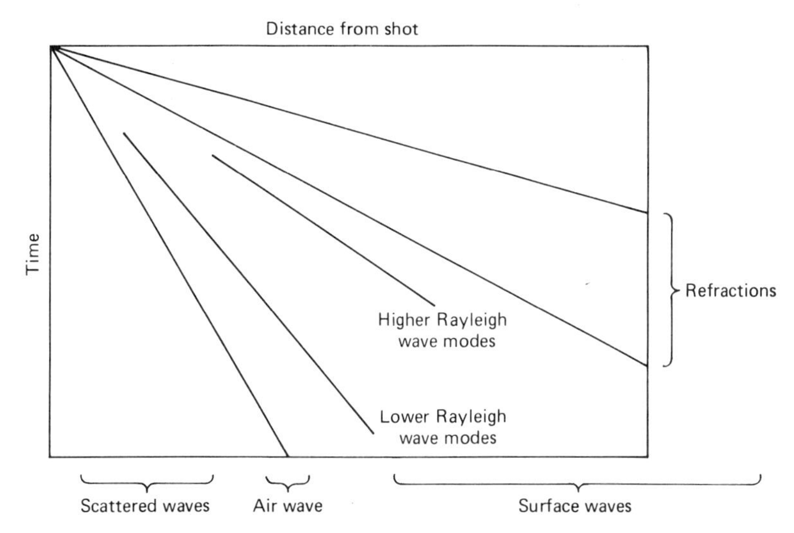

Figure 2.16 Interpretation of events on noise test panel.

limited to the geometric arrangement and to the relative weightings of the receivers within those arrangements. At this time it is uncommon in field recording practice to be able to introduce time delays between receivers, even though these effects could be easily simulated on the computer after the fact. The most common designs in field use today are based on equal linear spacing of the receiver elements with possibly variable weightings applied to each individual receiver before summing.

Let us now look at some of the mathematical equations that relate to array design using equally spaced linear array elements with weighting. We define the response of the array to be

$$A(k) = \int_{-\infty}^{\infty} w(k) \exp(-jkx)\, dx$$

where k is the apparent wavenumber across the array, $kx = \mathbf{k} \cdot \mathbf{x}$ with $\mathbf{k}$ the wavevector.

If the $N + 1$ array elements are located at equally spaced positions,

$$x_n = nd$$

with weights w_n for $-N \leq n \leq N$. Then we may write

$$w(k) = \sum_{n=-N}^{N} w_n \delta(x - nd)$$

and the response may be written

$$A(k) = \sum_{n=-N}^{N} w_n \exp(-jknd)$$

where d is the distance between receivers, and we have assumed that the receiver positions are symmetric about the origin. If the array is shifted with respect to the origin, then the response will be multiplied by an appropriate exponential, but this will not affect the magnitude response, $|A|$.

If the array weights are also symmetric, then we may write

$$|A(k)| = \left| w_0 + 2\sum_{n=1}^{N} w_n \cos knd \right|$$

and if, in addition, the array weights are constant, which we take to be one for convenience, then we have [238a]

$$|A(k)| = |\sin[(N + \tfrac{1}{2})kd]/\sin(kd/2)|$$

To obtain the above expressions in terms of apparent wavelength, λ, substitute $k = 2\pi/\lambda$. If the center weight w_0 is zero, then one must be subtracted from the last expression whose magnitude appears on the right above.

It is apparent from these equations that the response of a linear array with uniform spacing and equal weights is determined entirely by the spacing of the elements in the array and the apparent wavenumber of a wave component along the array. By choosing the spacing of the array element appropriately, we may, to a very limited degree, attempt to attenuate certain components within the recorded wave field. We may improve on the performance of our array by assigning different weights to the individual array elements, which would result in a weighted summing of the signal as opposed to a simple summing which we have described here. Since we have indicated earlier that an array should pass events with high apparent velocity and reject all others, or should at least approximate a low-pass filter for spatial frequencies, several improved designs have been suggested and have found use in practice. To mention a few of these, we have the *linearly tapered array*, in which the weights of the individual elements are tapered in a decreasing fashion away from the centre toward the end of the array, and an array in which the weights are based on a least-mean-square solution to the ideal low-pass filter response. Finally, one of the more popular and widely used arrays today was suggested by Parr and Mayne in 1955, and has come to be known as the *Chebyshev array* [128, 225]. The Chebyshev array is characterized by having equiripple characteristics in the reject band. This type of filter can achieve significantly improved performance over the arrays mentioned previously. To achieve these weights in the field, we must employ accurate resistive networks or, alternatively, employ equally weighted array elements in which the spacing between elements is nonuniform.

Unfortunately, practical implemention of array design in the field seldom lives up to theoretical expectations of performance. Indeed, there are many errors that enter into the implementation process which render the array response less than ideal. To begin with, we have based our theoretical discussion on a plane-wave formulation, whereas in fact we have to deal with curved wavefronts in the field which may, in fact, suffer amplitude decay as they cross the array. We have further assumed that our geophones will all respond identically to a given input, whereas in fact the response of the phone is sensitive to a number of factors. In particular, individual phones may or may not be coupled well to the ground and although velocity-sensitive phones are assumed to be planted vertically, they often deviate from this position in practice. It is further clear that the spacing of the individual phones is important in achieving the desired response and we have also assumed that the phones are responding to a homogeneous medium, whereas, in effect, the near-surface layer in which they are planted may be quite heterogeneous in character, resulting in a geophone response that may be quite different from its assumed response.

Finally, elevation differences in the terrain over which the geophones are planted may result in significant phase errors. An interesting analysis of the effect of the various types of errors in array response under actual field conditions may be found in the paper by Newman and Mahoney [216]. A good discussion of the causes and effects of imperfect geophone coupling to the earth will be found in [130].

An interesting alternative to the usual *weight-and-sum methods* used for geophone subarrays has been proposed by Biswell et al. [30] using a plane-wave stacking technique which removes differential moveout and directs the subarray peak response to the (apparent) incident angle of the seismic wavefront within a preselected range of angles of incidence. They report good results on marine data. The actual basis for discrimination here is apparent velocity, determined by the acceptable range of slowness (p) values.

Having taken a brief look at the problem of linear array design, let us now consider the possibility of combining areal arrays with the seismic experiment. Although this is not often done in practice, there are occasions when particularly difficult field conditions may warrant the use of areal arrays. Suppose that we have an array of sensors at spatial positions given by the vectors $\mathbf{r}_j$, $1 \le j \le N$. The response of this array to a plane wave propagating with vector wavenumber $\mathbf{k}$ is given by

$$A(\mathbf{k}) = \sum_{l=1}^{N} w_l \exp(-j\mathbf{k} \cdot \mathbf{r}_l)$$

where w_l is the weight applied to the lth sensor. It has been pointed out [154] that since $\mathbf{k}$ is a function of frequency and velocity, the spatial array will alter the characteristics of the propagating signal. We may now view the array design problem as a multidimensional filter design problem [79]. Needless to say, many of the problems that were mentioned with regard to the implementation of linear filter arrays are applicable to the problem of implementation of areal arrays, and indeed we may imagine that these problems are compounded by the additional degrees of freedom in placing the elements. In particular, if the areal array is laid out over an area with sufficient change in elevation, these elevation changes correspond precisely to phase errors at the individual elements, and it seems unreasonable to expect a high level of performance from a theoretically designed array under typical field conditions. It would appear that the only hope for implementing areal arrays with some degree of success, under field conditions, would be to take the theoretical design as a first approximation and then modify it under the actual field conditions until an acceptable approximation to the theoretical response is obtained. Although this type of procedure could become reality in the future, it has been somewhat impractical and costly to consider applying it in the past. As a result, areal arrays are not often used in exploration seismology, but it is certainly possible that this situation will change in the future.

The comments concerning array design to this point are largely applicable to either source or receiver arrays, and indeed we might contemplate the use of either type of array in the field. In particular, experiments have certainly been carried out with source arrays, both discrete and continuous (in the case of Primacord), to simulate sources that generate maximum compressional wave energy in the vertical direction and minimal compressional wave energy in the horizontal direction. In cases where field implementation of arrays has been compared with computer simulation of the arrays on the recorded data, the results have been that the computer simulations, in most cases, are as good as or better than the field attempt to implement the corresponding array. However, there have been a few instances where field tests have confirmed that the implementation of a desired array in the field is both successful and worthwhile, and it might be of interest to consider one such example.

In 1972, Brundrit and Van Wijnen of Shell Canada [39] carried out experiments using both source and receiver arrays at sea and their work led to a patented process consisting of a very long array of airgun sources together with a long receiver array for marine applications which has proved to be very effective in attenuating water bottom multiples (see also [300]). The basic idea of this work, as reported later by Lofthouse and Bennett [180], was to estimate the range of apparent primary wavelengths and the range of apparent multiple wavelengths observed in the area of interest and to attempt to design an array that could successfully discriminate between the two ranges of wavelengths (assuming that they do not overlap very much), attenuating the multiple wave energy and passing the primary energy. To gain some insight into this example, we shall follow the description of Ursin [300].

To estimate the desired response of the array and its characteristics, we make the following assumptions:

1. There is no dip, and moveout across source and receiver arrays is (approximately) linear.
2. The space-variant effect of surface reflections is ignored.
3. Each source subarray consisting of five airguns (there are five such subarrays) is considered a point source.

The (linear) moveout approximation is obtained from the hyperbolic approximation by the first terms of the Taylor series expansion. Let

y = distance between midpoint of source and midpoint of receiver arrays

T = travel time

Then we may write (ignoring higher-order terms)

$$T(y + \Delta y) \simeq T(y) + \frac{dT}{dy}\Delta y$$

The derivative is obtained from the hyperbolic moveout curve:

$$T(y + \Delta y) = \left[T^2(0) + \left(\frac{y + \Delta y}{V_s^2}\right)^2\right]^{1/2}$$

where V_s is stacking velocity. Thus

$$\frac{dT}{dy} = \frac{y}{T(y)V_s^2}$$

so the apparent velocity across the array is

$$V_a = \left(\frac{dT}{dy}\right)^{-1} = \frac{T(y)V_s^2}{y}$$

The apparent wavelength is, therefore, for any frequency f,

$$\lambda_a = \frac{V_a}{f} = \frac{T(y)V_s^2}{yf}$$

Assuming that the source signature is broadband, we replace these values by average values determined by

$$\lambda_{av}(T, y) = \frac{1}{N}\sum_{K=1}^{N}\frac{T(y)V_s^2}{f_k y} \equiv \frac{T(y)V_s^2}{f_{av}\, y}$$

where

$$f_k = f_1 + (f_N - f_1)\frac{k-1}{N-1}$$

is a relationship for choosing N discrete frequencies in the source frequency band and

$$\frac{1}{f_{av}} = \frac{1}{N}\sum_{k=1}^{N}\frac{1}{f_k}$$

Ursin gives a comparison of primary and multiple average wavelengths as functions of time and offset and the two wave fields do indeed separate very nicely after about 1 second of two-way travel time. This is to be expected since Ursin points out that the stacking velocity of the multiples is not very different from the velocity of sound in seawater, as we would expect. Lofthouse and Bennett point out that the most favorable geologic setting for a good separation is one in which a high-velocity layer underlies a shallow-water layer [180].

Using standard equations for the combined source and receiver arrays, Ursin estimates the "average" array response (to broadband reflections) by the equation

$$B(T, y) = \frac{1}{N}\sum_{k=1}^{N} A\left(\frac{T(y)V_s^2}{f_k y}\right)$$

where $A(\cdot)$ is the combined array response. This measure, $B(\cdot)$, gives some idea of the attenuation characteristics of the array for primaries or multiples at various offsets and travel times, and allows us to calculate the *primary-to-multiple ratio*, which is an indicator of array performance. Ursin goes on to suggest that we perform a partition of the χ–t plane based on primary and multiple attenuation and use only that portion in which primaries are attenuated by less than 6 dB and multiples are attenuated by more than 6 dB in forming a stack from the recorded data. He reports the use of symmetric weightings for the receiver array. Chebyshev weights and a set of weights given by Lofthouse and Bennett were also used, and very similar output sections were obtained with each set. The airgun (source) array was timed to give a very sharp pressure pulse with very small bubble energy. The spacing between subarrays is 20 to 56 m. He also gives some general rules for the design of the arrays in practice.

Lofthouse and Bennett report the use of this system to record 17,000 km of offshore data and Ursin's paper confirms its use by other companies as well. Arnold [21] reports on a land-based experiment using arrays of five and nine vibrators, respectively, but concludes in that study that computer-simulated arrays in processing produced results comparable to those obtained in the field.

It is probably safe to say that the greatest impact of array processing on seismic data is obtained in computer processing and not in field implementation. This author, however, predicts that the day is probably not far away when significantly improved field equipment will lead to improved array performance and use at the data acquisition stage.

In our discussion so far we have more or less passively assumed that our receiver arrays are composed of the standard-velocity phones, which measure primarily vertical *p*-wave components of ground motion. However, we have certainly available to us today multicomponent phones which can measure ground motion along three mutually perpendicular axes, and so are capable of distinguishing shear as well as compressional arrivals and indeed can be used to detect particle motion at the earth's surface and so provide a means of distinguishing Rayleigh waves or surface waves from body waves. It is probably true that not much effort has gone into the design of multicomponent arrays. However, some work has been reported and is worthy of comment here. We should point out that working with multicomponent phones carries with it economic and technical considerations, in that field equipment is necessarily limited in the number of available recording channels. As a result, any move to three component phones, which thereby reduces by a factor of 3 the number of recorded channels available for our receiver locations, must be weighed heavily against the potential advantage of using these phones. Since much work remains to be done in the area of multicomponent recording and filtering, it is not surprising that we do not see widespread use of multicomponent phones in the field today. However, as recording systems move in the direction of increased numbers of channels and as our array design procedures improve, and processing techniques improve, we will surely see the move to more and more multicomponent recordings, and in particular it is very likely that we will see increased emphasis on the recording and analysis of shear waves which they make possible. This additional information will hopefully enable us to take a very large step forward in moving from interpretation based on inferred structure to interpretation based on inferred lithology and ultimately to inferred presence or absence of hydrocarbons.

Much interesting work has been and is being carried out outside the field of exploration seismology in the areas of earthquake seismology and passive sonar which finds application in underwater acoustics. Although we shall not comment on the extensive literature related to underwater acoustics here, it is probably worthwhile to mention the interesting seismic experiments carried out with the large aperature seismic array (LASA). In particular, we mention an interesting study by Capon et al. [46] in which the objective was to identify procedures that would result in the successful detection and location of seismic events at as low a threshold magnitude as possible and to determine the effects of different types of array processing in improving the signal-to-noise ratio of the array. In this study three types of array processing were considered, the first being a delay and sum processing, in which a steering delay to align signals at each seismometer output was applied before summation. The second scheme that was tried was equivalent to the first but with variable weights applied to the seismometer outputs, which gives additional control over the side-lobe level as well as steering the main beam. Finally, they considered a filter-and-sum method that used a delay and amplitude weight for each seismometer at each resolvably different frequency, applied by filtering each trace and summing the outputs. In the latter method, maximum-likelihood filters were used. Capon et al. concluded that bandpass prefiltering of the data is very important and that the additional performance gained by maximum-likelihood filtering did not justify its use on-line, but in some cases was a valuable process for off-line processing (we shall return to this topic later).

An interesting refraction study, using source and receiver arrays, was published by Laster and Linville [175]. The source array consisted of 16 shots with spacing of 1000 ft plus three additional shots at the center and 3500 ft off each end. The receiver spread was located 11.5 miles away and consisted of 18 in-line groups and a six-group cross-spread each having a 500-ft interval. The shots were fired and recorded separately and the purpose of the experiment was to separate the refraction arrivals from several different interfaces. The normal procedures for carrying this out would be to design the experiment in such a way that each refracting interface would contribute the first arrival observed for an appropriate array design. In the Laster and Linville method, the data were collected, given the array layouts as described. Multichannel velocity filters were then used to process both the receiver and the shot array. For each shot the 18-element in-line receiver array was processed with a suite of velocity filters, each passing a different velocity. Each multichannel velocity filter produced a single output trace, and these traces, corresponding to a given velocity and different shot point, were gathered so that the resulting events seen on these gathers represent the propagation velocities under the source array.

An analogous filtering procedure was applied to the 16 equally spaced elements of the shot array, and multichannel velocity filters were applied to the 16 records recorded at each group. The output record for each group was then obtained by summing the 16 filtered records with appropriate time shifts. Laster and Linville point out that since the shot spacing was twice as large as the seismometer spacings, it was possible to center on closer velocity intervals than was possible with receiver F–K filtering and thereby they got slightly improved rejection between two early-arriving high-velocity events. They conclude that interfering refracted events can be separated in practice using arrays of shots and receivers combined with the F–K filtering process which proved to be effective in this application.

A very interesting study was published by Burg [42] in which he considered the multichannel and the multidimensional Wiener filtering theory applied to the output of a two-dimensional array of seismometers in order to detect a desired signal in noise. Burg essentially gives the extension of a single-channel Wiener filtering theory to the multichannel and multidimensional case represented by two spatial variables and one time variable and then applied the resulting theory to the outputs of a two-dimensional array of seismometers. He points out that the application of the three-dimensional Wiener theory to the outputs of the seismometer array before summation results in optimized linear array processing at all frequencies, whereas conventional processing techniques are usually optimum only over a very narrow range of frequencies. He goes on to point out that the improvement achieved by the optimal three-dimensional filtering depends on the number and geometry of seismometers in the array and on the separation of signal and noise power spectra in the three-dimensional frequency and vector–wavenumber space. This level of array processing is not generally practiced at the data acquisition stage and so we shall consider it in more detail later.

We mentioned earlier that the dimensionality of seismic array processing need not be restricted to the space and time domain. Indeed, we have pointed out that wave propagation presents us with a variety of parameters which we may use in the discrimination of signals and that these need not be restricted to the space and time parameters. In particular, there are two other signal properties which are easily identified as important in discriminating between certain types of wave motion. These are the dispersion characteristics of certain signals and the polarization characteristics which may be made apparent by the use of multicomponent phones, that is, phones that record wave motion in the direction of several orthogonal axes. However, as little use has been made of these possibilities at the data acquisition level in exploration seismology, we shall not delve into them further at this point, but shall return to this topic later.

Finally, we mention a paper by Brasel [34] in which a brief summary of array techniques which relate more or less to current exploration seismic practice is given, complete with a computer program for calculating amplitude response curves for linear arrays. Savit et al. [259] consider the design of field arrays which minimize the rms output over the reject band and point out that these arrays are useful for field implementation when noise power is distributed evenly over that band.

Although our discussion here has indicated that many factors weigh against the use of very sophisticated array procedures at the data acquisition level in exploration seismology, two comments are worth emphasizing. First, we have pointed out that in many cases, results almost as good as or better than those obtainable in the field can often be obtained by postprocessing data, using array procedures which we shall discuss later. However, some problems, such as dynamic range problems, indicate that this is a somewhat less than valid argument. Finally, we should emphasize that to overcome the many inherent problems with field applications of array processing procedures is very much an economic problem and will not be solved until exploration economics can justify the use of more sophisticated procedures at the field level, which in theory are already available.

2.7 DATA PROCESSING

Since all seismic data are effectively two- or three-dimensional array data, it is not unreasonable to expect that many of our processing procedures take advantage of the array aspect of the data and may be classified, themselves, as array processing procedures. This is not to say that the data could not be considered to be a collection of individual traces, and indeed for some purposes this is the way they are processed. In most events the extra dimensionality of the data supplied by the recording geometry affords us with the opportunity to extract additional information by making use of this extra dimensionality.

We have already mentioned that surface waves differ from body waves in several aspects, which express themselves in the multidimensional character of the recorded data. For example, surface waves seem to be more dispersive than body waves; they may be polarized differently, and they have different apparent velocity across an array. Any subset of these attributes may be used in a multidimensional sense to discriminate between body and surface waves, and this is one of the first major steps in our processing to improve signal-to-noise ratio (SNR).

While *deconvolution* is often treated as a time-series problem, that is, deconvolution operators are frequently applied to single traces, it is certainly possible to carry out multichannel or multidimensional deconvolution operations, and indeed there are cases in which this is desirable, even though it is often thought to be unnecessary in many cases in current practice. Certainly, velocity analysis utilizes the multidimensionality of the seismic data in exploiting the moveout characteristics at various arrivals of interest across the recording array. Without this array format of the data, it is difficult to see how velocity information would be obtained. Velocity information is important not only for the *inverse problem* which we are ultimately trying to solve, but it has other applications as well, including the identification and attenuation of multiple reflections, for example. Stacking is inherently a multidimensional process and whether it is done with or without weighting, it represents a multichannel filtering process.

Multichannel versus Multidimensional Processing

To begin our discussion of array processing of seismic data, we should be aware of the fact that there are basically two different approaches to describing operators to be applied to multidimensional data sets. In many exploration seismic applications, we treat an array of time series obtained at various geophone locations (depending on the application) as an array of input channels, singly indexed (in space) with time as the variable, and we may be interested in designing a linear system (filter) that will convert the input data into a desired output. In other applications, we may assume that a linear system has already imposed its effect on our data and we wish to estimate its parameters (or build a model) in order to invert or modify its effects.

Whatever the problem being considered, the multichannel model for a linear system with m input channels and n output channels is given by

$$\mathbf{y}(t) = \sum_{l=0}^{L} \mathbf{A}_l \, \mathbf{x}(t - l) \tag{2.19}$$

where $\mathbf{y}(t)$ is an $n \times 1$ vector, as shown by

$$\mathbf{y}^T(t) \equiv (y_1(t), \ldots, y_n(t))$$

and $y_k(t)$ are the output channels. Similarly, $\mathbf{x}(t)$ is an $m \times 1$ vector of m input channels, and the coefficients $\mathbf{A}_l$ are each $n \times m$ (constant) matrices. The superscript T indicates transposition. In the geophysical case, the subscript may be thought of as the spatial index or variable with t (time) as the other variable. If we denote the (i, j) element of $\mathbf{A}_l$ by a_{ij}^l, we may write

$$y_{kt} \equiv y_k(t) = \sum_{l=0}^{L} \sum_{j=0}^{m} a_{kj}^l \chi_j(t - l) \tag{2.20}$$

or, equivalently,

$$y_{kt} = \sum_{l=0}^{L} \sum_{j=0}^{m} a_{kj}^l \chi_{j,t-l} \tag{2.21}$$

We recall that in exploration seismology, t is taken to be a discrete variable (as is the spatial variable).

In the two-dimensional convolution representation for an input–output system, both variables referred to above enter into the convolution directly, that is, if we have one spatial variable and the time variable, then

$$y_{kt} = \sum_{l} \sum_{j} b_{lj} \chi_{k-l,t-j} \tag{2.22}$$

If we restrict our attention only to the m output channels $(y_1, \ldots, y_m)$ and to only past or present values (in time) of the n input channels, we may note that the two descriptions defined by Eq. (2.20) or (2.21) and (2.22) are quite similar, with different methods of indexing the spatial variable. However, we note that the multichannel description has many more degrees of freedom due to the larger number of coefficients available.

Equating the outputs of the two systems and requiring that the resulting relationship should hold for all possible inputs yields the result that the two-dimensional system with a fixed number of outputs is a multichannel system in which the coefficient matrices are *Toeplitz* (but not necessarily symmetric).

Requiring

$$\sum_{l=0}^{L} \sum_{j=0}^{m} a_{kj}^l \chi_{j,t-l} = \sum_{l} \sum_{j} b_{k-l,j} \chi_{l,t-j}$$

and choosing

$$\chi_{j,k} \equiv \delta_{p,q}^{j,k}$$

where

$$0 \leq p \leq m, \qquad 0 \leq q \leq L$$

we obtain

$$a_{kp}^{t-q} = b_{k-p,t-q}$$

or

$$a^s_{kp} = b_{k-p,s}; \qquad 0 \le k \le n, \quad 0 \le p \le m, \quad 0 \le s \le L$$

and it follows that $\mathbf{A}_s$ is Toeplitz (but not necessarily symmetric). Since $\mathbf{A}_s$ need not necessarily be Toeplitz, the class of multichannel systems is necessarily larger than the class of two-dimensional systems as described.

Of course, the study of multidimensional systems can be extended to higher-dimensional cases and to recursive descriptions, but then one could do the same for the multichannel description. In any event, it seems reasonable that the restricted degrees of freedom inherent in the two-dimensional linear system theory may be a major factor in explaining why certain single variable results do not extend to the two-dimensional case, but do extend to the multichannel description.

2.8 SPECTRAL ESTIMATION

One of the most widely used diagnostic tools in seismic exploration is the generation of *spectra* in one or more dimensions. In one dimension, spectra of individual seismic traces are generated and analyzed for possible resonances or "parasitic responses" of the individual geophone group, and to help identify other signal characteristics such as 60-Hz interference from power lines, probable usable signal bandwidth, the distribution of coherent or incoherent noise power, and estimates of signal-to-noise ratio. Some of these may be more accurately determined by going to two dimensions and considering the entire array as a function of spatial position and time, and performing a *spectral estimate* for this array. In this case, analyzing the spectrum in terms of *temporal frequency* and *wavenumber* (F–K domain), we find a much nicer separation of signal and noise characteristics determined by apparent velocity. That is, deep body-wave reflections at small offset distances tend to cluster near the infinite velocity line in the F–K domain (they exhibit very low spatial frequencies), whereas horizontally propagating events, such as surface waves and air waves, exhibit very low apparent velocities. Since straight lines through the origin represent constant-velocity lines in the F–K domain (see Fig. 2.13), we may separate a significant amount of signal energy from various types of noise energy in a very neat fashion in this domain. While the two-dimensional spectra obtained from the linear array data restricts us to frequency and wavenumber characteristics or, equivalently, apparent velocity of the arriving signals, it is easy to see that we could do even better if our data had been recorded with an areal array, in which case we could generate a three-dimensional spectral plot which could be used to discriminate on the basis of wave vector orientation as well as temporal frequency, and thereby obtain even more accurate discrimination of the individual plane waves that constitute our recorded wave field. This discrimination would still not be as complete as it could be if the third (vertical) spatial dimension could also be used.

In addition, if multicomponent recording has been employed, we may add the possibility of distinguishing signals whose arrivals exhibit different phase characteristics between the vertical and horizontal components. In particular, we recall that body *p*-wave arrivals appear in phase on each of the components of a multicomponent geophone, whereas Rayleigh waves traveling along the surface exhibit a phase shift between the vertical and horizontal channels. Filters that take advantage of this difference in phase between vertical and horizontal channels are called *polarization filters*. Note that this term has been given multiple meanings in seismology; the term has also been applied to distinguish plane waves whose wave vectors lie in a common plane [137]. Although polarization and dispersion analyses have seen little use in exploration geophysics, we shall make some reference to them later.

In view of the large number of potential applications resulting from the use of multidimensional spectral estimates in exploration seismology, we would expect that significant interest would have been shown in this problem. However, in most exploration applications, spectral estimation is carried out according to the very simplest of procedures and is seldom done in more than two dimensions, resulting in the familiar $F-K$ plots derived from linear seismic arrays with vertical velocity phones. Much more interest has been shown in multidimensional spectral estimation procedures in other areas of seismology and since they are applicable to exploration seismology, it seems appropriate to mention them here even though they have not found widespread use. This situation may change as time goes on and as computing power and exploration economics demand greater precision than has been required in the past.

Probably the most common method for obtaining spectral estimates in exploration seismology today is to weight the appropriate section of the data with a suitable *window*, apply the *fast Fourier transform* (*FFT*) *algorithm* (in one or two dimensions), and then calculate the square of the magnitude of the resulting Fourier transform. The result in the two-dimensional case is usually plotted as shown in Fig. 2.13, where we assume that we are dealing with real signals so that only two quadrants need be plotted and this is known as the $F-K$ plot. To represent the three-dimensional figure required for the two-dimensional spectral estimate, perspective plots, contour plots, and color plots are frequently used.

Until the late 1960s spectral estimation based on windowing data or autocorrelation sequences reached a high level of perfection and might have been the area of extensive research and development in an effort to extend these methods to higher dimensions as our array processing needs began to demand them. However, in the late 1960s new parametric (model-based) procedures for spectral analyses began to be developed, based on the work of Burg [41a], who conceived the *maximum-entropy procedure* for spectral estimation.

Whereas previous procedures had been concerned with either directly estimating the autocorrelation sequence of a partially sampled process, or attempting to remove the truncation effects of a finite data set, Burg simply estimated a model for the stochastic process which made the minimal assumption concerning the behavior of the process outside the observation interval. That is, Burg's method was equivalent to extending the autocorrelation sequence beyond a set of observed values, according to the *principle of maximum entropy*, which, in an information-theoretic sense, makes the least assumption concerning the structure of the underlying process. In this way, Burg was able to obtain an *autoregressive* (AR) model for the stochastic process from which the spectrum could be inferred, and indeed his method turned out to exhibit high-resolution characteristics. Although having some limitations, it did exhibit better spectral estimates, in general, than could be obtained from the windowing procedures, particularly for short data sets.

Burg constrained his algorithm, which could be applied directly to the observed time series, so that the autoregressive model was guaranteed to be stable. In some sense this led to a less-than-optimum model, and other researchers considered the possibility of removing the Burg constraints in an attempt to obtain increased resolution from the resulting spectral estimator [186, 299].

Since that time there has been considerable research on model-based spectral estimation algorithms, including *auto-regressive–moving average* (ARMA) models as well as the AR models. In any event, it is natural that we should attempt to extend these procedures, which do show improved performance in many of the single-channel cases, to the multichannel or multidimensional case. Multichannel extensions of the Burg algorithm were carried out fairly early and share many of its characteristics, including the existence of algorithms that resemble the single-channel algorithm (but which are much more complex). However, the attempt to extend the Burg algorithm or the maximum-entropy formalism to the multidimensional case met with much more difficulty; the maximum-entropy extension was carried out only recently, after many years of effort. The multidimensional analog of the Burg algorithm itself has defied extension.

Windowing procedurees, on the other hand, because they have fallen somewhat into disfavor based on the growing interest and popularity of the model-based algorithms, have not seen significant amounts of research and development, although this would seem to be still a valid area of inquiry, particularly since they are generally simple to calculate and apply. These arguments, of course, tend to lose their validity as computing power increases, but windows are still used in exploration seismology, and some interesting work has recently been reported which may allow further improvements in these algorithms.

In this section we consider window-based procedures and the model-based procedures for spectral estimation as they relate to exploration seismology.

Window-Based Methods

We begin the discussion of multidimensional spectral estimation by considering the simple window-based procedures which are commonly used in seismic processing systems today. We note that these procedures lack some of the sophistication that developed over many years and may be found, for example, in Jenkins and Watts [143]. For most diagnostic purposes, more sophisticated procedures derived from the *periodogram method* are not generally resorted to, as they may be frequently unjustified by the application, or we may do significantly better with less effort by moving to a model-based algorithm. Hence periodogram methods as used in exploration seismology are relatively simple and we shall dispense with them quickly here.

Periodogram methods in exploration seismology have been outlined above and are based on windowing the observed data set and calculating the square of the magnitude of the Fourier transform of the windowed data set using some appropriate window. Many exploration seismic processing systems employ some of the older windows, such as the *Hamming window*, but could probably benefit by replacing these windows with some of the more optimum windows that have been reported in recent years. In particular, we may mention the *Kaiser–Bessel*, *Dolph–Chebyshev*, and *Blackman–Harris* windows. The Hamming window, in fact, is very close to being the two-term Blackman–Harris window. The Kaiser–Bessel window is a very good approximation, which is easily calculated to the corresponding prolate–spheroidal wave function of order zero, which maximizes the energy in a given band of frequencies for some fixed energy level and which has finite support in the time domain. Using the Kaiser–Bessel approximation, we avoid an extremely difficult eigenvalue calculation to obtain the true optimal window. The Dolph–Chebyshev window is the closed-form solution to the problem of obtaining the window with minimum main-lobe width for a given sidelobe level. The Blackman–Harris windows are simply an attempt to optimize the window performance of a sum of weighted and shifted Dirichlet windows. The idea of using weighted and shifted Dirichlet windows was the basis for some of the better early windows, such as the Hamming window. To extend some of the peridogram-based methods, then, we shall have to consider the extension of some of these desirable windows to the multidimensional case.

One of the first papers published on the extension of windowing procedures to more than one dimension was a study by Huang [131], in which he considered extending good one-dimensional windows to the two-dimensional case by a simple transform. Huang considered the two-dimensional window W_2, obtained from the one-dimensional window W, given by $W_2(x, y) = W([x^2 + y^2]^{1/2})$, which results in a circularly symmetric two-dimensional window. He justified this as a potentially good procedure for certain filter design problems, but did not consider its significance for spectral estimation. Speake and Mersereau [275] considered the transformation above, as well as the separable transform $W_3(u, v) = W(u)W(v)$ (which has rectangular support) and again applied these windows to the filter design problem. An interesting comment on the Huang technique has been made by Kato and Furukawa [152], in which they point out that frequency characteristics of the good one-dimensional windows are essentially never preserved by applying the Huang transform and can result in a considerable increase in ripples and sidelobes. Kato and Furukawa suggest instead that a good one-dimensional window should be rotated in the frequency domain:

$$W_4(u, v) = W([u^2 + v^2]^{1/2})$$

A circularly symmetric two-dimensional window is then obtained by inverse Fourier transform. It follows that the one-dimensional window is related to the two-dimensional window by

$$W(x) = \int_{-\infty}^{\infty} W_4(x, y)\, dy$$

that is, a cross section of the two-dimensional frequency response through the origin is the Fourier transform of a one-dimensional integral of the two-dimensional window. Kato and Furukawa conclude that a good two-dimensional window cannot be constructed from a good one-dimensional window via Huang's procedure if it has a wide main lobe and/or large sidelobes.

The only real attempt to find a true multidimensional extension of a single-channel window has resulted from the work of Hesson [127]. Hesson shows that the I_0 sinh (Kaiser–Bessel) window is in fact a special case of the Green's function, G, for the one-dimensional dissipative wave operator, evaluated at a fixed time, T:

$$\nabla^2 G - a^2 \frac{\partial G}{\partial t} - \frac{1}{c^2}\frac{\partial^2 G}{\partial t^2} = 4\pi\, \delta(x - x_0)\, \delta(t - t_0)$$

with

$$G(x, t) = 0 \qquad \text{for } x \text{ on } \sigma,\ t > 0,\ \sigma \to \infty$$

$$G(x, 0) = \frac{\partial G}{\partial t}(x, 0) = 0 \qquad \text{for } x \text{ on } \sigma \text{ } (\sigma \text{ is the boundary of the region})$$

In higher dimensions, G_2 and G_3 can be obtained from $G = G_1$ by

$$G_3(R, \tau) = -\frac{1}{2\pi R}\frac{\partial[G_1(R, \tau)]}{\partial R}$$

where

$$R = |r - r_0|, \qquad \tau = t - t_0$$

and

$$G_2(R, \tau) = \int_{-\infty}^{\infty} G_3(R, \tau)\, dz$$

Formulas are given for G_2 and G_3 directly, and Hesson suggests that G_2 and G_3 are logical choices for higher-order windows in signal processing applications. Jain and Ranganath [142] have also reported multidimensional extensions of the prolate spheroidal wave functions which the Kaiser–Bessel windows approximate.

In summary, then, we may observe that nothing very sophisticated has been done to determine higher-dimensional windows for spectral estimation which are in some sense optimal, with the single exception of the possibility that the recent suggested extension of the Kaiser–Bessel windows will turn out to be good choices for higher-dimensional spectral estimation using periodogram methods. Nuttall et al. [220] consider the use of window and overlap methods for multidimensional spectral analysis.

Model-Based Methods

We now wish to consider model-based algorithms, and we begin with the *maximum-entropy method*, as formulated by Burg. Needless to say, it was not very long after the introduction of the single-channel Burg method that attempts were made to extend it to the higher-dimensional case. Several multichannel extensions of the Burg algorithm have been suggested, but the multidimensional extension is fraught with difficulty and only recently some reasonable first attempts have been made to achieve this generalization.

If a set of autocorrelation coefficients for a single-channel time series is known, the maximum-entropy method of spectral analysis may be viewed as a means of determining a spectral estimate by extending the given set of autocorrelation coefficients in such a way that the underlying time series is as random as possible. It was shown by A. van den Bos [302] that this is equivalent to fitting an *all-pole* or AR model to the data.

A standard method for fitting an all-pole model to data while honoring a set of autocorrelation coefficients $\{\phi_k\}$ for the data is to solve the so-called *normal* (or *Yule–Walker*) *equations* of some predetermined order shown by

$$\begin{bmatrix} \phi_0 & \phi_1 & \cdots & \phi_{n-1} \\ \phi_1 & \phi_0 & \cdots & \phi_{n-2} \\ \vdots & \vdots & & \vdots \\ \phi_{n-1} & \phi_{n-2} & \cdots & \phi_0 \end{bmatrix} \begin{bmatrix} a_1 \\ a_2 \\ \vdots \\ a_n \end{bmatrix} = \begin{bmatrix} \phi_1 \\ \phi_2 \\ \vdots \\ \phi_n \end{bmatrix}$$

This results in the spectral estimate

$$S(f) = \frac{\sigma^2}{\left| 1 - \sum_{k=1}^{n} a_k \exp(-2\pi jfk) \right|^2}$$

The filter coefficients $(1, -a_1, \ldots, -a_n)$ define the *prediction-error filter* for unit distance. It is also obtained by minimizing the mean-square value of the error between the values, χ_t, of the time series and its predicted values $\hat{\chi}_t$ based on the preceding n samples.

Burg suggested that a better spectral estimate might be obtained by passing a prediction-error filter over the data in both the forward and backward directions (without going off the data) and minimizing the average of the error power for these two predictors. This method has the advantage that estimates of the autocorrelation coefficients are not required, as the procedure is applied directly to the data set. Specifically, for a predictor $(a_1, \ldots, a_p)$ of order p (following Jones [145]), let

$$e_t^{(p)} = \chi_t - \sum_{k=1}^{p} a_k^{(p)} \chi_{t-k}, \qquad p + 1 \leq t \leq n$$

If the coefficients for the $(p - 1)$th-order predictor have been estimated, then use the

Levinson recursion,

$$a_k^{(p)} = a_k^{(p-1)} - a_p^{(p)} a_{p-k}^{(p-1)}, \qquad 1 \le k \le p-1$$

where $a_1^{(1)} = \phi_1/\phi_0$, to substitute for the pth-order predictor coefficient above.

$$e_t^{(p)} = \chi_t - \sum_{k=1}^{p-1} (a_k^{(p-1)} - a_p^{(p)} a_{p-k}^{(p-1)})\chi_{t-k} - a_p^{(p)} \chi_{t-p}$$

Define the *backward residuals* by

$$\eta_t^{(p)} = \chi_t - \sum_{k=1}^{p} a_k^{(p)} \chi_{t+k}, \qquad 1 \le t \le n-p$$

to obtain

$$e_t^{(p)} = e_t^{(p-1)} - a_p^{(p)} \eta_{t-p}^{(p-1)}, \qquad p+1 \le t \le n$$

and

$$\eta_t^{(p)} = \eta_t^{(p-1)} - a_p^{(p)} e_{t+p}^{(p-1)}, \qquad 1 \le t \le n-p$$

We then require the sum of the residual powers to be minimized in order to obtain $a_p^{(p)}$. That is,

$$\min_{\{a_p^{(p)}\}} \left[\sum_{t=p+1}^{n} [e_t^{(p)}]^2 + \sum_{t=1}^{n-p} [\eta_t^{(p)}]^2 \right]$$

which yields

$$a_p^{(p)} = \frac{2 \sum_{t=p+1}^{n} e_t^{(p-1)} \eta_{t-p}^{(p-1)}}{\sum_{t=1}^{n-p} [\eta_t^{(p-1)}]^2 + \sum_{t=p+1}^{n} [e_t^{(p-1)}]^2}$$

We then update the coefficient for the filter of order $p+1$ using the Levinson recursion above and generate our next estimates of the prediction error variance, σ^2, by

$$S^{(p)} = [1 - (a_p^{(p)})^2] S^{(p-1)}$$

with

$$S^{(0)} \equiv \phi_0 - \text{the zeroth autocorrelation lag}$$

The final estimate of the spectrum is given above with the filter coefficients of order n substituted for the set $\{a_k\}$. The requirement that the updated coefficients should satisfy the Levinson recursion is a constraint which guarantees the stability of the prediction-error filters. In the formulas above (indeed, throughout this chapter), we assume that the data available for analysis is real valued. We shall have more to say on the Levinson recursion and the Burg technique (for complex-valued data) in Chapter 4.

The extension of this algorithm to the multichannel case is not without difficulty, and the resulting algorithms are somewhat complex. The multichannel extension proceeds without difficulty as outlined above until it is observed that the matrices which appear in place of the term $a_p^{(p)}$ in the recursion equations for the forward and backward residuals are different, although the spectral estimates obtained from these must be identical.

Several successful multichannel extensions have been obtained that do guarantee stability of the predictor. Strand [282] gives one such generalization which reduces exactly to the Burg algorithm in the single-channel case. Another extension has been given by Morf et al. [209]. Since these derivations are quite involved, the reader is referred to these papers and to [53] for further details.

Although some difficulties were encountered in the extension of the Burg algorithm to the multichannel case, the extension to the multidimensional case is significantly more complicated for a variety of reasons. To begin with, if we attempt to formulate the multidimensional maximum-entropy problem as one that involves extrapolation of a given set of autocorrelation coefficients, then in a sense we are already in trouble. Indeed, it was shown by Rudin in 1963 [249] that if we have a function that is positive definite on some subset in n-dimensional Euclidean space, where n is greater than 1, there is no guarantee that an extension of the observed function to a positive definite function on the whole space exists, and indeed there are cases in which a positive-definite extension definitely does not exist. Even if we can be sure that we do have the sampled values of an autocorrelation function in a multidimensional space, the equations for an extension of that function, according to a maximum-entropy criterion, are highly nonlinear and resist solution (see Wernecke [307], etc.). Finally, even if there were no other difficulties, we would run into another problem in deriving an analog of the Burg algorithm in that the multidimensional equivalent of the Levinson recursion does not guarantee stability of the resulting predictor, or (following Marzetta [187]) may require the use of an infinite amount of computation to achieve the next successive approximation to the predictor of the required order. Although many schemes have been devised to approximate the performance of the one-dimensional maximum-entropy spectral estimator, for the multidimensional case, there is only one approach that to date has successfully generalized the maximum-entropy spectral estimation problem to the multidimensional case, and even this approach has not yielded a multidimensional analog of the Burg algorithm.

The successful treatment of the multidimensional spectral estimation problem has been reported by McClellan and Lang [194]. McClellan and Lang considered the problem of multidimensional maximum-entropy spectral estimation for signals received at a finite array of sensors at arbitrary locations. They assumed that the signal autocorrelation values are known on the coarray, Δ, which is a finite subset of the multidimensional Euclidean time–space domain. The problem is to find (or estimate) the (maximum entropy) spectrum that honors the observed autocorrelations and maximizes a measure of the entropy. Specifically, let the coarray, Δ, be the set

$$\mathbf{\Delta} = \{\mathbf{0}, \pm\boldsymbol{\delta}_1, \pm\boldsymbol{\delta}_2, \ldots, \pm\boldsymbol{\delta}_m\}$$

with

$$\boldsymbol{\delta}_0 \equiv 0$$

(we may write $\pm\boldsymbol{\delta}$ since the autocorrelation of a real signal is necessarily known at each of these lags if it is known at one of them). We are required to maximize the entropy

$$H(S) = \int_K \ln [S(\mathbf{k})] \, dv \tag{2.23}$$

where K is the spectral support (a compact subset of frequency–wavenumber space), v is a measure on this set, and S belongs to the set of continuous positive functions with support on K, subject to

$$r(\boldsymbol{\delta}) = \int_K S(k) \exp (j\mathbf{k} \cdot \boldsymbol{\delta}) \, dv, \qquad \boldsymbol{\delta} \in \mathbf{\Delta} \tag{2.24}$$

where r is the set of observed autocorrelations.

In order to proceed, it is necessary that r be actually extendable; that is, there must exist an acceptable S satisfying Eq. (2.24). McClellan and Lang derive a condition for *extendability* that can be shown to reduce to standard tests in particular cases. In particular, r is extendable if and only if $\langle r, p\rangle \geq 0$ for all positive δ-polynomials, p. That is, $p \in P$, where

$$P = \left\{ p(\chi) \,\middle|\, \sum_{\delta \in \Delta} p(\boldsymbol{\delta}) \exp (-j\mathbf{k} \cdot \boldsymbol{\delta}) \geq 0, \quad \text{for } k \in K \right\}$$

if we let

$$C = \left\{ S \in X \,\middle|\, \int_K S(\mathbf{k}) \exp (j\mathbf{k} \cdot \boldsymbol{\delta}) \, dv = r(\boldsymbol{\delta}) \right\}$$

where X is the Banach space of continuous functions on K and

$$D = \{S \in X \mid S(k) > 0\}$$

and suppose that r is extendable, then C and D are convex sets with nonempty intersection. The entropy $H(S)$ may be shown to be a concave functional, so we realize that this is a *constrained optimization problem.* That is, find

$$\max_{S \in C \cap D} H(S)$$

where C and D are convex sets. Using the *Finchel duality theorem* ([181, p. 201]), McClellan and Lang convert the optimization problem above into a dual problem via a direct application of the theorem. The dual problem is finite-dimensional and is guaranteed to have a solution in the set of Δ-polynomials, that is, polynomials of the form

$$P(\mathbf{k}) = \sum_{\boldsymbol{\delta} \in \boldsymbol{\Delta}} p(\boldsymbol{\delta}) \exp(-i\mathbf{k} \cdot \boldsymbol{\delta})$$

If the solution, $P(\mathbf{k})$ is positive on K, then $S(\mathbf{k}) = P^{-1}(\mathbf{k})$ is the maximum-entropy spectrum. Lang gives the necessary and sufficient condition for the existence of the solution to the maximum-entropy problem in the form

$$\int_K P^{-1}(\mathbf{k})\, dv = \infty \qquad \text{for all } p \in \partial P$$

where

$$\partial P = \{p \in P \mid p(\mathbf{k}) = 0 \text{ at some point in } K\}$$

If this is satisfied, the solution to the dual problem is in fact positive and its inverse is the maximum-entropy spectrum. Since the dual problem requires the optimization of a convex functional over a convex set, there are a number of algorithms that might be used to obtain the maximum-entropy solution provided that it exists.

We note that although this work solves the "classical" multidimensional maximum-entropy spectral estimation problem in some generality, it does not yield an analog of the Burg algorithm in its present form, so it remains to be seen whether such a generalization will be found. Because of the generality of Lang's approach, the methods are applicable to other spectral estimation problems as well. A new and different approach to the multichannel and multidimensional maximum-entropy problem has been suggested by Burg et al. [42a].

There is one more important class of estimators for power spectra which is probably worth mentioning with regard to array processing. These have been used with seismic arrays and are computationally quite convenient. In addition, they show increased resolution, as does the maximum-entropy method, over standard periodogram-based methods of spectral estimation referred to earlier. In fact, these *maximum-likelihood spectral estimation* [44] algorithms present no particular difficulties for nonequally spaced array data and have been around much longer than the recently derived extension of the maximum-entropy spectral estimator to the multidimensional case with nonequally spaced sensors. Several interesting papers have been written relating maximum-entropy spectral estimation to maximum-likelihood spectral estimation which clearly justify the statement that maximum-likelihood estimators show less resolution than maximum-entropy estimators. Indeed, Burg [41] showed that in the single-channel case, the reciprocal of the maximum-likelihood spectrum for an estimator of length N is simply the average of the reciprocals of the maximum-entropy spectra obtained by employing estimators of length 1 through N, respectively. Burg's result [41] may be expressed as

$$\frac{1}{\text{MLM}(k)} = \frac{1}{N} \sum_{n=1}^{N} \frac{1}{\text{MEM}(k, n)}$$

where MLM(k) is the N-length maximum-likelihood spectral estimate and MEM(k, n) is the n-length maximum-entropy spectral estimate.

Burg points out that whereas the maximum-entropy spectrum honors the observed autocorrelation values of the process, this is not true of the maximum-likelihood spectrum, although in the single-channel case, both spectra are expressed in autoregressive form. Another interesting relationship between the maximum-entropy and maximum-likelihood spectra was published by Pendrel and Smylie [226] in which they showed that the maximum-likelihood spectra is in fact obtained by maximizing a different entropy measure than the one which is used for the maximum-entropy spectral estimator. In particular, Pendrel and Smylie show that the M-length maximum-likelihood spectral estimator is obtained by maximizing the entropy H_L, where

$$H_L = \tfrac{1}{2} \ln |\mathbf{\Phi}_M|^{1/M}$$

and $\mathbf{\Phi}_M$ is the usual Toeplitz matrix of observed autocorrelations, $\{\phi_K\}$:

$$\mathbf{\Phi}_M(i, j) = \phi(i - j)$$

(see also Jones [144]). This implied relationship in the resolving power of the maximum-likelihood method (MLM) versus maximum-entropy method (MEM) spectral estimates is also supported in studies by Lacoss [167] and Baggeroer [22]. When we consider the multidimensional case, however, there has been no real contest in the past between MLM and MEM since the full multidimensional generalization of MEM spectral analysis was not available and MLM analysis was readily available [44] and easily computed. We will recall, however, that multichannel extensions of MEM have been available [209, 282, 145] but are computed with much greater difficulty.

Following Capon [44], we assume that the output of a sensor located at vector position $\mathbf{X}_j$ is a wide-sense-stationary discrete-time random process with zero mean given by the set $\{N_{jm}\}$, $-\infty < m < \infty$. The covariance of the noise is, as usual,

$$\rho_{kl}(m - n) = E(N_{km} N_{ln}^*)$$

where the asterisk denotes complex conjugation. The cross-power spectral density is

$$f_{kl}^{(\lambda)} = \sum_{m=-\infty}^{\infty} \rho_{kl}(m) \exp(jm\lambda)$$

We now require an estimate of the cross-power spectral density. Although many choices might be made here, Capon uses the following computation: Let

$$S_{in}(\lambda) = (N)^{-1/2} \sum_{m=1}^{N} a_m N_{i,m+(n-1)N} \exp(jm\lambda); \qquad 1 \le i \le K, \quad 1 \le n \le M$$

where K is the number of sensors to be used (the weights $\{a_n\}$ define a window). The number of data points, L, has been divided into M nonoverlapping blocks of N data points. We then define our estimate, $\hat{f}$, by

$$\hat{f}_{il}(\lambda) = \frac{1}{M} \sum_{n=1}^{M} S_{in}(\lambda) S_{ln}^{*}(\lambda); \qquad 1 \leq i \leq K, \quad 1 \leq l \leq K$$

The maximum-likelihood spectral estimate is given by

$$P(\lambda, \mathbf{k}) = \left\{ \sum_{i=1}^{K} \sum_{l=1}^{K} \hat{q}_{il}(\lambda) \exp\left[j\mathbf{k} \cdot (\mathbf{X}_i - \mathbf{X}_l) \right] \right\}^{-1}$$

where the matrix $[\hat{q}_{il}]$ is given by

$$[\hat{q}_{il}(\lambda)] = [\hat{f}_{il}(\lambda)]^{-1}$$

It is important to note that the maximum-likelihood "filter" whose power output is $P(\lambda, \mathbf{k}_0)$ passes a monochromatic plane wave with wave vector $\mathbf{k}_0$ without distortion, while attenuating in an optimum least-squares sense the power of waves with wave vectors different from $\mathbf{k}_0$ (see also [112]).

An interesting study by Woods and Lintz [319] considers the resolving power of the maximum-likelihood method. They show that the maximum-likelihood spectral estimate can achieve arbitrarily high resolution in the limit, as the background white noise tends to zero, provided that the wave field consists of a finite number (no more than the number of sensors, in the case of uniform arrays) of correlated plane waves.

In recent years, there has been much attention devoted to lattice implementations [183] of some of the familiar spectral estimation algorithms and we might expect that multichannel or multidimensional extensions of these methods would be sought. A number of papers [6, 111, 141, 184, 185, 237] have appeared carrying out this program for determining single-channel AR model coefficients, and Alam [3], Griffiths [113a], and Perry and Parker [227] consider multichannel applications. In particular, Alam [3] has presented a very general approach to this problem which can lead to AR, multichannel (MC), rational polynomial (RP), autoregressive orthonormal (ARON) models, and autoregressive–moving average (ARMA) models.

The approach used by Alam is to pose the spectral estimation problem as a problem of estimating the observation vector from a set of linearly independent vectors that form the estimation subspace. The procedure is basically to apply the observation vector to one channel and to feed the estimation vectors into other channels as the lattice is traversed. At each stage, inner products and error vectors are estimated, as linear combinations of the estimation vectors are fitted to the observation vector via orthogonal projection. The error vector inner products and norms are then used in a recursive scheme to estimate the desired model coefficients.

When the estimation vectors are delayed versions of the input (observation) vector, an AR model is generated. If the estimation vectors are a set of input AR vectors and a set of output AR vectors, an RP model results. If the estimation vectors are a mixture of autoregressive and orthonormal vectors, an ARON model results, and a multichannel model may be obtained from an arbitrary set of vectors. The actual filter structure will change with each type of model considered.

The actual calculations for the multichannel orthonormal lattice filter proceed as follows. Let the error vectors at the zeroth stage $\mathbf{X}_0^0, \mathbf{X}_0^1, \ldots, \mathbf{X}_0^N$ consist of the observation vector followed by the sequence of N estimation vectors. The inner products and reciprocal norms needed for the projections at stage j are generated by

$$\rho_i^j = -(\mathbf{X}_{i-1}^j, \mathbf{X}_{i-1}^i)$$

$$\mu_i^j = \frac{1}{[1 - (\rho_i^j)^2]^{1/2}}$$

where

$$\mathbf{X}_i^j = \mu_i^j \mathbf{X}_{i-1}^j + \mu_i^j \rho_i^j \mathbf{X}_{i-1}^j$$

for $1 \le i \le j - 1$. The filter coefficients, $A_0, \ldots, A_N$, are obtained from the recursions

$$P_0^1 = 1$$

$$Q_0^1 = 1$$

$$A_0 = 1$$

$$D_0 = 1$$

$$\mathbf{A}_j = \begin{bmatrix} A_{j-1} & 0 \\ 0 & D_{j-1} \end{bmatrix} \begin{bmatrix} \mu_j^0 & \\ \mu_j^0 & \rho_j^0 \end{bmatrix}$$

where D_j is obtained from the recursions

$$\mathbf{Q}_i^{j+1} = \begin{bmatrix} 0 \\ Q_i^j \end{bmatrix}; \qquad 0 \le i \le j - 1$$

$$\mathbf{P}_0^{j+1} = \begin{bmatrix} P_0^j \\ 0 \end{bmatrix}$$

$$P_i^{j+1} = \mu_i^{j+1} P_{i-1}^{j+1} + \mu_i^{j+1} \rho_i^{j+1} Q_{i-1}^{j+1}; \qquad 1 \le i \le j$$

$$Q_j^{j+1} = P_j^{j+1}$$

$$D_j = (P_j^{j+1})^R$$

where R denotes reversing the order of the coordinates in the vector. The coordinate lengths of these vectors increase by 1 at each iteration stage. The required spectral estimates are obtained from the resulting model in the usual way. Needless to say, these filters can be used as deconvolution filters, and so on, for the processes they model.

Several applications in exploration seismology depend on the multidimensional spectral estimates obtained from various choices of the methods mentioned above. Possibly the most important use made of these estimates is to identify strong noise components such as surface waves, air waves, 60-Hz pickup on our signals and to use these as a diagnostic guide for designing multidimensional filters to reject the noise components while retaining the signal components. As we have mentioned earlier, these components represent body waves, which tend to cluster around the line of infinite apparent velocity. In a simple case of discriminating solely on the basis of differences in apparent velocity, the resulting multidimensional filters are referred to as *velocity filters* or *moveout filters.*

Having designated regions in the F–K domain containing signal and, separately, noise components, we may calculate relative power levels for these different signal components in the F–K domain and use *Parseval's theorem* to equate the ratio of signal power to noise power to the corresponding ratio in the space–time domain. In this way, estimates of signal-to-noise ratios for various components of the recorded wave field can be easily obtained. In addition, these F–K plots are useful for detecting spatial and temporal aliasing problems on the data by observing folding in the F–K domain, and through these various calculations, may suggest proper field procedures which result in high-quality data for further processing.

2.9 MULTIDIMENSIONAL AND MULTICHANNEL FILTERING

Multidimensional and multichannel filtering of seismic data is used in a wide variety of applications in exploration seismology. Among these, we may briefly mention the attenuation of surface waves and other noise components on the recorded wave field using velocity as well as frequency filters, the attenuation of multiple reflections using a variety of methods, and a variety of deconvolution procedures. They also have applications in velocity analysis, stacking, and migration. We shall attempt to comment on all of these areas of application for multidimensional and multichannel filtering, and to begin our discussion, we consider the various possibilities for filtering array data.

We may distinguish between several different methods of processing array data and we may recall that any of these methods might be applied both at the data acquisition stage in the field as well as in the form of postprocessing on the recorded data after they have been collected. As we have already mentioned, the majority of array processing applications are carried out on the recorded data and are not implemented as a part of our field procedures except in the simplest cases which we have made reference to earlier.

At this point it is appropriate that we begin to consider the basic principles of array processing as it is applied in seismic data processing. Previous authors in this area have divided seismic array processing into roughly three levels of complexity. The first of these is known as *direct sum*, and has already been mentioned with regard to field procedures. This method relates to the simple concept of summing the output of an array of sensors in order to enhance their ability to distinguish between signals and noise. We have mentioned before that this simple procedure acts as a vertical beamformer with some improved noise rejection characteristics which depend entirely on the apparent wavelengths of the signals across the array. In particular, arrays of this type exhibit sidelobe characteristics that are not optimal in any sense.

The next improvement in the level of complexity in array processing consists of the *weight-and-sum* procedures, which are also reasonable for field implementation. In these, the geophones are assigned weights which are not necessarily equal. By using sets of weights such as are used in the Chebyshev arrays, for example, improved characteristics over the simple direct sum approach can be achieved. However, it is often felt that even this small step toward increasing complexity is somewhat impractical for field implementation, although this situation is rapidly changing as computer power comes to the field.

The obvious next step in complexity is to allow the ability to introduce delays. This clearly implies more sophisticated field devices which are capable of storing information, or in other ways delaying the time period from the arrival of a signal at the array until it is actually recorded. When combined with weighting and summing, we achieve full filtering capabilities for the array output and can apply sophisticated procedures based on multichannel or multidimensional techniques to process the arrays. In almost all cases today, this level of array processing is applied to the recorded data sets after they have been acquired in the field.

It is to this highest level of complexity that we now turn our attention. The most widely used class of filters in seismic data processing would be the class of *Wiener filters*, which find numerous applications in current data processing practice. Let us now consider the basic equations for multichannel Wiener filtering.

The relationship for multichannel filtering of the multi-input time series $\chi_1(t), \ldots, \chi_N(t)$ to obtain the multichannel output time series $y_1(t), \ldots, y_M(t)$ is given by

$$\mathbf{y}(t) = \sum_{k=0}^{L} \mathbf{A}_k \boldsymbol{\chi}(t-k)$$

where $\mathbf{y} = [y_1, \ldots, y_M]^T$, $\boldsymbol{\chi} = [\chi_1, \ldots, \chi_N]^T$, and $\mathbf{A}_k$ is an $M \times N$ matrix. If we determine a desired (multichannel) output, $\mathbf{d}$, we may seek to find coefficient matrices $\{\mathbf{A}_k\}$ for which the mean-square error between desired and actual output is a minimum. To be more specific, if we denote

$$\mathbf{A}_k = [a_{ij}^k]; \qquad 1 \leq i \leq M, \quad 1 \leq k \leq L, \quad 1 \leq j \leq N$$

and define

$$f_{ij}(k) = a_{ij}^k$$

then, in component form, we have

$$\begin{aligned} y_i(t) &= \sum_{k=1}^{L} \sum_{j=1}^{N} a_{ij}^k \chi_j(t-k) \\ &\equiv \sum_{k=1}^{L} \sum_{j=1}^{N} f_{ij}(k) \chi_j(t-k) \\ &= \sum_{j=1}^{N} f_{ij} * \chi_j(t) \end{aligned}$$

where $*$ denotes convolution. Our problem, then, is to determine the filters f_{ij} that minimize the mean-square error

$$\sum_{i=1}^{M} \sum_{t=1}^{N+L} \left[d_i(t) - \sum_{j=1}^{N} f_{ij} * \chi_j(t) \right]^2 \tag{2.25}$$

Solving this minimization problem, we arrive at the *normal equations* [293]

$$\mathbf{FR} = \mathbf{G} \tag{2.26}$$

where

$$\mathbf{F} = [\mathbf{f}_0, \ldots, \mathbf{f}_L]^T$$

$$\mathbf{f}_k = [f_{ij}(k)]$$

$$\mathbf{R} = \begin{bmatrix} \boldsymbol{\phi}_0 & \boldsymbol{\phi}_1 & \cdots & \boldsymbol{\phi}_M \\ \boldsymbol{\phi}_1^T & \boldsymbol{\phi}_0 & \cdots & \boldsymbol{\phi}_{M-1} \\ \boldsymbol{\phi}_M^T & \boldsymbol{\phi}_{M-1}^T & \cdots & \boldsymbol{\phi}_0 \end{bmatrix}$$

$$\mathbf{G} = [\boldsymbol{\Psi}_0, \boldsymbol{\Psi}_1, \ldots, \boldsymbol{\Psi}_M]^T$$

where

$$\boldsymbol{\phi}_k = \begin{bmatrix} \phi_{\chi_1\chi_1}(k) & \phi_{\chi_1\chi_2}(k) & \cdots & \phi_{\chi_1\chi_L}(k) \\ \phi_{\chi_2\chi_1}(k) & \phi_{\chi_2\chi_2}(k) & \cdots & \phi_{\chi_2\chi_L}(k) \\ \vdots & \vdots & & \vdots \\ \phi_{\chi_L\chi_1}(k) & \phi_{\chi_L\chi_2}(k) & \cdots & \phi_{\chi_L\chi_L}(k) \end{bmatrix}$$

and $\phi_{\chi_j\chi_l}(k)$ is the kth lag of the autocorrelation of χ_j with χ_l. Similarly, we have

$$\boldsymbol{\Psi}_k = \boldsymbol{\phi}_{\boldsymbol{\chi}\mathbf{d}}(k)$$

where $\boldsymbol{\phi}_{\boldsymbol{\chi}\mathbf{d}}(k)$ is the kth lag of the cross-correlation between $\boldsymbol{\chi}$ and $\mathbf{d}$. The normal equation (2.26) may be solved by a *multichannel generalization of the Levinson algorithm* [313].

Note that if the desired output d is the (multichannel) delta (situated at the origin), the filter, $\mathbf{F}$, is minimum phase [242], a property that generalizes the corresponding result in the single-channel case.* In the multidimensional case, however, this property does not generalize [109].

Let us now consider a two-dimensional analog of the Wiener filter. Following Justice [146], we consider the quarter-plane problem of finding an $(n + 1) \times (m + 1)$ array $\mathbf{a} = \{a_{kl}\}$ for which it is true that the convolution of $\mathbf{a}$ with an input array, $\mathbf{b} = \{b_{kl}\}$, $k \geq 0$, $l \geq 0$, yields the best least-squares approximation to the desired output array,

$$\mathbf{d} = \{d_{kl}\}; \qquad k \geq 0, \quad l \geq 0$$

*We say that a causal sequence is minimum phase if its z-transform has no poles or zeroes outside or on the unit circle.

The solution again results in a set of normal equations of the form

$$\mathbf{\Phi A} = \mathbf{G} \tag{2.27}$$

where **A** and **G** are vectors (obtained by concatenating other vectors), as shown by

$$\mathbf{A} = [\mathbf{a}_0, \mathbf{a}_1, \ldots, \mathbf{a}_n]^T$$

and

$$\mathbf{a}_k = [a_{k1}, \ldots, a_{km}]^T$$

$$\mathbf{G} = [\mathbf{G}_0, \ldots, \mathbf{G}_n]^T$$

where $\mathbf{G}_k = [g_{k1}, \ldots, g_{km}]$ and $g_{kj} = (k, j)$th lag of the cross-correlation of the input and desired output arrays. Finally, $\mathbf{\Phi}$ is the block-Toeplitz matrix of autocorrelations given by

$$\mathbf{\Phi} = \begin{bmatrix} \mathbf{\Phi}_0 & \mathbf{\Phi}_1 & \cdots & \mathbf{\Phi}_n^H \\ \mathbf{\Phi}_1 & \mathbf{\Phi}_0 & \cdots & \mathbf{\Phi}_{n-1}^H \\ \vdots & \vdots & & \vdots \\ \mathbf{\Phi}_n & \mathbf{\Phi}_{n-1} & \cdots & \mathbf{\Phi}_0 \end{bmatrix}$$

where the superscript H denotes Hermitian transposition (i.e., complex conjugation and ordinary transposition) and

$$\mathbf{\Phi}_{k-l} \equiv \mathbf{\Phi}_{k,l}$$

$$\mathbf{\Phi}_{k,l}(i, j) = \mathbf{\Phi}_{k-l,\, i-j}$$

The latter is the $(k - l, i - j)$th lag of the autocorrelation of the input array.

An algorithm for solving the system of normal equations (2.27) was given by Justice [146] (see also [2, 238, 305]). A more thorough discussion of the multidimensional Wiener filtering problem may be found in Ekstrom [84].

The Wiener filtering problem just discussed treats the (multichannel) problem of filtering input traces so that the output traces best approximate (in the least-squares sense) a desired output. In this formulation, the filter coefficients are unconstrained. Alternatively, we can formulate a *constrained adaptive filtering procedure* in which the filter design constantly adapts to meet some prescribed criterion while being constrained by some desired output characteristic.

An example of a filter of this type is an *adaptive* (*multichannel*) *noise canceler*, which is constrained to pass a particular type of signal while minimizing the output power (the error criterion). We may design an adaptive maximum likelihood filter in this way, for example. The term "adaptive" here (as opposed to the commonly used term "data adaptive") means *time adaptive* and implies that the filter coefficients are regularly updated (once per time step) to meet the error criterion as the data are processed. The concept of minimizing the array output power is equivalent to minimizing the average noise power if the array response to the desired signal (the constraint) is constant and the noise is uncorrelated with this signal [159a]. Since the first algorithms for solving the adaptive filtering problem were popularized in the late 1960s [Widrow, 310], several applications of these methods in geophysics have been reported.

Booker and Ong [32] formulate the problem as follows: Minimize the energy of the filter output

$$O_t = \sum_{i=1}^{N} \sum_{j=1}^{L} f_{ij} \chi_{i,t-j}$$

of a linear multichannel filter, f, subject to a system of linear constraints

$$\mathbf{MF} = \boldsymbol{\alpha}$$

where $\mathbf{M}$ and $\boldsymbol{\alpha}$ are given, N is the number of channels, and L is the filter length. In this system of equations $\mathbf{M}$ is a matrix, $\mathbf{f}$ and $\boldsymbol{\alpha}$ are vectors obtained by concatenating rows of what would otherwise be matrices. Booker and Ong derive the following gradient algorithm for updating the filter coefficients:

$$\mathbf{f}_{t+1} = \mathbf{f}_t - 2k_s O_t[\mathbf{I} - \mathbf{M}^T(\mathbf{M}\mathbf{M}^T)^{-1}\mathbf{M}]\chi_t$$

and the updating is done at each time increment. The *stepsize* parameter, k_s, prevents the filter from overreacting to transient events. They apply this procedure to several problems and indicate some improvement over the results obtained by conventional processing.

Gangi and Byun [100] investigate the use of the corrective gradient projection (CGP) method due to Frost [92] and compare this method with several other well-known methods for adaptive filter design. Shen [267] considers the design of an *adaptive maximum-likelihood beamformer* and applies it to short period array data from the Korean Seismic Research Station. Chang [50] suggests minimizing the l_1 norm of the filtered output rather than the usual output power. In [51], Chang relates Kalman filtering techniques to this approach and reports improved performance of the resulting filters.

It is clear that we are just beginning to see the application of adaptive multichannel filtering to seismic data processing and the gradient methods in use will surely supersede the former attempts to produce time-variant or adaptive filtering by gating the data, assuming local stationarity. The constraint that the desired signal shall be passed without distortion [112] will probably continue to be important for seismic data processing.

2.10 VELOCITY FILTERING

Certainly one of the earliest and most important applications of array processing procedures in exploration seismology was the design of filters for the purpose of discriminating among events with different apparent velocities. In particular, filters were designed to pass events that fell within a particular range of apparent velocities and to reject events whose apparent velocities were outside this range. These filters came to be known as *velocity filters*, *fan filters*, or *moveout filters* and were referred to in one early publication as *pie-slice filters.* A variety of techniques have been applied to the velocity filtering problem and, indeed, this is still an active area of research.

One of the earliest papers on the subject, published by Embree et al. [85], was concerned with the design of what was effectively a multichannel filter to pass certain apparent velocities and to reject others. Shortly after this publication appeared, Wiggins [311] considered the more general problem of designing two-dimensional filters that could pass or reject various portions of the F–K domain. Wiggins' approach was to calculate the required two-dimensional filter coefficients by doing a least-squares approximation to the desired response in the frequency domain. He also considered including a weighting factor so that the filtering scheme would apply more importance to certain areas of interest than to others. Burg [42] considered the velocity filtering problem for a two-dimensional array of seismometers and therefore considered the design of multichannel and multidimensional filters which exhibit desired pass and reject characteristics in the frequency–wave vector domain (see Fig. 2.17).

Because the original multichannel filtering algorithm, proposed by Embree et al., required a number of convolutions equal to the number of input channels, it was considered to be somewhat costly to apply in practice. Treitel et al. [295] suggested an improvement in this algorithm which consisted of weighting and shifting the input traces, summing the results, and then performing a single filtering or convolution operation on the resulting trace. In addition, the required filter was implemented in recursive form which resulted in a further improvement in the time required for the computation. Later, a more general design procedure was suggested for two-dimensional filters in which the form of the filter was recursive and the stability problem was resolved for the general class of all-pole multidimensional filters [266a] [148a].

In addition to the approaches mentioned above for designing velocity filters, other multichannel filter design methods were considered for application to this

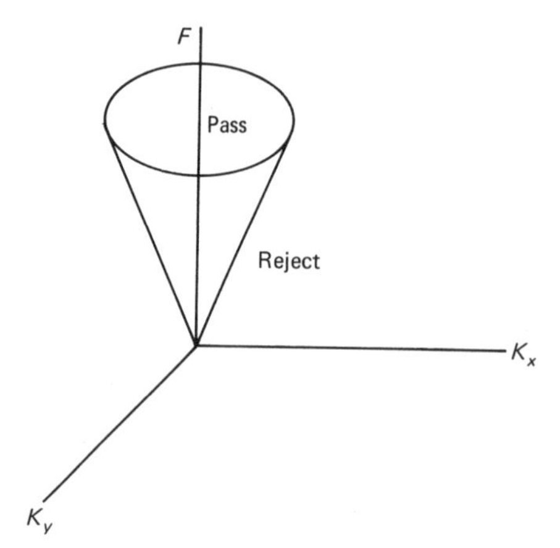

Figure 2.17 Example of filter design specifications in the frequency–wave vector domain for two spatial dimensions.

problem. Sengbush and Foster [266] considered a simple model for the seismic trace with offset given by

$$I_l(t) = S(t - \tau_l) + N(t - \tilde{\tau}_l) + U_l(t)$$

where I_l is the lth trace, S is the signal whose movement is τ_l on the lth trace, N is coherent noise with moveout $\tilde{\tau}_l$, and U is incoherent noise. S, N, and U are assumed to be independent second-order stationary processes. Finally, the moveouts τ_l and $\tilde{\tau}_l$ are assumed to be random variables with joint probability density function $f(\tau_1, \ldots, \tau_n; \tilde{\tau}_1, \ldots, \tilde{\tau}_n)$. Various choices for this probability density function relate to the degree of precision with which we know the values of the moveouts.

Using the Wiener multichannel filter design criteria together with the *harmonic orthogonal decomposition theorem* [113, p. 33], the problem of determining the filter, G_l, for the lth channel is reduced to solving a set of equations involving the power spectral densities of the signal and noise components, and expectations of terms involving the moveouts. Given a particular probability density, these terms are known and the corresponding equations can be solved to obtain the required (multichannel) Wiener filter:

$$\sum_{i=1}^{n} C_{li}(\omega)G_i(\omega) = E\left[\exp(-j\omega\tau_l)p_S(\omega)\right]; \qquad 1 \leq l \leq n$$

where E is the *expectation operator* and

$$C_{li}(\omega) = A_{li}(\omega)p_S(\omega) + B_{li}(\omega)p_N(\omega) + \delta_{li}\, p_u(\omega)$$

$$A_{li}(\omega) = E\left[\exp(j\omega\tau_i)\exp(-j\omega\tau_l)\right]$$

$$B_{li}(\omega) = E\left[\exp(j\omega\tilde{\tau}_i)\exp(-j\omega\tilde{\tau}_l)\right]$$

$$p_S(\omega) = \text{power spectral density of signal}$$

$$p_N(\omega) = \text{power spectral density of coherent noise}$$

$$p_u(\omega) = \text{power spectral density of incoherent noise}$$

$$\tau_l = \text{signal moveout on } l^{\text{th}} \text{ trace}$$

$$\tilde{\tau}_l = \text{coherent noise moveout on } l^{\text{th}} \text{ trace}$$

When the Sengbush and Foster paper appeared, another paper appeared beside it by Galbraith and Wiggins [99]. In this paper, multi-channel Wiener filters are derived for velocity filtering applications in which it is assumed that we have a knowledge of the relative (trace-to-trace) arrival times of signal and noise events, but no knowledge of the actual event shapes. Other parameters, such as variable gain between traces, can be accounted for, and the optimum Wiener filter can be computed to pass the "signal" events and attenuate the "noise" events. These filters are useful for velocity filtering, *deghosting* (removal of a signal echo generated by a reflection from the surface due to a buried source), and stacking.

Hubral [137] also considered a multichannel Wiener filter approach for two-dimensional seismic arrays, in which he assumed the same model used by Sengbush and Foster (above) for the seismic traces with offset. In a development similar to the Galbraith and Wiggins approach, the required multichannel normal equations are derived (in the frequency domain):

$$\sum_{i=1}^{N} A_i(f)[\Phi_{ss}(f) \exp [2\pi jf(\tilde{\alpha}_k - \tilde{\alpha}_i)] + \rho\Phi_{rr}(f) \exp [2\pi jf(\tilde{\alpha}_k - \tilde{\alpha}_i)] + \nu\Phi_{n_i n_k}(f)\delta_{ik}]$$

$$= \Phi_{ss}(f) \exp [2\pi jf\alpha_k]; \qquad 1 \le k \le N$$

where S = signal (common to all traces)
n_i = uncorrelated noise on ith trace
r = correlated noise
$\alpha, \tilde{\alpha}$ = time delays for signal and correlated noise between traces

and signal and noise are uncorrelated.

Hubral then posed several filtering "problems" (plane waves scattered around a given direction vector, "polarization" filter, and general velocity filter) and proceeded to generate a model and derive the necessary correlations for each of these problems based on expectations and probability density models. These filters are then designed to pass plane waves with certain characteristics and to reject all others.

Using a deterministic approach, McClellan and Parks [195] consider the Chebyshev design problem for *two-dimensional fan filters.* Making use of the special (frequency-domain) form of fan filters, a three-step mapping is obtained from the two-dimensional problem to a one-dimensional problem in this order. Let

$$H(f_1, f_2) = \sum_{k=0}^{N} \sum_{l=0}^{N} a(k, l) \cos (2\pi kf_1) \cos (2\pi lf_2)$$

be the two-dimensional symmetric (see [195]) amplitude response of the required filter. Define changes of variables

$$T_1: \begin{cases} u = \cos (2\pi f_1) \\ v = \cos (2\pi f_2) \end{cases}$$

$$T_2: \quad \chi = \tfrac{1}{2}(u - v)$$

$$T_3: \quad \chi = \cos (2\pi f)$$

The inverse transform is given by T_1 T_2 T_3 and this sequence preserves the equiripple (Chebyshev) character of the one-dimensional solution. Bandpass filters in one dimension transform approximately into fan filters in the two-dimensional case with some error in the pass and reject boundaries. Although McClellan and Parks point out that an equiripple two-dimensional design procedure would be an improvement, there is, to date, no known analog of the Haar condition in more than one dimension and so Chebyshev design procedures are lacking in these cases.

Other interesting approaches to velocity filter design have been given [222]. Of course, much of the extensive literature on two-dimensional filtering which exists today may be applied to this problem. Justice [148] points out that multidimensional filters may be used to design time- or space-variant filters of lower (multi) dimensionality.

Cassano and Rocca [48], again working with the model of the seismic trace suggested by Sengbush and Foster (above), suggest several errors which are inherent in these multichannel filter designs when applied to stacked data, and they suggest several modifications in the models to avoid these errors. Several interesting examples are shown for the application of these filters to actual stacked data.

A nonlinear multichannel stacking filter designed to enhance arrivals with a fixed phase velocity has been reported by Kanasewich et al. [151] based on a method due to Muirhead [211]. This technique is a simple delay-and-sum beamformer except that an Nth root is taken on each channel before summation followed by evaluating the Nth power of the sum. Specifically, let

$\chi_j(i)$ = ith sample on channel j

$Y_v(i)$ = ith output sample for a beamformer that passes phase velocity, v

G_j = gain of the jth channel

G = gain to which all channels are to be normalized

W_j = weighting coefficient

$\tau_j = D_j/V$

D_j = distance of jth receiver from center of array

V = phase velocity of desired wave

Then we have

$$Y_v(i) = R_v(i)\,|R_v(i)|^{N-1}$$

where

$$R_v(i) = \left(\sum_{j=1}^{K} W_j\right)^{-1} \sum_{j=1}^{K} \frac{\chi_j(i + r_j)}{|\chi_j(i + r_j)|} (GW_j\, G_j^{-1}\, |\chi_j(i + r_j)|)^{1/N}$$

Kanasewich et al. report that this filtering procedure exhibited superior performance to any linear multichannel filter tested.

Finally, we may recall that Biswell et al. [30] have reported a slant-stack procedure applied to geophone subarrays as an alternative to summing the outputs. This procedure is designed to produce an output trace consisting of events that exhibit apparent velocities in a preselected range (determined by the slowness parameter). Finally, we have mentioned earlier that we can do an even better job of wavefield discrimination if we increase the dimensionality of our recording arrays. As we begin to see increased interest in areal surveying geometries, we will necessarily see many techniques, such as velocity filtering, extended to the frequency-wavevector domain and, indeed, these methods are beginning to appear in the literature [39a].

2.11 POLARIZATION AND DISPERSION ANALYSIS

We have just discussed the discrimination of seismic wave fields on the basis of apparent velocity. Although it is certainly true that velocity filtering represents one of the most widely used and powerful tools for discriminating between different wave field components in current data processing practice, we indicated earlier that there are other potential discriminators available to us for this purpose. Two of these discriminators, *polarization* and *dispersion*, although certainly not widely used today in exploration seismology, have seen interest and development in the area of earthquake seismology and some thought at least is being given to the application of these methods in seismic data processing. Although characteristics such as dispersion will manifest themselves much more clearly in earthquake data, for example, than in exploration seismic data, due in part to the much shorter distances traveled in exploration seismology, dispersion may still often be a detectable characteristic and the same can be said for polarization effects exhibited by Rayleigh waves, for example. We can justify the consideration of these two additional wave parameters by the fact that they do represent applications of array processing procedures, and by the fact, to some extent at least, that each is deemed potentially important for exploration seismology.

To detect polarization in an acoustic wave we must have a multicomponent detector. That is, we must have a detector with capabilities for distinguishing particle motion or velocity, ideally along each of three mutually orthogonal axes. It appears to be generally true that body waves (i.e., those that are generated by reflection at boundaries between layers within the earth) tend to show *linear polarization*, that is, a zero-phase shift between vertical and horizontal components. This is another way of saying that the vertical and horizontal components for body wave reflections are in phase. On the other hand, Rayleigh waves tend to show *elliptic polarization* of particle motion in a plane orthogonal to the surface of the earth and to the propagating wavefront. The exact form of the particle motion depends on the mode of the Rayleigh wave being considered. For the fundamental mode, the motion is retrograde elliptic (Fig. 2.18).

In any event, these differences in polarization of the particle motion associated with each particular wave type suggest a means of distinguishing between these waves on the basis of their particle motion. As a result, if we have multicomponent detectors, we may conceive of the possibility of distinguishing between various components of the recorded wave field on the basis of linear or elliptic polarization of particle motion. Certainly the most significant attempts to distinguish between

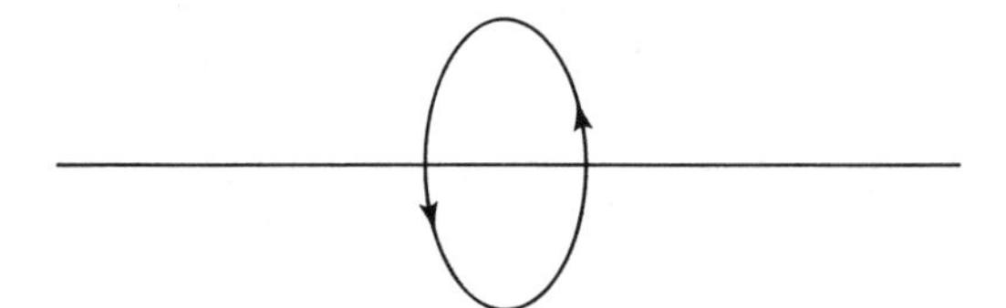

Figure 2.18 Retrograde elliptic particle motion for fundamental mode of Rayleigh wave.

wave field components on the basis of polarization have been made in the area of earthquake seismology. One of the earliest attempts to carry out this kind of analysis was published by Shimshoni and Smith [269]. In this study, Shimshoni and Smith define a cross-product over a finite time window of the horizontal and vertical components of motion and also define a normalized correlation coefficient using these two components, again over a finite time window.

If we let H_i and V_i be the horizontal and vertical components of motion at time $i\,\Delta t$, the cross-product M_j is defined by

$$M_j = \sum_{i=-n}^{n} H_{i+j} V_{i+j}$$

over the time window $[-n, n]$ and the correlation R_j is defined by

$$R_j = \frac{(2n+1)\sum H_{i+j} V_{i+j} - \sum H_{i+j} \sum V_{i+j}}{\{[(2n+1)\sum H_{i+j}^2 - (\sum H_{i+j})^2][(2n+1)\sum V_{i+j}^2 - (\sum V_{i+j})^2]\}^{1/2}}$$

Shimshoni and Smith point out that for a compressional wave whose vertical and horizontal components were exactly in phase (linear polarization), R_j would be $+1$, whereas for a vertically polarized shear wave, R_j would be -1. The cross-product, M_j, is then multiplied by the two components of the original signal to emphasize those events that are rectilinearly polarized. They also considered fitting ellipses to the particle motion and using the parameters of these ellipses as discriminators of polarization.

Subsequently, Mims and Sax [206] and others [18] considered the use of polarization filters to discriminate linearly polarized wave motion from elliptic and random polarizations. This type of filter is called a *rectilinear motion detector* (REMODE) and is again based on a cross-correlation between a vertical and horizontal component suitably rotated [150], so that the expected wave vector of the desired body wave bisects the vertical and horizontal component axes. The filter operator C is obtained from the equation

$$C_j = \sum_{i=-n}^{n} V_i H_{i+j}, \qquad j > 0$$

and C is made two-sided by the requirement

$$C_{-j} \equiv C_j$$

The filter C is now a zero-phase filter and tends to suppress nonlinearly polarized motion when convolved with the original component time series.

Flinn [87], and Archambeau and Flinn [18] use three-component recordings and calculate the eigenvalues of the quadratic form of the covariance matrix between the horizontal components and the vertical component. In practice, the data are filtered through narrowband Gaussian filters, then passed through the polarization filter, and the outputs are summed over the desired bandwidth to enhance signal-to-noise ratio.

Mercado [202] actually applies a two-channel velocity filter to the horizontal and vertical component channels to pass signals with a given phase delay between these channels and also applies a zero-phase narrow-bandpass filter prior to the velocity filter to refine further the (frequency-dependent) moveout between the two channels at a selected center frequency. These filtered outputs are, of course, summed over the desired bandwidth. The resulting process does not distort wave forms as do some of the nonlinear methods (above).

Flinn [87] and Simons [271] use all three components to design wave-vector filters which pass only waves with a given wave-vector orientation (using a single three-component phone as opposed to an areal array as discussed previously).

As an extension of the method proposed by Flinn [87], Montabetti and Kanasewich [207] obtain a measure of the rectilinearity and direction of particle motion from the covariance matrix of the three components over a small time interval. The covariance matrix

$$\mathbf{V} = \begin{bmatrix} \text{var}\,[R] & \text{cov}\,[R, T] & \text{cov}\,[R, Z] \\ \text{cov}\,[R, T] & \text{var}\,[T] & \text{cov}\,[T, Z] \\ \text{cov}\,[R, Z] & \text{cov}\,[T, Z] & \text{var}\,[Z] \end{bmatrix}$$

(where R is radial, T is transverse, and Z is vertical component) is diagonalized and a measure of rectilinearity is given by

$$F(\lambda_1, \lambda_2) = 1 - \left(\frac{\lambda_2}{\lambda_1}\right)^n$$

where λ_1 and λ_2 are the largest and next largest eigenvalues of $\mathbf{V}$, respectively. F is near unity for high rectilinearity and near zero for low rectilinearity. They go on to obtain *pointwise gain-control operators* to modulate the seismic records in order to enhance the desired signal component.

The inherent idea in polarization studies—that one component should be predictable from a knowledge of one or both of the other components—suggests that, for example, the vertical component of a Rayleigh wave may be estimated from its horizontal components, and the vertical estimate subtracted from the recorded vertical component to enhance compressional body reflections. This program was carried out by Potter and Roden [231] and experimentally by Claerbout [56a]. Potter and Roden give theoretical confirmation that a single station as used by Claerbout is insufficient to produce accurate estimates, justifying Claerbout's conclusion, and suggest that an aray of multicomponent phones must be used to effect the desired discrimination between wave-field components. Although much work clearly remains to be done, they do suggest that arrays of multicomponent phones can yield better discrimination than arrays of vertical phones.

A very good theoretical discussion of polarization and the design of operators in the frequency domain for the detection of polarization is given by Samson [252][252a]. This work should possibly be viewed as providing a mathematical basis for further polarization studies which would take into account the realities of noise, nonstationarity, and so on. Kanasewich [150] also provides a good discussion of developments in this area.

More recently, Harris [117] has considered the use of adaptive filtering techniques to estimate one of the vertical or radial (suitably rotated) channels from the other. Since pure Rayleigh wave motion would show a 90° phase shift between these channels, the adaptive filter will attempt to become a *Hilbert transformer.** By constraining the filter coefficient sequence to be even, it is prevented from becoming a Hilbert transformer. As a result, the output can only be an estimate of linearly polarized signal components (assuming that the input signal components are either linearly or elliptically polarized).

The question of whether polarization methods applied to multicomponent recordings will be applied to seismic exploration data is somewhat open in the opinion of this author, as many difficulties are foreseen in the application of these methods to seismic exploration. In particular, nonlinear filtering is frowned upon in exploration seismic data processing, since "wavelet character" shows promise of becoming an extremely important indicator of lithologic characteristics, so any distortion applied in a polarization filtering process may render us incapable of making these inferences in the final analysis. Second, it is not clear how sharp the distinction in polarization between desired body-wave components and other wave-field components really is. More work needs to be done in this area, and it is certainly true that almost no work in this area has taken place as far as exploration seismology is concerned. It seems quite possible that the proper use of polarization methods will be in conjunction with other methods of wave-field discrimination to refine further their abilities to discriminate between the various wave-field components. In any event, the increased capacity of field recording systems and the changing economics of seismic exploration, may very well allow us to consider the use of multicomponent recording techniques in the future and the probable increasing importance of shear-wave studies will tend to dictate that we will, in fact, move in this direction, if only for that reason.

The second potential discriminator of wave-field components, which is mentioned here, is dispersion. This is effectively a measure of the dependence of velocity of propagation on frequency. In particular, it has been noted for some time [119] that surface-wave components tend to be dispersive, whereas body-wave components tend not to show dispersion characteristics. As a result, dispersion becomes another potential tool for discriminating between wave-field components.

Some of the earliest recorded attempts to determine group velocity dispersion curves were made using earthquake data. The method used in each case was to pass the data through a bank of narrowband filters covering the frequency range of interest and to estimate the group velocity of an event at each of the center frequencies by examining the output from each of the narrowband filters. This concept, apparently originated with Alexander [10] and was refined by Archambeau and

*A Hilbert transformer is a device that has (1) a unit amplitude response for all frequencies, and (2) a phase response of $-90°$ for positive frequencies and $+90°$ for negative frequencies.

others [17–19] for use in determining the dispersion of both body and surface waves for determining the structure of the earth. Archambeau et al. used recursive filter design for the narrowband filters and a few years later Dziewonski et al. [82] considered the same method but introduced the use of Gaussian narrowband filters in order to obtain, in some sense, an optimal balance between frequency and time resolution. By truncating the Gaussian function, a reasonably good approximation was obtained to the optimal prolate–spheroidal wave function of zero order. They then calculated the *analytic signal* or pre-envelope* for each output from the narrowband filters and estimated the group arrival time for the center frequency of the narrowband filter, as the time at which the maximum of the envelope of the arrival in question occurred. To first-order accuracy, this is a reasonable approximation to the true group arrival time. A variation in group arrival time, at the center frequencies of the various bandpass filters, constitutes the estimate of the dispersion curve for that arrival. This method was applied to earthquake data to estimate the dispersion curve for Rayleigh waves. Later, Dziewonski et al. [83] proposed a technique whereby the observed seismogram is cross-correlated with a theoretical seismogram in which the dispersion closely approximates the observed dispersion and then carried out this program for estimating the dispersion curve. The dispersion estimate for the resulting pulse is obtained with higher precision and showed improved performance in the cases where velocity varied rapidly with frequency.

In the case of exploration seismology, one of the earliest studies of absorption effects on seismic data was carried out by McDonal et al. [196], who carried out experiments in the Pierre shale. Later, Wuenschel [320] analyzed these data and gave some credence to a dispersion relation which had been suggested. Robinson [247] designed a single-channel procedure for removing the effects of dispersion in order to bring seismic events into sharper focus and also suggested a way to introduce dispersion effects to synthetic seismic data. Most recently, the array processing procedure of transforming the time–distance representation of seismic data to the τ–p plane has been considered for dispersion analysis. The p (slowness) variable is fixed and a Fourier transform over the τ variable results in a slowness–frequency representation of the original data. It is in this domain that dispersion curves are then imaged.

The method for dispersion estimation suggested by McMechan and Yedlin [199] operates on common shot gathers which we denote by $V(x, t)$. Using the frequency–wavenumber representation suggested by Chapman [52], we write

$$V(x, t) = \iint \exp\left[j(Kx - \omega t)\right] \frac{N(K, \omega)}{D(K, \omega)} \, d\omega \, dK$$

where x = offset
t = travel time
K = wavenumber

*The pre-envelope of a real signal is complex. Its real part equals the given signal and its imaginary part equals the Hilbert transform of the signal.

N = function of the source
D = dispersion relation

Applying the τ–p transform [199] to both sides,

$$U(p, \tau) = \int V(x, \tau + px)\, dx$$

yields

$$U(p, \tau) = \int \frac{N(\omega p, \omega)}{D(\omega p, \omega)} \exp(-j\omega\tau)\, d\omega$$

If we now Fourier transform on τ, we obtain

$$U(p, \omega) = \frac{N(\omega p, \omega)}{D(\omega p, \omega)}$$

It is now a simple matter to determine the desired dispersion curve

$$D(\omega p, \omega) = 0$$

by simply searching for the largest values in the $U(p, \omega)$ plot. McMechan and Yedlin apply this analysis to marine data and obtain some very interesting results after verifying the method on synthetic data.

McMechan and Yedlin go on to suggest an alternative to the usual velocity filtering method for removal of dispersive surface waves by performing the τ–p transform and dispersive wave-field imaging, as suggested above, filtering (masking) the dispersion curves, and then performing inverse transforms to restore the nondispersed component of the recorded wave field.

2.12 DECONVOLUTION

To begin our discussion of deconvolution as it is interpreted in exploration seismology, we may do well to recall the basic convolution model considered earlier. We imagine that the seismic trace consists of the desired reflectivity sequence, which indicates the subsurface structural characteristics of the earth, convolved with or blurred by certain smearing functions, such as the seismic wavelet itself, the ghost signal, if there is one, and the effects of reverberation due to the multiple reflections of the seismic signal as it traverses the various layers of the earth. In addition, we may imagine that these effects also bear the imprint of the characteristics of the recording system, including everything from geophone coupling to filtering effects in the recording system, and, in addition to these effects, various types of noise that are added to the signal components. The purpose of *deconvolution* is to remove the undesirable effects of smearing by the seismic wavelet and redundancy produced by ghosting and multiple reflection, in order to derive some accurate estimate of the reflectivity sequence. This process should be carried out after the additive noise has been suitably attenuated and, ideally, the deconvolution operator should amplify the noise as little as possible.

In our discussion of deconvolution as it is applied to seismic data, we shall distinguish between *wavelet deconvolution*, that is, the attempt to remove the smearing effects of the seismic wavelet by filtering, from *deghosting* and *multiple attenuation procedures*. We will treat each of these separately.

Wavelet deconvolution is one of the most important problems considered in exploration data processing and much research effort has gone into devising successful procedures for carrying out this operation. It is interesting that whereas many seismic data processing procedures are multidimensional in nature, this particular problem tends to be treated today as a one-dimensional problem, and wavelet deconvolution operators are typically designed from and applied to individual traces. That is, the wavelet deconvolution problem is typically not treated as a multichannel or multidimensional problem. Probably the best reference in this area is a paper by Davies and Mercado [68] in which the standard multichannel filter design procedure is used, based on the assumption that the autocorrelations and cross-correlations of the input traces will approximate the corresponding correlation function for the input wavelet to be deconvolved. The cross-correlation between the input wavelet and the desired output wavelet is set equal to 1 at the zeroth lag position and is zero elsewhere. Filters designed in this way were compared with the operation of stacking, followed by single-channel deconvolution, and conversely, the operation of single-channel deconvolution followed by stacking. The three methods described were applied to both synthetic CDP data and to actual field recorded CDP data. The conclusion made from these studies was that the multichannel filter technique did not seem to offer significant improvement over the single-channel deconvolution and stacking procedures. Thus, although the literature of wavelet estimation and deconvolution is extensive and of great interest, it is largely devoid of multichannel and multidimensional methods which show improvement over the single-channel methods.

A simple procedure was published by Dash and Obaidullah [67] in which the seismic trace was modeled as a sum of a signal and noise component and the problem is to estimate the autocorrelation functions of each of the components. The assumption was made that two traces would be available with the same signal component but with different noise components. In that case, a fairly straightforward correlation analysis, which involves calculating the autocorrelations and cross-correlations of the two recorded traces, can yield estimates of both the signal autocorrelation and the noise autocorrelation using a simple differencing procedure with the autocorrelations and cross-correlations of the input traces. We assume that the signals are uncorrelated with the noise component and that the noise component between the two traces are uncorrelated. Of course this technique can be used to estimate the wavelet autocorrelation function which is needed for Wiener filter design for the purposes of deconvolution or wavelet shaping. In fact, a commonly used procedure in exploration seismology is simply to correlate a trace with itself, and under the assumption that the reflectivity sequence is white and that the noise component is uncorrelated, one immediately obtains an estimate of the wavelet autocorrelation which contains an additional multiplicative and additive factor at the zeroth lag due to the power of the reflectivity sequence and the noise component.

Aside from these few comments, the problem of wavelet deconvolution is primarily treated as a single channel procedure. Though volumes could be written on this topic, it is inappropriate to discuss it further here. We shall proceed to consider the other deconvolution problems that are more commonly treated using array processing techniques.

2.13 DEGHOSTING FILTERS

A *ghost* is essentially an echo of the source pulse which accompanies this pulse with a short time lag and is generated by buried sources due to the reflection of the upgoing energy from the surface which produces a slightly delayed downgoing version of the source pulse. The problem of removal of the ghost reflection is called *deghosting* and several procedures have been devised to perform this operation. While some of these techniques are single-channel techniques applied to individual traces, it is possible to treat the deghosting problem as a multichannel filter problem, and it is in this context that we shall consider the problem here.

Figure 2.19 shows a common cause of ghost reflections, although other effects in the near surface can generate ghost reflections as well. If we consider only a single trace, and if the primary–ghost delay time is known, a simple operator can attenuate the ghost event [242] provided that the ghost and the primary wave shapes are identical. Wiener filtering (or other deconvolution procedures) can also be used to attenuate this event if we have sufficient information.

A second possibility for ghost elimination would be to fire two shots at different depths into the same receiver spread. In this case, the ghosts from each shot will be displaced relative to each other when the primary events are aligned at each receiver. Considering the two traces obtained at a single receiver position, we may treat the ghost removal problem as a two-channel filtering problem.

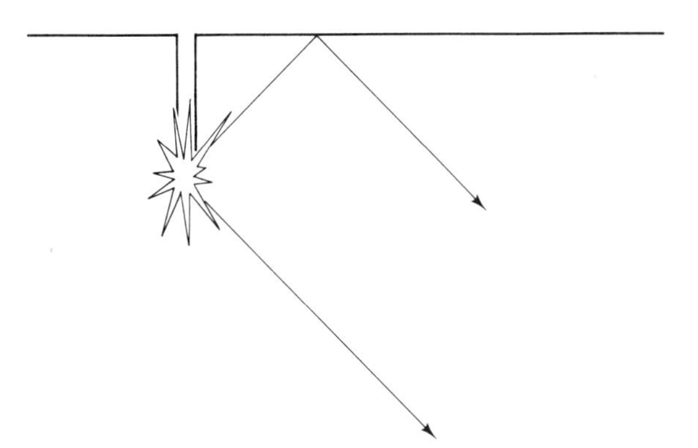

Figure 2.19 Generation of the ghost pulse from a buried source.

The method of Sengbush and Foster described earlier is clearly applicable to this problem by interpreting the ghost as correlated noise in their model. Indeed, this application was described by them in an earlier publication [88].

We consider here a multichannel method proposed by Schneider et al. [262a]. Schneider et al. indicate that their goals were to derive a procedure that would exhibit wideband ghost rejection properties while preserving the desired signal, and would require a minimum knowledge of the ghosting mechanism and the primary–ghost relationship. Consider the following model for the two channels corresponding to different shot depths:

$$f_K(t) = P_K(t) + G_K(t) + N_K(t); \qquad K = 1\text{: shallow shot, } K = 2\text{: deep shot}$$

where $P_K(t)$ = primary reflection sequence on channel K
$G_K(t)$ = ghost reflection sequence on channel K
$N_K(t)$ = random noise on channel K

We now time-shift the traces by ΔT so that $P_1 = P_2 = P$ and

$$G_1(t) = G(t)$$

$$G_2(t) = G(t - 2\Delta T)$$

where ΔT is the difference in uphole times.

We assume that noise is uncorrelated with signals or ghosts and neglect primary–ghost correlations as they are variable. The result is a slight loss in effectiveness of the process. If we assume further that the spectral content of primaries and ghosts is white over the frequency range of interest, we obtain the correlations

$$\phi_p(\tau) = P\ \delta(\tau)$$

$$\phi_g(\tau) = G\ \delta(\tau)$$

where the ϕ's represent autocorrelations, and P and G are the signal and ghost power levels, respectively. We now design a two-channel Wiener filter using the following autocorrelations and cross-correlations:

$$R_{11}(\tau) = \phi_p(\tau) + \phi_g(\tau) + N_0\ \delta(\tau)$$

$$R_{22}(\tau) = \phi_p(\tau) + \phi_g(\tau) + N_0\ \delta(\tau)$$

$$R_{12}(\tau) = \phi_p(\tau) + \phi_g(\tau - 2\Delta T)$$

$$R_{21}(\tau) = \phi_p(\tau) + \phi_g(\tau + 2\Delta T)$$

and the cross-correlation between signal and desired output is given by (for each channel)

$$D_1 = \phi_p(\tau)$$

$$D_2 = \phi_p(\tau)$$

N_0 is assumed to be the common (white) noise power on each channel (for convenience).

This simplified derivation was justified in view of the objectives and because of its practicality. It is not necessary to recalculate these filters continually to account for changing, near-surface conditions. Frequency-domain derivation of these filters is also considered, and in addition, an inverse procedure to estimate the parameters of the ghosting process is discussed. We defer further discussion of deconvolution processes as applied to multiple suppression until we have discussed velocity analysis.

2.14 VELOCITY ANALYSIS

We previously considered the shape of the moveout curve for a reflection generated by a stack of horizontal layers, each with its own constant velocity. We indicated that the curve is roughly hyperbolic and that the hyperbola is determined by the two-way normal incident travel time, the offset, and the rms velocity to the reflecting interface. It is also possible to derive a parabolic relationship for this curve, but Ursin [301] has shown that for a horizontally layered elastic medium, the hyperbolic function is more accurate than the parabolic function. Since this curve is determined in part by the rms velocity, we should be able to infer this velocity by determining the shape of this curve empirically. It is this concept that forms the foundation of one of the most widely practiced methods of seismic velocity analysis today.

We recall that the relationship between normal incidence two-way travel time, T_0, and the reflection time, T_x, at offset distance, x, for the nth layer in a flat-layered model is given by

$$T_x^2 = T_0^2 + \frac{x^2}{V_{\text{rms}}^2} + C_4 x^4 + C_6 x^6 + \cdots$$

where

$$V_{\text{rms}}^2 = \frac{\sum_{k=1}^{n} V_k^2 t_k}{T_0}$$

$$T_0 = \sum_{k=1}^{n} t_k$$

and V_K and t_K are the velocity and two-way normal incidence travel time in the kth layer. It seems reasonable that we attempt to estimate V_{rms} by plotting T_x^2 against x^2. Indeed, it is easy to see that at $x^2 = 0$,

$$\frac{d(T_x^2)}{d(x^2)} = \frac{1}{V_{\text{rms}}^2}$$

so the slope of the tangent line to this curve at the origin is the reciprocal of the rms velocity. One can infer the interval velocity structure of the model from a knowledge of the rms velocities [139].

Velocity analysis, as it is most commonly practiced today, is based on fitting the "best" hyperbola to a particular event and thereby estimating the velocity for that event. This velocity has been called [9] the *maximum coherency (MCS) velocity*. On the T_x^2 versus x^2 plot, this velocity is the reciprocal of the slope of the best least-squares straight-line fit to the T_x^2 versus x^2 curve for the event in question. It is this MCS or *stacking velocity* which is commonly estimated today and an analysis of the error between MCS and rms velocity is given by Al-Chalabi [9].

Our first problem is to consider how stacking velocities are obtained in practice. In general, this is done automatically by scanning across the traces of a gather looking for coherency in an event from trace to trace in an effort to determine whether an event is present and which hyperbola fits it best if an event is indicated. A number of schemes have been proposed for searching across seismic trace gathers for coherent energy.

Among the earliest techniques proposed were correlation methods seeking matching wavelets and their moveout positions across the set of traces. Schneider and Backus [262] suggested shifting the traces according to an assumed rms velocity (performing a *normal moveout* or NMO correction) and then performing trace-to-trace cross-correlations. If a primary event were aligned across the traces, an autocorrelation peak with maximum energy would appear at the origin. Any misalignment would reduce the power of the peak. Multiples would show up as peaks at a finite lag distance and uncorrelated noise would not show up at all. By analyzing the actual lag positions of the primary and multiple correlation peaks, t_p and t_m, respectively, the "true" primary and multiple rms velocities could be inferred from

$$t_p \simeq \frac{X_j^2 - X_i^2}{2T}\left(\frac{1}{V_p^2} - \frac{1}{V^2}\right)$$

$$t_m \simeq \frac{X_j^2 - X_i^2}{2T}\left(\frac{1}{V_m^2} - \frac{1}{V^2}\right)$$

where X_i and X_j indicate the offset of the two traces being correlated and V is the (assumed) rms velocity used for the NMO correction. V_p and V_m are primary and multiple rms velocities respectively, and T is the center of the time window over which the correlations are calculated. The cross-correlations are averaged and the method is refined from the suggested procedure above, in practice. Robinson [248] suggested an F–K domain version of this procedure, calculating the energy along constant velocity lines in the F–K domain for windowed segments of the data.

Taner and Koehler [284] suggested fixing the intercept time, T_0, and sweeping hyperbolas with fixed rms velocities across the gather and calculating a measure of the coherence of the events lying along each trajectory. The rms velocity corresponding to maximum coherence would be the estimated rms velocity for the event. The (three-dimensional) plot of coherency versus T_0 and V_{rms} was called the *velocity spectrum* (the accepted term today for this display). By picking the primary peaks on this display, we may infer the rms velocity as a function of two-way travel time. The coherence measure suggested was *semblance measure* defined by

$$S = \frac{\sum_{j=r}^{r+q} \left[\sum_{i=1}^{m} A_{ij(i)} \right]^2}{m \sum_{j=r}^{r+q} \sum_{i=1}^{m} A_{ij(i)}^2}$$

where $r = T_0 - q/2$ defines the window over which correlations are calculated, A_{ij} is the value of the jth sample on the ith trace, and $j(i)$ is usually the hyperbolic relationship determined by the trial value of rms velocity. This is a normalized measure, so that S lies in the range zero to one.

Using similar analyses, other measures have been suggested [9], but an interesting paper by Ursin [301] suggests that semblance is not a bad choice. Ursin considers a maximum-likelihood estimator and shows that, for his assumed model, this estimator is equivalent to a least-squares detector which differs from semblance only in the respect that semblance is essentially the same detector that has been normalized.

We may think of semblance as a kind of multichannel filter whose output is determined by the parameters T_0 and V_{rms}. When these parameters are chosen to correspond to an actual event, the output power is "maximum." Needless to say, this is a simplification of all that must really be considered in velocity analysis, but hopefully it gives the idea. Once the T_0 and V_{rms} parameters are selected for a primary event, the trace-to-trace moveout of this event is determined. We then perform a normal moveout (NMO) correction to each individual trace by shifting the event by an appropriate amount, so that the primary event on each trace is in alignment. For this procedure to have validity, any time shifts which may have been introduced by uneven terrain or other near-surface effects should have been removed (these are called *static corrections*). There may still remain some variation in the actual alignment of events after NMO corrections are made and much effort is made to analyze these *residual statics* [171, 312].

After all statics and NMO corrections are made for CDP gathers, our traces should all represent coincident source and receiver recordings at the common midpoint location. We would then normally sum ("stack") these traces to produce a single trace corresponding to coincident source and receiver at this point. The summing (stacking) process is a multichannel filter. If primaries have been properly aligned by the use of their rms velocities, we may expect that multiple and other events which generally have different rms velocities will not be aligned and will be attenuated by the stacking process.

An alternative procedure, *continuous velocity analysis*, simply performs an NMO correction for each entire trace in a gather for a fixed rms velocity and then stacks the result. The maximum amplitude for an event from the stacked traces will occur when the rms velocity is the correct one for that event [268].

2.15 STACKING FILTERS AND MULTIPLE SUPPRESSION

It is not always easy to distinguish between primary and multiple reflections on seismic data. For most purposes, however, we hope to have retained the primary energy in our seismogram and successfully attenuated or suppressed other forms of random or coherent signals including multiple reflections. Indeed, many of the algorithms that are used in the final stages in seismic data processing rely on the assumption that only primary energy is present in the traces being processed. The presence of multiple reflection energy will lead to inaccurate results and unreliable interpretations.

It becomes important, then, to establish criteria for the detection of multiple reflections so that suitable procedures can be devised for their removal. Two such criteria which are widely used are the periodicity relationships shown by certain types of multiples, particularly water bottom multiples in the marine case, and their rms velocity. In the case of water bottom multiples, for example, we can frequently estimate the period of successive multiple reflections from a knowledge of the water depth and the velocity of sound in water. When multiple reflections arrive with the same two-way normal incidence travel time as primary reflections, we can frequently distinguish between them by the apparently lower rms velocity of the multiple reflections. This lower rms velocity may be attributed to the fact that velocity tends to increase with depth in the earth and multiple reflections arriving at the same normal incidence time as a primary reflection will have spent more time at the shallower, lower-velocity regions of the earth than did the primary arrival. As a result, since the moveout hyperbola is determined by normal incidence time and rms velocity, the multiples will in general show a larger trace-to-trace moveout than will the corresponding primary reflection (Fig. 2.11) This differential moveout can be used as a criterion for the purpose of identifying the primary and multiple reflections, and can be used in design procedures which will preserve the primaries and attenuate or suppress multiple reflections. It is common practice today to identify these rms velocities using a velocity spectrum display. Thus, when several events are detected for a given normal incidence time but with different rms velocity, then, as a general rule of thumb, multiples will have the lower rms velocity and the primaries have the higher rms velocity. Although this rule should not be strictly followed in practice, it does at least suggest a way in which this distinction can be made.

In general, it is difficult to determine which events are primaries and which are multiples, which are diffractions, and so on, on a velocity display. This requires a great amount of skill and experience on the part of the interpreter, and if borehole velocity information is available, this can aid significantly in this process. In any event, if primary and multiple arrivals can be distinguished on the basis of apparent rms velocity, the resulting differential moveout between these two types of events can be used as the basis for procedures designed to preserve the primary events and to attenuate or suppress the multiple events.

A variety of methods have been suggested, based on the concept of *differential moveout* between primary and multiple events, in order to process them in such a way that the primaries are preserved and the multiples are suppressed. Perhaps the simplest and one of the most successful and widely used methods for CDP gathers is the concept of *stacking*. In this procedure we assume that we have accurately identified the primary rms velocity or its approximation, the stacking or maximum coherency velocity, and we also assume that this velocity differs from the stacking velocity for multiple reflections which may have the same normal incidence time. In this case, the differential trace-to-trace moveout for the primary stacking velocity is removed from the traces by shifting them an appropriate amount, so that the primary events are brought into horizontal alignment. The generally lower-velocity multiple events will still show differential moveout from trace to trace. If we then sum the traces to produce one output trace, the primary events that are in alignment will sum in phase, and if the alignment is accurate, they appear with little distortion on the summed output trace. The multiples, however, because they are summed out of phase, will suffer a degree of attenuation in this process. This method is known as the *CDP stack*, and since it also tends to attenuate random noise, it serves several useful purposes. Generally, it has the benefit that if the primary stacking velocities have been chosen accurately, the primaries will indeed be enhanced, and both multiples and incoherent noise will be suppressed. That is, the signal-to-noise ratio on the output trace will be improved.

The CDP stack is an attractive process because it requires a minimal amount of information to be successful, that is, we need only to have chosen the primary stacking velocities accurately. No information concerning the signal waveform, multiple delay times, or noise characteristics is necessary if our simple assumptions hold. Viewed in this light, stacking becomes a direct analogue of beam-forming. It is natural to raise the question whether we may do a better job if we could employ more complex multichannel filtering procedures instead.

Some of the first papers written on the subject of multichannel stacking filters for CDP data were directed to the problem of preserving primary reflections and attenuating coherent noise (multiple reflections) and incoherent noise. The model generally used was the following:

$$X_i(t) = S_i(t) + r_i(t) + N_i(t)$$

where X_i is the ith trace, S_i the signal component, r_i the correlated noise, and N_i the uncorrelated noise. When multichannel Wiener filtering is applied to this problem, for example, we may take the signal, S, to be the desired output and then we must estimate the required correlations based on our knowledge of the data.

One of the first papers to appear on this subject was written by Schneider et al. [262b] and the program above was suggested in that paper. Under the assumption that signal and noise are uncorrelated, we must estimate the signal and noise cross spectral densities, S_{ij} and N_{ij}, respectively, between channels i and j and solve the normal equations (in frequency-domain notation)

$$[S_{ij}^*(f) + N_{ij}^*(f)]Y_j(f) = S_{i0}^*(f)$$

where three channels were considered in the paper (the number would be much higher today). Allowing for statistically independent static errors between channels, with uniform probability distribution, we obtain the estimates

$$S_{ij}(f) = \Phi_{ss}(f) \frac{\sin (2\pi f \,\Delta S_i)}{2\pi f \,\Delta S_i} \frac{\sin (2\pi f \,\Delta S_j)}{2\pi f \,\Delta S_j}$$

where Φ_{ss} is signal autocorrelation and $\pm \Delta S_i$ is the expected range of signal misalignment on channel i. Similarly,

$$S_{io}(f) = \Phi_{ss}(f) \frac{\sin 2\pi f \,\Delta S_i}{2\pi f \,\Delta S_i}$$

and

$$N_{ij}(f) = \Phi_{mm}(f) \frac{\sin (2\pi f \,\Delta M_i)}{2\pi f \,\Delta M_i} \frac{\sin (2\pi f \,\Delta M_j)}{2\pi f \,\Delta M_j} \exp [-2\pi i f (M_i - M_j)]$$

where m stands for "multiple" and $\pm \Delta M_i$ is the expected range of misalignment of the multiple on channel i. Further, $N_{ii}(t) = \Phi_{mm}(f) + \Phi_{rr}(f)$, where r is uncorrelated noise. The autocorrelations, Φ_{mm}, Φ_{ss}, and Φ_{rr}, must be estimated in some way.

Galbraith and Wiggins [99] gave a similar, but perhaps more general analysis of the multichannel filtering problem, estimating the required correlations from assumed models of signal and noise in which some information about their moveout characteristics is known.

Meyerhoff [203] used a slightly different approach and considered filters designed to maximize a signal-to-noise ratio and pointed out that this is sometimes equivalent to Wiener multichannel filtering with the (primary) signal as the desired output. The signal-to-noise ratio for the mth trace is [203]

$$(\mathrm{SNR})_m = \frac{\Phi_{ss}(0)}{\Phi_{r_m r_m}(0) + \Phi_{nn}(0)}$$

where s is the signal, r the correlated noise, and n the uncorrelated noise. If we horizontally stack the m traces, the resulting signal-to-noise ratio is

$$(\mathrm{SNR})_A = \frac{\Phi_{ss}(0)}{\dfrac{1}{M^2} \sum_{m,\,p=1}^{M} \Phi_{r_m r_p}(0) + \dfrac{1}{M} \Phi_{nn}(0)}$$

If $\Phi_{r_m r_m}(0) = \Phi_{rr}(0)$, then, using *Schwarz's inequality*, we get

$$\frac{1}{M^2} \sum_{m,\,p=1}^{M} \Phi_{r_m r_p}(0) \le \Phi_{rr}(0)$$

which implies

$$(\mathrm{SNR})_A \ge (\mathrm{SNR})_m$$

If we weight each trace before stack,

$$X_{out}(t) = \sum_{m=1}^{M} a_m X_m(t)$$

then

$$(\mathrm{SNR})_{out} = \frac{A^2 \Phi_{ss}(0)}{\sum_{m,\,p=1}^{M} a_m a_p [\Phi_{r_m r_p}(0) + \Phi_{nn}(0)\delta_{mp}]}$$

where

$$A = \sum_{m=1}^{M} a_m$$

The weights $\{a_m\}$ are then determined by requiring that $(\mathrm{SNR})_{out}$ be maximized. Meyerhoff points out that if $\Phi_{sr_m}(0) = 0 \;\; 1 \leq m \leq M$, then this is equivalent to minimizing the mean-square error between the weighted sum and the signal component, S. Meyerhoff also considers linear filtering applied to each trace to maximize signal-to-noise ratio, where the filter is a pure delay operator, and finally he considers the multichannel Wiener filtering problem which Schneider et al. had considered [262b], but requires exact knowledge of the moveouts.

Cassano and Rocca [49] carried out a very similar analysis to the Galbraith and Wiggins approach, but derived their operators in the frequency domain. They show good performance of the filters on field data, particularly for long-period water-bottom multiples.

Since multidimensional filters can be analyzed in many cases by considering their response in the F–K domain, it is reasonable to ask whether this could be done for multichannel filters designed to pass events lying along hyperbolic curves. By parameterizing this class of moveout curves, Hubral [134] associated a velocity filter response with these multichannel filters. Analysis of the corresponding F–K diagram is useful for examining the pass and reject characteristics of multichannel filters.

Recently, Ozdemir [223] has considered calculating optimal multichannel least-squares (OMLS) filters over time gates rather than at fixed normal incidence travel time. He derives integral formulations for the required correlations and evaluates these numerically. Examples are given using synthetic and field data. Applications of multichannel maximum-likelihood filters to CDP gathers is considered by Mercado [202a], who comments on some of the problems in this process. He also gives examples with synthetic and field data.

More recent attempts to use differential moveout for multiple suppression have also made use of the NMO correction and a very simple two-dimensional filter [251]. The idea is to use a stacking velocity that will "overcorrect" primary events, but still leave multiple events "undercorrected" in the F–K domain. Primaries and multiples then show up in different quadrants and can be separated by a simple two-dimensional filter which passes one quadrant and rejects the other.

Predictive deconvolution is a single-channel technique frequently applied to single traces for multiple suppression. Taner [283] points out that for a reasonably flat sea bottom, energy reflecting at approximately the critical angle is influenced by a significant increase in the reflection coefficient. The result is that a high-amplitude reflection is generated and this reflection tends to continue to reflect at near the critical angle so that the water layer behaves as a wave guide for high-amplitude multiple energy. A multiple propagating in this mode, if detected at space–time coordinates (X_1, t_1), will reappear at $(2X_1, 2t_1)$, $(3X_1, 3t_1)$, and so on, and so forms a stationary time series along a radial line containing the origin and the point (X_1, t_1). It is reasonable, then, to design a multichannel Wiener prediction filter for this mode of propagation. The corresponding *prediction-error filter* (whose output is obtained by subtracting the predicted trace from the observed trace) should effectively suppress these multiple events. Taner calls this technique *radial multiple suppression* and demonstrates its effectiveness on marine data.

A variety of other multichannel methods have been proposed for multiple suppression [43, 204, 212]. White [309] considered the performance of optimum stacking filters which attempt to pass the primary signal undistorted, but attenuate multiples using the maximum-likelihood criterion. His conclusions are negative and he points out that we cannot expect our optimum stacking filters to perform better than a standard summation stack, due, in part, to the difficulties encountered in estimating signal-to-noise ratios accurately. In fact, even for simple weighted stacks, the performance is likely to exceed that of the straight stack only for very high signal-to-noise ratios, according to White. In an apparent direct response to this study, Rietsch [236] has considered the problem of estimating signal-to-noise ratios in great detail and does achieve improvements in weighted stacks over straight stacks in controlled experiments using both synthetic and field data. White's comments probably do pinpoint areas in which improvement is needed before the full benefits of optimum multichannel stacking filters can be realized.

2.16 MIGRATION

At this point, we may assume that we have gone to the field and collected seismic data, according to a two-dimensional or three-dimensional format and we assume that each geophone group was represented by a simple array designed to attenuate certain noise processes, inherent in the seismic experiment. We have then performed diagnostic procedures on the data and have performed further signal processing, including multichannel or multidimensional filtering to enhance signal-to-noise ratio, and have deconvolved the seismic wavelet from our traces. In addition, we have gathered the traces into some format which we may assume is the CDP format and have corrected these data for statics and have then carried out a standard velocity analysis to estimate the stacking velocity for primary reflections using a velocity spectrum display. At this point, we may have performed certain specific procedures to remove or attenuate multiple reflections and have then performed normal moveout corrections, followed by some stacking procedure which has resulted in a further attenuation of multiple reflections.

The single output trace from each CDP gather ideally should now represent the primary reflectivity sequence displayed as a function of two-way travel time as observed by a coincident source and receiver for normal incidence wave propagation. If these output traces are now displayed side by side for consecutive shot points, an image will appear, produced by the alignment of the primary reflections which appear on each individual trace. We assume that the picture which emerges is an accurate picture of the subsurface in terms of two-way travel time. However, this is true only in the case of an ideal flat earth model, and in any other case, our picture represents a distorted version of the true time section which we would like to reconstruct. If, indeed, we can reconstruct a correct picture of the subsurface in time, knowledge of the velocity structure of the subsurface will allow us to convert our picture to a corresponding picture of subsurface structure as a function of depth and this picture would accurately reflect the true structural characteristics of the subsurface. We have indicated, however, that the picture that results from the CDP stacked section is, in general, not a correct picture of the subsurface in terms of two-way travel time, so an additional procedure must now be applied to convert this distorted picture to a more accurate representation of the subsurface in terms of the two-way travel time. The procedure that carries out this operation is referred to as *migration.*

To gain an intuitive understanding of the error we have made in constructing our CDP stack section, let us recall that the single stacked output trace for each common depth point is imaged at the common depth point or common midpoint position. That is, it is imaged vertically beneath the common midpoint corresponding to the trace gather. However, in the case of the single dipping bed, for example, with constant velocity, it is clear that a reflecting point which is indicated on our single output trace was not located vertically beneath the common midpoint but rather at an offset position that must be determined. If we assume constant velocity, and coincident source–receiver location, which the CDP stack should now represent, it follows that the true reflecting point must lie somewhere on the arc of a circle with center at the common source–receiver location and radius determined by the time to the reflector.

Figure 2.20 shows the image of a reflecting point on a single CDP stacked trace. If the velocity is constant, and if the reflector is a dippling plane, the true reflecting point (at normal incidence) lies on a circle with center at the common source–receiver location and which is tangent to the reflector at the point of reflection.

If we construct circles for a given reflector, their envelope will, in fact, trace the true reflector and migration will have been accomplished (Fig. 2.21). This is a very simple way to imagine the migration problem and its solution in the case where velocity is constant. Alternatively, we may imagine that each reflecting point in the

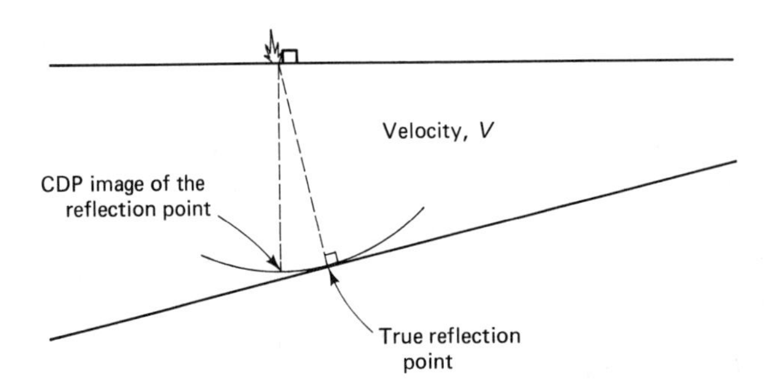

Figure 2.20 Migration moves the image point to the true reflection point.

subsurface acts as a *diffractor* and that our imaged section is constructed by *Huyghens' principle* as the sum of all diffraction patterns so produced. The CDP image of a diffraction is approximately hyperbolic and the diffracting point may be imagined to correspond to the apex of the hyperbola. As a result, we may conceivably image the reflecting points (diffractors) by a matched filtering operation in which the energy in the hyperbola is summed back to the apex. In more complex cases (variable velocity), the matched filter would have to be spatially (or temporally) variant, and the hyperbolic approximation for the diffraction curve may or may not be good.

Finally, we may view the migration problem as a wave propagation problem for some suitable experiment. While Schultz and Claerbout [265] point out that the CDP stacked section does not correspond to a physical experiment that can be carried out, there is a model that does describe the CDP stacked section as the image of a (nonrealizable) wavefield. This model was suggested by Loewenthal et al. [179] and is referred to as the *exploding reflector model.* According to the model, we may imagine a small source located at each point on each reflecting horizon and weighted by the corresponding reflection coefficient. At time $t = 0$, the sources are all fired and the resulting wave field, $U(X, Z, 0)$ propagates upward to the surface (at one-half the true velocity in order to arrive at the correct two-way travel time), where it is recorded in the form $U(X, 0, t)$ which represents the CDP stacked section. The migration problem, then, clearly becomes a problem of propagating the

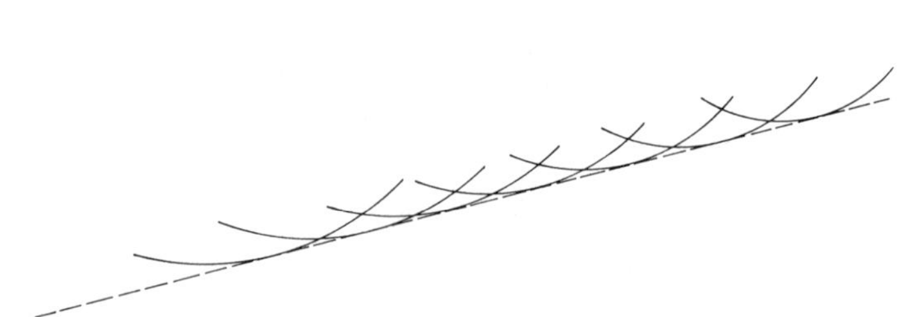

Figure 2.21 Construction of a reflector as the envelope of wavefront circles.

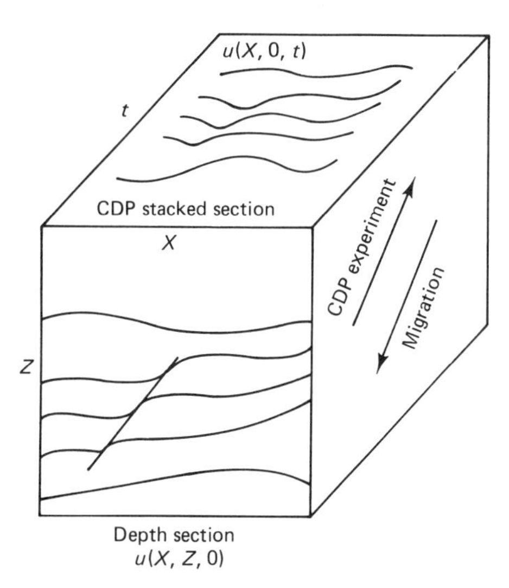

Figure 2.22 Exploding reflector model for a CDP stacked section. Migration is the process of propagating the resulting wave field backward in time.

wave field backward in time to reconstruct the subsurface image $U(X, Z, 0)$ (Figure 2.22).

Most modern migration procedures are based on this last (wave equation) model for the CDP stacked section and rely on methods that effectively propagate the recorded wave field backward in time to reconstruct the original reflectors. This process is carried out in a variety of ways based on finite difference approximations to the wave equation, the Kirchhoff integral formulation which relates $U(X, 0, t)$ to $U(X, Z, 0)$, and frequency-domain techniques to perform the same transformation of wave fields. In addition to these methods which are used most widely in practice in a variety of forms, other procedures, such as multidimensional filter approximations which are spatially variant, have been considered.

Although we shall not consider all of these methods in detail, we should at least have some familiarity with the concepts involved. These methods generally operate on primary (no multiples) wave fields only, and do not all tolerate lateral velocity variations well. They are fairly *robust* for vertical velocity variations, but generally do need modification where strong velocity contrasts are encountered. Some of the methods that operate well in the Gulf Coast would not perform as well in the North Sea, for example. In addition, shear waves are neglected (the acoustic wave equation is generally used) and two-dimensional methods assume that all energy is generated in the plane of the experiments.

Claerbout and his associates [57–59] pioneered the use of finite difference approximations to the wave equation in order to extrapolate the surface recorded wave field back into the earth. Claerbout considered an upward-moving coordinate frame and expressed the wave equation in this frame. By dropping one term from the resulting equation (which limits the accuracy of this approach), downgoing waves are eliminated and a parabolic equation results for the upgoing waves, which can be approximated by finite differences. It is then possible to extrapolate the upward-propagating wave field backward in time by increments to reconstruct the reflecting horizons (see [31] for a summary).

Schneider [261] derived an expression for the downward continuation using the appropriate Green's function and obtained the expression (three-dimensional case)

$$U(X, Y, Z, 0) = -\frac{1}{2\pi}\frac{\partial}{\partial Z}\iint \frac{U(x', y', 0, R/C)}{R}\,dx'\,dy'$$

where the velocity, C, is constant, and

$$R = [(X - x')^2 + (Y - y')^2 + Z^2]^{1/2}$$

Schneider gives examples in which this procedure remains robust when rms velocities are used for the term C, migrating dipping beds with good accuracy up to about 60° dip.

Stolt [280] considered improvements in Claerbout's finite difference approach and derived an F–K migration algorithm as well. In Stolt's notation, let

$$D = \frac{Ct}{2} + Z$$

where C is the velocity (normally a function of X and Z), t is time, $Z = d$ is the depth of the source–receiver midpoint (which is to be "pushed back into the earth") and D is the depth of some reflection point. Then the stacked section is represented by $\phi(X, 0, D)$, and the migrated section is represented by $\phi(X, D, D)$, where (X, Z) always represents the source–receiver midpoint coordinates. Transforming the surface field $\phi(X, 0, D)$ in both variables X and D results finally in the form (the migrated section)

$$\phi(X, D, D) = \frac{1}{2\pi}\iint B(p, k)\exp[-j(pX - kD)]\,dk\,dp$$

where

$$B(p, k) = \frac{1}{(1 + p^2/k^2)^{1/2}}\,A\left(p, \frac{kc}{2}\left(1 + \frac{p^2}{k^2}\right)^{1/2}\right)$$

$$k = \left(\frac{4\omega^2}{C^2} - p^2\right)^{1/2}$$

and p and ω (transform variables for X and D) are the spatial and temporal frequencies, respectively, and

$$A(p, \omega) = \frac{1}{2\pi}\iint \exp\left[j\left(pX - \frac{2\omega D}{C}\right)\right]\phi(X, 0, D)\,dD\,dX$$

In terms of full spatial representation of the stacked section $f(X, Z)$ as a function of offset, X, and depth, Z, we may obtain [55]

$$\bar{F}(K_x, K_z) = \frac{K_z}{(K_x^2 + K_z^2)^{1/2}} F(K_x, (K_x^2 + K_z^2)^{1/2})$$

where F is the (two-dimensional) Fourier transform of the stacked section, f, and $\bar{F}$ is the Fourier transform of the migrated section.

Berkhout [28a] unifies many of these ideas by considering the extrapolation of the surface wave field from depth z_0 to depth $z = z_0 + \Delta z$ using the Taylor expansion:

$$p_1 = p_0 + \frac{\partial p}{\partial z_0}\frac{\Delta z}{1!} + \frac{\partial^2 p}{\partial z_0^2}\frac{\Delta z^2}{2!} + \cdots$$

where $p \equiv p(x, z, t)$ is the pressure. By performing Fourier transforms on various variables, solving for unknown terms using the wave equation, and summing or approximating series, migration operators of various kinds operating in various domains (space–time, space–frequency, frequency–wavenumber, etc.) are obtained. These various approaches have suggested that migration may be viewed as a filter design problem, and some effort has been made to consider the design of approximate filters for migration [31, 104].

In addition to the approaches to migration mentioned above, Hubral [135] considers the problem in terms of ray tracing, and introduces the concept of the *image ray*, this is the ray from the reflecting point which emerges normal to the earth's surface. The emergence point is the point where the minimum travel time (apex of the travel time hyperbola) is observed. Although we indicated earlier that this is the (vertical) position to which the reflecting point will often be migrated, Hubral points out that the desired (true) migration position will be different from this for complex layering. As a result, time-to-depth conversions involving only scaling can be seriously in error, and a second migration process to move the time migrated point to the true depth position would be required. Sattlegger [255] presents an iterative approach to migration which takes into account the refraction of the image ray and attempts to correct for the resulting "over migration."

While many standard migration algorithms ignore velocity inhomogeneities, Stolt [280] suggests a transformation to render the problem in terms of constant velocity, appropriate for F–K migration. Berryhill [29] uses the Kirchhoff integral approach to derive an extrapolation operator for variable velocity, and Gazdag [107] gives a frequency domain approach in which velocity can vary both horizontally and vertically.

While migration has clearly been an active area of research, more remains to be done in other areas. Migration need not be applied after stack or to CDP gathers, and some studies have appeared that examine migration in these other cases [257] [258]. It is clear that pre-stack migration on CDP gathers (called "migration before stack" gathers by Hubral and Krey [139]) can be used for velocity analysis, for example, by finding the migration velocity which best sums the energy in a diffraction hyperbola to its apex at a given two-way image time. Indeed, the migration operator can perform the equivalent of NMO and stacking operations on CDP gathers, provided that the appropriate (variable) velocities are used (see also [72], [103], [221]).

The three earliest approaches to migration exhibited a variety of shortcomings that limited their usefulness. Finite difference migration, based on the parabolic approximation for the wave equation, could not accurately migrate steeply dipping events and exhibited dispersion, although it could handle velocity variations. The Kirchhoff integral approach could accurately migrate steeply dipping events, but lost accuracy in the case of velocity variations. The frequency-wavenumber migration algorithms were derived for constant velocity but also migrated steeply dipping events accurately.

Recent improvements in the various migration algorithms have overcome many of the earlier limitations, including lateral velocity variations, but still more work needs to be done in refining these methods. Many of these algorithms have been derived for both two-dimensional and three-dimensional cases.

An entirely different and very interesting approach to migration and velocity analysis has been reported by Cohen and Bleistein [63] using the Born approximation to the solution of a wave equation which is analogous to the Lipmann–Schwinger equation of quantum mechanics. In this approach, the velocity, v, is assumed to be a function of spatial position in a two-dimensional model, and is viewed as being a *perturbation* from a fixed reference velocity, C:

$$V^{-2}(X, Z) = C^{-2}(1 + \alpha(X, Z))$$

By employing the appropriate Green's function, an integral equation for α can be derived. Replacing the true wave field with the wave field for a constant-velocity medium (Born approximation) in the integral equation, it can be solved for α in closed form. (Actually, Cohen and Bleistein recommend solving for $\partial\alpha/\partial Z$.) The result is that the subsurface (migrated) section can be directly reconstructed, and if the observed data are free of gain control, velocity estimates (in X and Z coordinates) are obtained as well.

Even though the justification for this method requires small perturbations in the velocity field, Cohen and Bleistein indicate that the procedure is fairly robust and performed well in an example with 20 percent variation in velocity. This work has now become an important basis for further research into direct seismic inversion methods and interesting work is being reported [61, 232–234] which may lead to much more powerful procedures for inverting seismic data.

Finally, perhaps the most elegant and simple description of a migration procedure has been given by Hubral [138] using the slant stack to obtain an (approximate) plane-wave decomposition of the data. This procedure as described by Hubral consists of generating a plane-wave stack (see "τ–p Methods" in the next section) to isolate all plane waves which emerge at a specified emergence angle or apparent velocity. A plane-wave image is then formed by identifying the upgoing waves which generated these observed plane waves and locating them in the depth domain at time zero. The reconstructed plane waves, in depth, are then summed to reconstruct the migrated image. The second operation is trivially accomplished since plane waves emerging at angle α (to the horizontal) satisfy the equations

$$t = (X - X_0) \tan \frac{\beta}{V} \text{ in the time domain}$$

$$Z = (X - X_0) \tan \alpha \text{ in the depth domain}$$

where β is the angle between the wavefront and the horizontal in the time domain, V is the (constant) medium velocity, and X is the horizontal coordinate, and $\tan \beta = \sin \alpha$ is true. For each value of X_0, the amplitudes of the CDP stacked section are summed along the line $t = (X - X_0) \tan \beta$ (in X–t domain) and mapped to all points on the line $Z = (X - X_0) \tan \alpha$ (in X–Z domain). Summing all of the resulting plane waves in the X–Z domain results in the migrated section. Hubral points out that this procedure is really analogous to F–K migration, which also performs a plane-wave decomposition of the data in the F–K domain and then carries out operations analogous to those described above. He indicates that the slant-stack method may be more robust in the presence of noise, however.

2.17 ADDITIONAL TOPICS IN SEISMIC ARRAY PROCESSING

While our discussions to this point have outlined many of the areas in which array processing finds application in current seismic processing, several very interesting and relatively new areas of inquiry are starting to emerge which promise to significantly extend our capabilities to process and interpret seismic data in the future. Since these procedures are inherently multidimensional in character, it is appropriate that they should receive comment here as a potentially important part of seismic array processing techniques. In this section we wish to single out two of these important new contributions and discuss each of them briefly. The two topics are: *τ–p transform methods* in seismic data processing and the *vertical seismic profile* (*VSP*).

τ–p Methods

In our previous discussion of migration, we made reference to the fact that the CDP section represents data that do not correspond to a physically realizable experiment. If we view migration as the process of propagating an inferred wavefield backward in time, we must resort to the exploding reflector model in order to have a basis for this extrapolation when dealing with CDP stacked sections. Searching for a migration technique that accounts for strong lateral variations in velocity, Schultz and Claerbout [265] suggested a means of constructing a record section which corresponds to the illumination of a section by a plane wave. They referred to the resulting section as a *slanted plane-wave stack*. Hubral [138] also used the slant stack to formulate his elegant procedure.

The method suggested by Schultz and Claerbout for constructing a plane-wave stack is to combine many "single-shot, single-phone" experiments in the following way. A single output trace $P'(t)$ is obtained by summing a collection of shots with different offsets from a fixed receiver (in order to synthesize a plane wave). Appropriate "delays" between the shots give approximate plane waves at various angles of incidence.

Let g be the horizontal coordinate of the receiver, and let $\{S_i\}$ be the horizontal distances to the shots and define

$$f_i = S_i - g$$

$$\Delta f = f_{i+1} - f_i \qquad \text{(assumed constant)}$$

$$\Delta S = S_{i+1} - S_i = \Delta f$$

and

$$\Delta t = \frac{\Delta S}{V_H}$$

where V_H is the horizontal phase velocity of the wavefront. If we define the ray parameter, or *slowness*, p, by

$$p = \frac{1}{V_H}$$

then Snell's law implies that p is constant for a vertically stratified earth.

We note that $\Delta t = p\ \Delta f$. Given our field data in f and t coordinates, $P(f_i, t)$ we form our output trace P' by

$$P'(p, \tau) = \sum_{i=1}^{n} P(f_i, t = \tau + pf_i)$$

Schultz and Claerbout generate *slant-stack* sections by displaying the traces P' for successive geophone positions and note their similarity to standard CDP stacked sections. However, since a (approximate) single plane wave has generated the section, reflection coefficients which are sensitive to angle of incidence may show up prominently for certain plane waves, but may be missed in the CDP stacked section. They go on to consider the problem of migrating these slant-stack sections and discuss the need to consider both upgoing and downgoing waves for migration in the presence of strong lateral velocity variations, justifying the use of plane-wave illumination.

More recently, a number of papers have appeared extolling the virtues of plane-wave decompositions of seismic data since they tend to simplify many problems and also because the resulting decomposition tends to satisfy more accurately certain of our processing assumptions than do the CDP stacked sections.

Phinney et al. [228] give a good discussion of the τ–p (slant-stack) transform and its inverse and discuss some of its possibilities. The formal transform processes are: Let $\phi(x, t)$ be the observed record section with shot points at the origin and receivers evenly spaced in x, close enough that no aliasing occurs. If we assume that the wave field is continuous and twice differentiable in all arguments, the τ–p transform of the record section, ϕ, is given by the *Radon transform*

$$\psi(p, \tau) = \int_{-\infty}^{\infty} \phi(x, \tau + px)\, dx$$

and its inverse is given by

$$\phi(x, t) = \frac{1}{2\pi} \frac{d}{dt} H^{+} \int_{-\infty}^{\infty} \psi(p, t - px)\, dp$$

where H^{+} is the (forward) Hilbert transform.

The τ–p transform maps hyperbolas to ellipses, and straight lines to points, and neatly separates subcritical from postcritical reflections and helps to identify various signal components in this way. A simple masking procedure in the τ–p domain and an inverse transform can remove selected components from the record section. By windowing in the τ–p domain, various representations of the record section can be obtained which emphasize primaries only, multiples only, shear waves, backscattered signals (negative p), and subcritical reflections [228]. Phinney et al. also point out the probable value of the τ–p representation for the geophysical inverse problem.

Stoffa and others [73, 279] have pointed out that the hyperbolic moveout approximations referred to earlier are not valid for large offset distances and they derive accurate methods for velocity (versus depth) inversion using the slant stack. Clayton and McMechan [60] derive a migration algorithm for refraction data, using the τ–p transform, in order to directly invert the data to obtain velocity versus depth. McMechan and Yedlin [199] used the τ–p transform to develop a simple procedure for identifying dispersion (see "Polarization and Dispersion Analysis").

Alam and Austin [4] point out the advantages of autocorrelation and predictive deconvolution analysis carried out in the τ–p domain for multiple suppression. They comment that procedures of this kind seem to perform better in the τ–p domain as opposed to the t–x domain, since certain multiples become periodic for fixed values of p (recall also the radial multiple suppression technique referenced earlier [283]). Treitel *et al.* [295a] present a plane–wave decomposition technique and also comment on its value for predictive deconvolution. In addition, they present a plane–wave imaging technique for this process.

Chapman [52] first considered τ–p methods for synthetic seismogram generation. Stoffa and Wenzel [279a] suggest another procedure and point out the advantages of comparing synthetic with real data in the τ–p domain. In particular, iterative model building and response comparison with field data is computationally much less expensive in the τ–p domain. An interesting discussion of processing seismic data using the τ–p transform is given in [289].

Although these methods are relatively new, the physical meaning of the plane-wave decomposition and the correctness of using procedures, which are based on plane-wave analysis, to process data decomposed in this way, suggest that seismic processing and inversion studies will meet with reasonable expectations of success in the $\tau-p$ domain. Certainly the initial indications are promising.

The Vertical Seismic Profile

In our earlier discussions, we mentioned still another type of multidimensional display for certain kinds of seismic data, called the *vertical seismic profile* (VSP). The vertical seismic profile is generally obtained by firing surface shots into closely spaced geophone positions within a bore hole, which may span the entire length of the well from casing to total depth (Fig. 2.23). One advantage of the VSP is that much of the energy recorded has traveled directly from the surface through the earth to the geophone and so represents a one-way transmission path through the earth, thereby carrying potentially useful information which is not obtained in any other way. In addition, the VSP is extremely useful for identifying primary and multiple

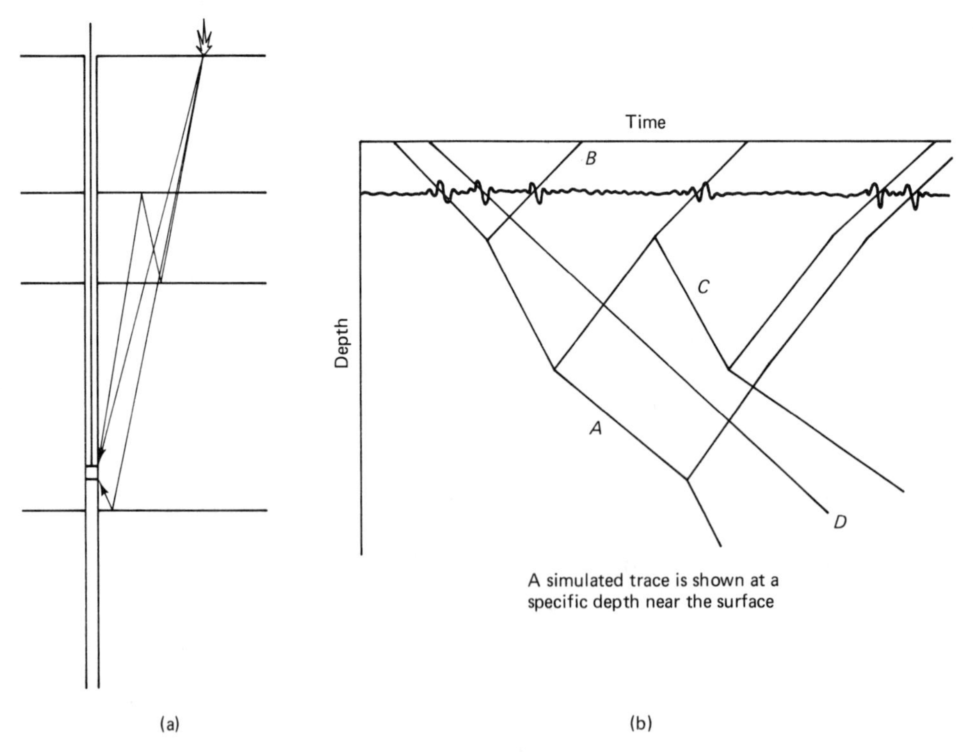

Figure 2.23 (a) VSP survey (showing some possible ray paths); (b) VSP profile: *A*, direct arrival; *B*, primary event; *C*, multiple reflection; *D*, tube wave.

events on seismograms and provides a very neat separation between upgoing and downgoing energy at the geophones.

Potential uses for the VSP include accurate in situ measurement of attenuation and dispersion characteristics of the seismic wavelet as it traverses through the various layers in the earth, information concerning mode conversion between compressional and shear waves, accurate interval velocity measurements, and the design of operators for deconvolving the seismic wavelet and for suppressing various types of multiple reflections. In addition, accurate ties can be established between well logs, seismic field data, and the VSP so that events from each of these kinds of data can be correlated and identified. VSP is also able to give us some information concerning the subsurface structure beneath the total depth of the well which can be useful information, whether it is obtained before the well has reached its final depth or after the drilling is completed. In addition, by comparing field-recorded VSPs with synthetic model-based VSPs, we can check the accuracy of models that we may be using.

A typical VSP is displayed as a series of seismic traces (recorded at geophone locations for a given depth in the well bore) versus time, and the time recording usually spans several seconds. The first arrival on each trace in a VSP can be expected to be the direct transmission path from the surface to the geophone, and the slope of these first arrivals is indicative of interval velocity. Primary reflections are recorded on all traces down to the depth of the reflector itself and are not recorded below that depth. The intersection of the curve corresponding to the direct arrival indicates the true depth position of the reflector. Downgoing energy will tend to parallel the direct arrival curve and upgoing energy will tend to align itself as a mirror image of these curves. Multiples will be observed on portions of VSP traces whose termination points are determined by the layers that give rise to them. If the direct arrival times from the VSP are doubled and the traces replotted, all upward-going energy will align itself vertically at its correct two-way travel time and a simple velocity filtering application could easily remove the downgoing energy if desired. On the other hand, if the data are replotted with the direct arrivals aligned vertically, the downgoing energy will align itself vertically and we could easily filter out the upgoing energy. An excellent outline of the basics of VSP interpretation can be found in Kennett et al. [157].

In addition to discussing the interpretation of VSP profiles, including primary and multiple identification, Kennett et al. discuss methods for determining the dip of dipping horizons, and discuss the use of downgoing waveforms to design deconvolution operators for surface seismic data. They compare seismic data processed in this way with conventionally processed data. Finally, they discuss the estimation of the acoustic impedance log *ahead of the drill bit*, by examining primaries whose presence is shown on the recorded traces, but which do not intersect the (truncated) direct arrival curve because the bit has not penetrated the reflecting horizon.

VSP data are not always as perfect as those discussed above. These surveys are particularly vulnerable to a certain kind of noise called the *tube wave*, which is a boundary wave propagating along the cylindrical fluid–solid boundary of the borehole. Tube waves may obscure significant portions of the VSP data and represent another potential application of array processing procedures to remove their effects. Various approaches to this problem, including velocity filtering, are discussed by Hardage [116]. A constructive use for tube waves has been suggested in [132].

Although numerous studies have been performed relating attenuation to fluid saturation, confining pressure, and so on, these experiments are carried out generally on very small rock samples at very high frequencies, and the correct extrapolation of these results back to the earth environment and seismic frequencies is poorly understood. The VSP offers an excellent opportunity to perform these experiments with meaningful parameters, and these experiments are just beginning to be reported.

Hauge [122] reports five VSP surveys in which attenuation measurements have been made. Assuming a linear dependence of attenuation on frequency, Hauge assumes a model for attenuation given by

$$A_x(f) = G_x A_0(f) \exp(-B_x f)$$

where f is the frequency, $A_0(f)$ the amplitude spectrum of the downhole pulse at reference depth X_0, and $A_x(f)$ the pulse amplitude spectrum for depth X. G_x is a gain factor which accounts for changes of pulse magnitude with depth and B_x is called the *cumulative attenuation* between depths X_0 and X. B_x and G_x can be determined by taking logarithms on both sides above:

$$\ln \frac{A_x(f)}{A_0(f)} = -B_x f + \ln G_x$$

and fitting a straight line to the data on the left.

After correcting for the effect of multiples on attenuation, good agreement was found between lithology and attenuation and, in general, porous sands were found to attenuate more than neighboring shales. The expected distortion of the seismic wavelet with attenuation was observed.

Newman and Worthington [215] carried out several VSP surveys with three component phones, and while acknowledging the limitations on the data, also support the types of conclusions drawn by Hauge, and conclude that attenuation is a sensitive indicator of lithology in porous, near-surface material. They also indicate that compressional wave attenuation may be more sensitive to these changes than shear-wave attenuation. Some interesting comments and results on the use of attenuation have been reported by Stone [281] and by Taner and Sheriff [285].

Spencer et al. [276] consider several problems inherent in VSP attenuation studies: interference effects, spatial resolution, and frequency-dependent attenuation. They conclude that attenuation estimates are very sensitive to noise and suggest that the *spectral ratio method* should probably be confined to the direct arrival.

Tatham and Stoffa [291] have suggested the use of the *compressional-to-shear velocity ratio* as a potential hydrocarbon indicator. In the past, shear waves have been generally assumed not to be significant in seismic data processing and interpretation. Lash [174], however, reports the results of VSP surveys carried out with three component phones to investigate the presence and importance of shear waves in seismic exploration. Lash concludes that shear waves are generated by explosive sources and that additional shear waves are generated by conversion of compressional waves at interfaces between elastic media in the earth. He suggests that we attach more importance to shear-wave studies for their potential use in estimating parameters of importance such as porosity and permeability.

Tatham [286] has suggested that crack or pore geometry may be the relevant factor in relating V_p/V_s values to lithology. That is, it may be a more important factor than the elastic parameters themselves. Experimental evidence for correlation of V_p/V_s with lithology would then imply an association of crack or pore geometry with lithology, using V_p/V_s as the indicator. Balch et al. [23] also discuss the possible correlation of VSP with lithology. In addition, they discuss the processing of VSP data.

These experiments are suggestive of important developments which may be now hovering on the horizon in seismic exploration. Recalling our earlier comments that structure alone will not be enough to satisfy our future interpretational requirements, these experiments and others like them are pointing the way to better methods, more information, and improved interpretations which will hopefully reduce the risks and costs inherent in the discovery of new hydrocarbon resources. In particular, we are beginning to accumulate the information and methods which may allow direct inference (at least in a probabilistic sense) of lithologic parameters such as porosity, saturation, rock type, and so on.

2.18 SUMMARY

In this chapter we have only scratched the surface of the many broad areas which may be classified as seismic array processing. There is still much more that may be said but which space and time will not allow. In particular, we have ignored the areas of seismic refraction surveys and processing, the very large and interesting area of multidimensional modeling of wave propagation through a stratified or heterogeneous elastic medium, and a problem that is growing in importance in hand with seismic modeling, that of using modeling to generate accurate synthetic data corresponding to the interpreter's model of the subsurface and comparing that data with the original field data in order to perform a nonlinear optimization on the model parameters, so that its response is finally in agreement with the observed field data. Although the latter is of great interest and growing importance, we are very far from having a good handle on its solution.

We have also made only passing reference to the important area of direct seismic inversion which is undergoing rapid development and shows significant promise for the future. Direct seismic inversion methods are designed to estimate velocities, densities, and/or acoustic impedance directly from the data, and some extremely interesting work is going on in this area [61, 63, 232–234]. Finally, we have not touched on many new areas of research still in the developmental stages, such as geotomography [65] [73a]. Many of these new areas of research represent potentially fruitful avenues of inquiry which will find important future applications in exploratory and developmental seismology.

With regard to seismic modeling, many sophisticated and accurate procedures have been and are being developed, from simple to highly sophisticated ray-tracing approximations, to propagator matrix formulations, and finally finite element or finite difference approximations to the elastic wave equation for heterogeneous media. Recent developments in computing power suggest that after years of development and limited use, these sophisticated modeling procedures are finally becoming practical as a part of our data processing and interpretation efforts, and it is reasonable to anticipate that they will grow significantly in importance and will become an integral part of the seismic exploration process.

Although some procedures have gone unmentioned here because it was felt that they were truly single-channel operations, they are by far outnumbered by the procedures that have been mentioned and which may truly be classified as array processing procedures in exploration seismology. We have, with few exceptions, given a fairly complete, if shallow, discussion of the broad spectrum of procedures which are employed in the processing and interpretation of seismic data and we have endeavored to indicate the status of these methods and areas for future research.

Seismic data processing today is probably on the verge of a revolution from which it will emerge in quite different form from that in which we have known it for the past two decades, during which computer processing has become the mainstay of seismic processing. In particular, we have indicated that the trend would be to new, high-resolution methods, which probably will account for many more parameters than have been considered in the past, using entirely new ways to represent the data, and new processing procedures which will accompany these. Indeed, this is an exciting time in the development of exploration seismology, and much remains to be done and many questions remain to be answered. In the future, we shall see increased use of three-dimensional seismic surveying with new procedures for processing these data, the further integration of geology and other geophysical methods with seismic methods, the increased integration of borehole information with seismic processing, and new developments in petrophysics which will assist this integration of borehole techniques with seismic processing.

Looking ahead to the future, we may say that seismic processing will begin to account for more and more parameters, enabling us to extract more information, which, through the use of pattern recognition procedures, for example, may allow us to make probabilistic inferences of the detailed structure and nature of the near-surface geology of the earth. Ideally, this would include the direct inference of the existence of hydrocarbons and the properties of the reservoir in which they are trapped. Once these hydrocarbons have been located and their commercial value established, we may look forward to increased use of seismology in conjunction with other methods to delineate reservoir parameters more accurately so that, in conjunction with reservoir engineers, we may learn to optimally design primary, secondary, and tertiary recovery procedures to maximize the recovery of valuable hydrocarbons from these reservoirs.

The past is prologue. . . .

REFERENCES AND BIBLIOGRAPHY*

1 J. D. Achenbach, in *Wave Propagation in Elastic Solids*, Vol. 16, ed. H. A. Lauwerier and W. T. Koiter, North-Holland, Amsterdam, 1973.

2 H. Akaike, "Block Toeplitz Matrix Inversion," SIAM J. Appl. Math., Vol. 24, No. 2, pp. 234–241, 1973.

2a K. Aki and P. Richards, *Quantitative Seismology*, Vols. 1, 2, W. H. Freeman, San Francisco, 1980.

3 M. A. Alam, "Orthonormal Lattice Filter—A Multistage Multichannel Estimation Technique," GE, Vol. 43, No. 7, pp. 1368–1383, 1978.

4 M. A. Alam and J. Austin, "Suppression of Multiples Using Slant Stacks," 51st Annu. Int. Mtg. SEG, Los Angeles, 1981, pp. 3225–3257.

5 M. A. Alam and A. P. Sage, "Sequential Estimation and Identification of Reflection Coefficients by Minimax Entropy Inverse Filtering," Comput. Electron. Eng., Vol. 2, pp. 315–338, 1975.

6 M. A. Alam and C. J. Sicking, "Recursive Removal and Estimation of Minimum-Phase Wavelet," GE, Vol. 46, No. 10, pp. 1379–1391, 1981.

7 M. Al-Chalabi, "An Analysis of Stacking, RMS, Average and Interval Velocities over a Horizontally Layered Ground," GP, Vol. 22, pp. 458–475, 1974.

8 M. Al-Chalabi, "Series Approximation in Velocity and Traveltime Computations," GP, Vol. 21, pp. 783–795, 1972.

9 M. Al-Chalabi, "Velocity Determination from Seismic Reflection Data," Dev. Geophys. Explor., Vol. 1, pp. 1–68, 1979.

10 S. S. Alexander, "Surface Wave Propagation in the Western United States," Ph.D. dissertation, California Institute of Technology, 1963.

11 M. I. Al-Husseini, J. B. Glover, and B. J. Barley, "Dispersion Patterns of the Ground Roll in Eastern Saudi Arabia," GE, Vol. 46, No. 2, pp. 121–137, 1981.

12 R. C. Anderson, "Use of Vertical Seismic Profiles to Supplement Geologic Interpretations Made from Surface Recorded Seismic Data," Preprint of a paper presented at the SEG/China Geophys. Soc., Beijing, 1981.

13 N. A. Anstey, "The Sectional Auto-correlogram and the Sectional Retro-correlogram," GP, Vol. 14, pp. 299–426, 1966.

14 API, *Basic Petroleum Data Book*, Vol. 2, No. 3, 1982.

15 S. P. Applebaum, "Adaptive Arrays," IEEE Trans. Antennas Propag., Vol. AP-24, No. 5, pp. 585–598, 1976.

*Abbreviations: GE, Geophysics; GP, Geophysical Prospecting.

16 C. B. Archambeau, "Fine Structure of the Upper Mantle," J. Geophys. Res., Vol. 74, No. 25, pp. 5825–5865, 1969.

17 C. B. Archambeau, J. C. Bradford, P. W. Broome, W. C. Dean, E. A. Flinn, and R. L. Sax, "Data Processing Techniques for the Detection and Interpretation of Teleseismic Signals," Proc. IEEE, Vol. 53, No. 12, pp. 1860–1865, 1965.

18 C. B. Archambeau and E. A. Flinn, "Automated Analysis of Seismic Radiation for Source Characteristics," Proc. IEEE, Vol. 53, No. 12, pp. 1876–1884, 1965.

19 C. B. Archambeau, E. A. Flinn, and D. G. Lambert, "Detection, Analysis, and Interpretation of Teleseismic Signals," J. Geophys. Res., Vol. 71, No. 14, pp. 3483–3501, 1966.

20 C. B. Archambeau, E. A. Flinn, and D. G. Lambert, "Fine Structure of the Upper Mantle," J. Geophys. Res., Vol. 74, pp. 5825–5865, 1969.

21 M. E. Arnold, "Beam Forming with Vibrator Arrays," GE, Vol. 42, No. 7, pp. 1321–1338, 1977.

22 A. B. Baggeroer, "Confidence Intervals for Regression (MEM) Spectral Estimates," IEEE Trans. Inf. Theory, Vol. IT-22, pp. 534–545, 1976.

23 A. H. Balch, M. W. Lee, J. J. Miller, and R. T. Ryder, "The Use of Vertical Seismic Profiles in Seismic Investigations of the Earth," GE, Vol. 47, No. 6, pp. 906–918, 1982.

24 A. Bamberger, G. Chavent, C. Hemon, and P. Lailly, "Inversion of Normal Incidence Seismograms," GE, Vol. 47, No. 5, pp. 757–770, 1982.

25 E. H. Bareiss, "Numerical Solution of Linear Equations with Toeplitz and Vector Toeplitz Matrices," Numer. Math., Vol. 13, pp. 404–424, 1969.

26 T. E. Barnard, "Two Maximum Entropy Beamforming Algorithms for Equally Spaced Arrays," IEEE Trans. Acoust. Speech Signal Process., Vol. 30, No. 2, pp. 175–189, 1982.

27 M. Becquey, M. Lavergne, and C. Willm, "Acoustic Impedance Logs Computed from Seismic Traces," GE, Vol. 44, No. 9, pp. 1485–1501, 1979.

27a A. J. Berkhout, "Wave Field Extrapolation Techniques in Seismic Migration, a Tutorial," GE, Vol. 46, No. 12, pp. 1638–1656, 1981.

28 A. J. Berkhout and B. A. De Jong, "Recursive Migrations in Three Dimensions," GP, Vol. 29, pp. 758–781, 1981.

28a A. J. Berkhout, *Seismic Migration Imaging of Acoustic Energy by Wave Field Extrapolation*, Elsevier, New York, 1980.

29 J. R. Berryhill, "Wave-Equation Datuming," GE, Vol. 44, No. 8, pp. 1329–1339, 1979.

30 D. E. Biswell, L. F. Konty, and A. L. Liaw, "A Geophone Subarray Beam Steering Process," 49th Annu. Int. Mtg. SEG, New Orleans, 1979.

31 G. Bolondi, F. Rocca, and S. Savelli, "A Frequency Domain Approach to Two-Dimensional Migration," GP, Vol. 26, pp. 750–772, 1978.

32 A. Booker and C. Y. Ong, "Multiple-Constraint Adaptive Filtering," GE, Vol. 36, No. 3, pp. 498–509, 1971.

33 J. C. Bradford, "Analysis of Array Noise by Coherence Functions," Proc. IEEE, Vol. 53, No. 12, pp. 1873–1874, 1965.

34 S. D. Brasel, "Patterns of Sources and Detectors," Dev. Geophys. Explor., Vol. 1, pp. 69–92, 1979.

35 T. M. Brocher and R. A. Phinney, "Inversion of Slant Stacks Using Finite-Length Record Sections," Preprint.

36 T. M. Brocher and R. A. Phinney, "A Ray Parameter-Intercept Time Spectral Ratio Method for Seismic Reflectivity Analysis," Preprint.

37 P. W. Broom and W. C. Dean, "Seismic Applications of Orthogonal Expansions," Proc. IEEE, Vol. 53, pp. 1865–1869, 1965.

38 R. J. S. Brown, "Normal-Moveout and Velocity Relations for Flat and Dipping Beds and for Long Offsets," GE, Vol. 34, No. 2, pp. 180–195, 1969.

39 D. R. Brundrit and J. C. Van Wijnen, "Elongated Seismic Source for Use in Marine Operations," Canadian Patent 1,002,647 (issued Dec. 28, 1976); "Super Long Detector Cable," Canadian Patent 1,002,648 (issued Dec. 18, 1976). Both patents assigned to Shell Canada Ltd.

39a L. T. Bruton and N. R. Bartley, "Highly Selective Three-Dimensional Recursive Beam Filters Using Intersecting Resonant Planes," IEEE Trans. on Circuits and Systems, Vol. CAS–30, No. 3, pp. 190–193, 1983.

40 H. Buchholtz, "A Note on Signal Distortion Due to Dynamic (NMO) Corrections," GP, Vol. 20, pp. 395–402, 1972.

41 J. P. Burg, "The Relationship between Maximum Entropy Spectra and Maximum Likelihood Spectra," GE, Vol. 37, No. 2, pp. 375–376, 1972.

41a J. P. Burg, "Maximum Entropy Spectral Analysis," Ph.D. dissertation, Stanford University, 1975.

42 J. P. Burg, "Three-Dimensional Filtering with an Array of Seismometers," GE, Vol. 24, No. 5, pp. 693–713, 1964.

42a J. P. Burg, D. G. Luenberger, and D. L. Wenger, "Estimation of Structured Covariance Matrices," Proc. IEEE, Vol. 70, pp. 963–974, 1982.

43 B. Buttkus, "Coherency Weighting—An Effective Approach to the Suppression of Long Leg Multiples," GP, Vol. 27, pp. 27–39, 1979.

43a J. A. Cadzow and K. Ogino, "Two-Dimensional Spectral Estimation," IEEE Trans. Acoust. Speech Signal Process., Vol. 29, No. 3, pp. 396–401, 1981.

44 J. Capon, "High-Resolution Frequency-Wavenumber Spectrum Analysis," Proc. IEEE, Vol. 57, pp. 1408–1418, 1969.

45 J. Capon, R. J. Greenfield, and R. T. Lacoss, "Long-Period Signal Processing Results for the Large Aperture Seismic Array," GE, Vol. 34, No. 3, p. 305, 1969.

46 J. Capon, R. J. Greenfield, and R. J. Kolker, "Multidimensional Maximum-Likelihood Processing of a Large Aperture Seismic Array," Proc. IEEE, Vol. 55, No. 2, pp. 192–211, 1967.

47 J. Capon, R. J. Greenfield, R. J. Kolker, and R. T. Lacoss, "Short-Period Signal Processing Results for the Large Aperture Seismic Array," GE, Vol. 33, No. 3, pp. 452–472, 1968.

48 E. Cassano and F. Rocca, "After-Stack Multichannel Filters without Mixing Effects," GP, Vol. 22, pp. 330–344, 1974.

49a E. Cassano and F. Rocca, "Multichannel Linear Filters for Optimal Rejection of Multiple Reflections," GE, Vol. 38, No. 6, pp. 1053–1061, 1973.

50 C. Y. Chang, "Adaptive Multichannel Filtering," Proc. ICASSP 80, Vol. 2, 1980, pp. 462–464.

51 C. Y. Chang, "Multichannel Adaptive Filtering with a Feedback Convergence Function," Proc. ICASSP 82, Vol. 2, 1982, pp. 667–670.

52 C. H. Chapman, "A New Method for Computing Synthetic Seismograms," Geophys. J. R. Astron. Soc., Vol. 54, pp. 481–518, 1978.

53 D. G. Childers, ed., *Modern Spectrum Analysis*, IEEE Press, New York, 1978.

54 J. H. Chun and C. A. Jacewitz, "The First Arrival Time Surface and Estimation of Statics," 51st Annu. Mtg. SEG, Los Angeles, 1981, pp. 1–46.

55 J. H. Chun and C. A. Jacewitz, "Fundamentals of Frequency Domain Migration," GE, Vol. 46, No. 5, pp. 717–733, 1981.

56 J. H. Chun and C. A. Jacewitz, "Structure Independent Iterative Automatic Statics," 47th Annual Mtg. SEG, Calgary, 1977.

56a J. F. Claerbout, "Detection of *P* Waves from Weak Sources at Great Distances," GE, Vol. 29, No. 2, pp. 197–211, 1964.

57 J. F. Claerbout, "Synthesis of a Layered Medium from Its Acoustic Transmission Response," GE, Vol. 33, No. 2, pp. 264–269, 1969.

58 J. F. Claerbout, "Toward a Unified Theory of Reflector Mapping," GE, Vol. 36, No. 3, pp. 467–481, 1971.

59 J. F. Claerbout and S. M. Doherty, "Downward Continuation of Moveout-Corrected Seismograms," GE, Vol. 37, No. 5, pp. 741–768, 1972.

60 R. W. Clayton and G. A. McMechan, "Inversion of Reflection Data by Wave Field Continuation," GE, Vol. 46, No. 6, pp. 860–868, 1981.

61 R. W. Clayton and R. H. Stolt, "A Born–WKBJ Inversion Method for Acoustic Reflection Data," GE, Vol. 46, No. 11, pp. 1559–1567, 1981.

62 B. Cochrane, K. P. Dawson, M. A. Fiddy, and T. J. Hall, "Sampling and Interpolation in Two Dimensions," Proc. ICASSP 82, Vol. 1, 1982, pp. 471–474.

63 J. K. Cohen and N. Bleistein, "Velocity Inversion Procedure for Acoustic Waves," GE, Vol. 44, No. 6, pp. 1077–1087, 1979.

64 K. S. Cressman, "How Velocity Layering and Steep Dip Affect CDP," GE, Vol. 33, No. 3, pp. 399–411, 1968.

65 W. D. Daily, R. J. Lytle, E. F. Laine, J. T. Okada, and F. J. Deadrick, "Geotomography in Oil Shale," J. Geophys. Res., Vol. 87, No. B7, pp. 5507–5515, 1982.

66 B. P. Dash and B. L. A. Hains, "Moveout Detection by an Autocorrelation Matrix Method," GE, Vol. 39, No. 6, pp. 794–810, 1974.

67 B. P. Dash and K. A. Obaidullah, "Determination of Signal and Noise Statistics Using Correlation Theory," GE, Vol. 35, No. 1, pp. 24–32, 1970.

68 E. B. Davies and E. J. Mercado, "Multichannel Deconvolution Filtering of Field Recorded Seismic Data," GE, Vol. 33, No. 5, pp. 711–722, 1968.

69 P. H. Delsarte, Y. Genin, and Y. Kamp, "Half-Plane Minimization of Matrix-Valued Quadratic Functionals," Soc. Ind. Appl. Math., Vol. 2, No. 2, pp. 192–211, 1981.

70 P. Delsarte, Y. V. Genin, and Y. G. Kamp, "Half-Plane Toeplitz Systems," IEEE Trans. Inf. Theory, Vol. IT-26, No. 4, pp. 465–474, 1980.

71 P. Delsarte, Y. Genin, and Y. Kamp, "A Survey of Two-Dimensional Toeplitz Systems," Int. Symp. Math. Theory Networks Syst., Vol. 3, Delft University of Technology, Delft, The Netherlands, pp. 17–30, 1979.

72 S. M. Deregowski and F. Rocca, "Geometrical Optics and Wave Theory of Constant Sections in Layered Media," GP, Vol. 29, pp. 374–406, 1981.

73 J. B. Diebold and P. L. Stoffa, "The Traveltime Equation Tau-*p* Mapping, and Inversion of Common Midpoint Data," GE, Vol. 46, No. 3, pp. 238–254, 1981.

73a Kris A. Dines and R. Jeffrey Lytle, "Computerized Geophysical Tomography." Proceedings of the IEEE, Vol. 67, No. 7, pp. 1065–1073, July 1979.

74 W. L. Dinstel, "Velocity Spectra and Diffraction Patterns," GE, Vol. 36, No. 2, pp. 415–417, 1971.

75 C. H. Dix, "Seismic Velocities from Surface Measurements," GE, Vol. 20, pp. 68–86, 1955.

76 S. M. Doherty and J. F. Claerbout, "Structure Independent Velocity Estimation," GE, Vol. 41, No. 5, pp. 850–881, 1976.

77 W. Dragoset and K. Larner, "Data Enhancement from a 500-Channel Streamer," 50th Annu. Int. Mtg. SEG, Houston, 1980.

78 A. A. Dubrulle and J. Gazdag, "Migration by Phase Shift—an Algorithmic Description for Array Processors," GE, Vol. 44, No. 10, pp. 1661–1666, 1979.

79 D. E. Dudgeon, "Fundamentals of Digital Array Processing," Proc. IEEE, Vol. 65, No. 6, pp. 898–904, 1977.

80 J. W. Dunkin and F. K. Levin, "Isochrons for a Three-Dimensional Seismic System," GE, Vol. 36, No. 6, pp. 1099–1137, 1971.

81 H. Durbaum, "Zur Bestimmung von Wellengeschwindigkeiten aus Reflexions-Seismischen Messungen," GP, Vol. 2, pp. 151–167, 1954.

82 A. Dziewonski, S. Bloch, and M. Landisman, "A Technique for the Analysis of Transient Signals," Bull. Seismol. Soc. Am., Vol. 59, No. 1, pp. 427–444, 1969.

83 A. Dziewonski, J. Mills, and S. Bloch, "Residual Dispersion Measurement—A New Method of Surface-Wave Analysis," Bull. Seismol. Soc. Am., Vol. 62, No. 1, pp. 129–139, 1972.

84 M. P. Ekstrom, "Realizable Wiener Filtering in Two Dimensions," IEEE Trans. Acoust. Speech Signal Process., Vol. ASSP-30, No. 1, pp. 31–40, 1982.

85 P. Embree, J. P. Burg, and M. M. Backus, "Wide-Band Velocity Filtering—The Pie-Slice Process," GE, Vol. 28, No. 6, pp. 948–974, 1963.

86 E. J. Farrell, "Sensor-Array Processing with Channel-Recursive Bayes Techniques," GE, Vol. 36, No. 5, pp. 822–834, 1971.

87 E. A. Flinn, "Signal Analysis Using Rectilinearity and Direction of Particle Motion," Proc. IEEE, Vol. 53, No. 12, pp. 1874–1876, 1965.

88 M. R. Foster, R. L. Sengbush, and R. J. Watson, "Design of Sub-optimum Filter Systems for Multi-trace Seismic Data Processing," GP, Vol. 12, pp. 173–191, 1964.

89 C. W. Frasier, "Discrete Time Solution of Plane *p*-SV Waves in a Plane Layered Medium," GE, Vol. 35, No. 2, pp. 197–219, 1970.

90 W. S. French, "Two-Dimensional and Three-Dimensional Migration of Model-Experiment Reflection Profiles," GE, Vol. 39, No. 3, pp. 265–277, 1974.

91 E. J. Frey, "Two-Dimensional Spectral Estimation: A Comparison of Current Techniques," M.Sc. dissertation, University of Colorado, Boulder, 1982.

92 O. L. Frost, III, "An Algorithm for Linearly Constrained Adaptive Array Processing," Proc. IEEE, Vol. 60, pp. 926–935, 1972.

93 R. L. Frost, "Recent Results in High-Resolution Astronomical Imaging," IEEE Int. Symp. Proc., Vol. 2, pp. 505–509, 1980.

94 W. I. Futterman, "Dispersive Body Waves," J. Geophys. Res., Vol. 67, No. 13, pp. 5270–5291, 1962.

95 R. A. Gabel and R. R. Kurth, "Digital Beamsteering with Recursive Multichannel Filters," Proc. ICASSP 82, pp. 803–806, 1982.

96 W. F. Gabriel, "Adaptive Arrays—An Introduction," Proc. IEEE, Vol. 64, No. 2, pp. 239–272, 1976.

97 W. F. Gabriel, "Nonlinear Spectral Analysis and Adaptive Array Superresolution Techniques," NRL Report 8345, 1980.

98 W. F. Gabriel, "Spectral Analysis and Adaptive Array Superresolution Techniques," Proc. IEEE, Vol. 68, No. 6, pp. 654–666, 1980.

99 J. N. Galbraith, Jr., and R. A. Wiggins, "Characteristics of Optimum Multichannel Stacking Filters," GE, Vol. 33, No. 1, pp. 36–48, 1968.

100 A. F. Gangi and B. S. Byun, "The Corrective Gradient Projection Method and Related Algorithms Applied to Seismic Array Processing," GE, Vol. 41, No. 5, pp. 970–984, 1976.

101 A. F. Gangi and D. Disher, "A Space–Time Filter for Seismic Models," GE, Vol. 33, No. 1, pp. 88–104, 1968.

102 D. C. Ganley, "A Method for Calculating Synthetic Seismograms Which Include the Effects of Absorption and Dispersion," GE, Vol. 46, No. 8, pp. 1100–1107, 1981.

103 G. H. F. Gardner, W. S. French, and T. Matzuk, "Elements of Migration and Velocity Analysis," GE, Vol. 39, No. 6, pp. 811–825, 1974.

104 G. Garibotto, "2-D Recursive Phase Filters for the Solution of Two-Dimensional Wave Equations," IEEE Trans. Acoust. Speech Signal Process., Vol. ASSP-27, No. 4, pp. 367–373, 1979.

105 R. Garotta, "Selection of Seismic Picking Based upon the Dip, Moveout and Amplitude of Each Event," GP, Vol. 19, pp. 357–373, 1971.

106 R. Garotta and D. Michon, "Continuous Analysis of the Velocity Function and of the Moveout Corrections," GP, Vol. 15, pp. 584–597, 1967.

107 J. Gazdag, "Wave Equation Migration with the Accurate Space Derivative Method," GP, Vol. 28, pp. 60–70, 1980.

108 J. Gazdag, "Wave Equation Migration with the Phase-Shift Method," GE, Vol. 43, No. 7, pp. 1342–1351, 1978.

109 Y. Genin and Y. Kamp, "Counterexample in the Least Square Inverse Stabilization of 2D-Recursive Filters," Electron. Lett., Vol. 11, pp. 330–331, 1975.

110 P. L. Goupillaud, "An Approach to Inverse Filtering of Near-Surface Layer Effects from Seismic Records," GE, Vol. 26, No. 6, pp. 854–760, 1961.

111 A. Gray and J. Markel, "Digital Lattice and Ladder Filter Synthesis," IEEE Trans. Audio Electroacoust., Vol. AU-21, pp. 491–500, 1973.

112 P. E. Green, Jr., E. J. Kelly, Jr., and J. J. Levin, "A Comparison of Seismic Array Processing Methods," Geophys. J. R. Astron. Soc., Vol. 11, pp. 67–84, 1966.

113 U. Grenander and M. Rosenblatt, *Statistical Analysis of Stationary Time Series*, Wiley, New York, 1957.

113a L. Griffiths, "Adaptive Structures for Multiple Input Noise Canceling Applications," Proc. 1979 Int. Conf. Acoust. Speech Signal Process., pp. 925–928, 1979.

114 S. F. Gull and G. J. Daniell, "Image Reconstruction from Incomplete and Noisy Data," Nature, Vol. 272, pp. 686–690, 1978.

115 O. S. Halpeny and D. G. Childers, "Composite Wavefront Decomposition via Multidimensional Digital Filtering of Array Data," IEEE Trans. Circuits Syst., Vol. CAS-22, No. 6, pp. 552–562, 1975.

116 B. A. Hardage, "An Examination of Tube Wave Noise in VSP Data," Midwestern Mtg. SEG, Tulsa, Okla., 1980.

117 D. B. Harris, "Recursive Least Squares with Linear Constraints," Proc. ICASSP 81, Vol. 2, pp. 526–529, 1981.

118 F. J. Harris, "On the Use of Windows for Harmonic Analysis with the Discrete Fourier Transform," Proc. IEEE, Vol. 66, No. 1, pp. 51–83, 1978.

119 N. A. Haskell, "The Dispersion of Surface Waves on Multilayered Media," Bull. Seismol. Soc. Am., Vol. 43, pp. 17–34, 1953.

120 L. Hatton, K. Larner, and B. S. Gibson, "Migration of Seismic Data from Inhomogeneous Media," GE, Vol. 46, No. 5, pp. 751–767, 1981.

121 R. A. Haubrich, "Array Design," Bull. Seismol. Soc. Am., Vol. 58, No. 3, pp. 977–991, 1968.

122 P. S. Hauge, "Measurements of Attenuation from Vertical Seismic Profiles," GE, Vol. 46, No. 11, pp. 1548–1558, 1981.

123 W. S. Hawes and L. Gerdes, "Some Effects of Spatial Filters on Signal," GE, Vol. 39, No. 4, pp. 464–498, 1974.

124 S. Haykin, ed., *Nonlinear Methods of Spectral Estimation*, Springer-Verlag, Berlin, 1979.

125 S. Haykin and S. Kesler, "The Complex Form of the Maximum Entropy Method for Spectral Estimation," Proc. IEEE, Vol. 64, pp. 822–823, 1976.

126 S. Haykin and S. Kesler, "Prediction-Error Filtering and Maximum-Entropy Spectral Estimation," in *Nonlinear Methods of Spectral Analysis*, Springer-Verlag, Berlin, 1979.

127 J. H. Hesson, "On Extremal Properties Statisfied by the I_0-sinh Window," Proc. ICASSP 82, Vol. 1, pp. 339–342, 1982.

128 M. Holzman, "Chebyshev Optimized Geophone Arrays," GE, Vol. 28, No. 2, pp. 145–153, 1963.

129 G. M. Hoover, "Acoustical Holography Using Digital Processing," GE, Vol. 37, No. 1, pp. 1–19, 1972.

130 G. M. Hoover and J. T. O'Brien, "The Influence of the Planted Geophone on Seismic Land Data," GE, Vol. 45, No. 8, pp. 1239–1253, 1980.

131 T. S. Huang, "Two-Dimensional Windows," IEEE Trans. Audio Electroacoust., pp. 88–89, 1972.

132 C. F. Huang and J. A. Hunter, "A Seismic 'Tube Wave' Method for In-Situ Estimation of Rock Fracture Permeability in Boreholes," 51st Annu. Mtg. SEG, Los Angeles, 1981.

133 T. P. Hubbard, "Deconvolution of Surface Recorded Data Using Vertical Seismic Profiles," 49th Annu. Int. Mtg. SEG, New Orleans, 1979.

134 P. Hubral, "Stacking Filters and Their Characterization in the (*f*-*k*) Domain," GP, Vol. 22, pp. 722–735, 1974.

135 P. Hubral, "Time Migration—Some Ray Theoretical Aspects," GP, Vol. 25, pp. 738–745, 1977.

136 P. Hubral, "Interval Velocities from Surface Measurements in the Three-Dimensional Plane Layer Case," GE, Vol. 41, No. 2, pp. 233–242, 1976.

137 P. Hubral, "Three-Dimensional Optimum Multichannel Velocity Filters," GP, Vol. 20, pp. 28–46, 1972.

138 P. Hubral, "Slant Stack Migration," in *Festschrift* Theodor Krey, Prakla–Seismos, Hannover, West Germany, pp. 72–78, 1980.

139 P. Hubral and T. Krey, "Interval Velocities from Seismic Reflection Time Measurements," ed. K. L. Larner, SEG, Tulsa, Okla. 1980.

140 P. Hubral, S. Treitel, and P. R. Gutowski, "A Sum Autoregressive Formula for the Reflection Response," GE, Vol. 45, No. 11, pp. 1697–1705, 1980.

141 F. Itakura and S. Saito, "A Statistical Method for Estimation of Speech Spectral Density and Formant Frequencies," Electron. Commun. (Jap.), Vol. 53-A, pp. 36–43, 1970.

142 A. K. Jain and S. Ranganath, "Extrapolation Algorithms for Discrete Signals with Application in Spectral Estimation," IEEE Trans. Acoust. Speech Signal Process., Vol. ASSP-29, No. 4, pp. 830, 845. 1981.

143 G. M. Jenkins and D. G. Watts, *Spectral Analysis and Its Applications*, Holden-Day, San Francisco, 1968.

144 R. H. Jones, "Estimation of Spatial Wavenumber Spectra and Falloff Rate with Unequally Spaced Observations," J. Atmos. Sci., Vol. 32, No. 2, pp. 260–268, 1975.

145 R. H. Jones, "Multivariate Autoregression Estimation Using Residuals," in *Applied Time Series Analysis*, ed. D. F. Findley, Academic Press, New York, 1978, pp. 139–162.

146 J. H. Justice, "A Levinson-Type Algorithm for Two-Dimensional Wiener Filtering Using Bivariate Szego Polynomials," Proc. IEEE Vol. 65, No. 6, pp. 882–886, 1977.

147 J. H. Justice, "The Step from One to Higher Dimensional Signal Processing—Case Histories," in *Applied Time Series Analysis, II*, ed. D. F. Findley, Academic Press, New York, 1981.

148 J. H. Justice, "Two-Dimensional Recursive Filtering in Theory and Practice," in *Applied Time Series Analysis*, ed. D. F. Findley, Academic Press, New York, 1978.

148a J. H. Justice and J. L. Shanks, "Stability Criterion for N–Dimensional Digital Filters, IEEE Trans. on Automatic Control, Vol. AC–18, No. 3, pp. 284–286, 1973.

149 J. F. Kaiser and R. W. Schafer, "On the Use of I_0-sinh Window for Spectrum Analysis," Trans. Acoust. Speech Signal Process., Vol. ASSP-28, No. 1, pp. 105–107, 1980.

150 E. R. Kanasewich, *Time Sequence Analysis in Geophysics*, University of Alberta Press, Edmonton, 1981.

151 E. R. Kanasewich, C. D. Hemmings, and T. Alpaslan, "*N*th-Root Stack Nonlinear Multichannel Filter," GE, Vol. 38, No. 2, pp. 327–338, 1973.

152 H. Kato and T. Furukawa, "Two-Dimensional Type-Preserving Circular Windows," IEEE Trans. Acoust. Speech Signal Process., Vol. ASSP-29, No. 4, pp. 926–928, 1981.

153 S. M. Kay and S. L. Marple, Jr., "Spectrum Analysis—A Modern Perspective," Proc. IEEE, Vol. 69, No. 11, pp. 1380–1419, 1981.

154 E. J. Kelly, "The Representation of Seismic Arrays in Frequency-Wave Number Space," Massachusetts Institute of Technology, Lincoln Laboratory Group Report 1964-15.

155 B. L. N. Kennett, "Slowness Techniques in Seismic Interpretation," manuscript.

156 P. Kennett and R. L. Ireson, "Some Techniques for the Analysis of Well Geophone Signals as an Aid to the Identification of Hydrocarbon Indicators in Seismic Processing," 43rd Annu. Int. Mtg. SEG, Mexico City, 1973.

157 P. Kennett, R. L. Ireson, and P. J. Conn, "Vertical Seismic Profiles: Their Applications in Exploration Geophysics," GP, Vol. 28, pp. 676–699, 1980.

158 A. K. Kerekes, "Seismic Array Design—A Practical Approach," 49th Annu. Int. Mtg. SEG, New Orleans, 1979.

159 T. A. Khan, M. D. McCormack, and D. L. Shafer, "Optimization of Source Arrays in the Recording Unit," 49th Annu. Int. Mtg. SEG, New Orleans, 1979.

159a H. Kobayashi, "Iterative Synthesis Method for a Seismic Array Processor," IEEE Trans. Geosci. Electron., Vol. 8, No. 3, pp. 168, 178, 1970.

160 O. Koefoed, "Reflection and Transmission Coefficients for Plane Longitudinal Incident Waves," GP, Vol. 10, pp. 304–351, 1962.

161 D. D. Kosloff and E. Baysal, "Forward Modeling by a Fourier Method," GE, Vol. 47, No. 10, pp. 1402–1412, 1982.

162 D. D. Kosloff and E. Baysal, "Migration with the Full Acoustic Wave Equation," Preprint.

163 T. Krey, *Festschrift*, Prakla-Seismos, Hannover, West Germany, 1980.

164 T. Krey, "Seismic Stripping Helps Unravel Deep Reflections," GE, Vol. 43, No. 5, pp. 899–911, 1978.

165 M. J. Kuhn, "Acoustical Imaging of Source Receiver Coincident Profiles," GP, Vol. 27, pp. 62–77, 1979.

166 R. Kumaresan and D. W. Tufts, "Improved Spectral Resolution III: Efficient Realization," Proc. IEEE, Vol. 68, No. 10, pp. 1354–1355, 1980.

167 R. T. Lacoss, "Data Adaptive Spectral Analysis, Methods," GE, Vol. 36, No. 4, pp. 661–675, 1971.

168 R. T. Lacoss, E. J. Kelly, and M. N. Toksoz, "Estimation of Seismic Noise Structure Using Arrays," GE, Vol. 34, No. 1, pp. 21–38, 1969.

168a S. W. Lang, "Spectral Estimation for Sensor Arrays," Ph.D. dissertation, Massachusetts Institute of Technology, 1981.

169 S. W. Lang and J. H. McClellan, "The Extension of Pisarenko's Method to Multiple Dimensions," Proc. ICASSP 82, Vol. 1, pp. 125–128, 1982.

170 S. W. Lang and J. H. McClellan, "Spectral Estimation for Sensor Arrays," Proc. First IEEE ASSP Workshop on Spectral Estimation, Hamilton, Ont., 1981.

171 K. L. Larner, B. R. Gibson, R. Chambers, and R. A. Wiggins, "Simultaneous Estimation of Residual Static and Crossdip Corrections," GE, Vol. 44, No. 7, pp. 1175–1192, 1979.

172 K. L. Larner, L. Hatton, B. S. Gibson, and I. Hsu, "Depth Migration of Imaged Time Sections," GE, Vol. 46, No. 5, pp. 734–750, 1981.

173 K. L. Larner and M. Rooney, "Interval Velocity Computation for Plane Dipping Multilayered Media," 42nd Annu. Int. Mtg. SEG, Anaheim, Calif., 1973.

174 C. C. Lash, "Shear Waves, Multiple Reflections and Converted Waves Found by a Deep Vertical Wave Test (Vertical Seismic Profiling)," GE, Vol. 45, No. 9, pp. 1373–1411, 1980.

175 S. J. Laster and A. F. Linville, "Preferential Excitation of Refracting Interfaces by Use of a Source Array," GE, Vol. 33, No. 1, pp. 49–64, 1968.

176 F. K. Levin, "Vertical Stacking as a Reflection Filter," GE, Vol. 42, No. 5, pp. 1045–1047, 1977.

177 A. I. Levorsen, *Geology of Petroleum*, W. H. Freeman, San Francisco, 1967.

177a J. S. Lim and N. A. Malik, "A New Algorithm for Two-Dimensional Maximum Entropy Power Spectrum Estimation," IEEE Trans. Acoust. Speech Signal Process., Vol. 29, No. 3, pp. 401–413, 1981.

178 D. Loewenthal, L. Lu, R. Roberson, and J. Sherwood, "The Wave Equation Applied to Migration," 36th Mtg. Eur. Assoc. Explor. Geophys., Madrid, 1974.

179 D. Loewenthal, P. R. Gutowski, and S. Treitel, "Direct Inversion of Transmission Synthetic Seismograms," GE, Vol. 43, No. 5, pp. 886–898, 1978.

180 J. H. Lofthouse and G. T. Bennett, "Extended Arrays for Marine Seismic Acquisitions," GE, Vol. 43, No. 1, pp. 3–22, 1978.

181 D. G. Luenberger, *Optimization by Vector Space Methods*, Wiley, New York, 1969.

182 J. Makhoul, "A Class of All-Zero Lattice Digital Filters: Properties and Applications," IEEE Trans. Acoust. Speech Signal Process., Vol. 26, pp. 304–314, 1978.

183 J. Makhoul, "Stable and Efficient Lattice Methods for Linear Prediction," IEEE Trans. Acoust. Speech Signal Process., Vol. 25, No. 5, pp. 423–428, 1977.

184 J. D. Markel, "Digital Inverse Filtering—A New Tool for Formant Trajectory Estimation," IEEE Trans. Audio Electroacoust., Vol. 20, pp. 129–137, 1972.

185 J. D. Markel and A. H. Gray, Jr., "On Autocorrelation Equations with Applications to Speech Analysis," IEEE Trans. Audio Electroacoust., Vol. 21, pp. 69–79, 1973.

186 L. Marple, "A New Autoregressive Spectrum Analysis," IEEE Trans. Acoust. Speech Signal Process., Vol. 28, No. 4, pp. 441–454, 1980.

187 T. L. Marzetta, "Additive and Multiplicative Minimum-Phase Decompositions of 2-D Rational Power Density Spectra," IEEE Trans. Circuits Syst., Vol. 29, No. 4, pp. 207–214, 1982.

188 T. L. Marzetta, "A Linear Prediction Approach to Two-Dimensional Spectral Factorization and Spectral Estimation," Ph.D. dissertation, Massachusetts Institute of Technology, 1972.

189 T. L. Marzetta, "Two-Dimensional Linear Prediction: Autocorrelation Arrays, Mimimum-Phase Prediction Error Filters, and Reflection Coefficient Arrays," IEEE Trans. Acoust. Speech Signal Process., Vol. 28, No. 6, pp. 725, 733, 1980.

189a J. A. Masters, "Deep Basin Gas Trap, Western Canada," AAPG Bulletin, Vol. 63, No. Z, pp. 152–181, 1979.

190 B. T. May and F. Hron, "Synthetic Seismic Sections of Typical Petroleum Traps," GE, Vol. 43, No. 6, pp. 1119–1147, 1978.

191 B. T. May and D. K. Straley, "Higher-Order Moveout Spectra," GE, Vol. 44, No. 7, pp. 1193–1207, 1979.

192 B. May and D. Straley, "Fourth Order Reflection Moveout," Amoco Production Company, Research Dept. Report F77-E-14, 1977.

193 J. H. McClellan, "Multidimensional Spectral Estimation," Proc. IEEE, Vol. 70, No. 9, pp. 1029–1039, 1982.

194 J. H. McClellan and S. W. Lang, "Multi-dimensional MEM Spectral Estimation," Institute of Acoustics, Spectral Analysis and Its Use in Underwater Acoustics, Underwater Group Conf., London, pp. 10.1–10.8, 1982.

195 J. H. McClellan and T. W. Parks, "Equiripple Approximation of Fan Filters," GE, Vol. 37, No. 4, pp. 573–583, 1972.

196 F. J. McDonal, F. A. Angona, R. L. Mills, R. L. Sengbush, R. G. Van Nostrand, and J. E. White, "Attenuation of Shear and Compressional Waves in Pierre Shale," GE, Vol. 23, No. 3, pp. 421–433, 1958.

197 R. N. McDonough, "Maximum-Entropy Spatial Processing of Array Data," GE, Vol. 39, No. 6, pp. 843–851, 1974.

198 A. E. McKay, "Review of Pattern Shooting," GE, Vol. 14, No. 3, pp. 420–437, 1954.

199 G. A. McMechan and M. J. Yedlin, "Analysis of Dispersive Waves by Wave Field Transformation," GE, Vol. 46, No. 6, pp. 869–874, 1981.

200 B. S. Melton and L. F. Bailey, "Multiple Signal Correlators," GE, Vol. 22, No. 3, pp. 565–588, 1957.

201 B. S. Melton and P. R. Karr, "Polarity Coincidence Scheme for Revealing Signal Coherence," GE, Vol. 22, No. 3, pp. 553–564, 1957.

202 E. J. Mercado, "Linear Phase Filtering of Multicomponent Seismic Data," GE, Vol. 33, No. 6, pp. 926–935, 1968.

202a E. Mercado, "Maximum Likelihood Filtering of Reflection Seismograph Data," GE, Vol. 43, No. 3, pp. 497–513, 1978.

203 J. H. Meyerhoff, "Horizontal Stacking and Multichannel Filtering Applied to Common Depth Point Seismic Data," 28th Mtg. Eur. Assoc. Explor. Geophys., Amsterdam, 1966.

204 D. Michon, R. Wlodarczak, and J. Merland, "A New Method of Canceling Multiple Reflection 'Souston'," GP, Vol. 18, pp. 615–645, 1971.

205 D. Middleton and J. R. B. Whittlesey, "Seismic Models and Deterministic Operators for Marine Reverberation," GE, Vol. 33, No. 4, pp. 557–583, 1968.

206 C. Mims and R. Sax, "Rectilinear Motion Detection (REMODE)," Teledyne, Inc., Alexandria, Va., Seismic Data Lab., Report 118, 1965.

207 J. F. Montalbetti and E. R. Kanasewich, "Enhancement of Teleseismic Body Phases with a Polarization Filter," Geophys. J. R. Astron. Soc., Vol. 21, pp. 119–129, 1970.

208 H. M. Mooney and B. A. Bolt, "Dispersive Characteristics of the First Three Rayleigh Modes for a Single Surface Layer," Bull. Seismol. Soc. Am., pp. 43–46, 1966.

209 M. Morf, A. Vieira, D. T. Lee, and T. Kailath, "Recursive Multichannel Maximum Entropy Spectral Estimation," IEEE Trans. Geosci. Electron., Vol. 16, No. 2, pp. 85–94, 1978.

210 P. M. Morse and H. Feshbach, *Methods of Theoretical Physics*, McGraw-Hill, New York, 1953.

211 K. J. Muirhead, "Eliminating False Alarms When Detecting Seismic Events Automatically," Nature, Vol. 217, pp. 533–534, 1968.

212 O. E. Naess, "Superstack–An Iterative Stacking Algorithm," GP, Vol. 27, pp. 16–28, 1979.

213 J. G. Negi and V. P. Dimri, "On Wiener Filter and Maximum Entropy Method for Multichannel Complex Systems," GP, Vol. 27, pp. 156–167, 1979.

214 N. S. Neidell and M. T. Taner, "Semblance and Other Coherency Measures for Multichannel Data," GE, Vol. 36, No. 3, pp. 482–497, 1971.

215 P. J. Newman and M. H. Worthington, "In-Situ Investigation of Seismic Body Wave Attenuation in Heterogeneous Media," GP, Vol. 30, pp. 377–400, 1982.

216 P. Newman and J. T. Mahoney, "Patterns—with a Pinch of Salt," GP, Vol. 21, pp. 197–219, 1973.

217 W. I. Newman "A New Method of Multidimensional Power Spectral Analysis," Astron. Astophys., Vol. 54, pp. 369–380, 1977.

218 A. Nilsen and B. Gjevik, "Inversion of Reflection Data," GP, Vol. 26, pp. 421–432, 1978.

219 A. H. Nuttall, "Some Windows with Very Good Sidelobe Behaviour," IEEE Trans. Acoust. Speech Signal Process., Vol. 29, No. 1, pp. 84–91, 1981.

220 A. H. Nuttall, G. C. Carter, and E. M. Montavon, "Estimation of the Two-Dimensional Spectrum of the Space–Time Noise Field for a Sparse Line Array," J. Acoust. Soc. Am., Vol. 55, No. 5, pp. 1034–1041, 1974.

221 K. Owusu, G. H. F. Gardner, and W. F. Massell, "Velocity Estimates Derived from Three-Dimensional Seismic Data," 51st Annu. Int. Mtg. SEG, pp. 2307–2354, 1981.

222 H. Ozdemir, "A Frequency Domain Mapping Approximation of Moveout Filters with Applications," GP, Vol. 30, pp. 292–317, 1982.

223 H. Ozdemir, "Optimum Hyperbolic Moveout Filters with Applications to Seismic Data," GP, Vol. 29, pp. 709–714, 1981.

224 T. W. Parks, C. F. Morris, and J. D. Ingram, "Velocity Estimation from Short-Time Temporal and Spatial Frequency Estimates," Proc. ICASSP 82, Vol. 1, pp. 399–402, 1982.

225 J. O. Parr, Jr., and W. H. Mayne, "A New Method of Pattern Shooting," GE, Vol. 20, No. 3, pp. 539–564, 1955.

226 J. V. Pendrel and D. E. Smylie, "The Relationship between Maximum Entropy and Maximum Likelihood Spectra," GE, Vol. 44, No. 10, pp. 1738–1739, 1979.

227 F. A. Perry and S. R. Parker, "Adaptive Solution of Multichannel Lattice Models for Linear and Nonlinear Systems," IEEE Int. Symp. Circuits Syst. Proc., Vol. 3, pp. 744–747, 1980.

228 R. A. Phinney, K. R. Chowdhury, and L. N. Frazer, "Transformation and Analysis of Record Sections," J. Geophys. Res., Vol. 86, No. B1, pp. 359–377, 1981.

229 R. A. Phinney and L. N. Frazer, "On the Theory of Imaging by Fourier Transform," 48th Annu. Mtg. SEG, San Francisco, 1978.

230 R. A. Phinney and D. M. Jurdy, "Seismic Imaging of Deep Crust," GE, Vol. 44, No. 10, pp. 1637–1660, 1979.

231 T. F. Potter and R. B. Roden, "Seismic Noise Estimation Using Horizontal Components," GE, Vol. 23, No. 4, pp. 617–632, 1967.

232 S. Raz, "A Procedure for Multidimensional Inversion of Seismic Data," GE, Vol. 47, No. 10, pp. 1422–1430, 1982.

233 S. Raz, "Direct Reconstruction of Velocity and Density Profiles from Scattered Field Data," GE, Vol. 46, No. 6, pp. 832–836, 1981.

234 S. Raz, "Three-Dimensional Velocity Profile Inversion from Finite-Offset Scattering Data," GE, Vol. 46, No. 6, pp. 837–842, 1981.

235 E. Rietsch, "Estimation of the Signal-to-Noise Ratio of Seismic Data with an Application to Stacking," GP, Vol. 28, pp. 531–550, 1980.

236 E. Rietsch, "Geophone Sensitivities for Chebyshev Optimized Arrays," GE, Vol. 44, No. 6, pp. 1142–1143, 1979.

237 D. C. Riley and J. P. Burg, "Time and Space Adaptive Deconvolution Filters," 42nd Annu. Int. Mtg. SEG, Anaheim, Calif., 1972.

238 J. Rissanen, "Algorithms for Triangular Decomposition of Block Hankel and Toeplitz Matrices with Application to Factoring Positive Matrix Polynomials," Math. Comput., Vol. 27, No. 121, pp. 147–154, 1973.

238a R. K. Ritt, *Fourier Series*, McGraw-Hill, New York, 1970.

239 E. A. Robinson, "Dynamic Predictive Deconvolution," GP, Vol. 23, pp. 780–798, 1975.

240 E. A. Robinson, "A Historical Perspective of Spectrum Estimation," Proc. IEEE, Vol. 70, No. 9, pp. 885–907, 1982.

241 E. A. Robinson, *Multichannel Time Series Analysis with Digital Computer Programs*, Holden-Day, San Francisco, 1967.

242 E. A. Robinson, "Multichannel z-Transforms and Minimum Delay," GE, Vol. 31, No. 1, pp. 482–500, 1966.

243 E. A. Robinson, *Random Wavelets and Cybernetics Systems*, Griffin, London/Hafner, New York, 1962.

244 E. A. Robinson, "Spectral Approach to Geophysical inversion by Lorentz, Fourier and Radon Transforms," Proc. IEEE, Vol. 70, No. 9, pp. 1039–1954, 1982.

245 E. A. Robinson and S. Treitel, "The Fine Structure of the Normal Incidence Synthetic Seismogram," Geophys. J. R. Astron. Soc., Vol. 53, pp. 289–310, 1978.

246 E. A. Robinson and S. Treitel, *Geophysical Signal Analysis*, Prentice-Hall, Englewood Cliffs, N.J., 1980.

247 J. C. Robinson, "A Technique for the Continuous Representation of Dispersion in Seismic Data," GE, Vol. 44, No. 8, pp. 1345–1351, 1979.

248 J. C. Robinson, "HRVA—A Velocity Analysis Technique Applied to Seismic Data," GE, Vol. 34, No. 3, pp. 330–356, 1969.

249 W. Rudin, "The Extension Problem for Positive-Definite Functions," Ill. J. Math., Vol. 7, pp. 532–539, 1963.

250 G. B. Rupert and J. H. Chun, "The Block Move Sum Normal Moveout Correction," GE, Vol. 40, No. 1, pp. 17–24, 1975.

251 J. V. Ryu, "Decomposition (DECOM) Approach Applied to Wave Field Analysis with Seismic Reflection Records," GE, Vol. 47, No. 6, pp. 869–883, 1982.

252 J. C. Samson, "Matrix and Stokes Vector Representations of Detectors for Polarized Waveforms: Theory, with Some Applications to Teleseismic Waves," Geophys. J. R. Astron. Soc., Vol. 51, pp. 583–603, 1977.

252a J. C. Samson, "The Spectral Matrix, Eigenvalues, and Principal Components in the Analysis of Multichannel Geophysical Data," presented at European Geophysical Society Meeting, Leeds, England, August, 1982.

253 J. W. Sattlegger, "Migration Velocity Determination: Part 1. Philosophy," GE, Vol. 40, No. 1, pp. 1–16, 1975.

254 J. W. Sattlegger, "A Method of Computing True Interval Velocities from Expanding Spread Data in the Case of Arbitrary Long Spreads and Arbitrarily Dipping Interfaces," GP, Vol. 13, pp. 306–318, 1965.

255 J. W. Sattlegger, "Migration of Seismic Interfaces," GP, Vol. 30, pp. 71–85, 1982.

256 J. W. Sattlegger, "Series for Three-Dimensional Migration in Reflection Seismic Interpretation," Geophysical Prospecting, Vol. 12, no. 1, pp. 115, 134, 1964.

257 J. W. Sattlegger and P. K. Stiller, "Section Migration before Stack, after Stack or In-between," GP, Vol. 22, pp. 297–314, 1974.

258 J. W. Sattlegger, P. K. Stiller, J. A. Echterhoff, and M. K. Hentschke, "Common Offset Migration (COPMIG)," GP, Vol. 28, pp. 859–871, 1980.

259 C. H. Savit, J. T. Brustad, and J. Sider, "The Moveout Filter," GE, Vol. 23, No. 1, pp. 1–25, 1958.

260 R. L. Sax, "Seismic Noise Models," Proc. IEEE, Vol. 53, No. 12, pp. 1870–1872, 1965.

261 W. A. Schneider, "Integral Formulation for Migration in Two and Three Dimensions," GE, Vol. 43, No. 1, pp. 49–76, 1978.

262 W. A. Schneider and M. M. Backus, "Dynamic Correlation Analysis," GE, Vol. 33, No. 1, pp. 105–126, 1968.

262a W. A. Schneider, K. L. Larner, J. P. Burg, and M. M. Backus, "A New Data-Processing Technique for the Elimination of Ghost Arrivals on Reflection Seismograms," GE, Vol. 29, No. 5, pp. 783–805, 1964.

262b W. A. Schneider, E. R. Prince, Jr., and B. F. Giles, "A New Data-Processing Technique for Multiple Attenuation Exploiting Differential Normal Moveout," GE, Vol. 30, No. 3, pp. 348–362, 1965.

263 M. Schoenberger and F. K. Levin, "Apparent Attenuation Due to Intrabed Multiples, II," GE, Vol. 43, No. 4, pp. 730–737, 1978.

264 P. S. Schultz, "A Method for Direct Estimation of Interval Velocities," 50th Annu. Mtg. SEG, Houston, 1980.

265 P. S. Schultz and J. F. Claerbout, "Velocity Estimation and Downward Continuation by Wavefront Synthesis," GE, Vol. 43, No. 4, pp. 691–714, 1978.

266 R. L. Sengbush and M. R. Foster, "Optimum Multichannel Velocity Filters," GE, Vol. 33, No. 1, pp. 11–35, 1968.

266a J. L. Shanks, S. Treitel, and J. H. Justice, "Stability and Synthesis of Two–Dimensional Recursive Filters," IEEE Trans. Audio and Electroacoustics, Vol. AU–20, No. 2, pp. 115–128, 1972.

267 W. W. Shen, "A Constrained Minimum Power Adaptive Beamformer with Time-Varying Adaptation Rate," GE, Vol. 44, No. 6, pp. 1088–1096, 1979.

268 J. W. C. Sherwood and P. H. Poe, "Continuous Velocity Estimation and Seismic Wavelet Processing," GE, Vol. 37, No. 5, pp. 769–787, 1972.

269 M. Shimshoni and S. W. Smith, "Seismic Signal Enhancement with Three-Component Detectors," GE, Vol. 29, No. 5, pp. 664–671, 1964.

270 S. A. Siddiqui, "Dispersion Analysis of Seismic Data," M.S. dissertation, University of Tulsa, 1971.

271 R. S. Simons, "A Surface Wave Particle Motion Discrimination Process," Bull. Seismol. Soc. Am., Vol. 58, No. 2, pp. 629–637, 1968.

272 S. M. Simpson, Jr., "Traveling Signal-to-Noise Ratio and Signal Power Estimates," GE, Vol. 32, No. 3, pp. 485–493, 1967.

273 M. K. Smith, "Noise Analysis and Multiple Seismometer Theory," GE, Vol. 21, No. 2, pp. 337–360, 1956.

274 A. Sommerfeld, *Mechanics of Deformable Bodies*, Academic Press, New York, 1950.

275 T. C. Speake and R. M. Mersereau, "A Note on the Use of Windows for Two-Dimensional FIR Filter Design," IEEE Trans. Acoust. Speech Signal Process., Vol. 29, No. 1, pp. 125–127, 1981.

276 T. W. Spencer, J. R. Sonnad, and T. M. Butler, "Seismic Q-Stratigraphy or Dissipation," GE, Vol. 47, No. 1, pp. 16–24, 1982.

277 J. Stewart, "Positive Definite Functions and Generalizations: An Historical Survey," Rocky Mt. J. Math., Vol. 6, No. 3, pp. 409–435, 1976.

278 P. L. Stoffa, P. Buhl, J. B. Diebold, and F. Wenzel, "Direct Mapping of Seismic Data to the Domain of intercept Time and Ray Parameter: a Plane Wave Decomposition," 49th Annu. Int. Mtg. SEG, New Orleans, 1979.

279 P. L. Stoffa, J. B. Diebold, and P. Buhl, "Velocity Analysis for Wide Aperture Seismic Data," GP, Vol. 30, pp. 25–57, 1982.

279a F. Wenzel, P. L. Stoffa, and P. Buhl, "Seismic Modeling in the Domain of Intercept Time and Ray Parameter," IEEE Trans. ASSP, Vol. ASSP–30, No. 3, pp. 406–423, 1982.

280 R. H. Stolt, "Migration by Fourier Transform," GE, Vol. 43, No. 1, pp. 23–48, 1978.

281 D. G. Stone, "Velocity and Bandwidth," Offshore Technol. Conf., Paper OTC 2270, 1975.

282 O. M. Strand, "Multichannel Complex Maximum Entropy (Autoregressive) Spectral Analysis," IEEE Trans. Autom. Control, Vol. Ac-22, No. 4, pp. 634–640, 1977.

283 M. T. Taner, "Long Period Sea-Floor Multiples and Their Suppression," GP, Vol. 28, pp. 30–48, 1980.

284 M. T. Taner and F. Koehler, "Velocity Spectra-Digital Computer Derivation and Applications of Velocity Functions," GE, Vol. 34, No. 6, pp. 859–881, 1969.

285 M. T. Taner and R. E. Sheriff, "Application of Amplitude, Frequency, and Other Attributes to Stratigraphic and Hydrocarbon Determination," AAPG, Seismic Stratigraphy—Applications to hydrocarbon exploration, Memoir 26, 1977.

286 R. H. Tatham, "V_p/V_s and Lithology," GE, Vol. 47, No. 3, pp. 336–344, 1982.

287 R. H. Tatham, D. V. Goolsbee, W. F. Massell, and H. R. Nelson, "Seismic Shear Wave Observations in a Physical Model Experiment," 51st Annu. Int. Mtg. SEG, los Angeles, 1981.

288 R. H. Tatham, J. Keeney, and W. F. Massell, "Spatial Sampling and Realizable 3-D Surveys," 51st Annu. Int. Mtg. SEG, Los Angeles, 1981.

289 R. H. Tatham, J. W. Keeney, and I. Noponen, "Application of the Tau-*P* Transform (Slant Stack) in Processing Seismic Reflection Data," 52nd Annu. Int. Mtg. SEG, Dallas, 1982.

290 R. H. Tatham, J. Keeney, C. D. T. Walker, D. Goolsbee, J. Wiley, W. F. Massell, and M. Parry, "Application of the Tau-*p* Transform (Slant Stack) in Processing Seismic Reflection Data," 52nd Annu. Int. Mtg. SEG, Dallas, 1982.

291 R. H. Tatham and P. L Stoffa, "V_p/V_s—A Potential Hydrocarbon Indicator," GE, Vol. 41, No. 5, pp. 837–849, 1976.

291a I. Tolstoy, *Wave Propagation*, McGraw–Hill, New York, 1973.

292 R. D. Tooley, T. W. Spencer, and H. F. Sagoci, "Reflection and Transmission of Plane Compressional Waves," GE, Vol. 30, No. 4, pp. 552–570, 1965.

293 S. Treitel, "Principles of Digital Multichannel Filtering," GE, Vol. 35, No. 5, pp. 785–811, 1970.

294 S. Treitel, P. R. Gutowski, and D. E. Wagner, "Plane-Wave Decomposition of Seismograms," GE, Vol. 47, No. 10, pp. 1375–1401, 1982.

295 S. Treitel, J. L. Shanks, and C. W. Frasier, "Some Aspects of Fan Filtering," GE, Vol. 32, No. 5, pp. 789–900, 1967.

295a S. Treitel, P. R. Gutowski, and D. E. Wagner, "Plane–Wave Decomposition of Seismograms," Geophysics, Vol. 47, No. 10, pp. 1375–1401, 1982.

296 A. W. Trorey, "Theoretical Seismograms with Frequency and Depth Dependent Absorption," GE, Vol. 27, No. 6, Part 1, pp. 766–785, 1962.

297 D. Tufekčív, J. F. Claerbout, and Z. Rašperíc, "Spectral Balancing in the Time Domain," 39th Mtg. Eur. Assoc. Explor. Geophys., 1977.

298 T. J. Ulrych and T. N. Bishop, "Maximum Entropy Spectral Analysis and Autoregressive Decomposition," Rev. Geophys. Space Phys., Vol. 13, No. 1, pp. 183–200, 1975.

299 T. J. Ulrych and R. W. Clayton, "Time Series Modeling and Maximum Entropy," Phys. Earth Planet. Inter., Vol. 12, pp. 188–200, 1976.

300 B. Ursin, "Attenuation of Coherent Noise in Marine Seismic Exploration using Very Long Arrays," GP, Vol. 26, pp. 722–749, 1978.

301 B. Ursin, "Seismic Velocity Estimation," GP, Vol. 25, pp. 658–666, 1977.

302 A. van den Bos, "Alternative Interpretation of Maximum Entropy Spectral Analysis," IEEE Trans. Inf. Theory, Vol. IT-17, pp. 493–494, 1971.

303 G. G. Walton, "Three-Dimensional Seismic Method," GE, Vol. 37, No. 3, pp. 417–430, 1972.

304 R. W. Ward and J. B. Reining, "Cubic Spline Approximation of Inaccurate RMS Velocity Data," GP, Vol. 27, pp. 443–457, 1979.

304a K. H. Waters, *Reflection Seismology*, John Wiley, New York, 1981.

305 G. A. Watson, "An Algorithm for the Inversion of Block Matrices of Toeplitz Form," J. ACM, Vol. 20, No. 3, pp. 409–415, 1973.

306 F. Wenzel, P. L. Stoffa, and P. Buhl, "Seismic Modeling in the Domain of Intercept Time and Ray Parameter," IEEE Trans. Acoust. Speech Signal Process., Vol. 30, No. 3, pp. 406–422, 1982.

307 S. J. Wernecke and L. R. D'Addario, "Maximum Entropy Image Reconstruction," IEEE Trans. Comput., Vol. 26, No. 4, pp. 351–364, 1977.

308 J. E. White and D. J. Walsh, "Proposed Attenuation–Dispersion Pair for Seismic Waves," GE, Vol. 37, No. 3, pp. 456–461, 1972.

309 R. E. White, "The Performance of Optimum Stacking Filters in Suppressing Uncorrelated Noise," GP, Vol. 25, pp. 165–178, 1977.

310 B. Widrow, P. Mantey, L. Griffith, and B. Goode, "Adaptive Antenna Systems," Proc. IEEE, Vol. 55, pp. 2143, 2159, 1967.

311 R. A. Wiggins, "ω–k Filter Design," GP, Vol. 14, No. 4, pp. 427–440, 1966.

312 R. A. Wiggins, K. L. Larner, and R. D. Wisecup, "Residual Statics Analysis as a General Linear Inverse problem," 45th Annu. Mtg. SEG, Denver, Colo., 1975.

313 R. A. Wiggins and E. A. Robinson, "Recursive Solution to the Multichannel Filtering Problem," J. Geophys. Res., Vol. 70, No. 8, pp. 1885–1891, 1965.

314 A. S. Willsky, "Relationships between Digital Signal Processing and Control and Estimation Theory," Proc. IEEE, Vol. 66, No. 9, pp. 996–1017, 1978.

315 J. W. Woods, "Markov Image Modeling," IEEE Trans. Autom. Control, Vol. 23, No. 5, pp. 846–850, 1978.

316 J. W. Woods, "Two-Dimensional Discrete Markovian Fields," IEEE Trans. Inf. Theory, Vol. 18, No. 2, pp. 232–240, 1972.

317 J. W. Woods, "Markov Image Modeling," IEEE Trans. Autom. Control, Vol. 23, No. 5, pp. 846–850, 1978.

318 J. W. Woods, "Two-Dimensional Markov Spectral Estimation," IEEE Trans. Inf. Theory, Vol. 22, No. 5, pp. 552–559, 1976.

319 J. W. Woods and P. R. Lintz, "Plane Waves at Small Arrays," GE, Vol. 38, No. 6, pp. 1023–1041, 1973.

320 P. C. Wuenschel, "Dispersive Body Waves—An Experimental Study," GE, Vol. 30, No. 4, pp. 539–551, 1965.

321 P. C. Wuenschel, "The Vertical Array in Reflection Seismology—Some Experimental Studies," GE, Vol. 41, No. 2, pp. 219–232, 1976.

321a G. B. Young and L. W. Braile, "A Computer Program For the Application of Zoeppritz's Amplitude Equations and Knott's Energy Equations," Bulletin of the Seismological Society of America, Vol. 66, No. 6, pp. 1881–1885, 1976.

322 R. Zavalishin, "Improvements in Constructing Seismic Images Using the Method of Controlled Directional Reception," 52nd Annu. Int. Mtg. SEG, Dallas, 1982.

3

NORMAN L. OWSLEY
U.S. Naval Underwater Systems Center
Newport, Rhode Island

Sonar Array Processing

3.1 INTRODUCTION

The descriptive terms *wet end* and *dry end* are frequently used when an overall sonar system is discussed. In a modern digital sonar, as illustrated in Fig. 3.1, the wet end consists of a spatial array of sound pressure sensing, electromechanical transducers (hydrophones), hereafter referred to as *sensors*, and an associated data link both of which are immersed in the underwater medium. The dry end, beginning with the data receiver, includes the functions of *array processing* and *postprocessing*. The preponderance of future sonar systems will utilize digital techniques beginning with the wet-end analog-to-digital conversion [1] of the sensor output data and terminating with the storage of digital display output data in the dry end. Within this context, sonar array processing includes the spatial filtering portion of the dry end signal processing and postprocessing incorporates the temporal operations. The space and time operations are rendered separable [2] either because the signal process is narrowband and the inverse signal bandwidth is sufficiently greater than the transit time of sound across the array, as in active sonar, or the signal is spectrally decomposed, as commonly done in passive sonar by performing a fast Fourier transform on the sensor output data to translate them into the discrete frequency domain. The discussion of the dry-end array processing to be presented in this chapter focuses on the various aspects of sensor array beamforming for spatial

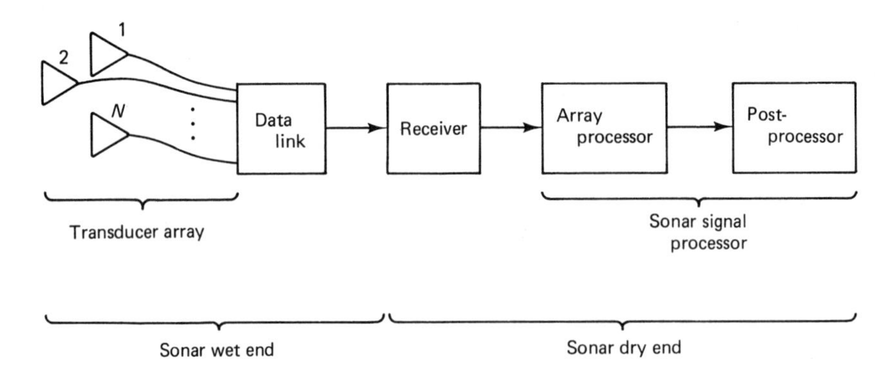

Figure 3.1 Generic digital sonar system.

signal field decomposition. A discussion of the dry-end postprocessing in a sonar would resemble any good discussion on modern *spectrum analysis* with the emphasis being more on spectral resolution capability for time-limited signals in active sonar systems and more on the simple detection of low signal-to-noise ratio signals of extended duration in passive, listening-only sonar. Extensive surveys of the techniques of spectrum analysis germane to the sonar postprocessing problem can be found elsewhere [3] and therefore are not presented herein. Rather, recent trends in the spatial filtering aspects of the dry-end signal processing with emphasis on data-dependent techniques are emphasized in this chapter.

The underwater sound environment is characterized by a multitude of *noise-like signals* and *signal-like noises.* In the face of this sometimes bewildering mix of signal and noise, two avenues for signal extraction are available, *temporal domain* processing and *spatial domain* processing. The most fundamental tools for temporal domain processing are correlation and spectrum analysis. Temporal domain correlation has a well-known justification based on the matched filter when a priori knowledge of the signal spectrum is available. Spectrum analysis is meant to include the complete range of techniques from traditional *Fourier power spectrum analysis* to data-dependent techniques such as the *maximum-likelihood method* (*MLM*), *maximum a posteriori* (*MAP*) *estimation*, and the *maximum-entropy method* (*MEM*) of spectrum estimation. It should, of course, not be surprising that nearly all of the temporal domain processing concepts have direct counterparts in the spatial domain. As a fundamental relationship, a parallel can be drawn between a sampled time interval of equally spaced, sampled data points and a one-dimensional (line) array of sensors which are equally spaced over some finite spatial aperture. With this simple analogy in mind, the temporal concept of *discrete frequency* (1/time) becomes interchangeable with the spatial frequency concept of *wavenumber* (1/distance). The lesser known discipline of sensor array processing by spatial multiple beamforming in wavenumber should now take on more meaning if the reader is able to relate to its discrete spectrum analyzer counterpart in temporal processing. In this chapter the *space–time analogy* pointed out above is exploited to help the reader obtain a firmer grasp of the spatial aspects of discrete sensor array processing.

In the early 1960s, the exploratory element of the sonar community became enamored with the idea of optimum processing for discrete sensor arrays [4–6]. Certainly, this initial theoretical work was motivated simply by the knowledge that, in many cases, the ambient noise environment for sonar arrays was highly nonuniform in the spatial frequency domain, allowing for the possibility of using spatial prewhitening filters. In this early development, little analysis was directed at performance prediction in realistic surroundings. In all likelihood, this occurred because the techniques of spatial coherence measurement and second-order statistical ambient noise modeling had not yet been sufficiently developed. Instead, the early theoretical studies were directed at so-called fully adaptive systems wherein all sensors in all beams were to be optimally and adaptively processed [7–10]. For the most part, this early work was evaluated using simulated data and did not initially have the benefit of either good noise models or actual recorded data from large operational sonar arrays. As the substantial hardware implications associated with a realization of a fully adaptive system became evident, the late 1960s and early 1970s saw an emphasis on hardware efficient configurations. The prospect of fabricating multibeam sonars that utilized 100 or more sensors and processing up to five octaves in bandwidth forced a trend to systems which either addressed a specific type of *directional noise*, namely, the point-source interference in the array far field [11–13], or were suboptimal by virtue of processing only a subarray but with little loss in performance [14]. Work on optimum array processor performance prediction using realistic models for ambient noise progressed significantly in the 1970s [15–17]. These models have provided insight into the minimal structures required for optimum processing against spatially coherent noise.

The *noise estimator–subtractor* structure has emerged in both the sonar [18, 19] and radar [20] disciplines as the most flexible implementation of an optimum-adaptive array processor in the sense of a *m*inimum *v*ariance *d*istortionless signal *r*esponse (MVDR) beamformer. This structure can be optimal and of minimum complexity provided that the spatial *c*ross-*s*pectral *d*ensity *m*atrix (CSDM) for the noise only is either known a priori or can be measured. In passive, listening-only sonar, it is possible to separate the signal and noise portions of the total CSDM for the array only if the signal wavefront and the receiving sensor array are of known shape and the wavefront is perfectly coherent. If these conditions are not satisfied, the potential exists for a condition of time delay and amplitude mismatch between the beamformer's model of the signal wavefront and the actual wavefront [21]. For sufficiently high signal-to-noise ratio, this model mismatch condition can significantly degrade the performance of an optimum array processor. Methods of avoiding the signal suppression effects resulting from wavefront mismatch involve the use of multidimensional filter linear and angular response pattern derivative constraints on the adaptive process [14, 18, 22–24]. Even these signal preservation constraints are not effective if medium propagation effects cannot be accounted for accurately.

As previously mentioned, adaptive MVDR beamforming can provide nearly optimum array processing given a reasonably accurate model for a wavefront in terms of its relative time delays and amplitudes at each sensor in the array. However, when either the wavefront or the array itself has unknown shape, a more general type of processing scheme than beamforming is perhaps indicated. In particular, array processing methods that incorporate medium propagation effects directly [25], as in the context of either wavefront shape measurement [26–28] for applications such as source localization or deformable array, unknown shape determination. Adaptive array focusing finds some application to this general problem and can be explained in terms of an orthogonal vector decomposition of the array CSDM. These same techniques may form the basis for investigations into Doppler-compensated coherent multipath recombination [29] and high-resolution wavenumber analysis [30, 31].

Finally, as mentioned above, the use of an array of sensors for the purpose of source location by time-delay estimation is an area of current interest [32, 33]. Theoretical bounds on performance have been established with respect to the basic time-delay measurement which is at the center of any source–location processor [34, 35].

Section 3.2 addresses the basic question: When is optimum-adaptive beamforming really useful? It should be realized that any kind of optimum-adaptive system requires expensive signal processing resources and that reasonable performance predictions should precede any decision to employ adaptive techniques. Section 3.3 derives the minimum variance distortionless response (MVDR) processor. Section 3.4 presents the MVDR estimator–subtractor configuration together with a discussion of the minimum complexity realization. Section 3.5 considers the degradation effects in an MVDR array processor due to either signal and noise correlation or mismatch of the beamformer wavefront shape model and actual signal wavefront geometry. Next, in Section 3.6, the various approaches to adaptive process realization are discussed in terms of both open-loop (feedforward) and closed-loop (feedback) algorithms. In Section 3.7, alternative approaches to the implementation of the MVDR processor are considered. In particular, the issue of time-domain versus frequency-domain techniques is briefly examined. Section 3.8 gives some operational examples of specific adaptive MVDR performance using different types of noise cancellation configurations. The cancellation of nonacoustic, high-wavenumber noise induced by the turbulent boundary layer at the surface of a moving array is discussed in Section 3.9. Finally, in Section 3.10, some alternative forms of array processing which have been considered to a lesser degree are surveyed. Specifically, two-dimensional Fourier transform methods and MEM and CSDM decomposition techniques are discussed with emphasis on high-resolution source range estimation.

The point of view reflected in this chapter is primarily that of low-frequency, passive sonar. This is true not only because of the author's recent background, but because the bulk of the development effort in the past decade has been with respect to passive techniques. Furthermore, active sonar considerations represent the easier of the two sonar operation modes with respect to the application of optimum-adaptive techniques. This is true for two reasons: first, in active sonar, the correlation characteristics of the noise can in many instances be measured directly without the signal (echo) present. Second, jammer and reverberation-type interference against which adaptive techniques may be considered tends to be at sound levels high enough so that as in radar, the simple rapidly converging sidelobe canceler noise suppression function is very effective.

3.2 SONAR OPTIMUM BEAMFORMING: IS IT NECESSARY?

When considering the use of an expensive adaptive implementation of an optimum beamformer for sonar array processing, a "quick-look" determination of the anticipated performance gains should first be made. Indeed, for arrays with as many as 100 sensors, the cost of digital signal processing to implement the optimum adaptive beamforming function relative to the cost of implementing a *delay-and-sum (DS) beamforming* system with a fixed, low-sidelobe-level spatial response pattern can be prohibitive. If the array gain improvement of an optimum sonar array beamformer over DS beamforming is potentially significant, further investigations into minimum complexity configurations are called for. In this section an example case study in optimum sonar array processor performance prediction is given. This study should serve as a guide for a rapid assessment of the utility of an optimum array, given either a point-source interference or a continuous angular distribution of coherent interference.

Point-Source Interference in the Array Far Field

Consider the far-field point signal and interference sources at discrete spatial frequencies ν_S and ν_I, respectively, as shown in Fig. 3.2. Spatial frequency is given by $\nu = (\omega \cos \theta)/v$, where ω is the temporal frequency, θ is the discrete angle of arrival of the respective planar wavefront at a linear array, and v is the speed of sound in water. Let the interference source sound pressure level SL_I be measured in units of power per unit area, called dB//1 μPa. This level is referenced to a 1-Hz band at 0.9 m (1 yd) from the assumed omnidirectional source. For simplicity, assume a low-frequency situation wherein the absorption loss for energy radiating from the interference can be neglected. For this case, the *transmission loss* (TL), which results from spherical spreading of the sound propagating from the interfering source to the receiving array at a distance r, is taken to be

$$\mathrm{TL} = 20 \log_{10} r \tag{3.1}$$

Let the ambient background noise level NL (dB//1 μPa) in a 1-Hz band at the receiving array be spatially uncorrelated. Let this be an *ambient noise-limited* situation in contrast to a *self-noise-limited* situation wherein either the array or the array platform is the dominant noise source. The level of this ambient noise at a

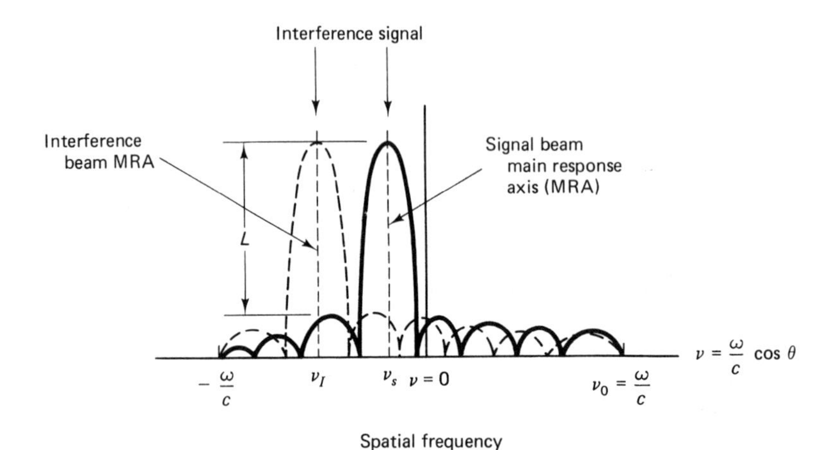

Figure 3.2 Sonar point-source interference geometry. Solid line: signal-directed beam response pattern; dashed line: point-source interference directed beam response pattern.

single sensor can be obtained from experimentally observed ambient noise spectra data and is typified by Fig. 3.3 for a deepwater environment. The *array gain* (AG) of a DS beamformer steered directly at the interfering source is the ratio of the *interference-to-noise* ratio, INR_{out}, at the beamformer output to the interference-to-noise ratio, INR_{in}, at a single sensor, and is given by

$$\begin{aligned} G &= \frac{\text{INR}_{\text{out}}}{\text{INR}_{\text{in}}} \\ &= 10 \log N \end{aligned} \tag{3.2}$$

where N is the number of sensors in the array. Finally, for a DS beam steered such that its *maximum response axis* (MRA) is exactly at an angle of interest for signal detection, let L $(0 \le L \le 1)$ be the relative level of this signal directed DS beam response in the direction of the interfering source. For example, $L = 1$ corresponds to the signal and interference at the same angle as the MRA of the signal beam, and $L = 0.5$ corresponds to the interference being at the -3-dB point of the signal beam main lobe.

With the definitions above, an expression for the interference-to-noise ratio at the output of the beam directed at the interference (INR_{out}) can be written using the *passive sonar equation* [36]:

$$\text{INR}_{\text{out}} = \text{SL}_I - \text{TL} - \text{NL} + \text{AG} \tag{3.3}$$

Furthermore, the *array gain improvement ratio*, which is the ratio of array gain using optimum beamforming G_0 to the array gain using DS beamforming G_{DS}, is given by

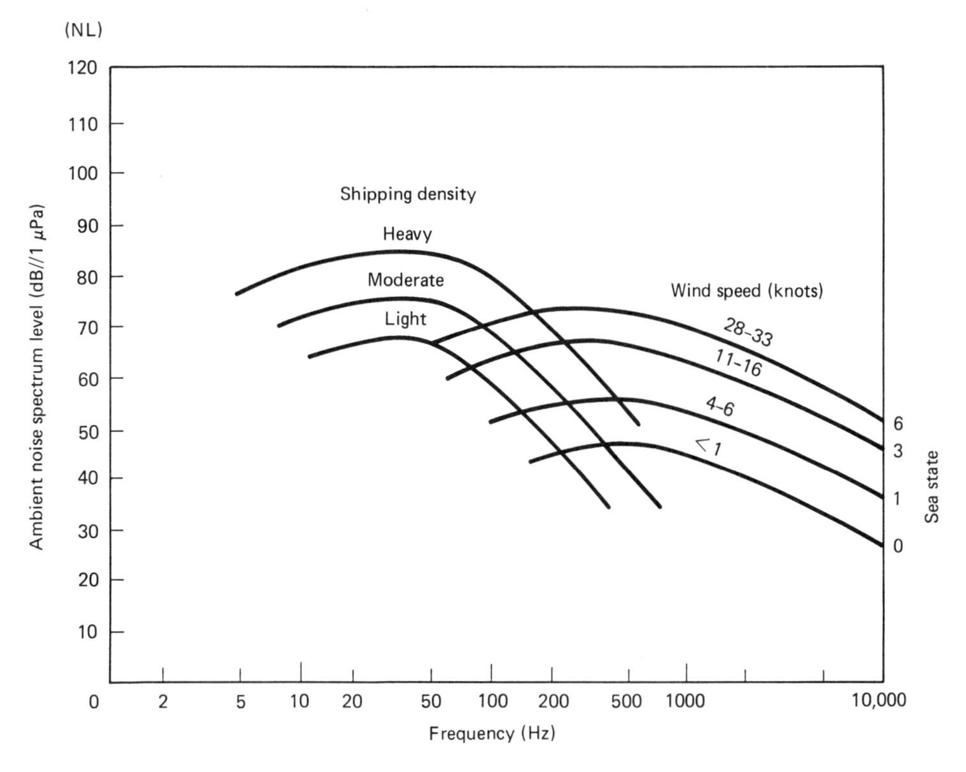

Figure 3.3 Typical deepwater ambient noise spectra.

the relation [37]:

$$R = \frac{G_0}{G_{DS}}$$

$$= 1 + \frac{(\text{INR}_{\text{out}})^2}{1 + \text{INR}_{\text{out}}} L(1 - L) \tag{3.4}$$

This expression for the gain improvement ratio R, which is derived in the following section, is plotted in Fig. 3.4 for transitional values of INR_{out} and L.

The relationships in Eqs. (3.1) to (3.4) allow a quick computation of the relative improvement in array processor performance in terms of the ratio of signal to total background interference plus noise power obtainable by using optimum versus conventional beamforming. For example, it is recalled that the interfering source has a sound source level SL_I and range r to a sensor array of N elements which is ambient-noise-limited by virtue of a spatially uncorrelated noise background of level NL at a single sensor. As a specific example of the above, let the interfering sonar contact have source level SL = 150 dB//1 μPa at 1 m and be at a range of r = 1 km

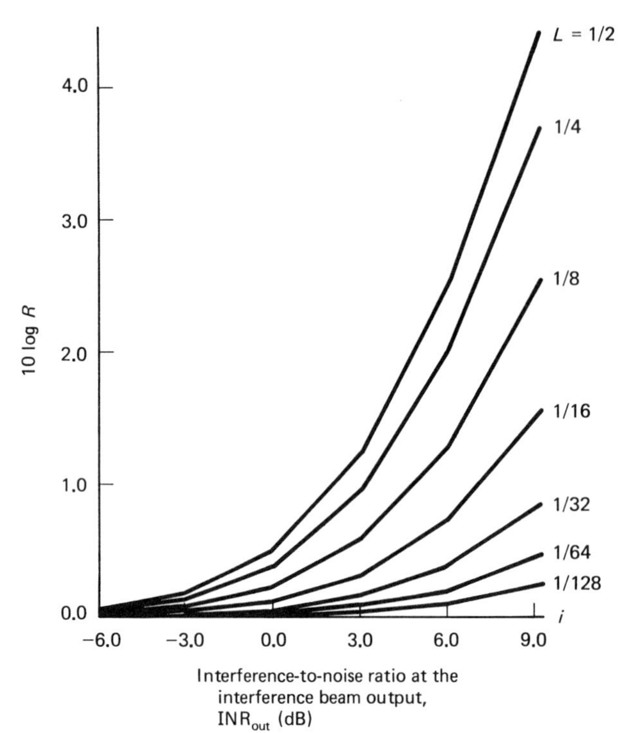

Figure 3.4 Array gain improvement ratio for a MVDR beamformer versus the interference-to-noise ratio at a beam output (INR_{out}) and sidelobe level (L).

from an array with $N = 50$ sensors, which is, in turn, located in a shipping zone with a moderate-to-heavy near shipping density. If a frequency of 100 Hz is considered such that $\text{NL} = 77$ dB//1 μPa and the interference is in the sidelobe region of the signal directed DS beamformer such that $L = 0.01(-20 \text{ dB})$, then Eq. (3.3) becomes

$$\begin{aligned}\text{INR}_{\text{out}} &= 150 - 20 \log 10^3 - 77 + 10 \log 50 \\ &= 30 \text{ dB}\end{aligned} \tag{3.5}$$

with a substantial resultant array gain improvement ratio obtained from Eq. (3.4) of approximately 10 dB. On the other hand, if the interfering source is either 10 dB quieter or the interfering source is at 10 km instead of 1 km or the ambient noise level is 10 dB higher, the array gain improvement ratio R reduces to approximately 3 dB. Given this predicted value of R or less, we may indeed have to question the value of an optimum beamformer and, at the very least, look for hardware-efficient realizations to make an optimum beamformer for a 3-dB signal-to-noise ratio improvement cost effective. Note that the array gain improvement is a maximum with respect to L for $L = 0.5$. This response level in the signal directed beam occurs at the -3-dB point in the main lobe of the response pattern, which is only one-half of the beamwidth from the main response axis (MRA).

If instead of a single interference as considered above, suppose that M spatially resolvable interferences occur simultaneously. For simplicity, let all the interfering sources have the same interference-to-noise ratio INR_{out} at the output of a beam steered directly at a particular source of interference. Also, let each interference occur at an average beam response level L with respect to the MRA level for a beam directed at the signal. The array gain improvement ratio for this case, which assumes that $L \ll 1$, is given by

$$G = 1 + \frac{(\text{INR}_{\text{out}})^2}{1 + \text{INR}_{\text{out}}} ML(1 - ML) \tag{3.6}$$

The maximum number of resolvable point-source interferences at frequency ω for an array with N uniformly spaced sensors at a distance $\pi v/\omega$ apart is $N - 1$. If L is on the order of $1/N$, then Eq. (3.6) indicates the performance of the optimum beamformer in terms of the gain improvement ratio improves with an increasing number of interferences M having the same power level until $M = N/2$ interfering sources are present. For $M > N/2$ the array gain improvement still indicates a net gain of optimum over DS beamforming but at a decreasing rate with increasing M. Equation (3.6) can be used to obtain a quick approximation of optimum array processor performance in the presence of an interfering noise which is distributed uniformly across an angular sector consisting of M beamwidths.

3.3 A SONAR ARRAY OPTIMUM BEAMFORMER

In this section the *constrained* minimum-variance distortionless response (MVDR) beamformer is derived as an optimum, linear processor for a hydrophone sensor array wherein all of the sensor outputs are individually filtered prior to linear combining. The structure for this particular configuration illustrated in Fig. 3.5 is formulated in the frequency domain. This has not been done simply to facilitate system analysis by virtue of the resulting separability of the spatial and temporal frequency domains. Rather, it reflects a current trend in the mode of implementation in both passive and narrowband active sonar systems. In particular, for a modern passive sonar system, it is typical to transform immediately the time-sampled hydrophone data to either the discrete spatial frequency or the discrete temporal frequency domain by a fast Fourier transform (FFT) operation. For the temporal frequency transformation, if T is the length of the FFT interval in seconds, the FFT resolution $\Delta = 1/T$ must be such that

$$\Delta T_A = \frac{L}{v} \ll 1.0 \tag{3.7}$$

where $T_A = L/v$ is the time required for sound traveling at a speed v to transit from one end of an array of length L to the other end. The reason for this bound on FFT resolution is related to the use of the complex phase-shift expression exp $[j2\pi f\tau(\theta)]$ in the discrete frequency domain as the approximation to a simple bulk time delay $\tau(\theta)$ which would be required in a time-domain realization [38]. Specifically, the

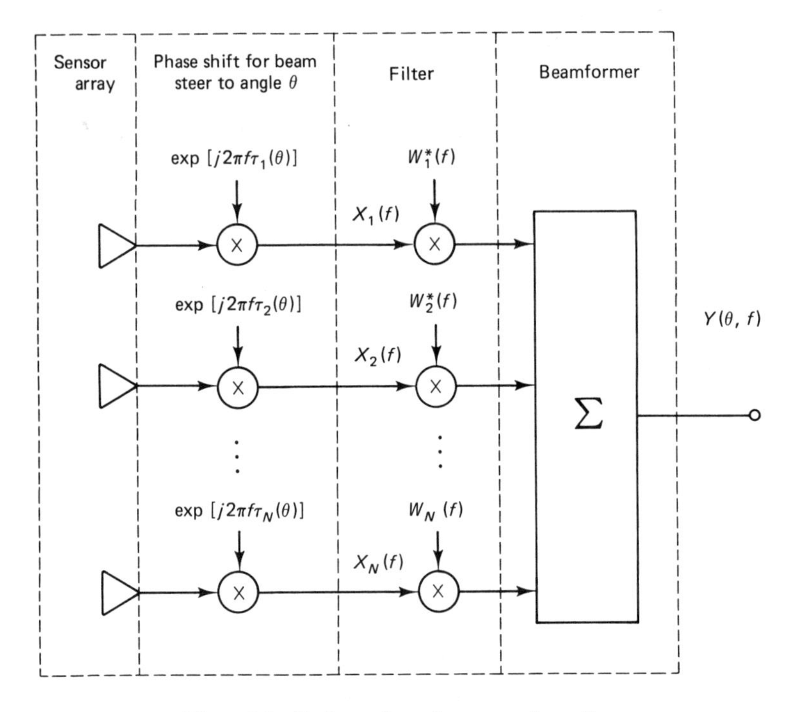

Figure 3.5 Optimum beamformer configuration.

Fourier transform pair relationships

$$x(t) \rightleftharpoons X(f) \tag{3.8}$$

and

$$x(t - \tau_k(\theta)) \rightleftharpoons X(f) \exp\left[-j2\pi f_k \tau(\theta)\right] \tag{3.9}$$

for a time delay $\tau(\theta)$ at the kth sensor are exact only at the center frequency f of a discrete FFT resolution cell, called a *bin.* For a worst-case signal at frequency $f \pm \Delta/2$ and a worst case (maximum) time delay of $\tau(\theta) = T_A$, use of Eq. (3.9) results in a phase error of $\Delta T_A/2$. Thus the bound expressed in Eq. (3.7) is necessary to reduce the attenuation of a signal with spectral content which is poorly matched to the FFT bin centers and with a wavefront that propagates for a significant period of time over the array. The worst-case maximum propagation time for a linear array, called the *endfire condition,* is for the direction of propagation to be parallel to the array axis. Clearly, for the direction of propagation perpendicular to the array, the *broadside condition,* Eq. (3.9), is exact for all frequencies. These "frequency quantization" effects associated with discrete frequency-domain beamforming following the FFT of sensor data are comparable to the corresponding time sample quantization effects which would accrue from a discrete time-domain beamformer realization [38].

After the sensor data have been phase-shifted for beam steering to the angle θ, each channel is then filtered. It is at this point that a frequency-domain realization can have significant impact on system hardware complexity. Specifically, if the sensor filtering operation is to be implemented in the time domain, the implied discrete convolution is typically implemented via a finite impulse response (FIR) filter. Alternatively, the discrete frequency-domain implementation of a linear convolution via circular convolution is a well-known, computationally efficient process [39]. The impact of discrete frequency-domain techniques on the hardware requirements for adaptive array processors [40] and filters has also been recognized [41, 42]. The only drawback associated with adaptive realizations in the frequency domain is the slower update rate and the potential for correspondingly slower convergence of an adaptive filter which receives input samples only at the rate at which the FFT operation is performed [43]. This issue will be considered again in a later section in the context of a convergence rate discussion.

Since the remainder of this chapter relies heavily on matrix notation, certain matrix definitions and operations must be introduced. The superscripts *, T, H, and -1 indicate the matrix operations of conjugation, transposition, Hermitian transposition, and inversion, respectively.

Let the output from an array of N sensors at frequency f be given by the $N \times 1$ vector

$$\mathbf{X}(f) = [X_1(f), X_2(f), \ldots, X_N(f)]^T \tag{3.10}$$

where

$$x_k(t) \rightleftharpoons X_k(f) \tag{3.11}$$

The kth sensor temporal output, $x_k(t)$, is assumed to be stationary and of zero mean. The *cross-spectral density matrix* (CSDM) for the array is defined as

$$\mathbf{R}(f) = E[\mathbf{X}(f)\mathbf{X}(f)^H] \tag{3.12}$$

$$= \sigma_s^2(f)\mathbf{P}(f) + \sigma^2(f)\mathbf{Q}(f) \tag{3.13}$$

where the matrices $\mathbf{P}(f)$ are the signal and noise cross-spectral density matrices, respectively, and are normalized so that trace $[\mathbf{P}(f)]$ = trace $[\mathbf{Q}(f)] = N$. The input signal and noise autospectral densities are σ_s^2 and σ^2.

An N-element filter vector to be applied to the beam steered array data at frequency f is defined as

$$\mathbf{W}(f) = [W_1(f), W_2(f), \ldots, W_N(f)]^T \tag{3.14}$$

where

$$w_k(t) \rightleftharpoons W_k(f) \tag{3.15}$$

is the corresponding impulse response for the kth-channel filter. The beamformer output for this vector filter is

$$Y(f) = \mathbf{W}^H(f)\mathbf{X}(f) \tag{3.16}$$

In the remainder of this chapter, the frequency f is deleted for expediency in all expressions unless it is needed explicitly. Let the signal cross-spectral density matrix $\mathbf{P}$ be of rank K and be expressible in diagonalized form as

$$\mathbf{P} = \mathbf{M\Lambda M}^H \tag{3.17}$$

where $\mathbf{M}$ is an $N \times K$ matrix of *orthonormal eigenvectors* of $\mathbf{P}$, $\mathbf{M}_k$, $k = 1, 2, \ldots, K$, as columns which form an orthonormal basis spanning a K-dimensional, complex signal subspace, Ω_s. The elements of the real $K \times K$ diagonal matrix $\mathbf{\Lambda}$ are the eigenvalues of $\mathbf{P}$ wherein the kth diagonal element λ_k gives the signal power associated with the kth signal subspace basis vector $\mathbf{M}_k$. It is now desired to form a beam output as in Eq. (3.16), which is defined so as to minimize the beamformer output power

$$E[|Y|^2] = \mathbf{W}^H\mathbf{RW} \tag{3.18}$$

but constrained to pass the multidimensional signal without spectral distortion. The symbol $E[\cdot]$ denotes the *expectation operator*. The distortionless multidimensional signal constraint on the beamforming vector can be expressed as a real-valued multiple linear constraint on $\mathbf{W}$ given by

$$\mathbf{M}^H\mathbf{W} = \mathbf{\Lambda}^{1/2}\mathbf{1} \tag{3.19}$$

where $\mathbf{1}$ is a vector of 1's with compatible dimension. Since Eq. (3.19) maintains the beamformer response at the proper level for each of the signal vector orthogonal components, minimization of Eq. (3.18) subject to Eq. (3.19) results in a minimization of the contributions to Eq. (3.18) due to the noise vector processes which have a projection into the subspace orthogonal to Ω_s. This constrained minimization is accomplished by minimizing the real functional

$$z = \mathbf{W}^H\mathbf{RW} + \mathbf{C}_R^T(\mathbf{M}^H\mathbf{W} + \mathbf{M}^T\mathbf{W}^* - 2\mathbf{\Lambda}^{1/2}\mathbf{1}) + j\mathbf{C}_I^T(\mathbf{M}^H\mathbf{W} - \mathbf{M}^T\mathbf{W}^*) \tag{3.20}$$

with respect to the real and imaginary parts of the filter vector $\mathbf{W}$. The vector $\mathbf{C} = \mathbf{C}_R + j\mathbf{C}_I$ is a K-element complex Lagrange multiplier vector. The minimum of Eq. (3.20) is found by solving the set of N simultaneous complex equations given by the gradient

$$\begin{aligned}\frac{\partial z}{\partial \mathbf{W}} &= \frac{\partial z}{\partial \operatorname{Re}[\mathbf{W}]} + j\frac{\partial z}{\partial \operatorname{Im}[\mathbf{W}]}\\ &= 2\mathbf{RW} - 2\mathbf{MC}^*\\ &= \mathbf{0}\end{aligned} \tag{3.21}$$

to yield

$$\mathbf{W} = \mathbf{R}^{-1}\mathbf{MC}^* \tag{3.22}$$

Application of the constraint in Eqs. (3.19) to (3.22) gives

$$\mathbf{C}^* = [\mathbf{M}^H\mathbf{R}^{-1}\mathbf{M}]^{-1}\mathbf{\Lambda}^{1/2}\mathbf{1} \tag{3.23}$$

with the resultant MVDR optimum filter vector

$$\mathbf{W}_{\text{opt}} = \mathbf{R}^{-1}\mathbf{M}[\mathbf{M}^H\mathbf{R}^{-1}\mathbf{M}]^{-1}\mathbf{\Lambda}^{1/2}\mathbf{1} \tag{3.24}$$

The use of the signal CSDM given by Eq. (3.17) is not always justifiable in the sense that the amount of information required to specify **P** could ever be known a priori. Rather, Eq. (3.17) should be interpreted as a definition of a vector subspace such that if a specific unknown signal direction vector has a significant projection into this subspace, that signal vector projection will suffer no suppression when the total array output power is minimized by using Eq. (3.24) as an array processor design algorithm.

An alternative approach to avoiding the suppression of a signal for which the signal direction vector is not perfectly known is to control the shape of the beamformer response mainlobe by constraining the derivatives of the response pattern on the MRA [44–45]. This type of constraint is expressed by the same linear constraint form as the K-dimensional constraint in Eq. (3.19), namely,

$$\mathbf{M}^H\mathbf{W} = \mathbf{E} \tag{3.25}$$

where **E** is a vector to be defined later.

However, now the mth column of the matrix **M** is given by the corresponding term-by-term $(m - 1)$th derivative of the elements of the nominal direction vector which has the signal arrival precisely on the MRA. This type of constraint can be most easily illustrated for the N-element line array with distance to the kth sensor given by x_k. The response of an optimum beamformer steered to angle θ due to a unit amplitude spatially uniform signal from angle ϕ is

$$Y(\phi) = \sum_{k=1}^{N} W_k^* \exp\left[\frac{j2\pi f x_k (\sin\theta - \sin\phi)}{v}\right] \tag{3.26}$$

The mth derivative of this response with respect to $\sin\theta$ evaluated on the MRA where $\sin\theta = \sin\phi$ is

$$\left.\frac{\partial^m Y(\phi)}{\partial \sin\phi^m}\right|_{\sin\phi = \sin\theta} = \sum_{k=1}^{N} W_k^*\left(j\,\frac{2\pi f x_k}{v}\right)^m \tag{3.27}$$

$$= e_{m+1} \tag{3.28}$$

such that e_{m+1} is the $(m + 1)$th element of the vector **E** in Eq. (3.25). If the constraints $e_1 = N$ and $e_n = 0$ for $1 < n \leq M$ are imposed on Eq. (3.28) with

$$[\mathbf{M}]_{km} = \left(-j\,\frac{2\pi f x_k}{v}\right)^m \tag{3.29}$$

defined as the kmth element in the matrix **M** in Eq. (3.25), then Eq. (3.25) is Eq. (3.28) placed in matrix form. Thus, if a vector **C** is defined as $\mathbf{C} = \mathbf{E}$ for the case of derivative constraints on the beamformer response pattern and $\mathbf{C} = \mathbf{\Lambda}^{1/2}\mathbf{1}$ for the case of multiple linear constraints, the general form of the optimum beamforming vector is

$$\mathbf{W}_{\text{opt}} = \mathbf{R}^{-1}\mathbf{M}[\mathbf{M}^H\mathbf{R}^{-1}\mathbf{M}]^{-1}\mathbf{C} \tag{3.30}$$

The derivative-type constraint is considered again in Section 3.6 in conjunction with the noise estimator–subtractor version of the MVDR beamformer.

The ratio of the MVDR beamformer output SNR to the SNR at the output of a single sensor is

$$G_0 = \frac{\mathbf{W}_{\text{opt}}^H \mathbf{P}\mathbf{W}_{\text{opt}}}{\mathbf{W}_{\text{opt}}^H \mathbf{Q}\mathbf{W}_{\text{opt}}} \tag{3.31}$$

Two special cases are of interest with respect to an evaluation of G_0, namely, a single plane-wave interference and a continuous, angularly extended noise.

Plane-wave Interference with Spatially Uncorrelated Noise

Consider the signal and noise cross-spectral density matrices for an N-element array given by, respectively,

$$\mathbf{P} = \mathbf{D}(\theta)\mathbf{D}(\theta)^H \tag{3.32}$$

and

$$\mathbf{Q} = \alpha\mathbf{D}(\phi)\mathbf{D}(\phi)^H + (1 - \alpha)\mathbf{I} \tag{3.33}$$

where $\mathbf{I}$ is the identity matrix. In Eq. (3.33), the parameter defined by $0 \leq \alpha < 1$ specifies the power contained in the plane-wave interference with direction vector $\mathbf{D}(\phi)$ relative to an uncorrelated noise component, which is of uniform level from sensor to sensor. Using Eq. (3.30) and the *matrix inversion identity*

$$[\mathbf{U}\mathbf{S}\mathbf{V}^H + \mathbf{B}]^{-1} = \mathbf{B}^{-1} - \mathbf{B}^{-1}\mathbf{U}[\mathbf{V}^H\mathbf{B}^{-1}\mathbf{U} + \mathbf{S}^{-1}]^{-1}\mathbf{V}^H\mathbf{B}^{-1} \tag{3.34}$$

to invert the array CSDM of Eq. (3.13) gives the optimum beamforming vector

$$\mathbf{W}_{\text{opt}} = g_1\mathbf{D}(\theta) - g_2\mathbf{D}(\phi) \tag{3.35}$$

where

$$g_1 = \frac{1}{N}\left(1 - \frac{|u|^2}{1 + i}\right)^{-1} \tag{3.36}$$

$$g_2 = g_1 \frac{u}{1 + (1/i)} \tag{3.37}$$

and

$$u = \frac{\mathbf{D}(\phi)^H\mathbf{D}(\theta)}{N} \qquad (|u| < 1) \tag{3.38}$$

$$i = \frac{N\alpha}{1 - \alpha} \tag{3.39}$$

The quantity u is the complex response of a conventional, delay-and-sum (DS) beamformer steered at angle θ to a planar interference wavefront of amplitude $1/N$ arriving from angle ϕ. Also, $i = \text{INR}_{\text{out}}$ is defined in Section 3.2 as the ratio of interfering noise power to the uncorrelated noise power at the output of DS beamformer steered directly at the interference angle ϕ. The optimum beamformer implementation dictated by Eq. (3.35) is quite interesting and, in fact, leads to the concept of a minimum complexity optimum beamformer, presented in the next section. Specifically, Eq. (3.35) indicates that a conventional DS beam is formed at a desired angle θ to receive the signal. The output of this signal-directed beam is then scaled by the real quantity g_1, which involves the response-level parameter $L = |u|^2$ introduced in Section 3.2. A second DS beam is formed at an angle ϕ which is the arrival angle of the interference. This interference beam output is then scaled and phase shifted before subtracting it from the scaled signal beam output. The result is the phase coherent subtraction of a portion of the interference waveform which appears in the signal DS beam output as a result of a finite response level L off the MRA. The array gain improvement ratio for the optimum MVDR beamformer relative to the DS beamformer is given by Eq. (3.4) in the preceding section, where the DS beamformer array gain is given by

$$G_{\text{DS}} = \frac{\mathbf{D}(\theta)^H \mathbf{P} \mathbf{D}(\theta)}{\mathbf{D}(\theta)^H \mathbf{Q} \mathbf{D}(\theta)} \tag{3.40}$$

Recall that Eq. (3.4) is plotted in Fig. 3.4 to answer the array designer's traditional question: When can I get at least a 3-dB improvement from optimum beamforming?

In the above, an interpretation of the optimum beamformer implementation was given which required an interference cancellation operation involving processing of only DS beamformer outputs. Anderson [46] presents an alternative interpretation of the optimum beamformer wherein the interference-directed DS beam output is used to cancel the interference waveform at the sensor level prior to DS beamforming at the signal. This sequence of operations can be represented by

$$Y(\theta) = \mathbf{W}_{\text{opt}}^H \mathbf{X} \tag{3.41}$$

$$= g_1 \mathbf{D}(\theta)^H \left[\mathbf{X} - \frac{g_3}{N} \mathbf{D}(\phi)\mathbf{D}(\phi)^H \mathbf{X} \right] \tag{3.42}$$

where

$$g_3 = \frac{1}{1 + (1/\text{INR}_{\text{out}})} \tag{3.43}$$

It is noted that for $g_3 = 1$, complete interference cancellation can be achieved. This corresponds to the case of infinite interference-to-noise ratio in the interference directed beam output. Theoretically, if all the gain levels in each sensor were known, which they seldom are in real sonar arrays, perfect cancellation of the interference could be obtained at the sensor level with an open-loop system implementing the beamforming process described in Eq. (3.42) with $g_3 = 1$. However, when the sensor gain levels are unknown, implementing the sensor-level interference cancellation as in Eq. (3.42) can lead to poor performance. This problem could be overcome by making the sensor-level cancellation process adaptive. However, this type of gain-level problem also can be addressed by a direct implementation of the noise cancellation process at the beam outputs as suggested by Eq. (3.35) [43].

Finally, it is noted that the general class of cancellation devices for interferences in the array far field are sometimes referred to as *null-steering systems.* This concept of interference cancellation by means of an estimator–subtractor circuit is developed further in the following section.

Angularly Extended Interference

Suppose that instead of the interference being a delta function in spatial frequency as in Fig. 3.2, the interfering noise source has angular extent with a spatial distribution of power given by

$$b(v) = \frac{2\beta}{v^2 + \beta^2} \tag{3.44}$$

as shown in Fig. 3.6. The half-power spatial bandwidth for this spatially extended interference is 2β. This type of directional interference is of considerable interest in sonar when multipath interference and random medium effects limit the distance over which correlation is exhibited along a propagating wavefront [37]. A simple expression of this spatial coherence is given in terms of the one-dimensional coherence function

$$q(x) = \exp(-\beta|x|) \tag{3.45}$$

where x is the separation of two points on a line array of sensors. Equations (3.44) and (3.45) are a Fourier transform pair for a first-order spatial Markov process with decorrelation distance $1/\beta$. As an example, consider a line array of $N = 32$ sensors

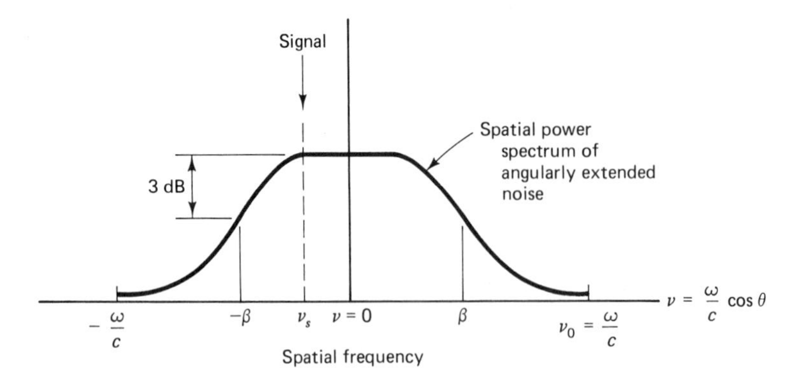

Figure 3.6 Angularly extended interference and discrete signal arrival.

which are uniformly spaced at a distance d. Let the parameter $\rho = \exp(-\beta d) < 1$ be defined. The kmth element of the noise-only CSDM, $\mathbf{Q}$, is given by

$$[\mathbf{Q}]_{km} = \exp(-\beta d\,|k-m|) \tag{3.46}$$

and the corresponding element in the signal CSDM is

$$[\mathbf{P}]_{km} = \exp[j(k-m)\,dv_s] \tag{3.47}$$

where v_s is the signal wavenumber. With the above, the array gain improvement ratio $R = G_0/G_{\mathrm{DS}}$ can be obtained in closed form [17]. This ratio is plotted in Fig. 3.7 as a function of $v_s d$ for various values of the noise coherence factor ρ. Note that for a modest noise coherence factor 15/16 the maximum array gain improvement possible is less than 2 dB. As an example, consider an array with the sensors spaced at one-half a wavelength at 200 Hz. That is, $d = \lambda/2 = 3.75$ m for a sound speed of 1500 m/s. This corresponds to an angularly extended interference with a 3-dB spatial bandwidth of 2.0°. The beamwidth for this array is given by $B = \lambda/L$, where L is the total array length. For this case $B = 3.8°$, which indicates that only a marginal array gain improvement of 2 dB is obtainable for this interference even though its 3-dB down angular spread is less than a beamwidth. In general, what is illustrated by this example is that only nominal gains are obtainable against noise of continuous angular extent greater than one-half a beamwidth. Moreover, the hardware costs for this situation are substantial. This aspect of the limitations of an optimum processor is discussed in some detail in [15–17].

To conclude this section, an optimum sonar array processor has been presented and its performance has been examined for both a point-source interference and angularly extended noise. The implication here is that the fully implemented opti-

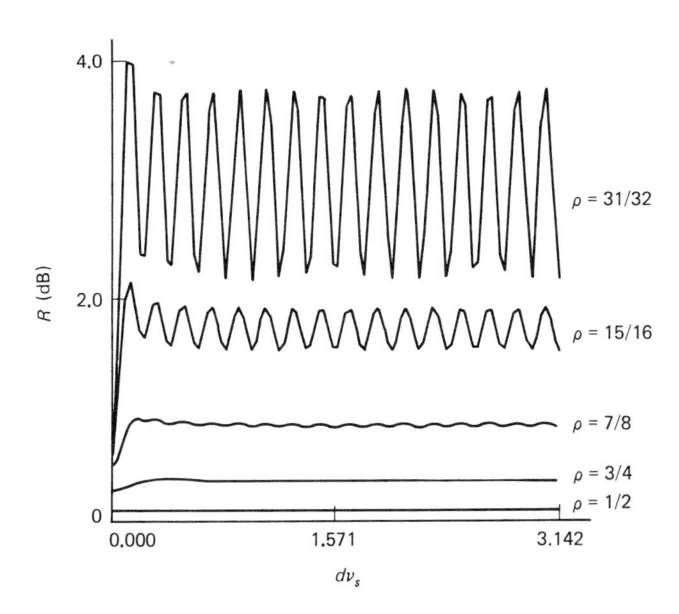

Figure 3.7 Array gain improvement ratio (R) of an MVDR. MVDR relative to a DS beamformer as a function of the sensor separation-signal spatial frequency product dv_s and adjacent sensor coherence ρ caused by an angularly extended noise source.

mum beamformer would necessarily be of considerable complexity [47–49] and that the resultant array gain improvement may be notable only in the point-source interference situation. With this in mind, the following section examines a noise-canceling alternative to the fully implemented beamformer for which the complexity could be minimized and kept in line with achievable and cost-effective performance.

3.4 MINIMUM COMPLEXITY IMPLEMENTATION OF AN OPTIMUM BEAMFORMER

In the preceding section it was demonstrated that plane-wave sources of interference, as contrasted to angularly extended interference, provide a major justification for the consideration of optimum sonar array processing techniques. Furthermore, the optimum array processor to discriminate against plane-wave interference was shown to reduce to single estimate-and-subtract operation for each interference performed at the output of each of the conventionally formed DS beams. Time domain suboptimum realization of this approach is discussed in [50]. This simple optimum beamformer realization can be related to the underlying dimensionality of the noise-only CSDM. The noise CSDM for a single plane-wave interference in uncorrelated noise from Eq. (3.33) is

$$\mathbf{Q} = \sigma^2\alpha\mathbf{D}(\phi)\mathbf{D}(\phi)^H + \sigma^2(1-\alpha)\mathbf{I} \tag{3.33}$$

Note that the directional noise component $\mathbf{D}(\phi)\mathbf{D}(\phi)^H$ is a *dyadic matrix* of rank 1. Now, let the form of Eq. (3.33) be generalized to

$$\mathbf{Q} = \alpha \sum_{i=1}^{L} b_i \mathbf{E}_i \mathbf{E}_i^H + (1-\alpha)\mathbf{I} \tag{3.48}$$

where the spatially correlated portion of the noise CSDM $\mathbf{Q}$ has the diagonalized form

$$\mathbf{Q}_c = \sum_{i=1}^{L} b_i \mathbf{E}_i \mathbf{E}_i^H \tag{3.49}$$

$$= \mathbf{EBE}^H \tag{3.50}$$

The matrix $\mathbf{Q}_c$ is a rank L matrix defined by its eigenvalues, b_i, and eigenvectors $\mathbf{E}_i$, which are the corresponding diagonal elements and columns of the diagonal matrix $\mathbf{B}$ and orthonormal matrix $\mathbf{E}$, respectively. The vector $\mathbf{E}_i$ can be envisioned as the *directional vector* for a generalized interference wavefront which is not necessarily planar. The signal plus noise CSDM for a single wavefront signal with direction vector $\mathbf{D}(\theta)$ is

$$\mathbf{R} = \sigma_s^2 \mathbf{D}(\theta)\mathbf{D}(\theta)^H + \sigma^2\alpha\mathbf{Q}_c + \sigma^2(1-\alpha)\mathbf{I} \tag{3.51}$$

for which the optimum filter structure is obtained by using Eq. (3.34) as

$$\mathbf{W}_0 = \beta\mathbf{D}(\theta) - \beta\mathbf{GF} \tag{3.52}$$

where

$$\beta = \frac{1}{\sigma^2(1 - \alpha)\mathbf{D}(\theta)^H\mathbf{R}^{-1}\mathbf{D}(\theta)} \tag{3.53}$$

and the $N \times (L + 1)$ matrix $\mathbf{G}$ is obtained from the coherent noise eigenvector matrix $\mathbf{E}$ by appending $\mathbf{D}(\theta)$ as the first column. The vector $\mathbf{F}$ is given by

$$\mathbf{F} = [\mathbf{G}^H\mathbf{G} + \sigma^2(1 - \alpha)\mathbf{C}^{-1}]^{-1}\mathbf{G}^H\mathbf{D}(\theta) \tag{3.54}$$

The matrix $\mathbf{C}$ is an $(L + 1) \times (L + 1)$ diagonal matrix obtained from the coherent noise eigenvalue matrix $\mathbf{B}$ by appending σ_s^2 as an additional diagonal element in the first location. Most important, we note that Eq. (3.52) can be written in the form

$$\mathbf{W}_0 = \beta_0\,\mathbf{D}(\theta) - \sum_{i=1}^{L} \beta_i\,\mathbf{E}_i \tag{3.55}$$

which is a complete generalization of the null steering operation prompted in the preceding section by Eq. (3.36). In simple terms, Eq. (3.55) indicates that the optimum beamformer first forms a conventional DS beam at the target angle θ and, in addition, forms L "eigenbeams" with direction vectors $\mathbf{E}_i$, $i = 1, 2, \ldots, L$. Then the ith eigenbeam output should be scalar filtered as specified by β_i and added to all of the other filtered eigenbeam outputs before subtracting this linear combination from the DS beam output filtered by β_0. This noise estimation by eigenbeam formation followed by cancellation of the noise from the signal beam output requires only L filters represented by the β_i coefficients. This L-filter, noise estimator–subtractor optimum beamformer realization is compared to the N-filter realization and is identified herein as a *minimum complexity implementation.*

In practice, it is questionable, first, that the signal direction vector $\mathbf{D}(\theta)$ would be known exactly, and second, that the diagonal form in Eq. (3.50) for the coherent portion of the noise only cross-spectral density matrix would be known a priori. The question then becomes: Can the minimal complexity demonstrated above be achieved without this detailed a priori knowledge and without a significant loss of performance? To address this question, consider the general adaptive noise estimator–subtractor illustrated in Fig. 3.8. Three basic functions are identifiable:

1. Signal waveform estimation by conventional time delay-and-sum (DS) beamforming
2. Adaptive spatial noise estimation implemented by adaptive filtering of a linear combination of hydrophone outputs and any auxiliary noise-only sensing device outputs
3. Coherent noise subtraction at the output of the DS beamformer

The DS beamformer output, Y_c, consists of signal, S, a component due to coherent noise, N_c, and incoherent noise N_I. The adaptive noise-estimation filter-beamformer provides an estimate of the coherent noise component, $\hat{N}_c$, and sub-

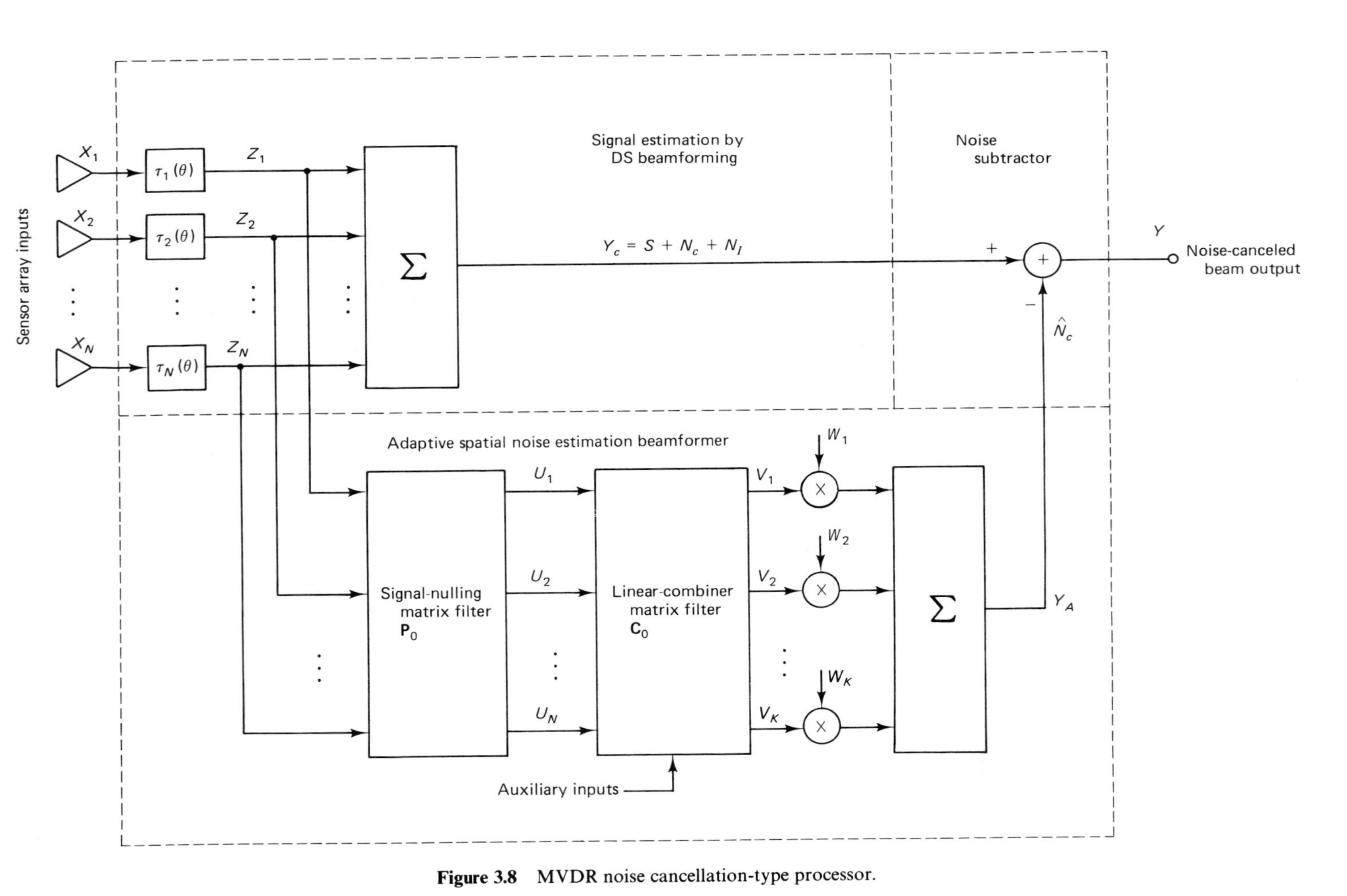

Figure 3.8 MVDR noise cancellation-type processor.

tracts it from the conventional beamformer output. To preclude subtraction of any portion of the signal when $\hat{N}_c$ is removed, the adaptive beamformer must remove the signal components from the steered hydrophone data, $\mathbf{Z}$. This signal removal is accomplished by the *signal-nulling matrix filter*, $\mathbf{P}_0$. The *linear-combiner matrix filter*, $\mathbf{C}_0$, is applied to the signal nulled data, $\mathbf{U}$, to reduce the number of adaptive filter channels required from N to K. Making K $(<N)$ as small as possible, commensurate with effective coherent noise subtraction, reduces implementation complexity.

The final adaptive noise estimation beamformer operation is that of multiplying the linear-combiner outputs, $\mathbf{V}$, by the corresponding filter weights, $\mathbf{W}$, and summing to give the adaptive beamformer output, $Y = \hat{N}_c$. The quantity $\hat{N}_c$ is a least-mean-square error (LMS) estimator of the coherent noise, N_c. This is because the filter weights, $\mathbf{W}$, are selected so as to minimize the average power of the noise-canceled beam output, Y. Since Y is formed as a difference between Y_c, which contains signal S, and Y_A, which contains no signal, the process of minimizing the power in Y with respect to the weights, $\mathbf{W}$, can only reduce N_c, which is coherent between the conventional beamformer output, Y_c, and the linear-combiner outputs, $\mathbf{U}$. The linear-combiner matrix filter output and the adaptive noise estimation beamformer filter weight vector are given by

$$\mathbf{V}^T = [V_1, V_2, \ldots, V_K] \tag{3.56}$$

and

$$\mathbf{W}^T = [W_1, W_2, \ldots, W_K] \tag{3.57}$$

A presteering matrix filter for a plane-wave nominal signal angle of arrival θ is defined as

$$\mathbf{D}_0(\theta) = \begin{bmatrix} \exp[-j2\pi f\tau_1(\theta)] & & & \\ & \exp[-j2\pi f\tau_2(\theta) & & \mathbf{0} \\ & & \ddots & \\ \mathbf{0} & & & \exp[-j2\pi f\tau_N(\theta)] \end{bmatrix} \tag{3.58}$$

where $\tau_n(\theta)$ is the time delay applied to the nth sensor for steering a DS beam at angle θ. The signal-nulling and linear-combiner matrix filters are given, respectively, by $\mathbf{P}_0$ and $\mathbf{C}_0$. As a result of the above, the DS beamformer output can be written as

$$Y = S + N_C + N_I \tag{3.59}$$

$$= \mathbf{1}^H \mathbf{D}_0(\theta)^H \mathbf{X} \tag{3.60}$$

and the adaptive noise estimation beamformer output as

$$Y_A = \hat{N}_C \tag{3.61}$$

$$= \mathbf{W}^H \mathbf{C}_0^H \mathbf{P}_0^H \mathbf{D}_0^H(\theta) \mathbf{X} \tag{3.62}$$

Let the output of the presteering matrix filter have the CSDM $\mathbf{R}$ given by Eq. (3.13) with the signal normalized CSDM $\mathbf{P}$ given by Eq. (3.17). Removal of all signal from the noise estimation adaptive beamformer output requires a signal nulling matrix filter given by

$$\mathbf{P}_0 = \mathbf{I} - \mathbf{M}\mathbf{M}^H \tag{3.63}$$

where $\mathbf{M}$ is defined in Eq. (3.17). This can be seen by computing the CSDM for the linear combiner matrix filter output $\mathbf{V}$. Specifically,

$$\mathbf{R}_{VV} = \mathbf{C}_0^H \mathbf{P}_0^H \mathbf{R} \mathbf{P}_0 \mathbf{C}_0 \tag{3.64}$$

$$= \mathbf{C}_0^H \mathbf{P}_0^H \mathbf{M} \boldsymbol{\Lambda} \mathbf{M}^H \mathbf{P}_0 \mathbf{C}_0 + \mathbf{C}_0^H \mathbf{P}_0^H \mathbf{Q} \mathbf{P}_0 \mathbf{C}_0 \tag{3.65}$$

where it is noted that $\mathbf{P}_0^H \mathbf{M}$ is an $N \times K$ null matrix of zeros which eliminates the contribution due to signal in the first term of Eq. (3.65).

The optimum filter weight vector that minimizes

$$E[|Y|^2] = E[|Y_c - Y_A|^2] \tag{3.66}$$

is given by

$$\mathbf{W} = \mathbf{R}_{VV}^{-1} \mathbf{P}_{VY_c} \tag{3.67}$$

where $\mathbf{R}_{VV}$ is given above and $\mathbf{P}_{VY_c}$ is the cross-spectral density vector between the linear combiner matrix filter output $\mathbf{V}$ and the DS beam output. Using these relationships, we may write

$$\mathbf{W} = [\mathbf{C}_0^H \mathbf{P}_0^H \mathbf{Q} \mathbf{P}_0 \mathbf{C}_0]^{-1} \mathbf{C}_0^H \mathbf{P}_0 \mathbf{Q} \mathbf{1} \tag{3.68}$$

which is the $K \times 1$ vector of optimum filter weights to be applied to the linear combiner output data.

With respect to the question of minimum complexity, it is desired to minimize the linear-combiner output filter dimensionality K. The number of adaptive filters that must be realized is exactly K[51]. The combined matrix filter $\mathbf{P}_0 \mathbf{C}_0$ is a fixed structure which does not require the complexity of any type of adaptive control. For reasons of economy, therefore, it is most desirable to make K as small as possible without drastically affecting the system performance in terms of canceling the significant coherent noise components represented in $\mathbf{Q}$. With this combined objective, the minimum value of K is L, as has been previously shown.

A necessary condition for $\mathbf{W}$ to process optimally against the coherent noise component $\mathbf{Q}$ of the CSDM $\mathbf{Q}$ is that the eigenvectors of the matrix $\mathbf{C}_0^H \mathbf{P}_0^H \mathbf{Q} \mathbf{P}_0 \mathbf{C}_0$ span the same vector subspace as the eigenvectors of $\mathbf{Q}$. An obvious choice of $\mathbf{P}_0 \mathbf{C}_0$ which satisfies this requirement is either $\mathbf{P}_0 \mathbf{C}_0 = \mathbf{E}$ or, more generally,

$$\mathbf{P}_0 \mathbf{C}_0 = \mathbf{E}\mathbf{U} \tag{3.69}$$

wherein the columns of $\mathbf{P}_0 \mathbf{C}_0$ are linearly independent and are expressed as a linear combination of the eigenvectors of the coherent noise CSDM $\mathbf{Q}$. This ensures that the L-dimensional vector subspace spanned by the eigenvectors $\mathbf{E}_i$, $i = 1, 2, \ldots, L$ is also spanned by the columns of the matrix $\mathbf{P}_0 \mathbf{C}_0$. Furthermore, if $\mathbf{P}_0 \mathbf{C}_0$ is formed as an $N \times L$ dimensional matrix with linearly independent columns, the coherent noise vector subspace is completely spanned by the columns of $\mathbf{P}_0 \mathbf{C}_0$ and optimum performance would result at least with respect to noise cancellation in the infinite time average case. Any variation of the form of $\mathbf{U}$ could affect the convergence rate of an adaptive realization of Eq. (3.68) by changing the eigenvalues of $\mathbf{R}_{VV}$ but not the steady-state noise cancellation performance. This convergence rate of a gradient descent based adaptive realization is determined by the spread in eigenvalues of the matrix $\mathbf{U}^H\mathbf{B}\mathbf{U}$ [52]. Even this dependence of convergence rate on the matrix $\mathbf{U}$ can be overcome with an open-loop realization of Eq. (3.68), as will be described in Section 3.6.

3.5 SIGNAL MODELS AND NOISE CANCELLATION PERFORMANCE

The deviation of assumed spatial signal models from an actual signal field as sensed by a hydrophone array beamformer in the underwater environment is frequently severe. Typically, this deviation, or signal model mismatch condition, results from a violation of either any or all of the following assumptions: (1) the signal wavefront and/or array shape is known, (2) the signal and noise are uncorrelated, and (3) the medium is distortionless with known relative signal level and phase at the outputs of all sensors in the array. The adaptive realization of an optimum beamformer, if it is to adapt itself to an arbitrary and potentially time-varying noise environment, should know with considerable accuracy, in terms of the assumptions listed above, what constitutes a signal. The basic effect which describes optimum/adaptive array processor degradation when the assumptions above are violated by the signal is that the actual signal appears as noise to the array processor and is rejected accordingly. The question of optimum/adaptive beamformer signal mismatch is considered for an array of full adaptivity as in Fig. 3.5 in references [21, 23, 44]. The treatment presented herein considers the question of signal mismatch in terms of the general estimator–subtractor structure in Fig. 3.8. This approach allows for explicit results in terms of the number of linear combiner output channels K, the signal-to-noise ratio (SNR), the interference-to-noise ratio (INR), and signal–interference correlation [53].

With respect to signal mismatch, the essence of the K-channel interference estimator–subtractor structure discussed in the preceding section is given in Fig. 3.9. In this figure the signal channel is equated to the DS beamformer output in Fig. 3.8 and the noise sample inputs are identical to the linear combiner outputs. The coherent noise component in Fig. 3.9 is the specific interference I. It is recalled that the function of the auxiliary array processor is to provide a least mean-square (LMS) error estimate $\hat{I}$ of the interference and subtract it from the DS beamformer output. The noise estimation processor provides this LMS error estimate of the interference by adapting the noise reference array filter weights W_k, $k = 1, 2, \ldots, K$, so that the noise-canceled output power is minimized. *Degradation* in this processor is said to occur when, in addition to cancellation of the interference, cancellation of the signal in the DS beam output also occurs. One way this degradation can be

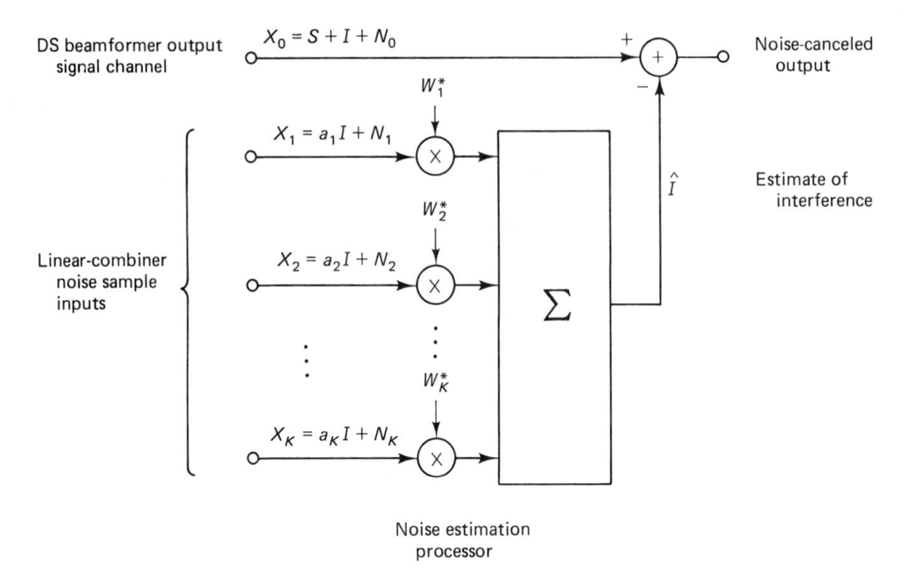

Figure 3.9 K-channel estimator–subtractor noise cancellation system.

analyzed is by allowing for correlation between the signal S and interference I. In an actual system, this correlation occurs when there is leakage of the signal through the signal nulling matrix filter of Fig. 3.8. This type of leakage occurs when the incorrect model of the signal structure is used, resulting in a violation of the assumptions relative to signal characteristics as stipulated above.

A common operational example of leakage occurs when the simple multipath situation illustrated in Fig. 3.10 is improperly modeled. If beams are formed with a horizontal line array which cannot resolve vertical from horizontal arrivals and the signal waveform arriving from any one vertical angle is assumed to be uncorrelated with the signal arriving on another path from any other angle, degradation can occur. Specifically, if, for example, the adaptive time constant on an optimum/adaptive beam which models only the direct path arrival is less than the decorrelation time between the direct path waveform and either of the other path waveforms, direct path signal and reflected path interference correlation can occur, interference in this case being any directional arrival other than that defined by the direct path model. Specifically, let the direction vectors describing the direct and surface reflected path wavefronts be denoted by $\mathbf{D}(\theta_d)$ and $\mathbf{D}(\theta_s)$, respectively. For a direct path signal model, the signal nulling matrix is

$$\mathbf{P}_0 = \mathbf{I} - \frac{1}{N}\mathbf{D}(\theta_d)\mathbf{D}(\theta_d)^H \tag{3.70}$$

For an assumed zero doppler differential on the two paths, perfectly correlated direct and surface-reflected signal arrivals result and the array output signal compo-

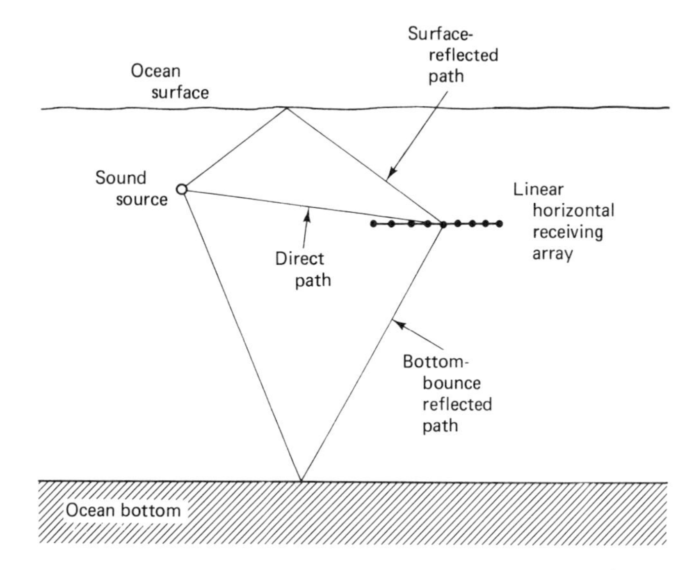

Figure 3.10 Simple example wherein the signal arrival on the direct path can be correlated with noise arrivals on the reflected paths.

nent is

$$\mathbf{S} = S[\mathbf{D}(\theta_d) + \mathbf{D}(\theta_s) \exp(-j2\pi f\tau)] \tag{3.71}$$

where τ is the difference in direct and surface-reflected path travel times such that the signal does not significantly decorrelate in time τ. The signal nulling matrix filter output due to this signal is

$$\mathbf{P}_0^H\mathbf{S} = S \exp(-j2\pi f\tau)[\mathbf{D}(\theta_s) - u\mathbf{D}(\theta_d)] \tag{3.72}$$

where $u = \mathbf{D}(\theta_d)^H\mathbf{D}(\boldsymbol{\theta}_s)/N$. Application of the linear combiner matrix filter to the signal nulling filter output of Eq. (3.72) gives a signal leakage component of the form

$$\mathbf{C}_0^H\mathbf{P}_0^H\mathbf{S} = \mathbf{A}S \tag{3.73}$$

where $\mathbf{A} = [a_1, a_2, \ldots, a_K]$ is a deterministic vector of complex elements which is straightforward to obtain. The form of Eq. (3.73) would be the same if both the surface-reflected and bottom-bounce arrival paths were included.

To facilitate a general analysis of cases where there is partial coherence between the multipath arrivals, the signal waveform is replaced by the interference waveform in Eq. (3.73) and the correlation

$$P_{SI} = E[S^*I] \tag{3.74}$$

$$= \sigma_S \sigma_I \rho \tag{3.75}$$

is defined wherein the zero-mean signal and interference powers are defined as

$$\sigma_S^2 = E[|S|^2] \tag{3.76}$$

and

$$\sigma_I^2 = E[|I|^2] \tag{3.77}$$

respectively. The factor ρ is the *complex coherence* between the signal and interference. Thus for $|\rho| < 1.0$, situations such as the extension of the example above to partially coherent multipath can be considered. On the other hand, the case for $|\rho| = 1$ allows the analysis of the important situation wherein sensor channel gain level and phase response are assumed to be known and, in fact, these properties may have modified substantially. For example, the direct path signal model in the matrix nulling filter in Eq. (3.70) assumes a uniform signal level over the array if the elements of $\mathbf{D}(\theta_d)$ are all of unity magnitude. Suppose that the kth diagonal element g_k of the diagonal matrix $\mathbf{G}$ gives the amplitude and phase response of the kth sensor channel relative to the assumed values. The signal leakage for the direct path arrival therefore becomes

$$\mathbf{V} = \mathbf{C}_0^H \mathbf{P}_0^H \mathbf{G} \mathbf{D}_0(\theta_d)\mathbf{S} \tag{3.78}$$

which is the signal component input into the noise cancellation process. As in the multipath case, Eq. (3.78) can also be placed in the form of Eq. (3.73). Clearly, this component can be highly correlated with the DS beamformer output since it also contains signal with the result that signal suppression will occur.

Rather than treat specific signal leakage situations, a general analysis is presented here. As shown in Fig. 3.9, and as represented in Eq. (3.73), the linear combiner output has the general K-vector form

$$\mathbf{V} = \mathbf{A}I + \mathbf{N} \tag{3.79}$$

where $\mathbf{A}$ is the leakage mechanism transfer function vector. The linear-combiner output which is not correlated with the DS beamformer output signal component is grouped into the vector $\mathbf{N} = [N_1, N_2, \ldots, N_K]^T$ with CSDM:

$$E[\mathbf{N}\mathbf{N}^H] = \gamma\sigma^2\mathbf{Q}_V \tag{3.80}$$

where $\sigma^2 = [|N_0|^2]$.

We previously pointed out the estimator–subtractor noise cancellation processor works by selecting the filter weight vector $\mathbf{W}^T = [W_1, W_2, \ldots, W_K]$ that minimizes the error power, that is, the noise-canceled output power

$$E[|Y|^2] = E[|\mathbf{Y}_c - \hat{I}|^2] \tag{3.81}$$

$$= \sigma_S^2 + \sigma_I^2 + \sigma^2 + 2\sigma_S\sigma_I \operatorname{Re}[\rho] - 2 \operatorname{Re}\left[\sigma_I^2\left(1 + \frac{\sigma_s}{\sigma_I}\rho\right)\mathbf{W}^H\mathbf{A}\right] + \mathbf{W}^H(\sigma_I^2\mathbf{A}\mathbf{A}^H + \gamma\sigma^2\mathbf{Q}_V)\mathbf{W} \tag{3.82}$$

The noise-canceled output power is minimized with respect to $\mathbf{W}$ for $\mathbf{W} = \mathbf{W}_{\text{opt}}$ such that if $[\mathbf{AA}^H + \sigma^2/\sigma_I^2 \mathbf{Q}_V]^{-1}$ exists, then

$$\mathbf{W} = \left(1 + \frac{\sigma_S}{\sigma_I}\rho\right)\left[\mathbf{AA}^H + \frac{\sigma}{\sigma_I}\mathbf{Q}_V\right]^{-1}\mathbf{A} \tag{3.83}$$

Using Eq. (3.83) in Eq. (3.82) gives the expression for the minimum noise-canceled output power

$$E[|Y|^2]_{W=W_{\text{opt}}} = \sigma_S^2\left[\frac{1 + (1 - |\rho|^2)q(\sigma_I^2/\sigma^2)}{1 + q(\sigma_I^2/\sigma^2)}\right] + \sigma^2 + \sigma_I^2\left[\frac{1 + 2(\sigma_S/\sigma_I)\text{ Re }[\rho]}{1 + q(\sigma_I^2/\sigma^2)}\right] \tag{3.84}$$

where

$$q = \frac{1}{\gamma}\mathbf{A}^H\mathbf{Q}_V\mathbf{A} \tag{3.85}$$

The performance of the noise cancellation processor can be expressed in terms of the *subtractor output signal-to-background noise ratio* (S/B). This performance metric is obtained directly from Eq. (3.84) as

$$\frac{S}{B} = \frac{\sigma_S^2}{\sigma^2}\frac{1 + (1 - |\rho|^2)q(\sigma_I^2/\sigma^2)}{1 + (1 + q + 2(\sigma_S/\sigma_I)\text{ Re }[\rho])(\sigma_I^2/\sigma^2)} \tag{3.86}$$

If the noise in the auxiliary noise sample inputs is uniform and uncorrelated, then

$$E[N_i N_j^*] = \gamma\sigma^2\delta_{ij} \tag{3.87}$$

where δ_{ij} is the Kronecker delta, equal to 1 for $i = j$ and zero otherwise, and the leakage gain, q, is defined as

$$q = |\mathbf{A}|^2 \tag{3.88}$$

$$= K|\bar{a}|^2 \tag{3.89}$$

such that $|\bar{a}|$ can be envisioned as the average magnitude of the elements in the leakage transfer function filter vector $\mathbf{A}$.

As a practical example, let the deterministic signal and interference be continuous-wave (CW) signals with complex envelopes in a narrow beamformer frequency band given by, respectively,

$$S = \sigma_S \exp(j2\pi ft) \tag{3.90}$$

and

$$I = \sigma_I \exp[j2\pi(f + \Delta)t] \tag{3.91}$$

If the estimator–subtractor form of an optimum beamformer is realized adaptively with an adaptive system time constant of T seconds, the signal and interference

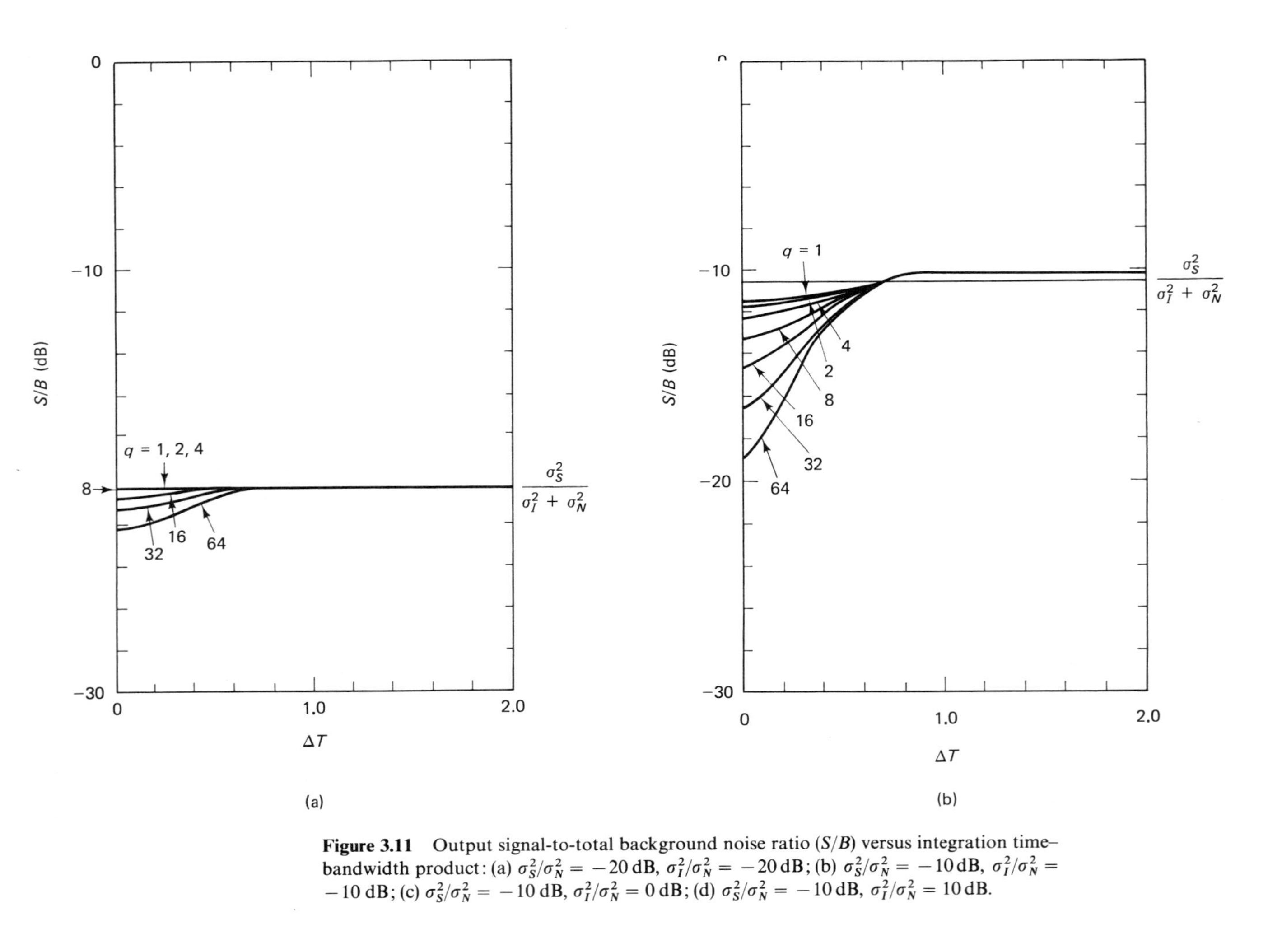

Figure 3.11 Output signal-to-total background noise ratio (S/B) versus integration time–bandwidth product: (a) $\sigma_S^2/\sigma_N^2 = -20$ dB, $\sigma_I^2/\sigma_N^2 = -20$ dB; (b) $\sigma_S^2/\sigma_N^2 = -10$ dB, $\sigma_I^2/\sigma_N^2 = -10$ dB; (c) $\sigma_S^2/\sigma_N^2 = -10$ dB, $\sigma_I^2/\sigma_N^2 = 0$ dB; (d) $\sigma_S^2/\sigma_N^2 = -10$ dB, $\sigma_I^2/\sigma_N^2 = 10$ dB.

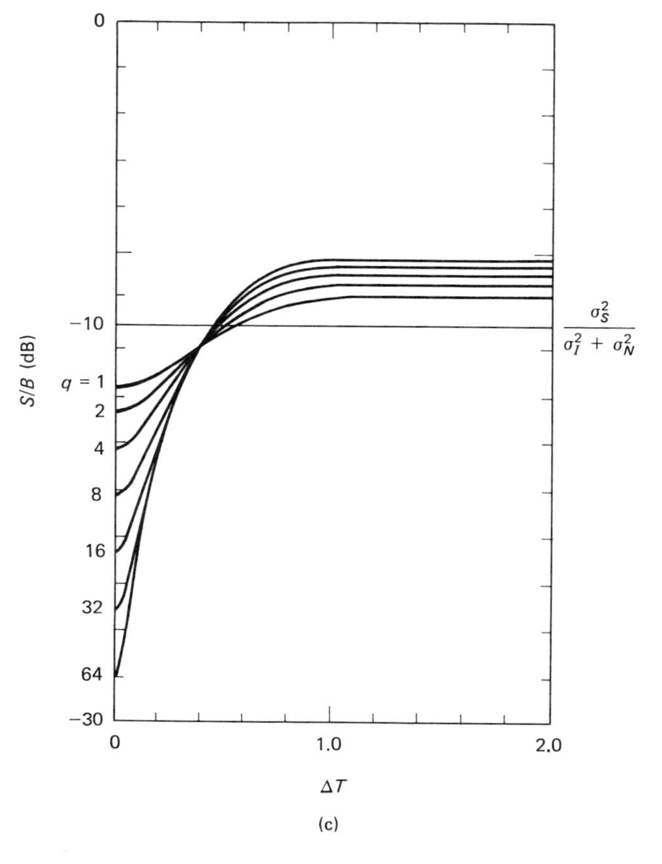

(c)

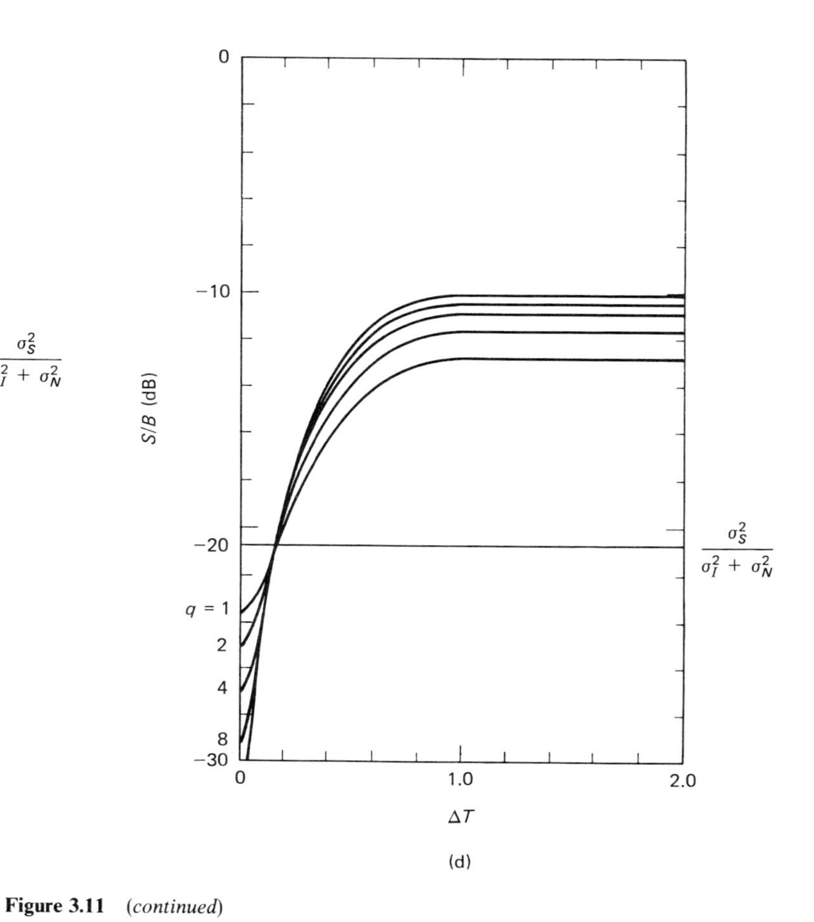

(d)

Figure 3.11 *(continued)*

correlation as measured by this system is approximated by

$$P_{SI} = \frac{1}{T}\int_{-T/2}^{T/2} S^*I\, dt \tag{3.92}$$

$$= \sigma_S \sigma_I \rho(\Delta T) \tag{3.93}$$

where

$$\rho(\Delta T) = \frac{\sin \pi\, \Delta T}{\pi\, \Delta T} \tag{3.94}$$

Using Eq. (3.94) in Eq. (3.86), the noise-canceled output signal-to-background noise ratio S/B is plotted in Fig. 3.11 for several values of σ_S^2/σ^2 and σ_I^2/σ^2 as a function of ΔT. The DS beamformer output signal-to-background noise ratio is $\sigma_S^2/(\sigma_I^2 + \sigma^2)$ and is also indicated to allow the relative performance of the interference cancellation system to be assessed. Note that for $\Delta T < 1.0$ degradation from optimum performance can occur. In fact, this degradation can lead to a performance worse than that for a system with no noise cancellation at all. This can be seen by letting $\rho = 0$ in Eq. (3.86), that is, no signal and interference coherence, to obtain

$$\left.\frac{S}{B}\right|_{\rho=0} = \frac{\sigma_S^2}{\sigma^2 + \left[\dfrac{1}{1 + q(\sigma_I^2/\sigma^2)}\right]\sigma_I^2} \tag{3.95}$$

At the other extreme, with $\rho = 1.0$ in Eq. (3.86), we obtain

$$\left.\frac{S}{B}\right|_{\rho=1} = \frac{\sigma_S^2}{\sigma^2 + [1 + q + 2(\sigma_S/\sigma_I)]\sigma_I^2} \tag{3.96}$$

Thus from Eq. (3.95) we see that degradation can never occur for either any level of interference-to-noise ratio, σ_I^2/σ^2, or any leakage gain, q. On the other hand, for $q(\sigma_I^2/\sigma^2)$ sufficiently large, essentially complete interference cancellation $(S/B) = (\sigma_S^2/\sigma^2)$ results. In contrast, noise canceler performance for fully coherent signal and interference, $\rho = 1.0$, can never be better than simple DS beamforming. Furthermore, it is clear from Eq. (3.96), degradation increases with $\rho = 1.0$ for any increasing level of either leakage gain q or signal-to-interference ratio (σ_S/σ_I).

This section has focused on another very important issue with regard to the utility of optimum/adaptive array processing systems in sonar. It has been shown that a complete lack of knowledge of the spatial CSDM of the array noise environment can be tolerated but that this advantage can be completely offset by a lack of the same knowledge about the signal! Experience with actual systems has shown the importance of this statement. Common effects such as multipath, spatial signal spreading, uncompensated hydrophone preamplifier drift, and medium inhomogeneities can lead to significant violations of the traditional assumptions about underwater signal fields. These violations result in performance degradation of operational adaptive array processors when oversimplified models of the array signal structure are assumed.

3.6 SONAR ARRAY PROCESSOR REALIZATION

The adaptive noise cancellation beamformer as defined in this section is a realization of the minimum-variance distortionless response (MVDR) beamformer discussed in Section 3.4 and illustrated in Fig. 3.8. Clearly, the need for adaptivity results from the fact that the sonar acoustic noise environment is neither easily predictable nor stationary. Thus to remain as nearly optimal as possible in the MVDR sense, the optimum filter weight vector for the noise estimation beamformer

$$\mathbf{W}_0 = \mathbf{R}_{VV}^{-1} \mathbf{P}_{VY_C} \tag{3.67}$$

must be approximated continuously in time. This adaptive estimation process can be realized in numerous ways. It is not the intent of this section to provide specific direction as to which technique should always be used because, in fact, at this time no general consensus exists, at least within the sonar array processing community. Rather, the type of adaptive realization which is in some sense best should first be a function of the character of the noise environment as defined in terms of spectral content, angular distribution, and degree of nonstationarity. Second, computational requirements should be a factor in the selection of a technique for the adaptive realization of Eq. (3.67).

As an initial issue, it is recognized that the MVDR sonar beamformer can be realized in either the frequency domain in terms of estimated cross-spectral density matrices, as in this chapter, or in the time domain, where estimated cross-correlation matrices are used [54]. However, the time-domain matrix filter expressions that result are identical in form to the frequency-domain counterpart. Moreover, *convergence rate* in number of iterations, filter channel dimensionality, and coherent noise-to-background noise power ratios are identical in either frequency or time-domain realizations. In this regard, one initial trend is observable in terms of the type of adaptive realization deemed appropriate. Specifically, when time-domain adaptation is selected, the closed-loop (error feedback) approach [9, 10, 50, 52] is more common in practice than an open-loop (feed forward) direct calculation approach [48–49, 55–61], which is increasingly favored in frequency-domain implementations. One reason for this observed trend, at least for multichannel systems, is a computational one. Direct estimation of the correlation matrix inverse required for an open-loop implementation of a K-channel adaptive FIR matrix filter with L points per filter could be a formidable numerical task. Given that a frequency-domain approach is to be used, results reported in the literature would seem to indicate a performance advantage for open-loop implementations [57–61].

In this section a partially historical overview of closed-loop and open-loop adaptive MVDR control strategies is presented, using a frequency-domain point of view. As such, this overview could apply to sonar or radar. Each structure discussed is related to the noise cancellation configuration of Fig. 3.8. For additional material on adaptive array control algorithm selection, Part II of Monzingo and Miller [47] is recommended.

Open-Loop Realizations

Open-loop adaptive realizations of the minimum complexity MVDR beamformer illustrated in Fig. 3.8 imply either a direct or an indirect computation of estimates for the CSDM

$$\mathbf{R}_{VV} = E[\mathbf{V}\mathbf{V}^H] \tag{3.97}$$

$= K \times K$ matrix of cross-spectral densities between all channels of the linear-combiner output

and the vector

$$\mathbf{P}_{VY_C} = E[\mathbf{V}Y_C^*] \tag{3.98}$$

$= K$-vector of cross-spectral densities between the linear-combiner output and the conventional beamformer output

Given the direct estimates of Eqs. (3.97) and (3.98), the corresponding estimate of the optimal filter $\mathbf{W}_{\text{opt}}$ in Eq. (3.67) is computed. At least three principal open-loop methods can be used to realize $\mathbf{W}_{\text{opt}}$ adaptively.

Estimate–invert–plug. The first method for adaptive realization of the MVDR beamformer classified as open-loop is based on the *estimate-and-plug* philosophy. In particular, this approach computes estimates of Eqs. (3.97) and (3.98) at time n by a continuous direct estimation process. For example, the continuous exponentially averaged estimates

$$\mathbf{R}_{VV}(n) = (1 - \nu)\mathbf{R}_{VV}(n - 1) + \nu\mathbf{V}(n)\mathbf{V}(n)^H \tag{3.99}$$

and

$$\mathbf{P}_{VY_C}(n) = (1 - \nu)\mathbf{P}_{VY_C}(n - 1) + \nu\mathbf{V}(n)Y_C^*(n) \tag{3.100}$$

can be implemented at intervals no greater than a time constant of the exponential integrator. The estimator of $\mathbf{R}_{VV}(n)$ is inverted and plugged, that is, substituted into the expression

$$\mathbf{W}_0(n) = \mathbf{R}_{VV}(n)^{-1}\mathbf{P}_{VY_C}(n) \tag{3.101}$$

and is then applied directly to the linear-combiner output. Instead of exponential averaging as in Eqs. (3.99) and (3.100), some versions of this estimate-and-plug approach would use uniformly weighted data (block averaging) over an interval of time n of length M, where M is sufficiently greater than K to ensure stability [62]. Numerous methods exist for inverting the estimator $\mathbf{R}_{VV}(n)$ [63]. Particular care must be taken in selecting a procedure which is robust to the limitations of finite-precision arithmetic and near singularity. In particular, it is necessary in cases of such ill-conditioning in $\mathbf{R}_{VV}(n)$ to inject a minimal level of synthetic white noise into the linear-combiner output vector to ensure full rank and resultant invertibility. The

Cholesky extension of the Crout decomposition method to Hermitian matrices is recommended from the standpoints of both computational efficiency and numerical accuracy [64].

Direct inverse–estimate–plug. The second open-loop method for the estimation of Eq. (3.67) is by an application of the matrix inversion indentity of Eq. (3.34) to Eq. (3.99) in the form

$$\mathbf{R}_{VV}(n)^{-1} = [(1-v)\mathbf{R}_{VV}(n-1) + v\mathbf{V}(n)\mathbf{V}(n)^H]^{-1} \tag{3.102}$$

$$= \frac{1}{1-v}\left[\mathbf{R}_{VV}(n-1)^{-1} - \frac{R_{VV}(n-1)^{-1}\mathbf{V}(n)\mathbf{V}(n)^H R_{VV}(n-1)^{-1}}{[(1-v)/v] + \mathbf{V}(n)^H\mathbf{R}^{-1}(n-1)\mathbf{V}(n)}\right] \tag{3.103}$$

The cross-spectral density vector $\mathbf{P}_{VY_C}$ can again be estimated as in Eq. (3.100). The open-loop, direct inverse approach is compared to both the estimate–invert–plug and closed-loop, gradient descent approach by Brennan et al. [57, 58] and Lunde [59] with performance results favoring open-loop methods. However, Brennan et al. point out the computational disadvantage of the open-loop techniques but then give several examples wherein the convergence time and adaptation noise are significantly less for the open-loop techniques than for the closed-loop approaches. Lunde states that for certain configurations where more beams are formed than there are adaptive channels, even the computational advantage of the closed-loop schemes diminishes. It is observed in [59] that the stability of Eq. (3.103) is heavily dependent on the linear combiner output word size, which, in turn, is related to the *conditional number*, $C = \lambda_{max}/\lambda_{min}$, the ratio of the maximum to minimum eigenvalues of $\mathbf{R}_{VV}(n)$. For a given word size, the ratio C can be controlled by the addition of white noise to the linear-combiner matrix filter output, as could also be required in the estimate–invert–plug technique discussed above. Finally, it should be clear that this approach is termed an open-loop, feedforward technique, because no feedback of beam output data is required in the adaptation process.

Orthonormalization. We now consider two methods based on the orthogonalization of the signal nulling matrix filter output N-vector. The first scheme transforms the nulling matrix filter output to orthogonal coordinates as suggested in Section 3.4 [65, 66]. Then a subset of these orthogonal filter outputs are selected as the linear combiner outputs for providing reference samples of the noise into the cancellation filter vector $\mathbf{W}$ in Fig. 3.8. The second approach is equivalent to a *Gram–Schmidt decomposition* of the signal nulling matrix filter output. This second scheme has the advantage of being decomposable into a simple two-input, one-output cancellation module [55, 60]. On the other hand, it is not clear that this configuration lends itself to either a minimum complexity structure or minimum convergence time.

Consider the first method, which involves the transformation of the signal nulling matrix filter output vector $\mathbf{U}^T = [U_1, U_2, \ldots, U_N]$ to orthogonal coordinates. The MVDR processor for the direct processing of $\mathbf{U}$ follows from Eq. (3.67) with $\mathbf{C} = \mathbf{I}$ and is given by

$$\mathbf{W}_0 = \mathbf{R}_{UU}^{-1}\mathbf{P}_{UY_C} \tag{3.104}$$

where

$$\mathbf{R}_{UU} = E[\mathbf{U}\mathbf{U}^H] \tag{3.105}$$

and

$$\mathbf{P}_{UY_C} = E[\mathbf{U}Y_C^*] \tag{3.106}$$

Let the signal-nulling matrix filter output CSDM have the orthogonal decomposition

$$\mathbf{R}_{UU} = \mathbf{M}\mathbf{\Lambda}\mathbf{M}^H \tag{3.107}$$

$$= \sum_{n=1}^{N} \lambda_n \mathbf{M}_n \mathbf{M}_n^H \tag{3.108}$$

where λ_n and $\mathbf{M}_n$ are the corresponding nth eigenvalue and eigenvector of $\mathbf{R}_{UU}$. The nth diagonal element and column of the diagonal matrix $\mathbf{\Lambda}$ and orthonormal matrix $\mathbf{M}$ are occupied by λ_n and $\mathbf{M}_n$, respectively. Furthermore, the eigenvalues are ordered such that $\lambda_1 \geq \lambda_2 \geq \cdots \geq \lambda_N$. The MVDR processor of Eq. (3.104) can now be rewritten by noting that

$$\mathbf{R}_{UU}^{-1} = \sum_{n=1}^{N} \lambda_n^{-1} \mathbf{M}_n \mathbf{M}_n^H \tag{3.109}$$

and

$$\mathbf{P}_{UY_C} = \sum_{n=1}^{N} \alpha_n \mathbf{M}_n \tag{3.110}$$

where α_n is the cross-spectral density between the nth orthonormal filter output and the DS beamformer output. Accordingly, there results from Eq. (3.104)

$$\mathbf{W}_{\text{opt}} = \sum_{n=1}^{N} \frac{\alpha_n}{\lambda_n} \mathbf{M}_n \tag{3.111}$$

Thus the suggested open-loop realization of $\mathbf{W}_{\text{opt}}$ by orthogonal transformation requires a continuous process of estimating the eigenvectors of the signal nulling matrix filter output $\mathbf{U}$ corresponding to the K largest coefficients (α_n/λ_n) and the corresponding correlations with the DS beamformed output. The linear combination of Eq. (3.111) is truncated at K terms to provide a computationally acceptable approximation to the optimum filter. The coefficient α_n/λ_n in Eq. (3.111) may be viewed as, first, a normalization by the power in the nth normal mode eigenvalue λ_n and, second, a rephasing of the nth modal filter noise output to allow coherent subtraction of this noise estimate from the DS beamformer output. Computationally, the estimation of the eigenvectors and eigenvalues of $\mathbf{R}_{UU}(n)$ would seem to be a disadvantage of this approach. However, this relatively large computing load is characteristic of all open-loop methods. However, it is pointed out that the dimensionality, K, of the eigenvector linear-combiner matrix can be made as small as possible, consistent with the effective cancellation of the dominant correlated noise modes. Efficient computational methods do exist for estimating just the dominant eigenvalue modal vectors [65].

The second open-loop method of noise cancellation by orthogonalization is illustrated in Fig. 3.12. This matrix filter could perform a Gram–Schmidt orthogonalization (GSO) of either the DS beam output and the signal nulling matrix filter output or the DS beam output and the linear combiner matrix filter output [60]. If the linear combining intermediate process is used to reduce the dimensionality of the GSO process, the linear combiner should reflect a priori knowledge of the dominant noise modes. This is because the GSO process has no capability for isolating these modes automatically, as in the transformation to orthogonal coordinates as described above [29]. The GSO network of Fig. 3.12 consists of two modules: a spectrum normalizing automatic gain control module, and a noise cancellation module. The basic orthogonalization operation in the GSO network is one of decorrelating the two inputs such as G_1 and G_2 by generating a new output, G_0, which is a linear combination of these inputs. It is required also that G_0 be uncorrelated with G_1. Accordingly, we have the linear combination of inputs

$$G_0 = G_2 - W^* G_1 \tag{3.112}$$

and $E[G_0^* G_1] = 0$ for

$$W = \frac{E[G_2^* G_1]}{E[|G_1|^2]} \tag{3.113}$$

where $E[G_2^* G_1]$ is the cross-spectral density between the two inputs and $E[|G_1|^2]$ is autospectral density of the G_1 input. The value of W given by Eq. (3.113) also minimizes the variance of the residual, G_0, by cancellation of that portion of G_2 which is coherent with G_1. We see from Eq. (3.113) that this operation can be performed in two steps, an *automatic gain control* (AGC) process and a *noise cancellation* (NC) process. These two processes are represented by the AGC and NC functions in the GSO network in Fig. 3.12. The noise inputs to the AGC function are whitened therein by the AGC self-spectrum normalization action. Then the statistical independence between the normalized channel and all other channels, including the DS beamformer output, is realized as a bank of noise cancellation (NC) functions. This process is implemented in cascade fashion as shown in Fig. 3.12. The result is that each signal free noise reference channel is made statistically independent from the DS beamformer output. One of the main advantages of the GSO network realization relative to a direct matrix inversion approach is its amenability to the modularization of simple arithmetic operations, [60] with obvious impact on computational requirements.

The GSO module considered above is termed the *first-order module.* A similar approach using a *second-order* GSO module could also be considered, and would reduce the number of stages in the overall cascaded network. In particular, the

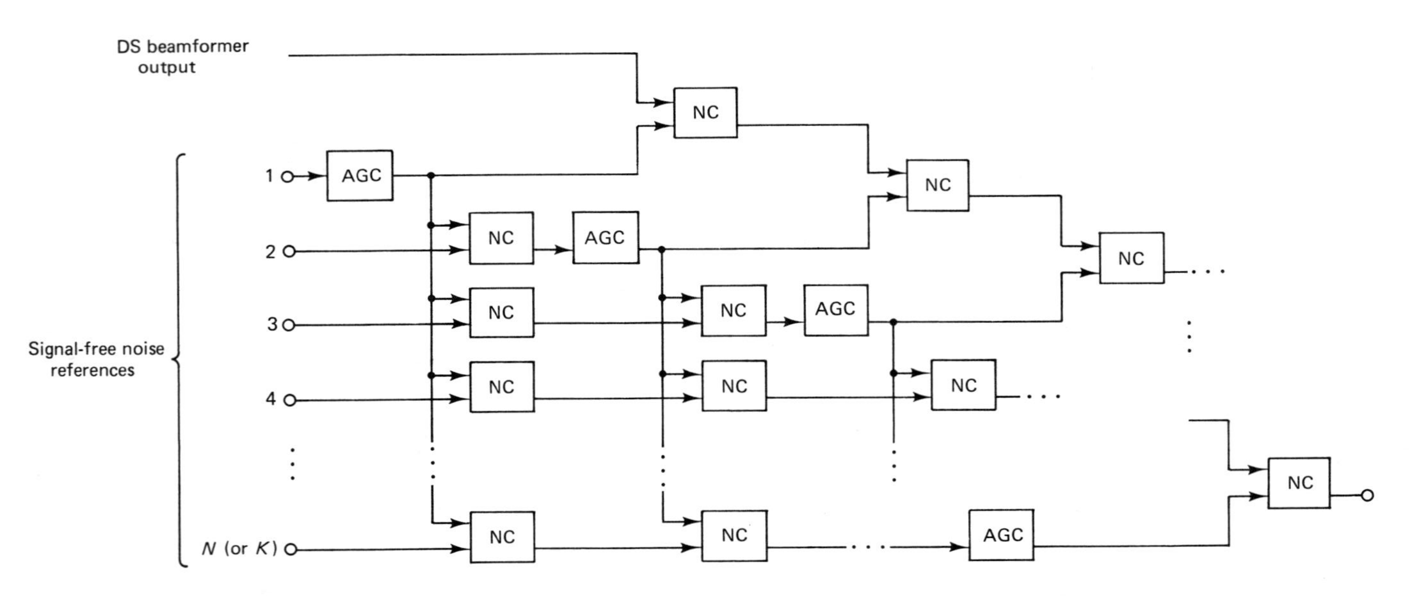

Figure 3.12 The Gram–Schmidt orthogonalization (GSO) network for noise cancellation.

orthogonalization of the three-module inputs G_1, G_2, and G_3 is realized by requiring that for the linear combination

$$G_0 = G_1 - W_1^* G_2 - W_2^* G_3 \tag{3.114}$$

the cross-spectral densities $E[G_0^* G_2]$ and $E[G_0^* G_3]$ be identically zero. These requirements give the matrix equation

$$\begin{bmatrix} W_1 \\ W_2 \end{bmatrix} = \begin{bmatrix} E[|G_2|^2] & E[G_2 G_3^*] \\ E[G_3^* G_2] & E[|G_3|^2] \end{bmatrix}^{-1} \begin{bmatrix} E[G_1^* G_3] \\ E[G_1^* G_3] \end{bmatrix} \tag{3.115}$$

which is still amenable to a modular realization.

Closed-Loop Realizations

The closed-loop methods permit the immediate computation of an estimate of the filter vector $\mathbf{W}$ defined by Eq. (3.67). The adaptive least-mean-square (LMS) algorithm [9, 14, 50, 52, 67–71] is used to minimize the noise-canceled beam output power in Fig. 3.8. The adaptive weight vector estimate $\mathbf{W}(n)$ at time n is updated to $\mathbf{W}(n+1)$ by adding the negative of the multidimensional derivative of the noise-canceled beam output power at time n with respect to the elements in the adaptive weight vector. In the limit, as n approaches infinity, this update process leads to a mean value of $\mathbf{W}(n)$ equal to the optimum weight vector $\mathbf{W}_{\text{opt}}$. This correction algorithm is readily stated in terms of the linear-combiner matrix filter output vector $\mathbf{V}(n) = [V_1(n), V_2(n), \ldots, V_K(n)]^T$ and the noise-canceled beam output $Y(n)$ at time n. Specifically, the gradient

$$\mathbf{J}(n) = \frac{\partial |Y(n)|^2}{\partial \mathbf{W}(n)} \tag{3.116}$$

$$= -2\mathbf{V}(n)Y(n)^* \tag{3.117}$$

is used to update the weight vector $\mathbf{W}(n)$ according to

$$\mathbf{W}(n+1) = \mathbf{W}(n) + \mu \mathbf{J}(n) \tag{3.118}$$

It is well known that for stationary noise the update step-size factor controls both the convergence rate of the mean value of $\mathbf{W}(n)$ and the steady-state misadjustment noise level associated with the closed-loop feedback. For a discussion of convergence and stability, see Widrow et al. [67] for the case of real input data. The adaptive LMS algorithm is regarded as a closed-loop control process because the gradient $\mathbf{J}(n)$ serves as a feedback vector error signal. If the control process is stable, the mean value of the gradient will go to zero as n approaches infinity.

A major difference between the various closed-loop algorithms which have been proposed over the past 20 years is in the approach to maintaining the response of the adapted beamformer to the signal. In particular, the early correlation cancellation systems can be visualized as in Fig. 3.8 but with the signal nulling matrix filter deleted (i.e., $\mathbf{P}_0 = \mathbf{I}$). The lack of signal preservation constraints was acceptable because these early systems addressed the radar problem wherein received signals of interest were of a duration which was short compared to the response time of the adaptive process and of a level that was low relative to any clutter or jamming noise [20]. This type of unconstrained sidelobe cancellation system was considered for active sonar applications. However, some passive and active sonar systems, on the other hand, had the requirement to preserve signals which could be of significant levels and of comparable duration relative to the coherent noise which was to be rejected. However, some important applications of the unconstrained process exist in passive sonar wherein a signal-free auxiliary noise reference is available [50]. Such auxiliary noise reference sensors as indicated in Fig. 3.8 are typically not in the actual acoustical sensor array itself. This need to prevent either cancellation or distortion of signal in the main beam gave rise to an evolution of signal preservation constraints, which is summarized below. The pilot signal approach [9] gave way to the P-vector, *soft-constraint* implementation [10], which, in turn, was replaced by the direct imposition of *linear hard constraints* on the updated filter weight vector to produce a signal null [14, 22]. Most recently, as illustrated in Fig. 3.8, the use of a signal nulling preprocessor matrix filter applied directly to the presteered array data has been proposed [17, 18, 40, 51].

The signal-nulling matrix filter preprocessor, which, if properly matched to the signal null space, provides signal free-noise references, has two significant implementation advantages over filter weight vector constrained algorithms. First, it can have an extremely simple realization which may consist of only add/subtract and scale operations. For example, an effective null on the main beam MRA requires only that the signal nulling matrix $\mathbf{P}_0$ have elements in any column that add to zero. This ensures that, for a signal which is in compliance with the signal wavefront model implicit in the $\mathbf{P}_0$ matrix filter, the linear combiner outputs will be signal-free. Two of the most useful spatial nulls are produced by $(1, -1)$ and $(-1, 2, -1)$ entries in a column of that $\mathbf{P}_0$ matrix, with the remainder of the elements in the column being zero. For sensors equally spaced at a distance d, these two nulling functions, which are referred to as the *sine* and *Hann* nulls, respectively, have normalized wavenumber amplitude response patterns given by

$$n_S(vd) = \left| \sin \frac{vd}{2} \right| \tag{3.119}$$

$$n_H(vd) = \sin^2 \frac{vd}{2} \tag{3.120}$$

Each order of the constraint used eliminates one potential spatial degree of freedom available to the noise cancellation process. These two nulling functions are plotted in Fig. 3.13. If the reader is familiar with the aspects of FFT interval window functions, the specification of higher-order nulls which reflect increasing uncertainty with respect to the signal wavefront model are simple extensions of such windows to the spatial transform domain. The realization of signal-nulling constraints in a time-domain implementation is also straightforward, albeit computationally more

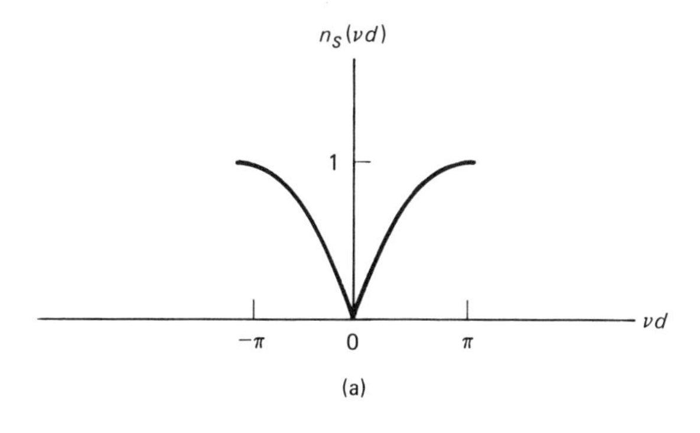

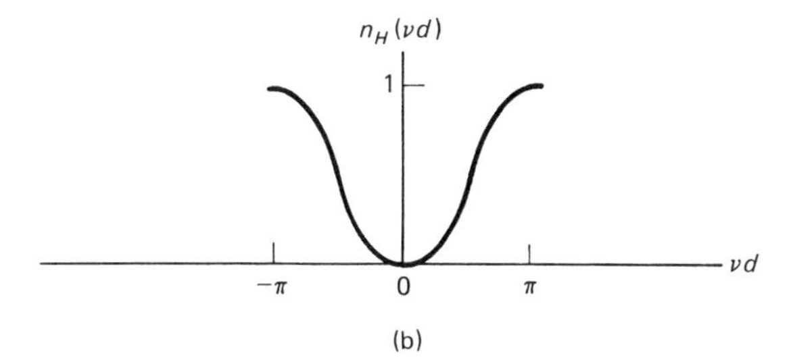

Figure 3.13 Normalized spatial wavenumber amplitude response functions for the (a) sine and (b) Hann spatial-nulling functions.

demanding. It is necessary only to observe that the signal-nulling filters considered above are expressed as two- or three-point convolution operators in the wavenumber domain and have full aperture multiplicative equivalents in the spatial domain. As a practical matter, when a high-level directional signal occurs at an angle within the spatial notch of a wavenumber-nulling filter but not at the null itself, then the filter weights will become very large in an attempt to cancel this signal. This can be seen from Eq. (3.67) where the elements of $\mathbf{R}_{VV}$ are small and those of $\mathbf{P}_{VY_C}$ are large. If this cancellation is not desired, it can be limited by a corresponding upper limitation on the magnitude of the elements in the filter weight vector. However, the phase of the magnitude constrained filter weight must be maintained to preserve the correct relative directional response.

It should be made clear that the type of spatial-nulling constraints discussed above apply equally well to both open- and closed-loop realizations. The discussion on higher-order nulling constraints is presented with the closed-loop system discussion because it was in this context that these techniques evolved historically. Similarly, the GSO network modules have a simple closed-loop implementation based on the minimization of the spectral density $E[|G_0|^2]$ in Eq. (3.112) by the adaptive LMS algorithm.

The issue of convergence rate deserves some consideration in a discussion of system realization because it is basic to the differences between the open- and closed-loop approaches presented above. The estimate–invert–plug approach is considered here as representative, in terms of convergence rate, of the open-loop schemes. It is desired to compare the rate of convergence of the weight vector for this open-loop approach to the adaptive LMS, closed-loop approach. In particular, let the mean value of the estimated CSDM $\mathbf{R}_{VV}(n)$ after J iterations of Eq. (3.99) be given by $\mathbf{G}(J)$. By iteration of this mean value expression there results

$$\mathbf{G}(J) = (1 - v)^J \mathbf{G}(0) + [1 - (1 - v)^J]\mathbf{R}_{VV} \tag{3.121}$$

The first term in Eq. (3.121) is the transient term wherein an exponential envelope $\exp[-J/\tau]$ can be fitted to the geometric expression $(1 - v)^J$. For large τ, the convergence time in iteration cycles, $\exp(-1/\tau) \cong 1 - 1/\tau$ and $\tau = 1/v$. Thus, for v selected small enough to give acceptable variance in the estimator $\mathbf{R}_{VV}(J)$, it is clear that convergence rate for this open-loop estimate is only a function of v and not a property of the data. On the other hand, for the adaptive LMS closed-loop approach the convergence occurs independently along each principal axis defined by the eigenvectors of the CDSM $\mathbf{Q}$ in Eq. (3.49). The time constant associated with the pth axis convergence process is

$$\tau_p = \begin{cases} \dfrac{1}{2\mu\sigma^2[\alpha(b_p - 1) + 1]}, & p \leq L \\ \dfrac{1}{2\mu\sigma^2(1 - \alpha)}, & p > L \end{cases} \tag{3.122}$$

where b_p is the pth eigenvalue associated with the coherent noise CSDM $\mathbf{Q}_c$. Thus it is seen that convergence rate for the closed-loop approach is dependent on the noise level such that the individual modal time constants are an inverse function of σ^2 and the relative amount of coherent noise power, b_p, in the pth mode. The τ_p are referred to simply as the *modal closed-loop time constants.* For the closed-loop system to converge in the mean, we require that the step-size parameter μ satisfy the condition

$$\frac{1}{\sigma^2[\alpha(b - 1) + 1]} < \mu < 0 \tag{3.123}$$

where the ordering $b_1 \geq b_2 \geq \cdots \geq b_L$ of the eigenvalues of $\mathbf{Q}_c$ is assumed. Therefore, convergence for closed-loop control can either be very rapid and correspondingly noisy for high coherent noise levels or slow and stable for low coherent noise levels. On the other hand, in open-loop control, the convergence will be of predetermined rate and noisiness. When the coherent noise is dominated by a strong mode, rapid convergence may be desirable if this mode represents an interfering source with a high angular velocity for example. If the increased adaptation noise due to rapid convergence succeeds in masking low-level signals of interest, rapid convergence in the mean is of little benefit. A resolution of this issue for a particular sonar system would in all likelihood depend on the mission to be performed. A number of configuration issues are currently of some interest to designers of adaptive signal processing systems in sonar. Several of these diverse issues are presented in the following section.

3.7 IMPLEMENTATION CONSIDERATIONS

The typical frequency band of interest in medium-range active and passive sonar systems seldom extends above 15 kHz. Also, it is characteristic of most long-range sonars to be concerned with the frequency band less than 5 kHz. As such, array processors with throughput capabilities equivalent to commercially available units are having a significant impact on the implementation of sonar signal processing. This is particularly true of systems implemented in the discrete frequency domain wherein high-speed FFT processing, complex arithmetic, and floating-point precision are at least beneficial if not necessary.

The majority of available literature on sonar-related adaptive array processing has presented a time-domain point of view. Until now, this chapter has dealt primarily with a frequency-domain formulation. This approach is intended to provide both a more balanced perspective as well as an indication of current implemention trends. In the discrete frequency domain, the basic adaptive element has been denoted as in Fig. 3.14(a), with the complex weight of the form

$$W = |W| \exp(j \arg W) \tag{3.124}$$

A separate complex filter weight is applied in each of the L FFT output bins. An equivalent implementation of the frequency-domain adaptive filter element for an analog narrowband process is also illustrated in Fig. 3.14(b). The traditional time-domain adaptive filter element has been the L-tap *transversal filter* illustrated in Fig. 3.14(b). However, much attention has been given recently to alternatives to the adaptive transversal filter as the principal time-domain adaptive system building block. In particular, the adaptive multichannel lattice noise-canceling structure [70, 71] with faster convergence than the transversal filter has been formulated. The adaptive lattice structure achieves a more rapid convergence as a result of cascaded partial orthogonalization operations. Full-mode isolation by complete orthogonalization of the multichannel time data probably is not practical in a time-domain implementation as it is in a frequency-domain implementation because of the prohibitive dimensionality which would characterize a K-channel $\times$ L-point version of the lattice filter. It is recalled that such mode isolation is possible in a frequency-domain implementation because the frequency bins are processed independently, thereby separating the temporal and spatial adaptation functions. Conclusive work using the adaptive lattice structures for sonar noise cancellation applications in the time domain has yet to be reported. This is primarily because the lattice structure has evolved principally from single-channel linear predictive filtering applications to speech processing and high signal-to-noise ratio spectrum analysis.

In terms of operational sonar adaptive hardware implemented to date, no distinct trend has evolved that, in general, favors implementation in the time domain over the frequency domain. Relatively small-aperture arrays have typically exploited time-domain techniques. Historically, small-aperture arrays have operated

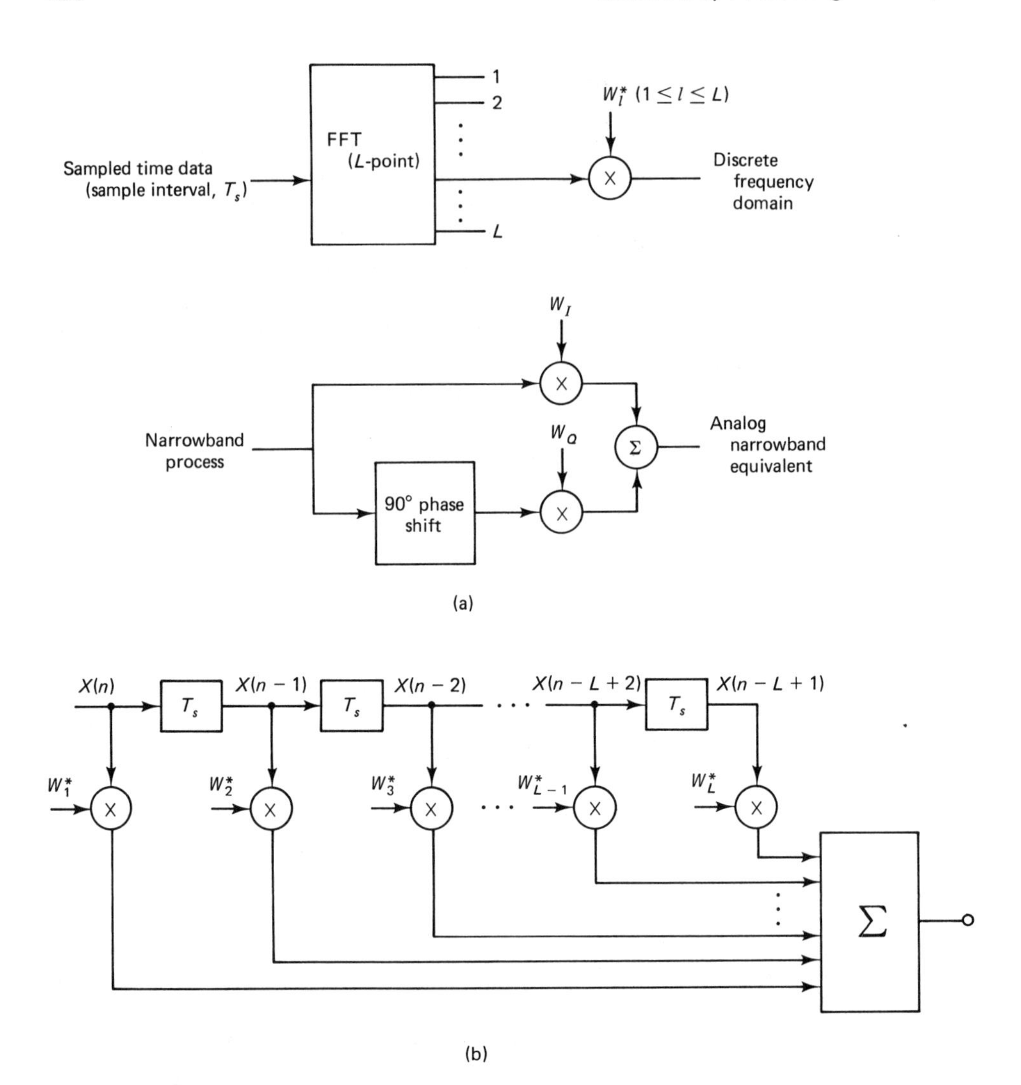

Figure 3.14 (a) Frequency-domain adaptive filter element implementation; (b) transversal filter time-domain adaptive element implementation.

on the basis of broadband-radiated signal energy and, therefore, no requirement for frequency partitioning has existed. However, because of linearity, complex spectrum analysis at the beam output and individual sensor level are, of course, interchangeable. Furthermore, for a given number of adaptive degrees of freedom, complex spectrum analysis at the sensor level followed by frequency-domain adaptive processing can offer certain advantages over time-domain processing, if spectrum-analyzed beam-output data are required. Specifically, as mentioned above, frequency-domain systems provide separability of the spatial and spectral adaptive processes. A fixed number of adaptive degrees of freedom, therefore, remain assigned to each spectral bin. This feature is a form of spectral self-normalization. It assures that no portion of the band with high-level spatially coherent noise will "capture" an inordinate share of the available degrees of freedom at the expense of spatial gain improvement in a local spectral region that might be strategically important. However, for a given number of adaptive weights, the time-domain approach, in effect, will distribute these weights in frequency so that the performance averaged over the entire processed band is optimized. Thus, if only either broadband energy or short duration signals are to be used for beamformer output signal detection, the time-domain techniques would seem to be appropriate. On the other hand, if spectrum analysis of the beamformer output is to be performed in the postprocessing sequence, adaptive beamformer implementation in the frequency domain should be given consideration.

3.8 SONAR ARRAY PROCESSING EXAMPLES

A useful display format for presentation of both active and passive sonar multi-beamformer output data is referred to as the *bearing time recorder* (BTR). In this section, several examples from passive sonar array processing are presented in terms of the BTR information presented to a sonar operator on such a display. The array considered consists of 18 sensors uniformly spaced on a horizontal line at one-half a wavelength at the highest frequency of interest, f.

The first example consists of five different signal and interference conditions wherein various types of beamforming configurations are utilized. For each condition, a 5-min time history for a BTR generated from 31 simultaneously formed digital beams is shown in Fig. 3.15. The BTR gray-scale intensity modulation provides an indication of apparent acoustic energy level arriving at the sensor array as a function of time on the vertical axis and horizontal arrival angle (azimuth) on the horizontal axis. The signal arrival angle is denoted as θ_s and the interference arrival angle as θ_I. The signal-to-uncorrelated noise ratio at a sensor is SNR $= -27$ dB in the processed band. The interference-to-uncorrelated noise ratio at a sensor in the same band is INR $= -5$ dB. Part (a) of Fig. 3.15 consists of the signal arrival only with an MVDR beamformer operational. In part (b), the signal remains at the same angle and SNR and the MVDR beamformer is still operational; however, the interference is turned on. The interference-to-uncorrelated noise ratio at the output of a DS beam pointed directly at the interference is

$$\text{INR}_{\text{out}} = \text{INR}_{\text{in}} + 10 \log 18 \tag{3.125}$$

$$= 8 \text{ dB} \tag{3.126}$$

which, from Section 3.2, indicates the potential for a considerable array gain improvement. However, we note that, even though the signal is easily detected at θ_S in part (b), it is actually slightly suppressed relative to part (a). This signal suppression is a result of only partial decorrelation of the signal and interference waveforms

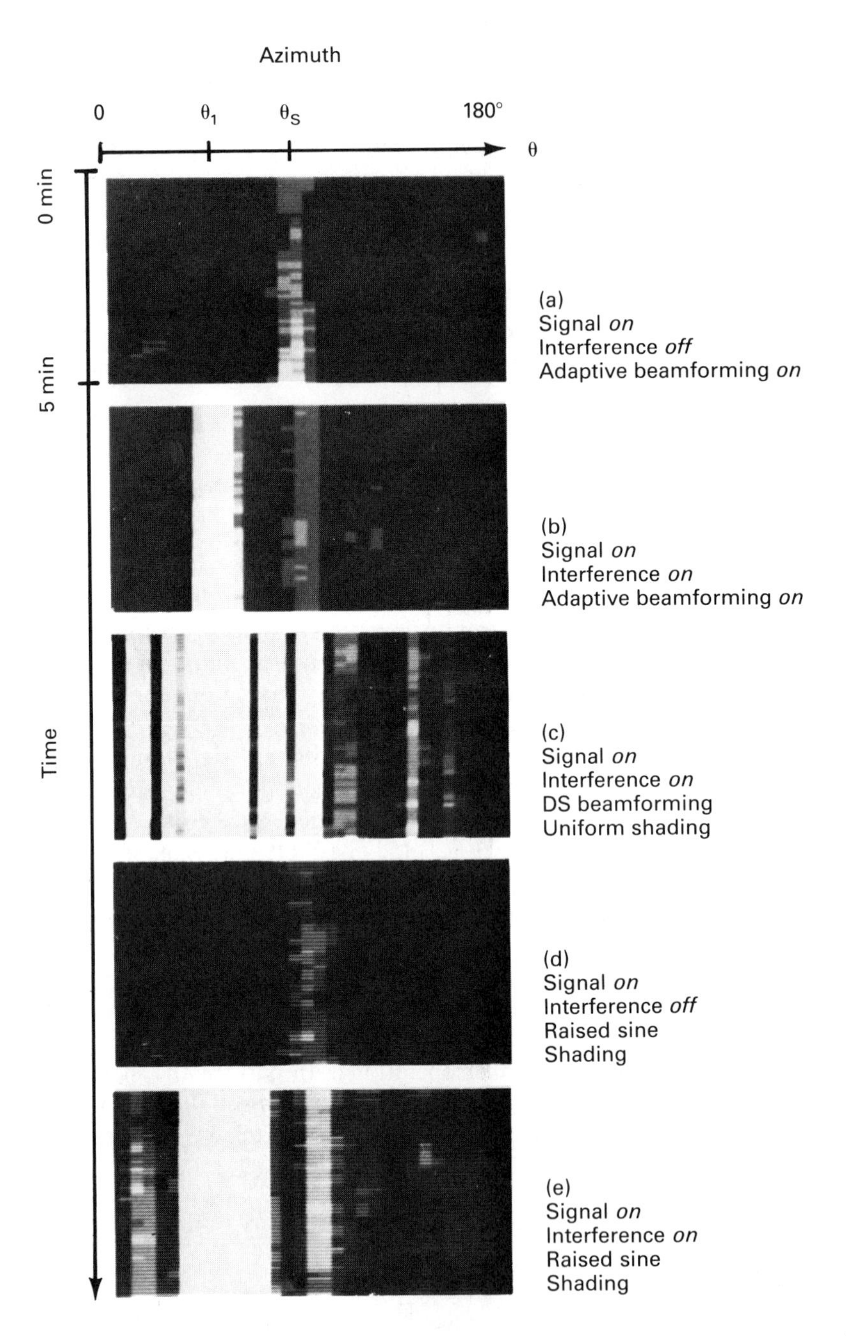

Fig. 3.15 Five different signal and interference array processing conditions displayed on a bearing-time recorder (BTR).

within the adaptive time constant of the beamformer as discussed in Section 3.5. As can be seen by comparison with the interference pattern in Fig. 3.15(c) without adaptive MVDR beamforming, the interference is substantially controlled. Furthermore, the signal is still completely detectable in part (b) in spite of the signal suppression. In part (c), with fixed constant sensor output weighting (uniform shading), the signal is completely masked by the interference on the second spatial sidelobe of the signal-directed beam. In order to reduce the spatial sidelobes and thereby achieve some measure of interference rejection, the kth sensor weight called raised sine shading

$$a(k) = \left[0.1 + 0.5 \sin \frac{(k-1)\pi}{17}\right], \qquad k = 0, 1, \ldots, 17 \tag{3.127}$$

is applied to the kth sensor output in the 18-sensor array. This raised sine array spatial *shading* window widens the main-lobe width relative to that in part (c) of Fig. 3.15 by a factor of approximately 1.3 and produces a maximum sidelobe response level of not more than -22 dB relative to the response of a signal on the MRA. Figure 3.15(d) gives the BTR with raised sine shading with the signal only turned on. One should note the slight loss of signal detectability due to the loss of directivity resulting from the widening of the main lobe. Finally, in Fig. 3.15(e) both the signal and interference are turned on with raised sine shading. In this part, it should be noted that the angular resolution of the interference is worse, by a factor of at least 1.3, than either the adaptive sequence (b) or the uniformly shaded condition of part (c). The interference still remains on a sidelobe of the signal directed beam to the extent that detection of the constant bearing signal from the BTR would be doubtful. The symmetry of the BTR for part (e) about the main interference response is the feature that casts doubt on the response to the right of the interference, at the known azimuth angle of the signal arrival.

The second example of a passive sonar beamformer uses a different type of linear-combiner matrix filter. In particular, when a sonar environment contains a source of far-field interference, a DS beamformer directed at the interference provides an essentially signal-free reference of the interference waveform. This interference directed beamformer operation is implemented with a properly selected row of phase-shift elements in the linear-combiner matrix filter. This type of beamformer output processing is referred to as *postbeamformer interference cancellation* (*PIC*). An example of PIC processing is shown in Fig. 3.16. Part (a) of this figure shows a multibeam BTR display with DS beamforming and no PIC processing. The bottom display [Fig. 3.16(b)] uses PIC processing in the signal azimuth sector and the two signal bearing "tracks" are easily detected. This type of beam space processing [43] can have advantages over element space processing. Specifically, if the hydrophone sensitivities are unknown and the corresponding gain levels cannot be incorporated into the element space processor correctly, signal leakage can occur, with attendant signal suppression. Furthermore, when the dominant interference direction is known, which is usually the case for a single strong interfering sonar contact, the

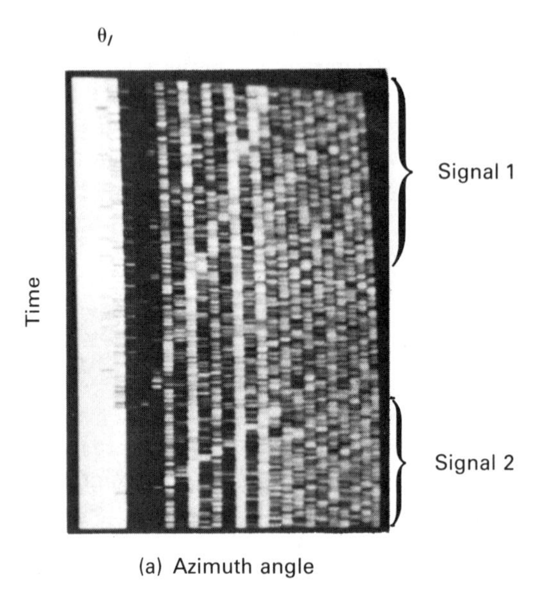

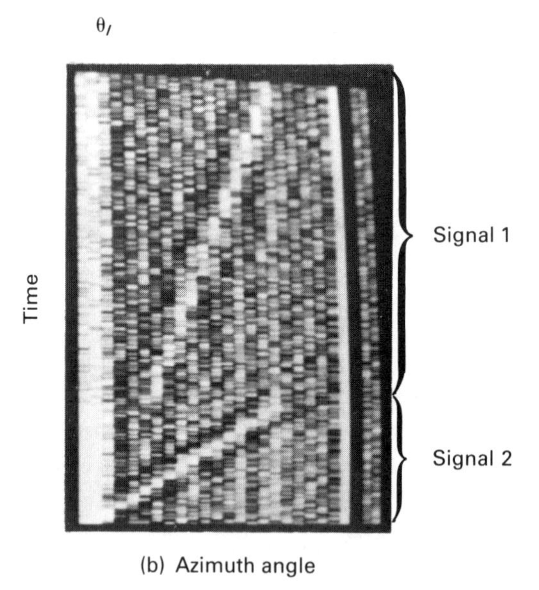

Figure 3.16 (a) Experimental BTR display with a strong interference at azimuth angle θ_I and two moving low-level signals. Postbeamformer interference cancellation (PIC) used. (b) Rerun of interference and signal geometry of Fig. 3.15(a) but with PIC operational in the signal azimuth sector only.

interference component of the noise only CSDM has a single dominant eigenvalue. The corresponding eigenvector can be used as the linear combiner matrix filter of order 1.

On the other hand, not all sonar noise environments consist of far-field interference. In fact, in the presence of complex multipath structures, reflections in the near-field and high-wavenumber noise caused by turbulent flow fluctuation mechanisms [72], beam space processors would be minimally effective. For these more complex noise environments, some form of either element space processing or higher-order, noise CSDM eigenvector preprocessor to achieve both minimum complexity and orthogonal noise reference channels could be the most effective. Adaptive noise cancellation for a noise generation mechanism unique to the sonar array is discussed in the following section.

3.9 HIGH-WAVENUMBER NOISE

Previously, we considered the minimization of the number of (spatial) channels that must be adaptively filtered, K, relative to the number of hydrophones in the array N. In this section we consider an analogous question in terms of spectral resolution channels. Before adaptive beamforming is considered for a particular array, the typical spatial and spectral characteristics of the actual coherent noise should be measured. In many applications, a study of intersensor noise coherence may lead to the conclusion that using fully optimum/adaptive processing may be totally futile. On the other hand, other noise mechanisms may be identified as noise cancellation candidates. For example, Fig. 3.17 presents one portion of an analysis of the spectral content of spatially coherent ambient noise and self-noise [36] for a mobile sonar array. This figure gives the percentage of the total processed acoustic bandwidth that exhibits spatial coherence above a specified value. The data pertain to the coherence between two adjacent sensors in the array moving through the water at speeds of 5 and 10 knots. As the speed increases, the percentage of bandwidth that exhibits significant coherence decreases.

Since the extent of coherent noise appears to be a decreasing function of speed, we may question the basic requirement for an optimum/adaptive array processor if this array is to be used primarily for high-speed operation. In fact, what underlies the measured results illustrated by Fig. 3.17 is that as the array speed increases, the relative power level of the turbulent boundary layer (TBL) induced noise increases, thereby masking out the other signal and noise components which exhibit coherence between adjacent sensors [72]. The result is a net reduction in the coherent signal and noise bandwidth.

It is well known that hydrophones in various types of mobile sonar arrays encounter a nonradiating noise component induced by TBL pressure fluctuations. This type of random structural excitation is characterized by wavelengths and coherence decorrelation distances that are short compared with those for the waterborne radiating acoustic component of sonar array coherent noise. Furthermore, as the sonar platform increases speed, the TBL induced-noise component increases in level accordingly and becomes a limiting factor in mobile-sonar performance.

One approach to reducing the effect of TBL-induced noise is the use of clusters of closely spaced hydrophones instead of single hydrophones. The distance between

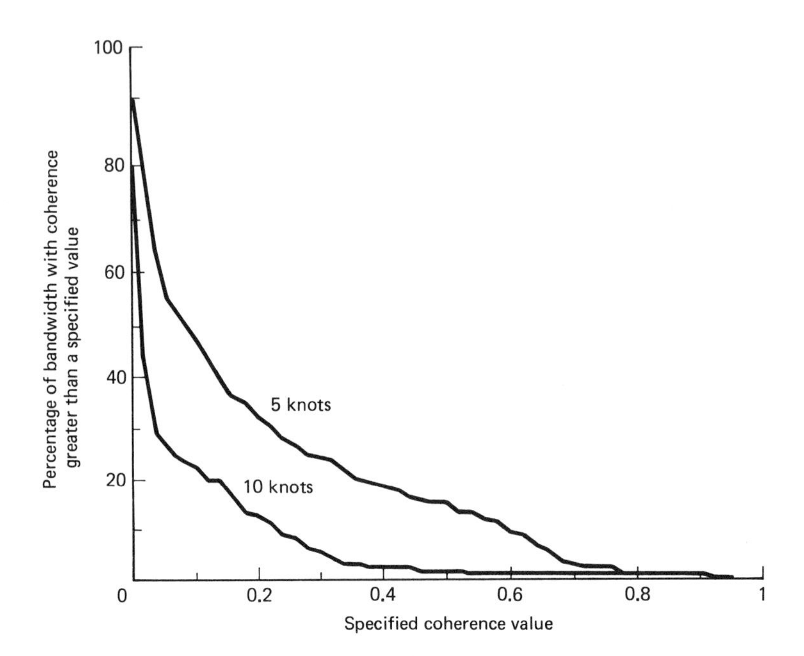

Figure 3.17. Coherence bandwidth between two adjacent sensors in a mobile sonar passive array.

sensors within a cluster is small compared with the acoustic wavelengths of radiating signals of interest. However, if this distance is not greater than the TBL-induced noise component decorrelation distance, the use of such clusters with appropriate signal processing can provide a measure of discrimination against this noise that is characterized by high-wavenumber spectral content. Various TBL-induced noise discrimination schemes have been proposed that require forming weighted linear combinations of the hydrophone outputs within a cluster before beamforming of the combined outputs [72–74]. The basic approach is to apply the noise cancellation process as presented in Fig. 3.8 to the processing of the small clusters of closely spaced hydrophones in MVDR fashion. The CSDM for a linear M-hydrophone cluster given a radiating signal in the presence of both TBL-induced noise and uncorrelated noise is modeled as an $M \times M$ matrix

$$\mathbf{R} = \sigma_S^2 \mathbf{1}\mathbf{1}^T + \sigma^2 \alpha \mathbf{Q} + \sigma^2(1 - \alpha)\mathbf{I} \tag{3.128}$$

Because the cluster dimensions are assumed small with respect to a radiating signal wavelength, the signal component is assumed to be the same at all sensors within a given cluster. Thus the signal direction vector is a vector of all 1's denoted by $\mathbf{1}$. The TBL-induced noise is described by the CSDM $\mathbf{Q}$ with mnth element

$$[\mathbf{Q}]_{mn} = \rho^{m-n} \tag{3.129}$$

where ρ is the TBL-noise coherence between two adjacent sensors in a cluster. This coherence is assumed to be of the form

$$\rho = r \exp (j\phi) \tag{3.130}$$

where r is the magnitude of the TBL-induced noise coherence function between the sensors at distance d_0. The coherence amplitude r, $0 \leq r < 1.0$, is a function of angular frequency $\omega = 2\pi f$ and has the form

$$r = \exp \left(\frac{-\beta \omega d_0}{u_T} \right) \tag{3.131}$$

In Eq. (3.131), β is a positive constant that is a function of the array structure and u_T is the *convective* velocity of flow in the TBL. The *convective* velocity u_T is a fractional part of the free-stream velocity, u_0, with which the array is moving through the water. Typically, we have $0.6 < u_T/u_0 < 0.8$. The parameter ϕ is given by $\phi = \omega d_0 / u_T$.

In Fig. 3.18, the gain is presented for both two- and three-element clusters

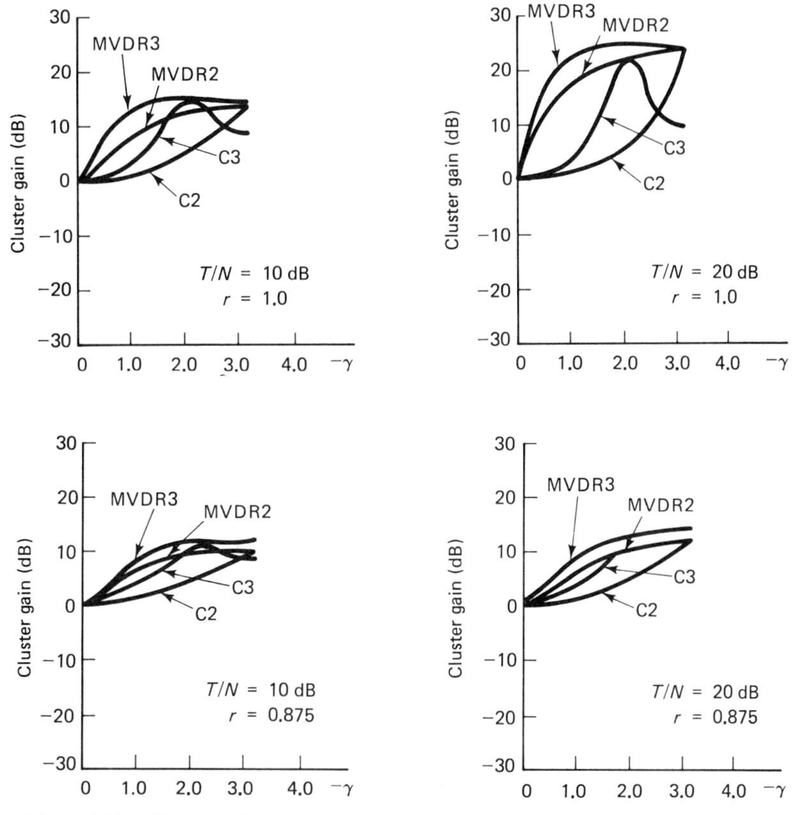

Figure 3.18 Cluster gain against TBL-induced noise versus $\gamma = -v_T d_0$, where v_T is the TBL-induced noise wavenumber and d_0 is the sensor spacing within the cluster.

using both conventional, equally weighted sensor output combining (C2, C3) and MVDR noise cancellation (MVDR2, MVDR3). The parameter γ is defined as

$$\gamma = \omega d_0\left(\frac{1}{v} - \frac{1}{u_T}\right) \tag{3.132}$$

$$\simeq -\nu_T d_0 \tag{3.133}$$

where $\nu_T \simeq \omega/u_T$ is the wavenumber of the TBL-induced noise. We see that the TBL-induced noise-to-uncorrelated noise ratio

$$\text{TNR} = \frac{\alpha}{1-\alpha} \tag{3.134}$$

and the coherence magnitude r are the critical parameters in the performance of this type of high-wavenumber noise cancellation process.

What has been demonstrated by a detailed examination of sensor-to-sensor coherence presented above is that for a specific sonar array under different operating conditions, the bandwidth over which adaptive processing is required can be determined. Because adaptivity is an expensive resource, it is conceivable that a fixed, affordable amount of this resource could be allocated to portions of the spectrum in real time as determined by on-line coherence level measurements. Certainly, if adaptivity is not needed, it should not be used! This is because it can only result in signal suppression if signal leakage occurs and it will certainly result in adaptation noise caused by steady-state filter misadjustment [9].

3.10 HIGH-RESOLUTION ARRAY PROCESSING

The measurement of intersensor time delays for a wavefront propagating in an array of sensors is a process receiving considerable attention in relation to two problems. First, to locate the position of a point source in the near field of an array, it is necessary to measure, with considerable accuracy, the arrival angle and curvature of the supposed spherically radiating wavefront generated by the source. Clearly, if these two parameters can be measured, the relative location of the source is determined. As the dual problem, if a point source is known to be in the far field of the array with resultant plane-wave insonification of that array, and if the propagation conditions are known, a situation exists wherein unknown array sensor coordinates can be estimated by measuring interelement time delays. The source location problem and the acoustic sensor array shape estimation problem have in common the observation of the signal at the multiple, spatially separated sensor locations in the array. The signals received by the array sensors ideally would consist of time-delayed replicas of the source signal in additive uncorrelated noise. Knowledge of this time-delay structure is equivalent to knowing the relative wavefront-array shape geometry. The estimation of time delay between the arrivals of a signal at just two sensors has been analyzed extensively [75–79]. Time-delay estimation in an array of sensors is the logical extension of the two-sensor process to the general delay vector estimation problem of interest in the practical problems of high-resolution source location and array shape estimation [26–28, 32, 33, 80.] In the array time-delay estimation problem, however, the problem is best formulated in terms of source location parameters [81–83] rather than in terms of the possible pairwise time delays.

Included in this group of proposed source location array processing techniques are schemes with direct counterparts in high-resolution temporal frequency analysis. The maximum-entropy method (MEM), discussed in Chapter 2, is also under consideration for sonar array processing for arrays of equally spaced sensors. In particular, it would appear that the desirable stability of the maximum-likelihood wavenumber spectrum estimator can be balanced against the desirable high resolution and undesirable instability [84] of the MEM by using a combination of each method [85].

Other high-resolution techniques based on an analytic expansion of the first column of the estimated CSDM using Prony's method or the equivalent, for complex polynomial solution, has been attempted by Yen [87] with good results for high SNR far-field signals. Similarly, the recent work on frequency parameter estimation by Tufts and Kumaresan [88–90] which provides an improved method of spectral resolution, could also be applicable to the multiple source wavenumber resolution problem. (See Chapter 4.)

High-resolution power spectrum analysis methods have well-known applications to single-channel frequency analysis [91] and multiple-channel sensor array frequency–wavenumber analysis. More recently, sensor array simultaneous frequency–wavenumber–range estimation has provided a further extension of fundamentally the same power spectrum analysis procedures [92]. Principal among these techniques are, first, the previously discussed minimum-variance, distortionless response (MVDR) filter, which is closely related to the maximum-likelihood procedure of Capon [49]. This process includes a bias term in the multiple-source case as a penalty for not having to know the signal spectrum. Second, the class of least-squares linear smoothing/prediction techniques is relevant and is referred to herein loosely as the maximum-entropy (ME) method [85, 93–94]. It is typical of the MVDR technique to assume no prior knowledge of the vector space dimensionality of either the signal or noise components of the data to be analyzed. In contrast, ME techniques assume a parametric model for the signal generation. Such prior knowledge of the signal takes the general form of bandwidth, angular, and radial extent of the signal for frequency, wavenumber, and range analyses, respectively. Prior knowledge that the noise can be prewhitened and the signal is known to have a spectral dimensionality which is small relative to the analysis data dimensionality also can be exploited to enhance the spectrum estimator resolution capabilities to an extent limited only by the observation time. Accordingly, an approach to data adaptive spectral analysis is described herein which can be characterized by an orthonormalization of the observation data covariance matrix. This approach, which can be expressed alternatively in terms of either orthonormal [65, 95–103], singular value [88], or Cholesky [101] decompositions of the array CSDM, is in contrast to either the Gram–Schmidt orthogonalization procedures or related schemes as embodied in the lattice filter structure [104].

The use of eigenvector orthonormal decompositions for frequency–wavenumber spectrum analysis has evolved from nonparametric and adaptive array processing schemes [14, 95, 105] to applications as high-resolution spectral estimators [96, 97, 100–102]. Frequently, these high-resolution techniques are viewed as ad hoc methods based on the notion of separability of signal and noise processes into orthogonal vector subspaces. In fact, two distinct versions of these high-resolution spectrum estimators are derived from the formal expressions for the MVDR and ME spectrum estimators. The additional prior information which allows the extension of the MVDR and ME estimators to their so-called enhanced high-resolution, dominant-mode forms is either that the noise is uncorrelated or can be prewhitened and the signal components in terms of the number of sources of the analyzed data are known and narrow in wavenumber bandwidth.

In this section the dominant-mode form of the estimated CSDM is presented together with the expression for the readily derived inverse of this matrix. Next, the MV and ME spectrum estimators in modal decomposition form are developed together with 3-dB down-resolution expressions. Finally, the modal MVDR and ME estimators are applied to passive sensor array processing for source range estimation.

Signal Model

Let the covariance matrix for the stationary, zero-mean complex data N-vector $\mathbf{X}(t)$ at discrete time t be given by an expression similar to Eq. (3.12):

$$\mathbf{R} = E\{\mathbf{X}(t)\mathbf{X}(t)^H\} \tag{3.135}$$

Of course, in practice, a time-averaged estimate of $\mathbf{R}$ can be used instead of $\mathbf{R}$ in terms of the ensemble average as defined in Eq. (3.135). This covariance matrix is assumed to consist of the signal and noise components

$$\mathbf{R} = \sigma_s^2\mathbf{P} + \sigma^2\mathbf{I} \tag{3.136}$$

where $\mathbf{P}$ is a rank $K \leq L \ll N$ signal covariance matrix, σ^2 the uncorrelated noise variance, and $\mathbf{I}$ an $N \times N$ unit diagonal matrix. The signal covariance matrix has two representations. First, in terms of actual signal descriptors, we can write

$$\sigma_s^2\mathbf{P} = \sum_{p=1}^{L} \sum_{q=1}^{L} \sigma_{pq}^2 \mathbf{D}(\boldsymbol{\theta}_p)\mathbf{D}(\boldsymbol{\theta}_q)^H \tag{3.137}$$

and second, in terms of a dominant-mode(s) decomposition of Section 3.3, Eq. (3.17) we have

$$\sigma_s^2\mathbf{P} = \sum_{k=1}^{K} \lambda_k \mathbf{M}_k \mathbf{M}_k^H \tag{3.138}$$

The signal is represented in Eq. (3.137) on an N-dimensional vector subspace by the spanning set of signal vectors $\{\mathbf{D}(\boldsymbol{\theta}_p): p = 1, 2, \ldots, L\}$. These L signal vectors are not necessarily linearly independent. The quantity σ_{pq}^2 is the cross-correlation between the pth and qth signal wavefront complex envelopes, [106] which, in turn, are assumed to be stationary and zero mean. In Eq. (3.138), the signal covariance matrix is written as an orthonormal expansion in terms of its dominant eigenvectors and K associated eigenvalues, where $\lambda_1 \geq \lambda_2 \geq \cdots \geq \lambda_k$ are the rank-ordered eigenvalues of $\sigma_s^2\mathbf{P}$ and $\mathbf{M}_1, \mathbf{M}_2, \ldots, \mathbf{M}_K$ are the corresponding orthonormal eigenvectors.

A narrowband (i.e., frequency-domain) representation of the signal is assumed which allows separability of the time and space description of the signal [107]. Two important examples of signal representations in each of these domains are given by defining the nth element of the signal vector $\mathbf{D}(\boldsymbol{\theta}_p)$ to be

$$\mathbf{D}(\boldsymbol{\theta}_p)\Big]_n = \exp\,(-j2\pi f_p n\Delta) \tag{3.139}$$

for spectral analysis of a time series in terms of the discrete Fourier frequency parameter f_p, with a uniform temporal sampling interval Δ. The spatial spectrum analysis counterpart to Eq. (3.133) for a one-dimensional (linear) spatial sensor array on the x-axis is

$$\mathbf{D}(\boldsymbol{\theta}_p)\Big]_n = \exp\,(-j2\pi f\tau_{pn}) \tag{3.140a}$$

$$= \exp\,[-j2\pi f(r_p^2 + x_n^2 - 2r_p x_n \cos\beta_p)^{1/2}/v] \tag{3.140b}$$

$$\cong \exp\left[-j2\pi f\left(r_p - x_n\cos\beta_p + \frac{\sin^2\beta_p}{2r_p}x_n^2\right)\Big/v\right] \tag{3.140c}$$

The spatial model for the pth signal component at frequency f represented in Eqs. (3.140a)–(3.140c) expresses the propagation time τ_{pn} for wavefront propagation and speed v in a homogeneous medium from the pth source at range r_p and bearing β_p to the nth of N sensors in a linear array. This geometry is illustrated in Fig. 3.19. Equation (3.140c) is obtained from Eq. (3.140b) by taking a Taylor series expansion

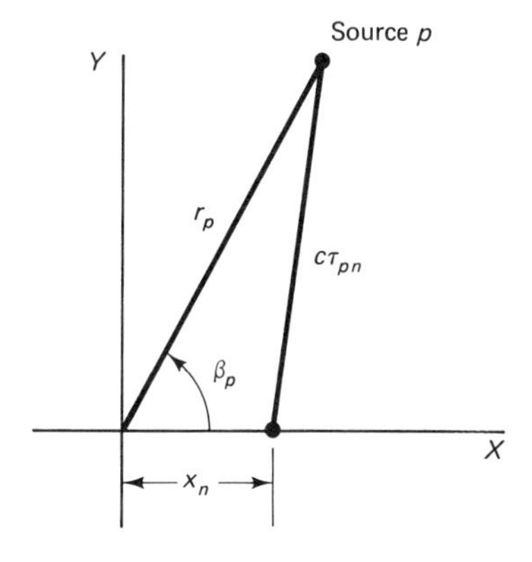

Figure 3.19 Linear array of N discrete sensors located at points X_n, $n = 1, 2, \ldots, N$, with a source at range r_p and bearing β_p.

of the propagation time

$$\tau_{pn} = (r_p^2 + x_n^2 - 2r_p x_n \cos \beta_p)^{1/2}/v \tag{3.141}$$

with respect to x_n assuming that the $r_p \gg x_n$. As mentioned above, *estimation of the time delay*

$$D_{m,n}(p) = \tau_{pm} - \tau_{pn} \tag{3.142}$$

between the mth and nth sensors $\{m, n = 1, 2, \ldots, N\}$ and the corresponding set of L source range and bearing parameters is a problem of considerable current interest [108–111]. This problem is equivalent to estimating the geometric parameters β_p and r_p, $p = 1, 2, \ldots, L$, for the signal sources [80].

The estimation of either frequency or time delay is discussed herein in terms of operations on the estimated data covariance matrix given, for example, simply by

$$\hat{\mathbf{R}} = \frac{1}{T} \sum_{t=1}^{T} \mathbf{X}(t)\mathbf{X}(t)^H \tag{3.143}$$

However, in the following, the ensemble average form of the data covariance matrix as given in Eq. (3.136) is assumed and the effect of finite averaging time is addressed empirically in the section that considers specifically the range estimation example. Before proceeding, two forms of the data covariance matrix $\mathbf{R}$ and the inverse $\mathbf{R}^{-1}$ are presented. If the $N \times K$ modal matrix $\mathbf{M}$ of $\mathbf{P}$ with $\mathbf{M}_k$ as the kth column and the $K \times K$ diagonalization matrix $\mathbf{\Lambda}$ with λ_k as the kth diagonal element are defined, an enhanced data covariance matrix, $\mathbf{R}(e)$, is defined as

$$\mathbf{R}(e) = e\mathbf{M\Lambda M}^H + \sigma^2\mathbf{I} \tag{3.144}$$

with

$$\mathbf{R}(e)^{-1} = \frac{1}{\sigma^2}\left[\mathbf{I} - \mathbf{M}\left(\mathbf{I} + \frac{\sigma^2}{e}\mathbf{\Lambda}^{-1}\right)^{-1}\mathbf{M}^H\right] \tag{3.145a}$$

$$= \frac{1}{\sigma^2}\left[\mathbf{I} - \sum_{k=1}^{K}\left(1 + \frac{\sigma^2}{e\lambda_k}\right)^{-1}\mathbf{M}_k\mathbf{M}_k^H\right] \tag{3.145b}$$

The scalar parameter e is hereafter referred to as the modal enhancement factor. With respect to the original definition of $\mathbf{R}$ in Eq. (3.136) we have $\mathbf{R}(1) = \mathbf{R}$. However, once the signal modes are identified in terms of estimates of $\mathbf{M}$ and $\mathbf{\Lambda}$ obtained from $\hat{\mathbf{R}}$ in Eq. (3.143), an enhancement factor $e > 1$ can readily be inserted. Next, consider the alternative form

$$\mathbf{R}(e) = e\mathbf{CC}^H + \sigma^2\mathbf{I} \tag{3.146}$$

wherein $\mathbf{P} = \mathbf{CC}^H$ is a Cholesky factorization of the signal covariance matrix $\mathbf{P}$ obtained from $\mathbf{R}$ after the diagonal, uncorrelated noise matrix $\sigma^2 I$ has been removed. An algorithm for this procedure is discussed in a subsequent section. The form

$$\mathbf{R}(e)^{-1} = \frac{1}{\sigma^2}\left[\mathbf{I} - \mathbf{C}\left(\mathbf{C}^H\mathbf{C} + \frac{1}{e}\mathbf{I}\right)^{-1}\mathbf{C}^H\right] \tag{3.147}$$

for the inverse of the data covariance matrix with signal enhancement is referenced in subsequent discussions.

Spectrum Analysis

Two approaches to high-resolution spectrum analysis are presented in the following. Both methods result from a linearly constrained quadratic minimization problem formulation in conjunction with the additional information that, first, the signal vector subspace is of dimension K; second, the noise process is zero mean, independent, and identically distributed; and third, the spectral resolution is to be maximized. The maximization of resolution is equivalent to saying that the enhancement factor, e, is to be made large. This is because a large e is equivalent to a high signal-to-noise ratio, which is a principal factor in resolution improvement.

Minimum-variance distortionless response. For this procedure it is desired to find a linear filter vector $\mathbf{W}$ for the analysis data vector which is a solution to the following constrained minimization problem.

- Minimize:

$$\sigma^2_{\text{MV}} = E\{|\mathbf{W}^H\mathbf{X}(t)|^2 : e\} \tag{3.148a}$$

$$= \mathbf{W}^H\mathbf{R}(e)\mathbf{W} \tag{3.148b}$$

- Maximize: Spectral resolution (i.e., $e \to \infty$)
- Constraints: (a) Distortionless (unit) response requires

$$1 = \mathbf{W}^H\mathbf{D}(\boldsymbol{\theta}) \tag{3.148c}$$

  (b) Signal space of dimension K and uniform independent noise requires Eq. (3.144).

The solution to this problem using Eq. (3.145b) for $\mathbf{R}(e)^{-1}$ is

$$\mathbf{W} = \mathbf{W}_{\text{MV}} \tag{3.149a}$$

$$= \frac{\mathbf{R}(e)^{-1}\mathbf{D}(\boldsymbol{\theta})}{\mathbf{D}(\boldsymbol{\theta})^H\mathbf{R}(e)^{-1}\mathbf{D}(\boldsymbol{\theta})} \tag{3.149b}$$

$$= g(\boldsymbol{\theta}, e)\left[\mathbf{D}(\boldsymbol{\theta}) - \sum_{k=1}^{K} b(k, e)\mathbf{M}_k^H\mathbf{D}(\boldsymbol{\theta})\mathbf{M}_k\right] \tag{3.149c}$$

where

$$g(\boldsymbol{\theta}, e) = \left[N - \sum_{k=1}^{K} b(k, e) \,|\, \mathbf{M}_k^H \mathbf{D}(\boldsymbol{\theta}) \,|^2 \right]^{-1} \tag{3.150}$$

$$b(k, e) = \left(1 + \frac{\sigma^2}{e\lambda_k} \right)^{-1} \tag{3.151}$$

The MV power specuum estimator is defined by Eq. (3.148a) with $\mathbf{W} = \mathbf{W}_{\text{MV}}$, which yields

$$P_{\text{MV}}(\boldsymbol{\theta}) = \frac{1}{\mathbf{D}(\boldsymbol{\theta})^H \mathbf{R}(e)^{-1} \mathbf{D}(\boldsymbol{\theta})} \tag{3.152a}$$

$$= \sigma^2 g(\boldsymbol{\theta}, e) \tag{3.152b}$$

The enhanced minimum-variance (EMV) spectrum estimator is obtained from Eq. (3.152) by taking

$$\lim_{e \to \infty} b(k, e) = 1 \tag{3.153}$$

which gives

$$P_{\text{EMV}}(\boldsymbol{\theta}) = \frac{\sigma^2}{N - \sum_{k=1}^{K} |\mathbf{M}_k^H \mathbf{D}(\boldsymbol{\theta})|^2} \tag{3.154}$$

for maximum resolution. It is noted that Eq. (3.149b) with $e = 1$ is the standard MVDR solution to the minimization requirement subject only to the distortionless response constraint of Eq. (3.148c) and uniform uncorrelated noise assumption. When the signal and noise vector space dimensionality information is exploited, Eqs. (3.149c) and (3.150) result. Finally, the resolution maximization requirement is satisfied when Eq. (3.153) is implemented and results in the estimator of Eq. (3.154).

Maximum entropy. For smoothing and prediction, it is desired to find a filter vector $\mathbf{W}$ which is to be applied linearly to the data vector $\mathbf{X}(t)$ in such a way that the nth element in $\mathbf{X}(t)$, say $X_n(t)$, is estimated in a least-squares sense by a linear combination of the other $N - 1$ elements of $\mathbf{X}(t)$. This objective, in conjunction with both resolution maximization and the a priori knowledge that the data covariance matrix has the structure specified by Eq. (3.144), can be stated as the following constrained optimization problem.

- Minimize: $\sigma_{\text{ME}}^2 = E\{|\,\mathbf{W}^H \mathbf{X}(t)\,|^2 : e\}$ (3.155a)
- Maximize: Spectral resolution (i.e., $e \to \infty$)
- Constraints: (a) Smoothing/prediction at point n in data window requires

$$1 = \mathbf{W}^H \mathbf{1}_n \tag{3.155b}$$

where $\mathbf{1}_n$ is a real N-vector consisting of all zeros except a 1 for the nth element
(b) Signal space of dimensionality K and uniform independent noise requires Eq. (3.122).

The solution to this problem, invoking only the constraint of Eq. (3.155b), is

$$\mathbf{W}_{\text{ME}} = \frac{\mathbf{R}(e)^{-1}\mathbf{1}_n}{\mathbf{1}_n^H\mathbf{R}(e)^{-1}\mathbf{1}_n} \tag{3.156}$$

and with the addition of constraint (b),

$$\mathbf{W}_{\text{ME}} = g(n, e)\left[\mathbf{1}_n - \sum_{k=1}^{K} b(k, e)\mathbf{M}_k^H \mathbf{1}_n \mathbf{M}_k\right] \tag{3.157}$$

where

$$g(n, e) = \left[1 - \sum_{k=1}^{K} b(k, e)\,|\,\mathbf{M}_k^H\mathbf{1}_n\,|^2\right]^{-1} \tag{3.158}$$

Now let the data vector $\mathbf{X}(t)$ contain samples which are uniformly spaced at an interval Δ for either a time sequence as in frequency analysis or a homogeneous spatial signal field as in wavenumber analysis. Both processes are stationary. Spatial stationarity requires strictly that there be no near-field sources for the spatial spectrum analysis application. This apparently limits the present discussion to wavenumber analysis (i.e., bearing estimation) and preclude range spectrum analysis. With these restrictions and $\hat{X}(t-n)$ defined as a least-squares estimate of $X_n(t) = X(t-n)$, the equation

$$\epsilon_n(t) = X(t-n) - \hat{X}(t-n) \tag{3.159}$$

$$= X(t-n) + \sum_{\substack{k=0 \\ k\neq n}}^{N-1} W_k^* X(t-k) \tag{3.160}$$

gives the smoothing error $\epsilon_n(t)$ for $0 < n \leq N-1$ and prediction error if $p = 0$. It is the entropy (uncertainty) in the residual error process $\epsilon_n(t)$ which is maximized as a function of $\mathbf{W}$. The term w_k is the kth element of the filter weight N-vector $\mathbf{W}_{\text{ME}}$. Moreover, if $X(t)$ is modeled as an autoregressive (AR) process generated by an all-pole filter excited by white noise, $\epsilon_n(t)$, with minimum variance $\sigma_{\text{ME}}^2 = [\mathbf{1}_n^H\mathbf{R}^{-1}(e)\mathbf{1}_n]^{-1}$, then the difference equation (3.160) has the z-transform transfer function

$$H_n(z) = z^{-n}\left(1 + \sum_{\substack{k=0 \\ k\neq n}}^{N-1} W_k^* z^{n-k}\right) \tag{3.161a}$$

$$= z^{-n}\sum_{k=0}^{N-1} W_k^* z^{n-k} \qquad (W_n = 1) \tag{3.161b}$$

$$= A_n(z^{-1}) \tag{3.161c}$$

The corresponding ME power spectrum estimate for $X(t)$ is, therefore,

$$P_{\mathrm{ME}}(\boldsymbol{\theta} = \omega) = \left. \frac{[\mathbf{1}_n^H \mathbf{R}(e)^{-1} \mathbf{1}_n]^{-1}}{|A_1(z^{-1})|^2} \right|_{z = e^{j\omega}} \tag{3.162a}$$

$$= \frac{[\mathbf{1}_n^H \mathbf{R}(e)^{-1} \mathbf{1}_n]^{-1}}{|\mathbf{W}_{\mathrm{ME}}^H \mathbf{D}(\omega)|^2} \tag{3.162b}$$

$$= \frac{\mathbf{1}_n^H \mathbf{R}(e)^{-1} \mathbf{1}_n}{|\mathbf{1}_n^H \mathbf{R}(e)^{-1} \mathbf{D}(\omega)|^2} \tag{3.162c}$$

The kth element in $\mathbf{D}(\omega)$ is given by

$$d_k(\omega) = \exp\left[-j\omega(n-k)\Delta\right] \tag{3.163}$$

for frequency analysis with $\omega = 2\pi f$ and

$$d_k(\omega) = \exp\left[-j\omega(n-k)d\right] \tag{3.164}$$

for wavenumber analysis with d equal to a uniform sensor spacing interval and $\omega = 2\pi f \cos \beta/v$. Note that the ME power spectrum estimation process is not strictly realizable without accepting a delay of n samples. This is because the signal model generation process is as illustrated in Fig. 3.20. The z^{-1} term represents the unit-delay operation, which is realizable, and z term represents a unit advance which is not realizable without a process delay. In actuality, a delay is implicit in the ME process realization, which requires that $\mathbf{R}(e)^{-1}$ be estimated with a delay for time averaging.

The final step in the implementation of the enhanced ME process is to substitute the dominant-mode signal covariance matrix constraint expressed by Eq. (3.145b) into Eq. (3.162c) to yield

$$P_{\mathrm{ME}}(\theta) = \frac{1}{g(n, e)\left|1 - \sum_{k=1}^{K} b(k, e)\mathbf{1}_n^H \mathbf{M}_k \mathbf{M}_k' \mathbf{D}(\theta)\right|^2} \tag{3.165}$$

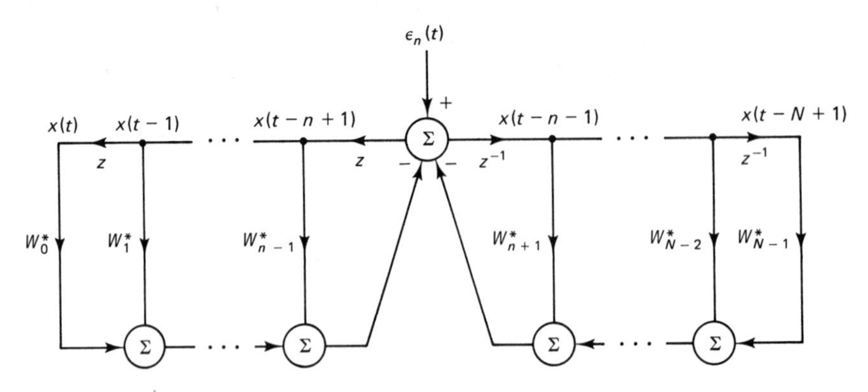

Figure 3.20 Generation of a nonrealizable all-pole process $x(t)$.

with the enhanced version of Eq. (3.165), $P_{EME}(\theta)$, given by taking $b(k, e) = 1$.

We can now generalize the use of $P_{MV}(\theta)$ and $P_{ME}(\theta)$ to less restrictive forms for $\mathbf{D}(\theta)$. Clearly, for $P_{ME}(\theta)$ to be a frequency (wavenumber) spectrum estimator in the strict sense, the uniform sampling interval Δ specified above is required. However, this does not preclude the use in general of Eq. (3.157) for $\mathbf{W}_{ME}$ as a filter for least-squares estimation of $X_n(t)$. Moreover, the use of Eq. (3.165) as a measure of the orthogonality between $\mathbf{W}_{ME}$ and $\mathbf{D}(\omega)$ is still perfectly valid. This orthogonality test point of view is useful whether the exponents in the elements of $\mathbf{D}(\omega)$ exhibit either a uniform sampling interval structure characterized by a linear dependence on the transform variable as for frequency spectrum analysis and wavenumber analysis or a nonlinear dependence on the transform variable. Such nonlinear dependence on the transform variable occurs in the range sepctrum analysis application and with arrays of nonuniformly spaced sensors. To illustrate this, we take the simple case of $K = 1$, where it can be shown that $\mathbf{M}_1 = \mathbf{D}(\theta)/\sqrt{N}$. Using this in Eq. (3.165) gives

$$P_{ME}(\theta) = \frac{1}{g\left|\mathbf{1}_n^H\left[I - \frac{b(1, e)}{N}\mathbf{D}(\theta_0)\mathbf{D}(\theta_0)^H\right]\mathbf{D}(\theta)\right|^2} \tag{3.166}$$

which exhibits a peak value at $\theta = \theta_0$. In fact, Eq. (3.166) becomes infinite when $\theta = \theta_0$ if the $\mathbf{M}_1$ mode is enhanced [i.e., $b(1, e) = 1$]. This is because

$$\begin{aligned}\lim_{e\to\infty} \mathbf{R}(e)^{-1} &= \frac{1}{\sigma^2}\left[\mathbf{I}_N - \frac{1}{N}\mathbf{D}(\theta_0)\mathbf{D}(\theta_0)^H\right] \\ &= \mathbf{R}(\infty)^{-1}\end{aligned} \tag{3.167}$$

and with $\theta = \theta_0$ it is easy to see that the signal transform vector $\mathbf{D}(\theta_0)$ is in the null space of $R(\infty)^{-1}$ causing an infinite response in the power spectrum estimator. This result in no way depends on the linearity of the complex exponent for the elements of $\mathbf{D}(\theta)$.

3.11 RESOLUTION PERFORMANCE

The resolution performance of the MVDR and MEM procedures can be examined at high signal-to-noise ratio, SNR, and long averaging time, T, by using Eq. (3.144) for the estimated data covariance matrix. For a single Gaussian signal source with SNR $= a$, the enhanced covariance matrix can be expressed as

$$\mathbf{R}(e) = ae\mathbf{D}(\theta_0)\mathbf{D}(\theta_0)^H + \mathbf{I} \tag{3.168}$$

with θ_0 equal to either the signal frequency (f), wavenumber (k), or range (r), depending on the particular application. Following the procedures of references [91] and [107], the width of the spectrum response at the 3-dB down points due to this single component signal in a white noise background can be obtained. Specifi-

TABLE 3.1 3-dB RESPONSE WIDTHS FOR HIGH-RESOLUTION ANALYSIS OF FREQUENCY, WAVENUMBER, AND RANGE SPECTRA ASSUMING HIGH SIGNAL-TO-NOISE RATIO (SNR) AND LONG AVERAGING TIME (T)

Type of analysis	Domain of application		
	Frequency	Wavenumber (line array)	Range ($N = 3$)
Conventional (EC)	$\sqrt{6}\,\dfrac{1}{N\pi\Delta}$	$\beta = \pi/2$: $\sqrt{6}\,\dfrac{\lambda}{N\pi d}$; $\beta = 0$: $2\sqrt{\sqrt{6}\,\dfrac{\lambda}{N\pi d}}$	$3\,\dfrac{\lambda}{\pi}\left(\dfrac{r}{d \sin \beta}\right)^2$
Minimum variance (EMV)	$2\sqrt{\dfrac{3}{Nea}\,\dfrac{1}{N\pi\Delta}}$	$\beta = \pi/2$: $2\sqrt{\dfrac{3}{Nea}\,\dfrac{\lambda}{N\pi d}}$; $\beta = 0$: $2\sqrt{2\sqrt{\dfrac{3}{Nea}\,\dfrac{\lambda}{N\pi d}}}$	$\sqrt{\dfrac{6}{ea}\,\dfrac{\lambda}{\pi}}\left(\dfrac{r}{d \sin \beta}\right)^2$
Maximum entropy (EME)	$\dfrac{2}{Nea}\,\dfrac{1}{N\pi\Delta}$	$\beta = \pi/2$: $\dfrac{2}{Nea}\,\dfrac{\lambda}{N\pi d}$; $\beta = 0$: $2\sqrt{\dfrac{2}{Nea}\,\dfrac{\lambda}{N\pi d}}$	$\dfrac{1}{ea}\,\dfrac{\lambda}{\pi}\left(\dfrac{r}{d \sin \beta}\right)^2$

cally, the values of δ for which

$$\frac{P_{\mathrm{EMV}}(\theta_0 + [\delta/2])}{P_{\mathrm{EMV}}(\theta_0)} = 0.5 \tag{3.169a}$$

and

$$\frac{P_{\mathrm{EME}}(\theta + [\delta/2])}{P_{\mathrm{EME}}(\theta_0)} = 0.5 \tag{3.169b}$$

are referred to as the *3-dB response widths* for the enhanced MV and ME spectral analysis processes, respectively. A third estimator,

$$P_{\mathrm{EC}}(\theta_0 + [\delta/2]) = \mathbf{D}(\theta_0 + [\delta/2])^H \mathbf{R}(e) \mathbf{D}(\theta_0 + [\delta/2]) \tag{3.169c}$$

referred to as the *enhanced conventional spectrum estimator*, is also considered. Equation (3.169c) comes from the uniformly weighted data transform

$$P_{\mathrm{EC}}(\theta_0 + [\delta/2]) = E\{|\,\mathbf{D}(\theta_0 + [\delta/2])^H \mathbf{X}(t)\,|^2 : e\} \tag{3.170}$$

in conjunction with the signal and noise model embodied in Eq. (3.136). The 3-dB response widths for the three spectrum estimators given above are summarized in Table 3.1. Table 3.1 includes δ for spectrum analysis, wavenumber analysis, and range analysis. In the range analysis case, the results are restricted to an $N = 3$ sensor array with uniform spacing d between the sensors.

The response width δ is presented herein as a measure of a given spectral analysis method to resolve two closely spaced signal components of equal strength. For a single component in the presence of uncorrelated noise, neither MV nor ME processing of any type can provide a better estimate of the signal frequency (wavenumber, range) location than conventional Fourier processing in terms of unbiased estimator variance. In fact, it is possible for a high-resolution estimator to have a higher variance than the conventional estimator in the single-component case when a limited averaging time is required. It is only in the multiple-component case that the lack of resolution capability can lead to superior performance of the EMV and EME techniques because increased resolution reduces the component of total rms error due to bias. These concepts are illustrated by an example in the final section.

Finally, the dependence of δ on the SNR $= a$ for each of the three spectral analysis techniques is noted. The conventional process response width is independent of SNR, while the MV and ME widths vary inversely as $(\text{SNR})^{1/2}$ and SNR, respectively.

Computation of the Enhanced Spectrum Estimators

Two fundamental approaches are available for the realization of the enhanced minimum-variance (EMV) and maximum-entropy (EME) estimators [65]. The first approach suggests a stochastic gradient search algorithm for adaptive estimation of the K largest eigenvalues and corresponding eigenvectors. This method involves feedback of the modal filter outputs defined by

$$y_k(t) = \mathbf{M}_k^H(t)\mathbf{X}(t), \qquad k = 1, 2, \ldots, K \tag{3.171}$$

which are then correlated with the data vector $\mathbf{X}(t)$ in an attempt to maximize the average value of the instantaneous eigenvalue estimate [8]

$$|y_k(t)|^2 = \mathbf{M}_k^H(t)\mathbf{X}(t)\mathbf{X}^H(t)\mathbf{M}_k \tag{3.172}$$

This objective is accomplished by adjustment of the eigenvector estimates at time t, $\mathbf{M}_k(t)$, $k = 1, 2, \ldots, K$, to maximize Eq. (3.172) subject to the orthogonality

$$\mathbf{M}_k'(t)\mathbf{M}_m(t) = \delta_{km} \qquad (1 \le k, m \le K) \tag{3.173}$$

and normality

$$|\mathbf{M}_k(t)|^2 = 1 \qquad (1 \le k \le K) \tag{3.174}$$

constraints [14]. This approach is computationally superior to the second approach to be outlined subsequently because both storage and multiply/add requirements are proportional only to NK. However, the performance of this gradient search scheme degrades drastically with a decrease in SNR, as might be anticipated with noisy feedback as a component in the realization.

The second scheme is an open-loop, feedforward realization utilizing a direct computation of the data covariance matrix estimate $\mathbf{R}$ using, for example, Eq. (3.143). Given $\mathbf{R}$ variations of subroutine EIGCH for Hermitian matrices contained in the IMSL Library, [112] can be used to obtain all N eigenvalues and eigenvectors of $\mathbf{R}$. Alternatively, because N can be large and the dimensionality K of the signal space is typically small relative to N, numerical techniques based on the power method for eigensystem computation can be used to obtain only the first K eigenvalues and eigenvectors of $\mathbf{R}$ [65]. However, the storage and computation requirements in this direct method are proportional to N^2.

As a variation on the feedforward approach, $\mathbf{R}$ can be assumed to be of the form given by Eq. (3.14b). An estimate of the diagonal component $\sigma^2\mathbf{I}$ can be obtained and subtracted from $\mathbf{R}(e)$. Then the residual matrix

$$\mathbf{P}(e) = \mathbf{R} - \sigma^2\mathbf{I} \tag{3.175}$$

can be factorized using a Cholesky decomposition. This factorization is used in Eq. (3.147), which, in turn, is used in the MV and ME spectral estimators in the form

$$\mathbf{R}^{-1}(e) = \frac{1}{\sigma^2}(\mathbf{I} - \mathbf{C}[\mathbf{C}^H\mathbf{C}]^{-1}\mathbf{C}^H) \tag{3.176}$$

(i.e., with $e = \infty$). Although computationally expedient, the difficulty with this method is that the estimation and removal of the assumed diagonal matrix $\sigma^2\mathbf{I}$ can result in problems when the assumption is invalid. One algorithm, which follows that given in [101], suggests iteratively subtracting an increasingly larger term from the diagonal of $\mathbf{R}$ until the resulting matrix becomes singular. This singular matrix is used as an estimate of $\mathbf{P}$, its rank is determined and Cholesky factorization is performed. In reality, an estimate of $\mathbf{R}$ obtained from a finite time average will not be characterized by the diagonal noise component and multiple eigenvalue $\lambda_k = \sigma^2$ for $K + 1 \le k \le N$ as would be true of the assumed model in Eq. (3.146). Such a modeling error would lead to a premature termination of the iterative diagonal removal procedure above. The result is poor relative enhancement of the dominant signal mode eigenvectors and excessive computation because of the greater than necessary dimensionality of the factorization. In other words, the number of columns in the matrix $\mathbf{C}$ is excessive resulting in signal modes that have not been sufficiently isolated. This shortcoming is typical of all such methods that attempt to enhance the signal by subtracting the uncorrelated noise component on the diagonal of the estimated covariance matrix.

A final variation on the direc. feedforward method is to estimate the orthonormal complement $\{\mathbf{M}_k: k = K + 1, K + 2, \ldots, N\}$ of the dominant modal vectors $\{\mathbf{M}_k: k = 1, 2, \ldots, K\}$. The complement form is based on the fact that

$$\sum_{k=1}^{N} \mathbf{M}_k\mathbf{M}_k^H = \mathbf{I} \tag{3.177a}$$

so that

$$\mathbf{I} = \sum_{k=1}^{K} \mathbf{M}_k \mathbf{M}_k^H + \sum_{k=K+1}^{N} \mathbf{M}_k \mathbf{M}_k^H \tag{3.177b}$$

and

$$\sum_{k=1}^{K} \mathbf{M}_k \mathbf{M}_k^H = \mathbf{I} - \sum_{k=K+1}^{N} \mathbf{M}_k \mathbf{M}_k^H \tag{3.178}$$

With mode enhancement, Eq. (3.145b) with $e = \infty$ and substituting Eq. (3.178) becomes

$$\mathbf{R}(e = \infty)^{-1} = \frac{1}{\sigma^2}\left(\mathbf{I}_N - \sum_{k=1}^{K} \mathbf{M}_k \mathbf{M}_k^H\right) \tag{3.179a}$$

$$= \frac{1}{\sigma^2} \sum_{k=K+1}^{N} \mathbf{M}_k \mathbf{M}_k^H \tag{3.179b}$$

and the EMV and EME estimators can be written as [99–103, 113]

$$P_{\text{EMV}}(\boldsymbol{\theta}) = \frac{\sigma^2}{\sum_{k=K+1}^{N} |\mathbf{D}(\boldsymbol{\theta})^H \mathbf{M}_k|^2} \tag{3.180}$$

and

$$P_{\text{EME}}(\theta) = \frac{\sum_{k=K+1}^{N} |\mathbf{1}_n^H \mathbf{M}_k|^2}{\left|\sum_{k=K+1}^{N} \mathbf{1}_n^H \mathbf{M}_k \mathbf{M}_k^H \mathbf{D}(\boldsymbol{\theta})\right|^2} \tag{3.181}$$

after Eqs. (3.149) and (3.162), respectively. Equations (3.180) and (3.181) are intuitively satisfying when it is noted that the orthonormal complementary set of eigenvectors spans a vector subspace which is orthogonal to the subspace spanned by the dominant eigenvectors. Thus, when $\mathbf{D}(\boldsymbol{\theta})$ lies in the signal subspace, we have $\mathbf{M}_k^H \mathbf{D}(\boldsymbol{\theta}) = 0$ for $K + 1 \leq k \leq N$ and the response functions given in Eqs. (3.180) and (3.181), theoretically at least, can become infinite. The mechanism that precludes such an infinite response is the inability to estimate either the eigenvectors for the dominant modes or their orthonormal complements exactly due to finite averaging time. This issue is examined more completely in the following section.

High-Resolution Range Estimation

To passively locate a source with a linear array of sensors, it is desired to estimate the two-dimensional, range-bearing power spectrum of the array obtained by processing the array sensor output data over the sequence of T samples. Three power spectrum estimators are considered: the uniformly weighted averaged transform,

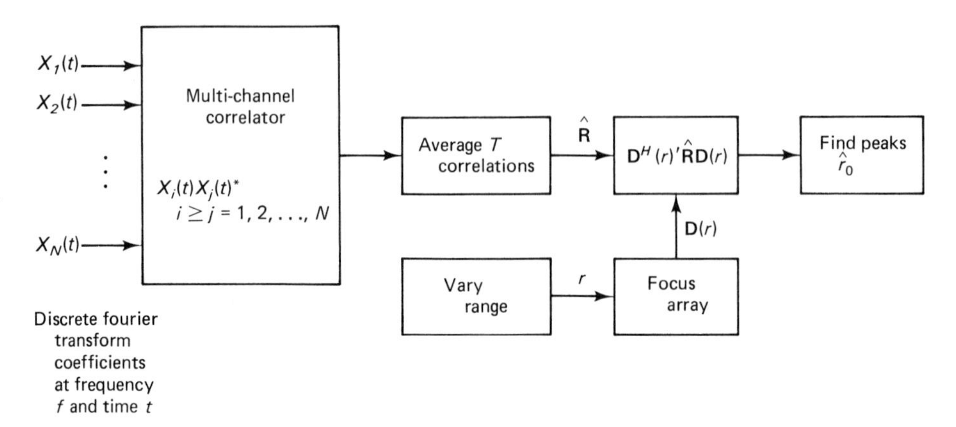

Figure 3.21 Realization of the focused beamformer for estimating the range r of a source.

referred to as the focused beamformer (FB); and the minimum-variance distortionless response (MV) and maximum-entropy (ME) estimators [92, 96, 97]. The array focusing vector $\mathbf{D}(\boldsymbol{\theta}) = \mathbf{D}(r)$ at frequency f and bearing β is defined by Eq. (3.140b). The conventional focused beamformer is formed as illustrated in Fig. 3.21. The MV and ME response functions are generated in a functionally similar manner, with the only difference being in the operations performed on the estimated correlation matrix $\mathbf{R}$ in the beamformer. The input data $X_k(t)$, $n = 1, 2, \ldots, N$, are the discrete Fourier coefficients for a transform interval centered at time t.

The range spectrum estimator is emphasized exclusively in this discussion, with similar results obtainable for the bearing angle dimension in terms of the wavenumber spectrum. The FB, MV, and ME range spectrum estimators are examined herein together with their enhanced dominant-mode decomposition spectra, referred to as the EFB, EMV, and EME, respectively. The issue of primary interest is a comparison of the various range estimator standard deviations to the minimum bound in the presence of both one and two sources.

For a single Gaussian source in a background of spatially uncorrelated and uniform Gaussian noise with zero mean and variance σ^2, the minimum bound on the variance of the range estimator is given by [82],

$$\sigma_r^2 = \left[2N\left(\frac{2\pi}{\lambda}\right)^2 \frac{\sigma_s^2}{\sigma^2} \frac{1}{1 + N(\sigma_s^2/\sigma^2)} \left(\sum_{n=1}^{N} \alpha_n^2 - N\bar{\alpha}^2\right)T\right]^{-1} \tag{3.182}$$

where N = number of sensors
λ = signal wavelength
σ_s^2/σ^2 = signal-to-noise ratio (SNR) at the output of a sensor,

$$\alpha_n = \frac{r - x_n \cos \beta}{(r^2 + x_n^2 - 2x_n r \cos \beta)^{1/2}} \tag{3.183}$$

$$\bar{\alpha} = \frac{1}{N} \sum_{n=1}^{N} \alpha_n \tag{3.184}$$

T = number of Fourier transform samples in the observation interval. As a simple example for simulation of a linear array, the parameters $r = 4000\lambda$, $\beta = 90°$, and $N = 3$ are selected with a uniform spacing between sensors of $d = 75\lambda$. Figure 3.22 gives a comparison of the minimum range estimator standard deviation from Eq. (3.182) and the standard deviation of the location of the spectrum peak for the FB, MV, ME, EMV, and EME range power spectrum estimators as a function of signal-to-noise ratio. It is noted that the actual simulated performance results using a CW signal are slightly better than minimum-bound results. This occurs for two reasons. First, Gaussian complex signal envelope model is used both to obtain the minimum bound result of Eq. (3.182) and to derive the processor structure, whereas in the simulation performed, a constant-frequency coherent signal was used. Second, a range gated estimator was used. This results in the simulation results being consistent in trend with the predicted minimum bound at either high SNR or long averaging time T. However, for low SNR and short averaging time, the range estimator variance appears to saturate. This occurs because the range gate region of allowable spectrum peaks is selected to contain only the correct response lobe for a representative high-SNR, large-T-response function, as illustrated in Fig. 3.23. With a gate width q and either a low SNR or small T, the range estimate histogram generated in the simulation becomes nearly flat with resultant range estimator variance approaching $q^2/12$. A similar problem which uses a time-delay gate to prevent selection of ambiguous lobes of a cross-correlation function for a bandlim-

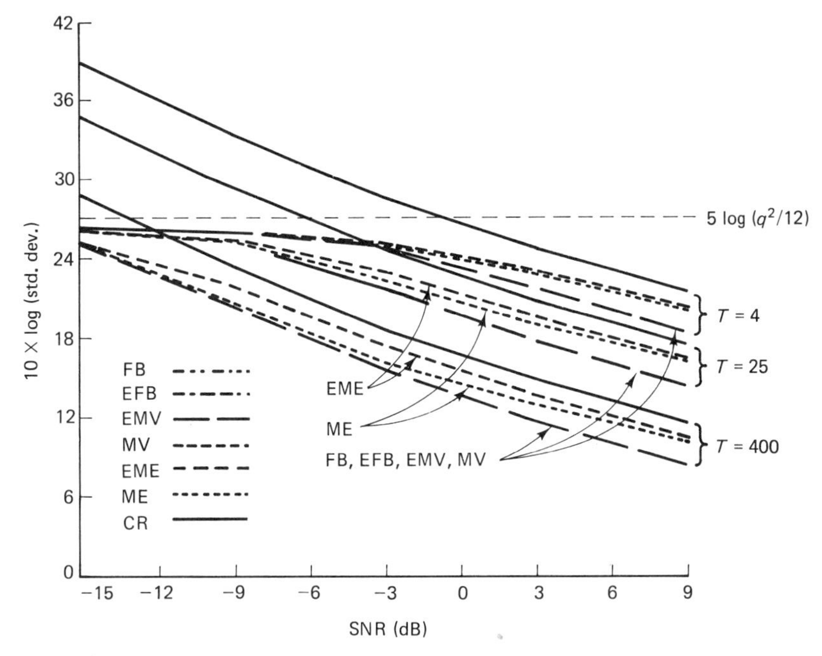

Figure 3.22 Range estimate standard deviation versus SNR for various range power spectrum estimators; $r = 4000\lambda$, $\beta = 90°$, $N = 3$.

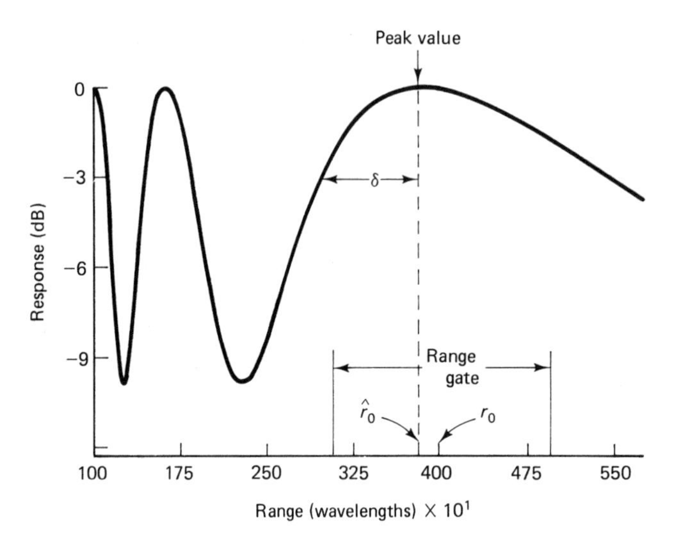

Figure 3.23 Typical focused beamformer range response with a range gate from 3000λ to 5000λ. ($q = 2000\lambda$) and a source located at $r = 4000\lambda$.

ited signal is considered in [111, 114]. The use of a range gate in this case is justified by noting that the location of the true response peak is invariant with respect to λ. This result says that, by frequency diverse range response function estimation, the ambiguous response peaks can be resolved. The most significant feature about the simulation results shown in Fig. 3.22 is that, in terms of range estimator rms error, none of the high-resolution spectrum estimators perform either better or worse than the focused beamformer with the exception of the maximum entropy process. Both the maximum entropy (ME) and enhanced maximum entropy (EME) do marginally worse at high SNR with performance approaching that for the Gaussian signal when, in fact, a coherent signal is present. As SNR decreases, the ME performance approaches that for the FB and MV processes. This is consistent with predictions made by Baggeroer [84] for relative ME performance.

To evaluate the ability of any range power spectrum estimation technique to resolve the spectrum peak resulting from one source and a second peak from a nearby source, two metrics are appropriate. First, the separation of the 3-dB-down points, δ, on the peak due to a single source as shown in Fig. 3.23 is considered. Second, the rms range error for one source in the presence of a second source is established. The 3-dB-down range response peak widths for the techniques under consideration are as given in Table 3.1. The effective dominant-mode enhancement, e, for range estimation is a function of the SNR and observation time T. Figure 3.24(a) ($T = 400$) and 3.24(b) ($T = 25$) give the range response 3-dB-down peak width for the simulation experiment previously discussed. The theoretical ME and MV results from Table 3.1 are included for reference and are seen to be in excellent agreement with experimental results for SNR > -3 dB and long averaging time ($T \geq 25$).

The ability of either an enhanced MV or ME process to improve resolution as indicated by 3-dB response peak width is clearly demonstrated in Fig. 3.24 to be a

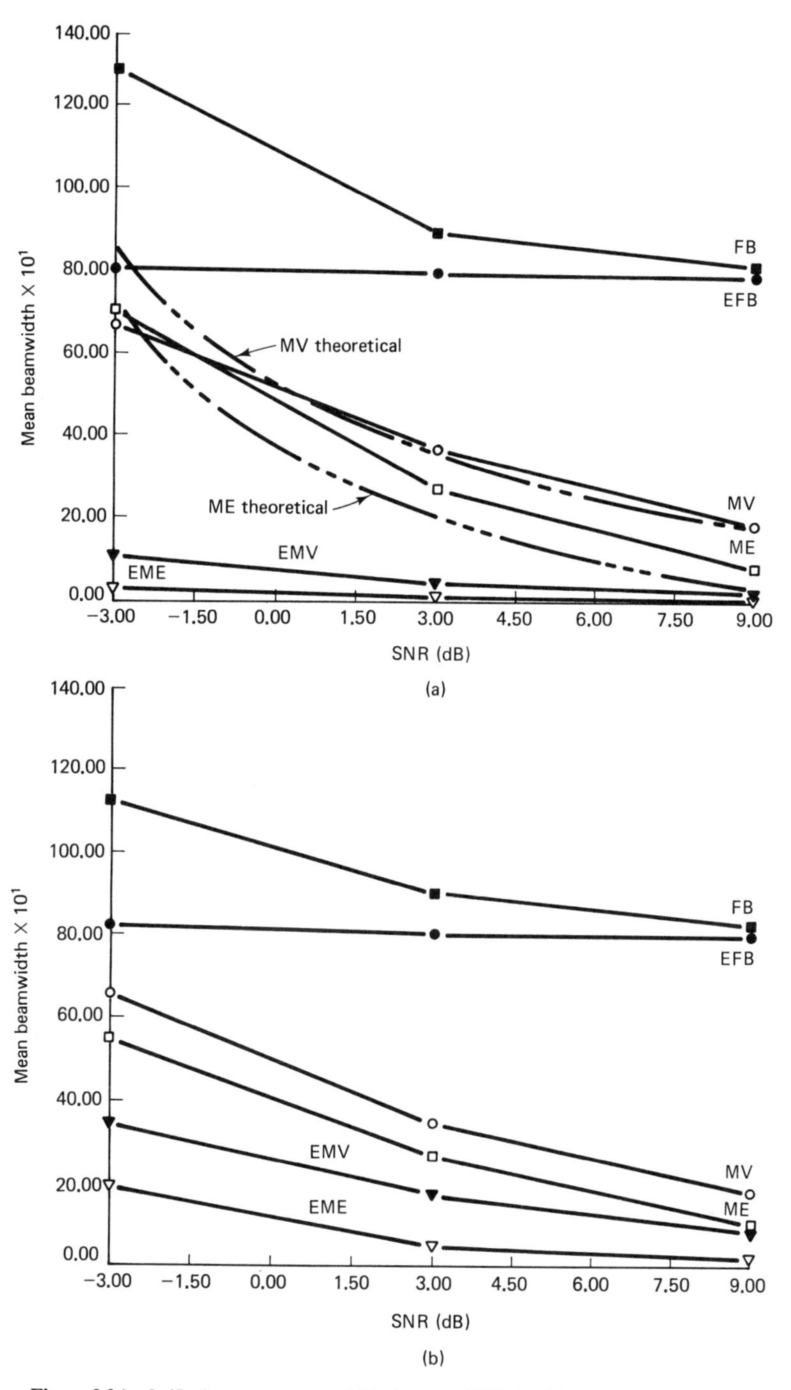

Figure 3.24 3-dB-down response width δ versus SNR for (a) $T = 400$, (b) $T = 25$.

TABLE 3.2 EFFECTIVE ENHANCEMENT FACTOR FOR THE MV AND ME PROCESSES

(a) MV process ($r = 2000\lambda$)

T	SNR 3	SNR 6	SNR 9
4	1.68	1.74	1.74
25	10.12	12.96	6.76
400	156.25	125.00	90.25

(b) ME process ($r = 2000\lambda$)

N	SNR 3	SNR 6	SNR 9
4	1.16	1.39	1.29
25	5.20	4.86	4.00
400	18.66	12.33	9.00

function of both averaging time T and SNR. Table 3.2 gives the empirically determined effective enhancement factors for the MV and ME processes as a function of these two parameters. The results in Table 3.2 indicate that effective enhancement increases with averaging time T. Furthermore, the results indicate that the obtainable relative enhancement of the MV process is greater than that for the ME process. However, the absolute 3-dB peak widths for the ME and EME processes are still marginally less than for the MV and EMV processes, respectively. There is slightly better range estimator rms error performance exhibited by the MV and EMV processes compared to the corresponding ME approaches. Finally, it is important to note that both enhancement processes continue to exhibit impressive resolution capabilities with sufficient averaging time. This remains true even for SNR levels below -3 dB.

For the analysis of range estimator performance when two sources are present, let two sources be located at ranges r_1 and r_2. Let $\mathbf{X}(t)$ be the N-vector of array complex Fourier coefficient time series data. It is assumed that the signals and noise are uncorrelated. The covariance matrix of $\mathbf{X}(t)$ is

$$\mathbf{R} = E[\mathbf{X}(t)\mathbf{X}^H(t)] \tag{3.185}$$

$$= \sigma_s^2[\alpha\mathbf{D}(r_1)\mathbf{D}(r_1)^H + (1-\alpha)\mathbf{D}(r_2)\mathbf{D}(r_2)^H] + \sigma^2\mathbf{I} \tag{3.186}$$

where $0 < \alpha < 1$ indicates the relative power of the sources. Because $\mathbf{X}(t)$ is assumed to be zero mean, complex, and Gaussian, the probability density function of $\mathbf{X}(t)$ is

$$P\left(\frac{\mathbf{X}(t)}{r_1, r_2}\right) = \frac{1}{\pi^H|\mathbf{R}|} \exp\,[-\mathbf{X}(t)^H\mathbf{R}^{-1}\mathbf{X}(t)] \tag{3.187}$$

The minimum variance bound for an unbiased estimator of the range parameter vector $\mathbf{r}$ defined by the transpose $\mathbf{r}^T = [r_1, r_2]$ is cov $(\mathbf{r}) = \mathbf{J}(\mathbf{r})$, where $\mathbf{J}(\mathbf{r})$ is the 2×2 Fisher information matrix [106]. The (i, j)th entry of $\mathbf{J}(\mathbf{r})$ is given by

$$[\mathbf{J}(\mathbf{r})]_{ij} = -E\left[\frac{\partial^2 L(\mathbf{X})/\mathbf{r})}{\partial r_i \, \partial r_j}\right] \tag{3.188}$$

where

$$L(\mathbf{x}/\mathbf{r}) = -T \log \pi - T \log |\mathbf{R}| - \sum_{t=1}^{T} \mathbf{X}(t)'\mathbf{R}^{-1}\mathbf{X}(t) \tag{3.189}$$

Using the identity

$$-E\left[\frac{\partial^2 L(\mathbf{X}/\mathbf{r})}{\partial r_i \, \partial r_j}\right] = \text{tr}\left(\mathbf{R}^{-1} \frac{\mathbf{R}}{\partial r_i} \mathbf{R}^{-1} \frac{\mathbf{R}}{\partial r_j}\right) \tag{3.190}$$

expressions can be derived for J_{11}, J_{22}, and J_{12}, which are functions of the focusing vectors, signal-to-noise ratio, number of sensors, and number of samples. For high signal-to-noise ratio and $x_n \ll r_i \, (i = 1, 2)$ [115].

$$J_{11} \simeq 2N\left(\frac{2\pi}{\lambda}\right)^2 \left(\frac{\alpha\sigma_s^2}{\sigma^2}\right)^2 \left[N \sum_{n=1}^{N} t_{n1}^2 - \left(\sum_{n=1}^{N} t_{n1}\right)^2\right] \frac{1 + \dfrac{(1-\alpha)M\sigma_s^2}{\sigma^2}(1 - |g|^2)}{\text{DV}} \tag{3.191}$$

$$J_{22} \simeq 2N\left(\frac{2\pi}{\lambda}\right)^2 \left[\frac{(1-\alpha)\sigma_s^2}{\sigma^2}\right]^2 \left[N \sum_{n=1}^{N} t_{n2}^2 - \left(\sum_{n=1}^{N} t_{n2}\right)^2\right] \frac{1 + \dfrac{\alpha M\sigma_s^2}{\sigma^2}(1 - |g|^2)}{\text{DV}} \tag{3.192}$$

$J_{12} \simeq 0$

where

$$t_{ni} = \frac{r_i - X_n \cos \beta_i}{\sqrt{r_i^2 + X_n^2 - 2r_i X_n \cos \beta_i}}, \qquad n = 1, 2, \ldots, N \quad i = 1, 2 \tag{3.193}$$

$$g = \frac{1}{N} \sum_{n=1}^{N} \exp\left[-j2\pi \frac{c}{\lambda}(\tau_{1n} - \tau_{2n})\right] \tag{3.194}$$

$$\text{DV} = \left(1 + \frac{\alpha N\sigma_s^2}{\sigma^2}\right)\left[1 + \frac{(1-\alpha)N\sigma_s^2}{\sigma^2}\right] - \frac{\alpha(1-\alpha)M^2|g|^2}{\sigma^4} \tag{3.195}$$

The remaining terms have previously been defined for the minimum-variance bound for a single source. Using J_{11} and J_{22}, it follows that the minimum-variance bounds for estimators of r_1 and r_2 are J_{11}^{-1} and J_{22}^{-1}, respectively. These expressions provide insight into the performance of a multiple-source range estimation process. For example, if $g = 0$, the minimum-variance bounds for r_1 and r_2 are the same as they would be in the single-source case in Eq. (3.152). If $|g| > 0$, it can be shown that the minimum-variance bounds for r_1 and r_2 will be larger. In fact, J_{11}^{-1} and J_{22}^{-1} are monotonically increasing functions of $|g|$.

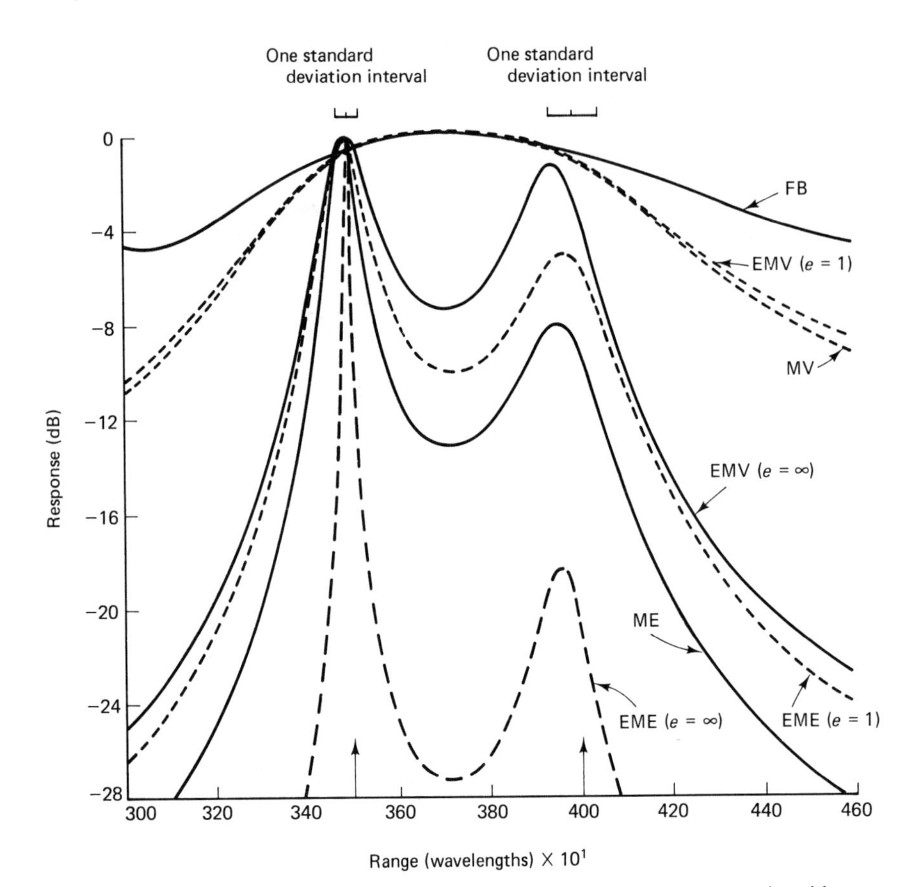

Figure 3.25 Typical range spectrum response for a two-source example with SNR = 9 dB, $T = 100$, $N = 5$, $d = 75\lambda$, $r_1 = 3500\lambda$, $r_2 = 4000\lambda$.

As a final illustration of range high-resolution performance with two sources present, a typical set of range response functions from a simulation run is presented in Fig. 3.25. The simulation parameters are SNR = 9 dB, $T = 100$, $N = 5$, $d = 75\lambda$, $r_1 = 3500\lambda$, and $r_2 = 4000\lambda$. The *single-source* range estimate, one-standard-deviation interval is shown above the individual source ranges centered at the true range value. All response functions resolve the two sources except the FB, MV, and EMV with $e = 1$. The enhanced processes used $K = 2$ for the number of dominant modes for which the eigenvalues and eigenvectors are estimated. The response level for the source at 4000λ is lower than for that at $r_1 = 3500\lambda$ because the reduced resolution at r_2 manifests itself as a range-distributed source. Clearly, the enhanced processes with $e = \infty$ exhibit substantial resolution capability. The inability to resolve the two sources by the FB, MV, and EMV ($e = 1$) processes results in a critical rms bias error of 200λ to 300λ suggesting that only a single source is present.

We have presented two high-resolution spectral analysis procedures based on minimum variance (MV) and maximum entropy (ME). These procedures exhibit superior resolution performance relative to either the MV or ME spectrum estimators, with ultimate performance limited only by observation time. Such performance should have substantial impact in spectral analysis applications that are estimator bias limited as opposed to estimator random fluctuation limited.

3.12 TWO-DIMENSIONAL, SPACE–TIME FOURIER TRANSFORM

The simplest array geometry context within which to introduce two-dimensional Fourier processing is that of the line array of N sensors which are uniformly spaced at a distance d. Let $x_n(t, r)$ be the tth time sample in the rth time interval of T samples from the nth sensor output. The two-dimensional space-time (TDST) Fourier transform for the space–time sequence $x_n(t, r)$, $t = 0, 1, \ldots, T-1$ and $n = 1, 2, \ldots, N$, is given by

$$X(k, \omega: r) = \frac{1}{NT} \sum_{t=0}^{T-1} \sum_{n=1}^{N} W_T(t) W_S(n) x_n(t, r) \exp\left[-j2\pi\left(\frac{\omega t}{T} + \frac{kn}{N}\right)\right] \tag{3.196}$$

where $W_T(t)$ = temporal sequence window function ($0 \le t \le T-1$)
$W_S(n)$ = spatial sequence window function ($1 \le n \le N$)
ω = temporal frequency index
k = spatial frequency index

Two-dimensional fast Fourier transform (2D-FFT) techniques [116] are applicable to the computation of Eq. (3.178).

As time advances, it is desirable for detection purposes to estimate the TDST power spectral density $P(k, \omega)$ by an operation such as exponential averaging

$$P_r(k, \omega) = (1 - v)P_{r-1}(k, \omega) + v\,|X(k, \omega: r)|^2 \tag{3.197}$$

for example. This data could then be displayed to an operator as a function of wavenumber number index, k, and frequency index, ω. An example of such a wavenumber, frequency display-generated from an actual sonar line array data, is presented in Fig. 3.26. The region inside the $\theta(k) = \pm 90°$ cone in this figure corresponds to acoustically radiating energy in the medium for which the relationship

$$\frac{k}{N} = \frac{fd \cos \theta(k)}{v} \tag{3.198}$$

$$= \frac{dv(k)}{2\pi} \tag{3.199}$$

is valid. f is the discrete frequency. The region outside the cone corresponds to the region of nonacoustically radiating, mechanically transmitted energies in the array which are in the high-wavenumber region by virtue of lower propagation velocities.

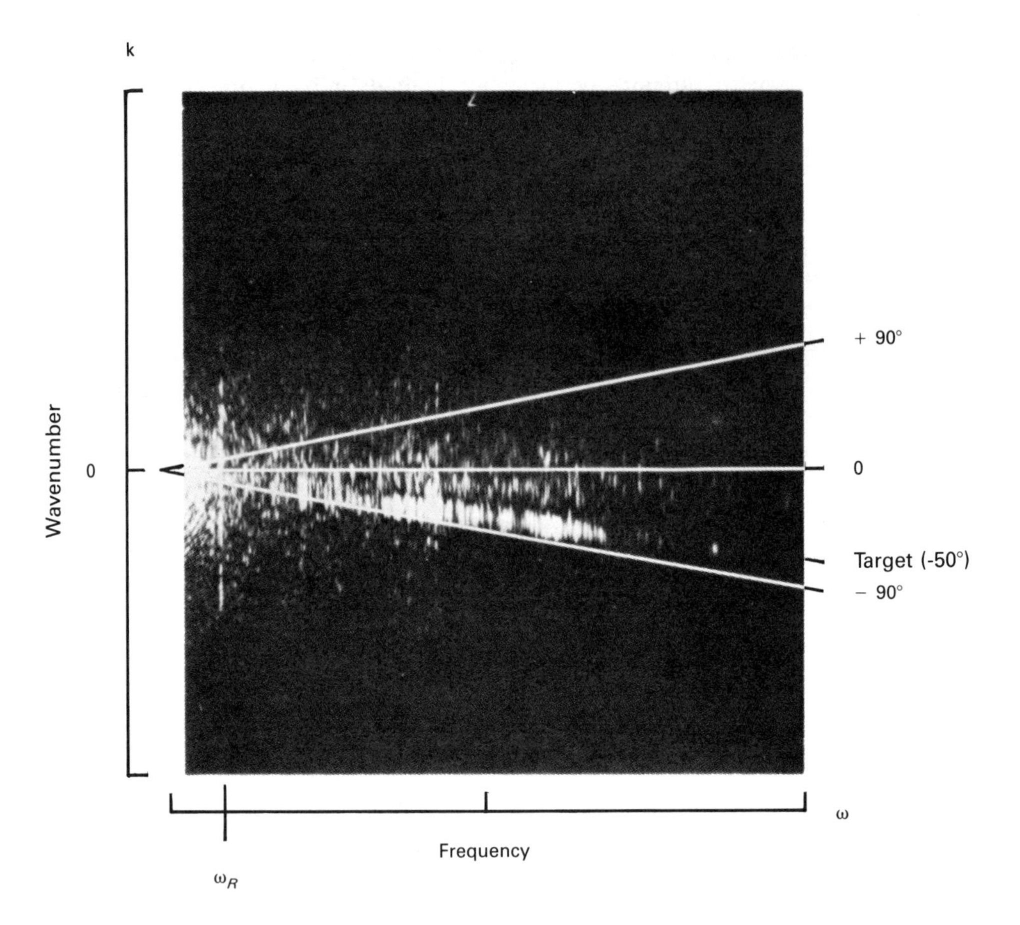

Figure 3.26 Wavenumber frequency display from a sonar array two-dimensional space–time array processing system.

Evidence of a high-wavenumber-noise, discrete frequency component resulting from a mechanical resonance in the array structure can be seen in Fig. 3.26 at $\omega = \omega_R$. It is also possible to see an indication of TBL-induced high-wavenumber-noise "stripes" which vary in roughly a linear fashion in the lower-frequency range. Both of these high-wavenumber components exhibit energy which is aliased into the radiating energy k–ω region and, thereby, degrade detection.

The restriction of TDST Fourier processing techniques to uniformly spaced arrays has been considered by Mucci and Lasky [117], and an approach to noise cancellation within the k–ω process is discussed by Scharf [86]. Moreover, Tufts and Kumaresan have considered the application of high-resolution spectrum analysis to the two-dimensional array geometry [89, 90]. Wax, Shan and Kailath also have recently extended the orthogonal eigenvector decomposition techniques to simultaneous frequency wavenumber spectrum analysis [118].

3.13 SUMMARY

The advent of digital signal processing in sonar has resulted in capabilities for the simultaneous formation of hundreds of directional beams from arrays with several hundred sensors over bandwidths spanning more than five octaves. Digital techniques have allowed the spatial counterpart of the frequency filter to become ever more sophisticated in its ability to spatially locate signal sources. It can be said that one of the most significant developments that digital techniques have made possible is the multibeam sonar. The increased time for which a beam is directed at a source provided by multibeam sonars compared with scanned systems has resulted in such a tremendous increase in detection sensitivity that the masking of low-level sources by either high-level interfering sources or other coherent-noise mechanisms has become an issue of concern. In this regard, the recent trend has been to relax the assumptions required concerning the array noise environment. This process has been made possible by advances in both digital beamformer fixed sidelobe-level design techniques and the advent of adaptivity for data-dependent sidelobe-level control. The difficulty with these seemingly superior array processing schemes is that a similar relaxation of the definition of a signal has not been permitted. In fact, the less prior knowledge that is assumed about the noise, the more it is necessary to know a priori about the signal so that it can be extracted from a background about which potentially nothing is known. In this requirement of more complete knowledge of signal spatial structure, nature does not always cooperate. Nature provides the sonar array with an inhomogeneous medium which, in conjunction with the surface and bottom interfaces, yields both multipath propagation and nonplanar wavefronts. To complicate the situation further, the sensitivities in hydrophone components and preamplifiers are seldom uniform, to say nothing of the complications introduced by the deformable shapes which characterize some types of large aperture sonar arrays.

At this point in time, the technology exists to implement the fully optimum and adaptive array processor. However, in many cases it is difficult to justify this ultimate processor as a general approach to sonar array processing, particularly if less expensive suboptimum noise cancellation techniques serve adequately. Finally, if the intent of future sonar systems is to make ever more precise measurements on signal wavefront characteristics, it will be expedient to incorporate ever-more-sophisticated mechanisms in terms of underwater sound propagation characteristics into the spatial models which are implicit in a sonar array processing system. Accordingly, it could very well be that sonar array processors which are more adaptive to signal field structure variations than noise uncertainties may well define the next step in sonar array processor development.

REFERENCES

1 B. Blesser and J. M. Kates, "Digital Processing in Audio Signals," in *Application of Digital Signal Processing*, ed. A. V. Oppheim, Prentice-Hall, Englewood Cliffs, N.J., 1978.

2 S. Pasupathy and A. N. Venetsanopoulos, "Optimum Active Array Processing Structure and Space–Time Factorability," IEEE Trans. Aerosp. Electron. Syst., Vol. AES-10, No. 6, pp. 770–779, Nov. 1974.

3 A. B. Baggeroer, "Sonar Signal Processing," in *Applications of Digital Signal Processing*, ed. A. V. Oppenheim, Prentice-Hall, Englewood Cliffs, N.J., 1978.

4 F. Bryn, "Optimal Signal Processing of Three-Dimensional Arrays Operating on Gaussian Signals and Noise," J. Acoust. Soc. Am., Vol. 34, pp. 289–297, 1962.

5 P. L. Stocklin, "Space–Time Sampling and Likelihood Ratio Processing in Acoustic Pressure Fields," J. Br. IRE, pp. 79–91, July 1963.

6 H. Mermoz, "Filtrage adapté et utilisation optimale d'une antenne," Signal Processing with Emphasis on Underwater Acoustics, Proc. NATO Advanced Study Institute, 1964, pp. 163–294.

7 S. W. W. Shor, "Adaptive Technique to Discriminate against Coherent Noise in a Narrowband System," J. Acoust. Soc. Am., Vol. 39, pp. 74–78, 1966.

8 H. L. Van Trees, "A Unified Approach to Optimum Array Processing," Proc. 1966 Hawaii Symp. Signal Process.

9 B. Widrow et al., "Adaptive Antenna Systems," Proc. IEEE, Vol. 55, pp. 2143–2159, 1967.

10 L. Griffiths, "A Simple Adaptive Algorithm for Real-Time Processing in Antenna Arrays," Proc. IEEE, Vol. 57, pp. 1696–1704, 1969.

11 V. C. Anderson, "DICANNE, a Realizable Adaptive Process," J. Acoust. Soc. Am., Vol. 45, pp. 398–405, 1969.

12 V. C. Anderson and P. Rudnick, "Rejection of a Coherent Arrival at an Array," J. Acoust. Soc. Am., Vol. 45, p. 406–410, 1969.

13 H. Cox, "Array Processing against Interference," Proc. Symp. Inf. Process., Purdue University, Lafayette, Ind., Apr. 1969, pp. 453–464.

14 N. L. Owsley, "A Recent Trend in Adaptive Spatial Processing for Sensor Arrays: Constrained Adaptation," in *Signal Processing*, ed. J. W. R. Griffiths et al., Academic Press, New York, 1973, pp. 591–604.

15 J. B. Lewis and P. M. Schultheiss, "Optimum and Conventional Detection Using a Linear Array," J. Acoust. Soc. Am., Vol. 49, pp. 1083–1091, 1971.

16 S. Pasupathy, "Optimum Spatial Processing of Passive Solar Signals," IEEE Trans. Aerosp. Electron. Syst., Vol. 14, No. 1, pp. 158–164, Jan. 1978.

17 N. L. Owsley, "An Overview of Optimum Adaptive Control in Sonar Array Processing," in *Applicctions of Adaptive Control*, ed. K. S. Narendra and R. V. Monopoli, Academic Press, New York, 1980, pp. 131–164.

18 L. J. Griffiths, "An Adaptive Beamformer Which Implements Constraints Using an Auxiliary Array Processor," in *Aspects of Signal Processing*, ed. G. Tacconi, D. Reidel, Boston, 1977.

19 S. Pasupathy, "Optimal and Conventional Array Processing," Can. Electr. Eng. J., Vol. 3, No. 1, pp. 27–31, 1978.

20 S. P. Applebaum, "Adaptive Arrays," Syracuse University Research Corp., Report Spl. TR 66-1, 1966; reprinted in IEEE Trans. Antennas Propag., Vol. AP-24, No. 5, pp. 585–598, Sept. 1976.

21 H. Cox, "Resolving Power and Sensitivity to Mismatch of Optimum Array Processors," J. Acoust. Soc. Am., Vol. 54, No. 3, pp. 771–785, 1973.

22 O. L. Frost, "An Algorithm for Linearly Constrained Adaptive Array Processing," Proc. IEEE, Vol. 60, pp. 926–935, Aug. 1972.

23 H. Cox, "Sensitivity Considerations in Adaptive Beamforming," in *Signal Processing*, ed. J. W. R. Griffiths et al., Academic Press, New York, 1973.

24 A. M. Vural, "Effects of Perturbation on the Performance of Optimum/Adaptive Arrays," IEEE Trans. Aerosp. Electron. Syst., Vol. AES-15, No. 1, pp. 76–87, Jan. 1979.

25 H. Mermoz, "Complementarity of Propagation Model Design with Array Processing," in *Aspects of Signal Processing*, ed. G. Tacconi, D. Riedel, Boston, 1977.

26 W. R. Hahn and S. A. Tretter, "Optimum Processing for Delay-Vector Estimation in Passive Signal Arrays," IEEE Trans. Inf. Theory, Vol. IT-19, No. 5, pp. 608–614, Sept. 1973.

27 W. R. Hahn, "Optimum Signal Processing for Passive Sonar Range and Bearing Estimation," J. Acoust. Soc. Am., Vol. 58, pp. 201–207, July 1975.

28 N. L. Owsley, "Wavefront-Array Shape Matching by Adaptive Focusing," Proc. Time Delay Estimation Appl. Conf., Naval Post Grad. School, Monterey, Calif., May 1979.

29 N. L. Owsley, "Spectral Set Extraction," in *Aspects of Signal Processing*, ed. G. Tacconi, D. Reidel, Boston, 1977.

30 G. Bienvenu, "Influence of the Spatial Coherence of the Background Noise on High Resolution Passive Methods," Proc. IEEE ICASSP, Washington, D. C., Apr. 1979, pp. 306–309.

31 R. Kumaresan and D. Tufts, "Estimating the Angles of Arrival of Multiple Plane Waves," IEEE Trans. Aerosp. Electron. Syst., Vol. AES-19, No. 1, pp. 134–139, Jan. 1983.

32 IEEE Proc. 1983 Int. Conf. Acoust. Speech Signal Process., Session 19, Time Delay Estimation and Source Location, Boston, Apr. 15, 1983.

33 P. M. Schultheiss and E. Weinstein, "Estimation of Differential Doppler Shifts," J. Acoust. Soc. Am., Vol. 66, No. 5, pp. 1412–1419, Nov. 1979.

34 E. Weinstein, "Optimal Source Localization and Tracking," IEEE Trans. Acoust. Speech Signal Process., ASSP-30, No. 1, pp. 69–76, Feb. 1982.

35 E. H. Scanell, Jr. and G. C. Carter, "Confidence Bounds for Magnitude Squared Error Coherence Estimates," Proc. IEEE ICASSP, Tulsa, Okla., pp. 665–673, Apr. 1978.

36 R. Urick, *Principles of Underwater Sound*, 2nd ed., McGraw-Hill, New York, 1975, Chap. 11.

37 H. Cox, "Line Array Performance When the Signal Coherence Is Spatially Dependent," J. Acoust. Soc. Am., Vol. 54, No. 6, pp. 1743–1746, 1973.

38 R. G. Pridham and R. A. Mucci, "Digital Interpolation Beamforming for Low-Pass and Bandpass Signals," Proc. IEEE, Vol. 67, No. 6, pp. 904–919, June 1979.

39 A. V. Oppenheim and R. W. Schafer, *Digital Signal Processing*, Prentice-Hall, Englewood Cliffs, N.J., pp. 110–115, 1975.

40 W. G. Fish, J. F. Law, A. A. Lesick, N. L. Owsley, and E. L. Walters, "A Frequency Domain Adaptive Beamformer Implementation," Naval Underwater Systems Center Technical Report 5255, New London, Conn., June 21, 1976.

41 M. Dentino, J. McCool, and B. Widrow, "Adaptive Filtering in the Frequency Domain," Proc. IEEE, Vol. 66, No. 12, pp. 1658–1659, Dec. 1978.

42 N. J. Bershad and P. L. Feintuch, "Correlation Function for the Weight Sequence of the Frequency Domain Adaptive Filter," IEEE Trans. Acoust. Speech Signal Process., Vol. ASSP-30, No. 5, pp. 801–804, Oct. 1982.

43 A. M. Vural, "A Comparative Performance Study of Adaptive Array Processors," Proc. IEEE ICASSP, Hartford, Conn., pp. 695–700, Apr. 1977.

44 A. M. Vural, "Effects of Perturbations on the Performance of Optimum/Adaptive Arrays," IEEE Trans. Aerosp. Electron. Syst., Vol. AES-15, No. 1, pp. 76–87, Jan. 1979.

45 S. Applebaum and D. J. Chapman, "Adaptive Arrays with Main Beam Constraints," IEEE Trans. Antennas Propag., Vol. 24, No. 5, pp. 650–662, Sept. 1976.

46 V. Anderson, "DICANNE, a Realizable Adaptive Process," J. Acoust. Soc. Am., Vol. 45, pp. 398–405, 1969.

47 R. A. Monzingo and T. W. Miller, *Introduction to Adaptive Arrays*, Wiley-Interscience, New York, 1980.

48 J. Capon, "High-Resolution Frequency-Wavenumber Spectrum Analysis," Proc. IEEE, Vol. 57, pp. 1408–1418, Aug. 1969.

49 J. Capon, R. Greenfield, and R. J. Kolker, "Multidimensional Maximum—Likelihood Processing of a Large Aperture Seismic Array," Proc. IEEE, Vol. 55, No. 2, pp. 192–211, Feb. 1967.

50 B. Widrow, J. R. Glover, and J. M. McCool, "Adaptive Noise Cancelling: Principles and Applications," Proc. IEEE, Vol. 63, pp. 1692–1716, Dec. 1975.

51 C. W. Jim, "A Comparison of Two LMS Constrained Optimal Array Structures," Proc. IEEE, Vol. 65, No. 12, pp. 1730–1731, Dec. 1977.

52 B. Widrow, "Adaptive Filters," in *Aspects of Network and System Theory*, ed. R. Kalman and N. DeClaris, Holt, Rinehart and Winston, New York, pp. 563–587, 1971.

53 N. L. Owsley, "Noise Cancellation in the Presence of Correlated Signal and Noise," Naval Underwater Systems Center Technical Report 4639, New London, Conn., Jan. 11, 1974.

54 W. S. Hodgkiss, "Adaptive Array Processing: Time vs. Frequency," Proc. IEEE ICASSP, Washington, D.C., pp. 282–285, Apr. 1979.

55 C. Giraudon, "Results on Active Sonar Optimum Array Processing," in *Signal Processing*, ed. J. W. R. Griffiths et al., Academic Press, New York, 1973.

56 H. Mermoz, "Modularité du traitement adaptif d'antenne," Ann. Telecommun., Vol. 29, No. 1-2, 1974.

57 I. S. Reed, J. D. Mallet, and L. E. Brennan, "Rapid Convergence Rate in Adaptive Arrays," IEEE Trans. Aerosp. Electron. Syst., Vol. AES-10, No. 6, pp. 853–864, Nov. 1974.

58 L. E. Brennan, J. D. Mallet, and I. S. Reed, "Adaptive Arrays in Airborne MTI Radar," IEEE Trans. Antennas Propag., Vol. AP-24, No. 5, pp. 607–615, Sept. 1976.

59 E. B. Lunde, "The Forgotten Algorithm in Adaptive Beamforming," in *Aspects of Signal Processing*, ed. G. Tacconi, Part 1, D. Reidel, Boston, 1977.

60 C. Giraudon, "Optimum Antenna Processing: A Modular Approach," in *Aspects of Signal Processing*, ed. G. Tacconi, Part 2, D. Reidel, Boston, 1977.

61 L. E. Brennan, E. L. Pugh, and I. S. Reed, "Control Loop Noise in Adaptive Array Antennas," IEEE Trans. Aerosp. Electron. Syst., Vol. AES-7, No. 2, pp. 254–262, Mar. 1971.

62 J. Capon and N. R. Goodman, "Probability Distributions for Estimators of the Frequency-Wavenumber Spectrum," Proc. IEEE, pp. 1785–1786, Oct. 1970.

63 M. Marcus, *Basic Theorems in Matrix Theory*, NBS Applied Mathematics Series 57, National Bureau of Standards, Washington, D.C., pp. 15–18, Jan. 22, 1967.

64 G. E. Forsythe, "Today's Computational Methods of Linear Algebra," SIAM Rev., Vol. 9, No. 3, July 1967.

65 N. L. Owsley, "Adaptive Data Orthogonalization," Proc. IEEE ICASSP, Tulsa, Okla., pp. 109–112, Apr. 1978.

66 W. D. White, "Cascade Processors for Adaptive Antennas," IEEE Trans. Antennas Propag., Vol. AP-24, No. 5, pp. 575–581, Sept. 1976.

67 B. Widrow et al., "Stationary and Nonstationary Learning Characteristics of the LMS Adaptive Filter," Proc. IEEE, Vol. 64, No. 8, pp. 1151–1162, Aug. 1976.

68 J. McCool, "A Constrained Adaptive Beamformer Tolerant of Array Gain and Phase Errors," in *Aspects of Signal Processing*, ed. G. Tacconi, D. Reidel, Boston, 1977.

69 P. Feintuch, "An Adaptive Recursive LMS Filter," Proc. IEEE, Vol. 64, pp. 1622–1624, Nov. 1976.

70 L. J. Griffiths, "Adaptive Structures for Multiple-Input Noise Cancelling Applications," Proc. IEEE ICASSP, Washington, D.C., pp. 925–928, Apr. 1979.

71 L. J. Griffiths, "A Adaptive Lattice Structure for Noise-Cancelling Applications," Proc. IEEE ICASSP, Tulsa, Okla., pp. 87–91, Apr. 1978.

72 D. W. Jorgensen and G. Maidanik, "Response of a System of Point Transducers to Turbulent Boundary-Layer Pressure Field," J. Acoust. Soc. Am., Vol. 43, pp. 1390–1394, 1968.

73 R. M. Kennedy, "Cancellation of Turbulent Boundary-Layer Pressure Fluctuations," Naval Underwater System Center Technical Report 3006, Sept. 10, 1970, New London, Conn.

74 N. L. Owsley, and J. W. Fay, "Optimum Turbulent Boundary-Layer Noise Suppression with Suboptimum Realizations," J. Acoust. Soc. Am., Vol. 66, No. 5, pp. 1404–1411, Nov. 1979.

75 G. C. Carter and C. H. Knapp, "Time-Delay Estimation," Proc. IEEE ICASSP, pp. 357–361, Apr. 1976.

76 H. J. Young, "Underwater Sound Arrival Angle Estimation by Multiple Cross-Correlation Measurements," Proc. IEEE ICASSP, Tulsa, Okla., pp. 659–664, Apr. 1978.

77 Y. T. Chan, R. B. Hattin, and J. B. Plant, "The Least Squares Estimation of Time Delay and Its Use in Signal Detection," Proc. IEEE ICASSP, Tulsa, Okla., pp. 665–669, Apr. 1978.

78 G. C. Carter, "The Role of Coherence in Time Delay Estimation," in *Aspects of Signal Processing*, ed. G. Tacconi, D. Reidel, Boston, pp. 251–257, 1977.

79 W. R. Hahn, "Target Location by Wavefront Curvature Measurement," Proc. Time Delay Estim. Appl. Conf., U.S. Post Graduate School, Monterey, Calif., May 1979.

80 N. L. Owsley and G. R. Swope, "Time Delay Estimation in a Sensor Array," IEEE Trans. Acoust. Speech Signal Process., Vol. ASSP-29, No. 3, pp. 519–523, June 1981.

81 V. H. MacDonald and P. M. Schultheiss, "Optimum Passive Bearing Estimation in a Spatially Incoherent Noise Environment," J. Acoust. Soc. Am., Vol. 46, pp. 37–43, 1969.

82 W. J. Bangs and P. M. Schultheiss, "Space–Time Processing for Optimal Parameter Estimation," in *Signal Processing*, ed. J. W. R. Griffiths et al., Academic Press, New York, 1973.

83 P. M. Schultheiss, "Locating a Passive Source with Array Measurements: A Summary of Results," Proc. IEEE ICASSP, Washington, D.C., pp. 967–970, Apr. 1979.

84 A. B. Baggeroer, "Confidence Intervals for Maximum Entropy Spectral Estimates," in *Aspects of Signal Processing*, ed. G. Tacconi, D. Reidel, Boston, pp. 617–630, 1977.

85 J. P. Burg, "The Relationship between Maximum Entropy Spectra and Maximum Likelihood Spectra," Geophysics, Vol. 37, No. 2, pp. 375–376, Apr. 1972.

86 L. Scharf, "Optimum and Adaptive Array Processing in Frequency-Wavenumber Space," Proc. Underwater Sound Advisory Group (USAG) Workshop on Multidimensional Analysis of Acoustic Fields, Woods Hole, Mass., Oct. 1973.

87 N. Yen, "Analytic Expansion Technique for Ambient Noise Directionality Determination," Proc. 96th Mtg. Acoust. Soc. Am., 1978.

88 D. W. Tufts and R. Kumaresan, "Singular Value Decomposition and Spectral Analysis," Proc. First IEEE ASSP Workshop on Spectral Estimation, McMaster University, Hamilton, Ont., Aug. 18, 1981, paper 6.4.1.

89 D. W. Tufts and R. Kumaresan, "Estimation of Frequencies of Multiple Sinusoids: Making Linear Prediction Perform Like Maximum Likelihood," Proc. IEEE, Vol. 70, No. 9, pp. 975–989, Sept. 1982.

90 R. Kumaresan and D. W. Tufts, "Estimating the Angles of Arrival of Multiple Plane Waves," IEEE Trans. Aerosp. Electron. Syst., Vol. AES-19, No. 1, pp. 134–138, Jan. 1983.

91 R. T. Lacoss, "Data Adaptive Spectral Analysis Methods," Geophysics, Vol. 36, No. 4, pp. 661–675, Aug. 1971.

92 N. L. Owsley and G. Swope, "High Resolution Range Estimation with a Linear Array," Proc. IEEE EASCON '82, Washington, D.C., Sept. 1982.

93 A. Van den Bos, "Alternative Interpretation of Maximum Entropy Spectral Analysis," IEEE Trans. Inf. Theory (Corresp.), Vol. 17, pp. 493–494, July 1971.

94 J. P. Burg, "Maximum Entropy Spectrum Analysis," 37th Annu. Mtg. Soc. Explor. Geophysicists, Oklahoma City, 1967.

95 H. Mermoz, "Complementarity of Propagation Model Design with Array Processing," in *Aspects of Signal Processing*, Part 2, ed. G. Tacconi, D. Reidel, Boston, pp. 463–468, 1977.

96 N. L. Owsley, "Modal Decomposition of Data Adaptive Spectral Estimates," Yale University Workshop on Applications of Adaptive Systems Theory, ed. K. S. Narendra, May 1981.

97 N. L. Owsley and J. F. Law, "Dominant Mode Power Spectrum Estimation," Proc. IEEE ICASSP, Paris, Apr. 1982.

98 D. Tufts and R. Kumaresan, "Data-Adaptive Principal Component Signal Processing," Proc. 19th IEEE Conf. Decis. Control, Albuquerque, N. Mex., pp. 949–954, Dec. 1980.

99 A. Cantoni and L. Godara, "Resolving the Directions of Sources in a Correlated Signal Field Incident on an Array," J. Acoust. Soc. Am., Vol. 67, No. 4, pp. 1247–1255, Apr. 1980.

100 G. Bienvenu and L. Kopp, "Adaptive High Resolution Spatial Discrimination of Passive Sources," in *Underwater Acoustics and Signal Processing*, ed. L. Bjorno, D. Reidel, Boston, 1981.

101 R. Klemm, "High-Resolution Analysis of Nonstationary Data Ensembles," in *Signal Processing: Theories and Applications*, ed. G. Kunt and G. Coulon, pp. 711–714, Oct. 1980.

102 D. Johnson, "Improving the Resolution of Bearing in Passive Sonar Arrays by Eigenvalue Analysis," Proc. First IEEE ASSP Workshop on Spectral Analysis, McMaster University, Hamilton, Ont., Aug. 1981, paper 5.6.

103 R. Schmidt, "Multiple Emitter Location and Signal Parameter Estimation," Proc. RADC Spectral Estimation Workshop, Rome Air Development Center, Rome, N.Y., 1979.

104 L. J. Griffiths, "A Continuously Adaptive Filter Implemented As a Lattice Structure," Proc. IEEE ICASSP, Hartford, Conn., pp. 683–689, 1977.

105 W. S. Liggett, "Passive Sonar: Fitting Models to Multiple Time Series, in *Signal Processing*, ed. J. W. R. Griffiths et al., Academic Press, New York, 1973.

106 H. L. Van Trees, *Signal Detection and Estimation and Modulation Theory*, Part I, Wiley, New York, 1968.

107 O. B. Gammelsaeter, "Adaptive Beamforming with Emphasis on Narrowband Implementation," in *Underwater Acoustic and Signal Processing*, ed. Leif Bjorno, NATO ASI Series C, D. Reidel, Boston, pp. 307–326, 1981.

108 Special Issue on Time Delay Estimation, Part II, IEEE Trans. Acoust. Speech Signal Process., Vol. ASSP-29, No. 3, June 1981.

109 G. C. Carter, "Time Delay Estimation for Passive Sonar Signal Processing," IEEE Trans. Acoust. Speech Signal Process., Vol. ASSP-29, No. 3, pp. 463–470, June 1981.

110 L. Ng and Y. BarShalom, "Time Delay Estimation in a Multitarget Environment," Proc. 21st IEEE, Conf. Decis. Control, Orlando, Fla., Dec. 1982.

111 J. P. Ianniello, "Threshold Effects in Time Delay Estimation Using Narrowband Signals," Proc. IEEE ICASSP 82, Paris, pp. 375–379, May 3, 1982.

112 International Mathematical and Statistical Libraries, Inc., *Eigensystem Analysis*, 9th ed., Vol. 2, 1982, Chap. E.

113 W. Gabriel, "Adaptive Superresolution of Coherent RF Spatial Sources," Proc. First IEEE ASSP Workshop on Spectral Analysis, McMaster University, Hamilton, Ont., Aug. 1981.

114 J. P. Ianniello *et al.*, "Comparison of the Ziv-Zakai Bound on Time Delay Estimation with Correlator Performance," IEEE ICASSP, Boston, Mass., pp. 875–878, April 14–16, 1983.

115 G. Swope, "Spectrum Analysis Techniques Applied to Multi-source Range Estimation," IEEE ASSP Spectrum Estimation Workshop II Proceedings, Tampa, Fla., Nov. 10–11, 1983. Thesis, Rensselaer Polytechnic Institute, Aug. 1982 (to be published).

116 A. V. Oppenheim and R. W. Schafer, *Digital Signal Processing*, Prentice-Hall, Englewood Cliffs, N.J., 1975.

117 R. A. Mucci and M. L. Lasky, "Digital Beamformer Implementation Considerations," Proc. IEEE EASCON '81, pp. 104–113.

118 M. Wax, T. J. Shan and T. Kailath, "Covariance Eigenstructure Approach to Two-Dimensional Harmonic Retrieval," IEEE ICASSP, Boston, Mass., pp. 891–894, April 14–16, 1983.

4

SIMON HAYKIN
Communications Research Laboratory
McMaster University, Hamilton, Ontario

Radar Array Processing for Angle of Arrival Estimation

4.1 INTRODUCTION

Radar is an electronic device that is used for the detection and location of targets of interest [1]. It operates by transmitting a particular type of waveform (e.g., a pulse-modulated sine wave) and then analyzing the nature of the resultant echo signal. In noise-limited situations, the radar signal may consist of an echo and additive receiver noise. In more complicated situations, it may also contain a *clutter* component due to backscatter from the surrounding environment. A further source of complication may be the presence of interferences produced by jammers beamed at the radar site from directions unknown to operators at the site. When the clutter and/or the interferences are of a continually changing nature, the use of an *array antenna* coupled with some form of *adaptive control* offers an attractive solution to the design of the radar system required to deal with situations of this kind. The collection of papers on array processing edited by Haykin [2], the book on adaptive arrays written by Monzingo and Miller [3], and the book by Hudson [4] give detailed treatments of adaptive array antennas useful for radar applications. Indeed, many of the adaptive techniques discussed in Chapters 2 and 3 have a corresponding use in radar.

In this chapter we study a radar system in which a transmitting antenna is used to *floodlight* the environment surrounding the radar site and a receiving array antenna is used to analyze the received signal. In particular, we study the use of a linear array antenna to solve the problem of estimating the angle(s) of arrival of plane wave(s) impinging on the array from unknown direction(s). The results of this study are of particular interest in two areas:

1. *Direction-finding systems* [5]: The traditional method of improving the angle-measurement accuracy of navigation aids and/or surveillance systems has been to increase the physical aperture of the receiving antenna. Alternatively, we may use a *sampled aperture*, combined with sophisticated signal processing techniques. This approach has become possible by virtue of remarkable improvements in the cost as well as capability of digital signal processing hardware.
2. *Low-angle tracking radar* [6–11]: In a marine environment, the conventional method of tracking a target has been to use a monopulse radar [1]. However, the operation of such a system can be severely limited by the presence of multipath (i.e., two or more signal paths in close proximity to each other), particularly at low elevation angles. Here, again, we find that by using a sampled aperture with appropriate signal processing, it is possible to resolve a target in the presence of multipath, even though the separation between the signal paths incident on the array may be a small fraction of a beamwidth.

We begin this study by describing some basic signal and noise models for an environment that contains a multiplicity of noncoherent and/or coherent sources.

4.2 BASIC SIGNAL AND NOISE MODELS

Consider a uniformly spaced, linear array antenna consisting of N elements, and with M *plane waves* arriving at the array from distinct directions. We assume that all these plane waves are *narrowband*, with the same *carrier frequency* f_c. Figure 4.1 illustrates the configuration of a linear array with a single incident wave coming from a direction defined by the vector $-\mathbf{v}_m$. We may thus express the noiseless signal produced at the nth element of the array by this wave, denoted as the mth plane wave, as follows:

$$s(n, m, t) = A_m \cos\left[2\pi f_c t + 2\pi\left(n - \frac{N+1}{2}\right)\mathbf{v}_m \cdot \mathbf{z} + \alpha_m\right] \tag{4.1}$$

where $n = 1, 2, \ldots, N$
$m = 1, 2, \ldots, M$
f_c = carrier frequency
t = time
$\mathbf{v}_m$ = *vector wavenumber* of the mth incident plane wave
$\mathbf{z}$ = *unit vector* along the line of the array

The parameters A_m and α_m denote the *amplitude* and *phase* of the signal $s(n, m, t)$, measured at the center of the array for which $n = (N + 1)/2$.

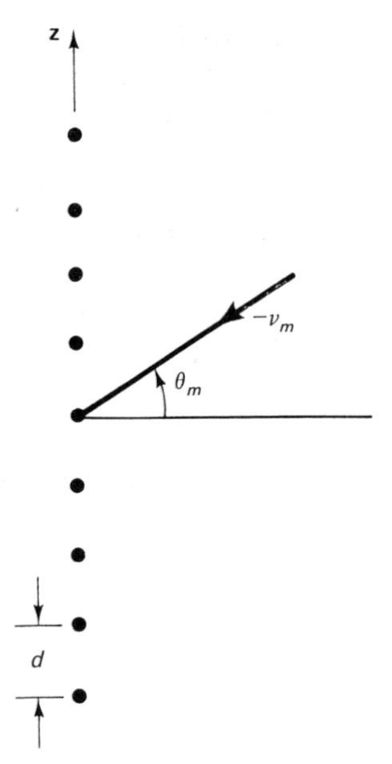

Figure 4.1 Geometry of linear array, showing a ray of an incident plane wave.

Let θ_m denote the *angle of arrival* of the mth plane wave, measured with respect to the normal to the array, as illustrated in Fig. 4.1. Then we may express the *dot product* of the two vectors $\mathbf{v}_m$ and $\mathbf{z}$ as follows:

$$\mathbf{v}_m \cdot \mathbf{z} = \frac{d}{\lambda} \sin \theta_m \tag{4.2}$$

where d = interelement spacing
λ = received radar wavelength

Accordingly, we may rewrite Eq. (4.1) as

$$s(n, m, t) = A_m \cos\left[2\pi f_c t + \frac{2\pi d}{\lambda}\left(n - \frac{N+1}{2}\right) \sin \theta_m + \alpha_m\right] \tag{4.3}$$

We may simplify the representation of the signal $s(n, m, t)$ by adopting the complex notation for narrowband signals [12]. In particular, we may represent the signal $s(n, m, t)$ by its *complex amplitude*, defined by

$$s(n, m) = A_m \exp\left[\frac{j2\pi d}{\lambda}\left(n - \frac{N+1}{2}\right) \sin \theta_m + j\alpha_m\right] \tag{4.4}$$

Define the *electrical phase angle* from element to element along the array as

$$\phi_m = \frac{2\pi d}{\lambda} \sin \theta_m \tag{4.5}$$

Also, let a_m denote the complex amplitude of the signal $s(n, m, t)$, measured at the center of the array; that is,

$$\begin{aligned} a_m &= s(n, m)\Big|_{n=(N+1)/2} \\ &= A_m \exp(j\alpha_m) \end{aligned} \tag{4.6}$$

Note that if the number of elements N is odd, the center of the array represented by $n = (N + 1)/2$ corresponds to the center element. On the other hand, when N is even, the center of the array is a fictitious point. In any event, we may use Eqs. (4.5) and (4.6) to rewrite Eq. (4.4) in the simplified form

$$s(n, m) = a_m \exp\left[j\left(n - \frac{N+1}{2}\right)\phi_m\right] \tag{4.7}$$

The signal measured at the output of any element in the array differs from the signal actually received by that element by an amount attributed to *noise.* In practice, this noise is modeled as a *white, ergodic random process.* To limit the effects of the noise, it is customary to filter the output of each element in the array to the same narrow band of frequencies occupied by the actual received signal. As a result of this operation, we have an *observed signal,* which for each element of the array, consists of the actual received signal plus a *narrowband noise.* For element n we may describe this narrowband noise by

$$w(n, t) = B_n \cos(2\pi f_c t + \beta_n) \tag{4.8}$$

where the amplitude B_n is *Rayleigh distributed* and the phase β_n is *uniformly distributed* over the range $(0, 2\pi)$. Representing the narrowband noise $w(n, t)$ by its complex amplitude, $w(n)$, we have

$$w(n) = B_n \exp(j\beta_n) \tag{4.9}$$

This is a complex-valued random variable that is Gaussian-distributed with a mean that is typically zero.

Equation (4.7) defines the complex amplitude of the signal received by element n due to the mth plane wave. Summing the contributions produced by a set of M plane waves, and then adding to the result the complex amplitude of the narrowband noise, we may express the complex amplitude of the observed signal as

$$\begin{aligned} x(n) &= s(n) + w(n) \\ &= \sum_{m=1}^{M} s(n, m) + w(n) \\ &= \sum_{m=1}^{M} a_m \exp\left[j\left(n - \frac{N+1}{2}\right)\phi_m\right] + w(n), \qquad n = 1, 2, \ldots, N \end{aligned} \tag{4.10}$$

We may put Eq. (4.10) into a more compact form by using matrix notation.

We first introduce the following definitions:

1. An $N \times 1$ vector

$$\mathbf{x} = \begin{bmatrix} x(1) \\ x(2) \\ \vdots \\ x(N) \end{bmatrix} \tag{4.11}$$

called the *observed signal vector.*

2. An $N \times 1$ vector

$$\mathbf{s} = \begin{bmatrix} s(1) \\ s(2) \\ \vdots \\ s(N) \end{bmatrix} = \begin{bmatrix} \sum_{m=1}^{M} s(1, m) \\ \sum_{m=1}^{M} s(2, m) \\ \vdots \\ \sum_{m=1}^{M} s(N, m) \end{bmatrix} \tag{4.12}$$

called the *received signal vector.*

3. An $M \times 1$ vector

$$\mathbf{a} = \begin{bmatrix} a(1) \\ a(2) \\ \vdots \\ a(M) \end{bmatrix} \tag{4.13}$$

called the *signal-in-space vector.*

4. An $N \times M$ matrix

$$\mathbf{B} = \begin{bmatrix} \exp\left(j\dfrac{1-N}{2}\phi_1\right) & \exp\left(j\dfrac{1-N}{2}\phi_2\right) & \cdots & \exp\left(j\dfrac{1-N}{2}\phi_M\right) \\ \exp\left(j\dfrac{3-N}{2}\phi_1\right) & \exp\left(j\dfrac{3-N}{2}\phi_2\right) & \cdots & \exp\left(j\dfrac{3-N}{2}\phi_M\right) \\ \vdots & \vdots & & \vdots \\ \exp\left(j\dfrac{N-1}{2}\phi_1\right) & \exp\left(j\dfrac{N-1}{2}\phi_2\right) & \cdots & \exp\left(j\dfrac{N-1}{2}\phi_M\right) \end{bmatrix} \tag{4.14}$$

called the *direction matrix*.

5. An $N \times 1$ vector

$$\mathbf{w} = \begin{bmatrix} w(1) \\ w(2) \\ \vdots \\ w(N) \end{bmatrix} \tag{4.15}$$

called the *noise vector*.

Then, using these definitions, we may rewrite Eq. (4.10) in the form

$$\mathbf{x} = \mathbf{s} + \mathbf{w} \tag{4.16}$$

or, equivalently,

$$\mathbf{x} = \mathbf{Ba} + \mathbf{w} \tag{4.17}$$

where

$$\mathbf{s} = \mathbf{Ba} \tag{4.18}$$

The observed signal vector $\mathbf{x}$ represents a *snapshot* of data corresponding to a particular instant of time. Ordinarily, several independent measurements are made, so that the data available for processing may be expressed in the form

$$\mathbf{x}(k) = \mathbf{Ba}(k) + \mathbf{w}(k), \qquad k = 1, 2, \ldots, K \tag{4.19}$$

where K is the total number of snapshots taken.

Based on such a set of data, we may use *temporal averaging* to improve the estimation of the angles of arrival of the plane waves incident upon the array. The rational here is that the processing interval, represented by the duration of the K snapshots, is usually short enough to ensure that the angles do not change significantly. It is for this reason that we have assumed that the direction matrix $\mathbf{B}$ in Eq. (4.19) is essentially constant during the processing interval. On the other hand, the signal-in-space vector typically varies with time; this accounts for the use of $\mathbf{a}(k)$ in Eq. (4.19). The signal-in-space vector $\mathbf{a}(k)$ is modeled as a stochastic process because the behavior of the sources responsible for its generation is, in general, unpredictable.

The *spatial correlation matrix* of the observed signal vector $\mathbf{x}$ is defined by

$$\mathbf{R} = E[\mathbf{x}^*(k)\mathbf{x}^T(k)] \tag{4.20}$$

where E is the *expectation operator*, the asterisk denotes complex conjugation, and the superscript T denotes transposition. Substituting Eq. (4.19) in (4.20), and recognizing that the signal-in-space vector $\mathbf{a}$ and the noise vector $\mathbf{w}$ are statistically independent, we get

$$\begin{aligned} \mathbf{R} &= E[\mathbf{B}^*\mathbf{a}^*\mathbf{a}^T\mathbf{B}^T] + E[\mathbf{w}^*\mathbf{w}^T] \\ &= \mathbf{B}^*\mathbf{R}_a\mathbf{B}^T + \mathbf{R}_w \end{aligned}$$

where $\mathbf{R}_a$ = *spatial correlation matrix of the signal-in-space vector* $\mathbf{a}$
$= E[\mathbf{a}^*\mathbf{a}^T]$
$\mathbf{R}_w$ = *spatial correlation matrix of the noise vector* $\mathbf{w}$
$= E[\mathbf{w}^*\mathbf{w}^T]$

The matrix product

$$\mathbf{R}_s = \mathbf{B}^*\mathbf{R}_a\mathbf{B}^T$$

represents the spatial correlation matrix of the signal component of the received signal vector $\mathbf{s}$. With the noise vector $\mathbf{w}$ assumed white, we have

$$\mathbf{R}_w = \sigma_w^2\mathbf{I}$$

where σ_w^2 is the variance of the elemental noise $w(n)$ for all n.

Depending on the structure of the spatial correlation matrix $\mathbf{R}_a$, we may distinguish two different cases:

1. The sources responsible for the signal-in-space vector $\mathbf{a}$ are *jointly uncorrelated* or *noncoherent* so that the off-diagonal elements of the corresponding spatial correlation matrix $\mathbf{R}_a$ are all zero. Then we find that the spatial correlation matrix of the received signal vector, $\mathbf{R}_s = \mathbf{B}^*\mathbf{R}_a\mathbf{B}^T$, is a *Toeplitz matrix.* We say that a square matrix is Toeplitz if all the elements on its main diagonal are equal, and likewise for the elements on any other diagonal parallel to the main diagonal. We say that the received signal vector $\mathbf{s}$ is *spatially stationary* if its spatial correlation matrix is Toeplitz, or vice versa. Clearly, if the spatial correlation matrix $\mathbf{R}_s$ is Toeplitz, the spatial correlation matrix of the observed signal vector, $\mathbf{R}$, is likewise Toeplitz.
2. The sources responsible for the received signal vector $\mathbf{s}$ are *correlated* or *coherent.* The result is that the spatial correlation matrix $\mathbf{R}_s$ is *non-Toeplitz,* and the received signal vector $\mathbf{s}$ is said to be *spatially nonstationary.* Clearly, if $\mathbf{R}_s$ is non-Toeplitz, the spatial correlation matrix, $\mathbf{R}$, pertaining to the observed signal vector, is also non-Toeplitz.

4.3 MULTIPATH MODEL

When a target lies in close proximity to a reflecting surface (e.g., ground, sea surface), the received signal reaches the antenna via a multiplicity of paths (i.e., two or more), even though there may exist only one target in the environment. This is known as the *multipath* phenomenon. Figure 4.2(a) illustrates one special case of the multipath problem that arises when (1) the reflecting surface is perfectly smooth, and (2) the target is at a long range from the receiving antenna. The target radiates or reflects signals in all directions. The radar receives two signals: (1) a signal from angle θ_1, via the *direct path* from the target to the antenna; and (2) a *specular* component from angle θ_2, due to reflection from the surface. We may also view the

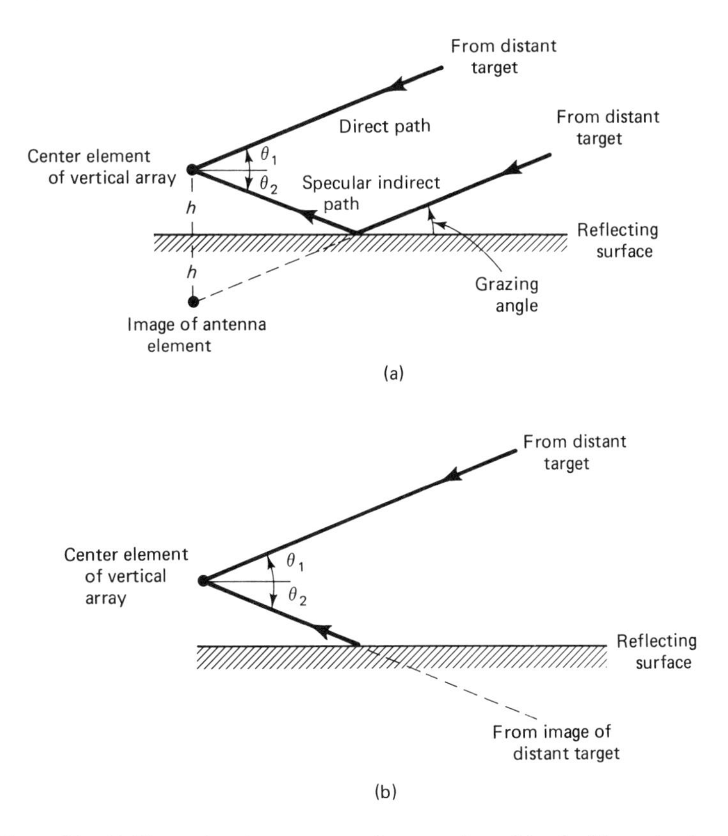

Figure 4.2 (a) Illustrating the geometry of symmetric multipath; (b) another interpretation of symmetric multipath.

specular component as being due to the presence of an *image target*, as illustrated in Fig. 4.2(b). Thus the estimation of the angular position of the target becomes a problem similar to that of resolving two closely spaced *coherent sources*.

For the two-path model shown in Fig. 4.2, we may express the complex amplitude of the observed signal for element n of the array as

$$x(n) = a_1 \exp\left[j\left(n - \frac{N+1}{2}\right)\phi_1\right] + a_2 \exp\left[j\left(n - \frac{N+1}{2}\right)\phi_2\right] + w(n),$$

$$n = 1, 2, \ldots, N \tag{4.21}$$

where ϕ_m is defined by Eq. (4.5) for $m = 1, 2$. The first term on the right-hand side of Eq. (4.21) represents the direct signal, the second term represents the specular component, and the third term represents the noise component. The amplitude of the specular component is related to that of the direct component as follows:

$$|a_2| = \rho\,|a_1| \tag{4.22}$$

where ρ is the magnitude of the *surface reflection* coefficient. This coefficient is a function of the *grazing angle* [equal to angle θ_1 in Fig. 4.2(a)], type of reflecting surface, carrier frequency, and polarization of the received radar wave. For a perfectly calm sea surface, for example, the surface reflection coefficient ρ typically equals 0.9.

The phase of the specular component differs from that of the direct component by a largely deterministic amount, denoted by ψ, as shown by

$$\arg [a_2] = \arg [a_1] + \psi \tag{4.23}$$

The phase difference ψ is the result of the difference between the path lengths of the direct and specular components. From the geometry of Fig. 4.2(a), pertaining to a symmetric situation with $\theta_1 = -\theta_2 = \theta$, we find that this path-length difference equals

$$\Delta R = 2h \sin \theta \tag{4.24}$$

where h is the height of the center of the array above the reflecting surface. Hence we may express the phase difference ψ as

$$\begin{aligned} \psi &= \frac{2\pi}{\lambda} \Delta R + \zeta \\ &= \frac{4\pi h}{\lambda} \sin \theta + \zeta \end{aligned} \tag{4.25}$$

where ζ is the phase shift induced on the radar wave by the reflecting surface. We thus find that, except for ζ, the phase difference ψ is determined by the geometry of the environment. Combining Eqs. (4.22) and (4.23), we get

$$a_2 = \rho a_1 \exp (j\psi) \tag{4.26}$$

The multipath model described above assumes a perfectly smooth reflecting surface. When the reflecting surface is rough, the geometry of the multipath signal configuration takes on a more complicated form, as shown in Fig. 4.3. Power from

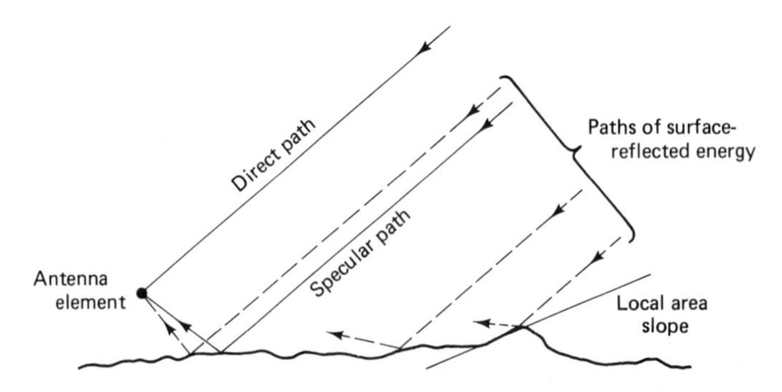

Figure 4.3 Illustrating the multipath phenomenon in the presence of a rough surface.

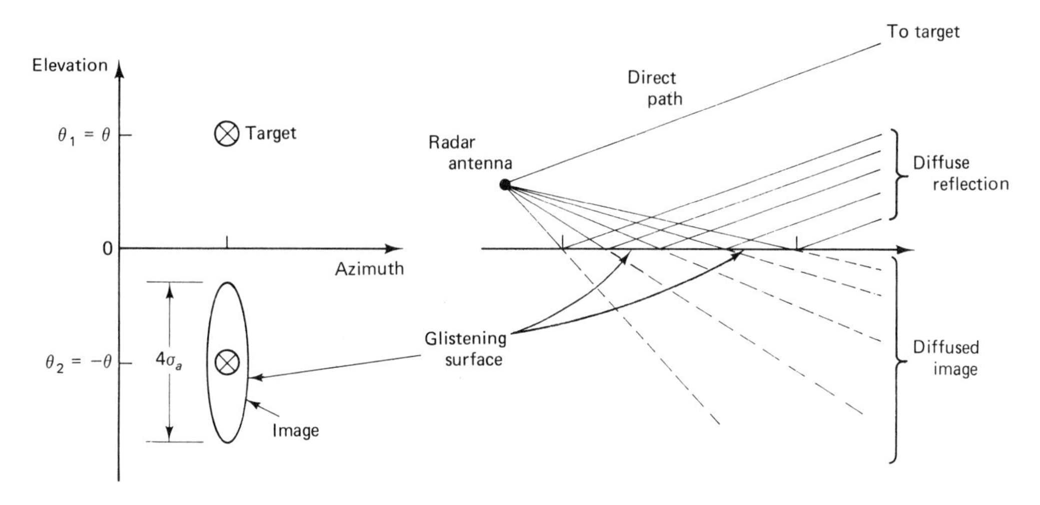

Figure 4.4 Diffuse reflection from a glistening surface.

the target which reaches the rough surface and which is not absorbed or reflected specularly will be scattered at all other angles. This form of scattering is referred to as *diffuse multipath.* Most of the diffuse power, from the homogeneous surface (usually assumed Gaussian distributed), will reach the radar from the region within the *glistening surface* (Barton [13]). The glistening surface, shown in Fig. 4.4, represents the area from which power can be reflected to the radar by facets having slopes less than or equal to σ_a. Facets are small flat plates that are used to approximate a rough reflecting surface. The quantity σ_a is the root-mean-square (rms) slope of the small surface facets.

For a more detailed discussion of specular and diffuse forms of multipath, the reader is referred to the paper by Barton [13].

Effect of Multipath on the Spatial Correlation Matrix

Consider a linear array of N elements embedded in the multipath environment shown in Fig. 4.2. We wish to evaluate the effect of the specular component on the spatial correlation matrix of the observed vector $\mathbf{x}$.

We find it convenient to define the index n so that $n = 0$ corresponds to the center of the array. Accordingly, we may express the observed signal at the output of element n as follows:

$$\begin{aligned} x(n) &= a_1 \exp(jn\phi_1) + a_2 \exp(jn\phi_2) + w(n) \\ &= a_1 \exp(jn\phi_1) + \rho a_1 \exp(jn\phi_2 + \psi) + w(n) \end{aligned} \tag{4.27}$$

where $n = -(N-1)/2, \ldots, (N-1)/2$. The $N \times N$ correlation matrix $\mathbf{R}$ is the matrix of correlations of $x(n)$ and $x(m)$ over the complete ranges of n and m. The (n, m)th element $R(n, m)$ of this matrix is defined by

$$R(n, m) = E[x(n)x^*(m)]$$

Using this definition, performing some algebraic manipulations and normalizing $R(n, m)$ with respect to the mean-square value of a_1, we obtain

$$\frac{R(n, m)}{E[|a_1|^2]} = \exp[j(n-m)\phi_1] + \rho^2 \exp[j(n-m)\phi_2] + 2\rho \exp\left[\frac{j}{2}(n-m)(\phi_1+\phi_2)\right] \cos\left[\tfrac{1}{2}(n+m)(\phi_1-\phi_2)-\psi\right] + \frac{1}{\text{SNR}}\delta_{nm} \tag{4.28}$$

where δ_{nm} is the *Kronecker delta* equal to 1 for $n = m$ and zero otherwise, and SNR is the *signal-to-noise ratio* defined by

$$\text{SNR} = \frac{E[|a_1|^2]}{E[|w_n|^2]} \tag{4.29}$$

Note that the signal-to-noise ratio is defined for a single element of the array in free space (i.e., in the presence of the direct path only).

For the sequence described by the spatial samples $x(n)$, $n = -(N-1)/2, \ldots, (N-1)/2$, to be wide-sense stationary in space, the autocorrelation function $R(n, m)$ must only be a function of the spatial lag $(n - m)$. For this to occur, the dependence on $(n + m)$ contained in the third term of Eq. (4.28) must be removed. There are two ways in which this requirement may be satisfied:

1. The phase difference ψ is an odd multiple of $\pi/2$ radians. For such a value of ψ, we find that the cosine term

$$\cos\left[\tfrac{1}{2}(n+m)(\phi_1+\phi_2)-\psi\right] = \pm\sin\left[\tfrac{1}{2}(n+m)(\phi_1+\phi_2)\right]$$

averages to zero over all the values of $(n + m)$ that are possible when $(n - m)$ is held constant.

2. The direct and specular components are separated by an integer multiple of standard beamwidths. The definition of a standard beamwidth is given in Eq. (4.35). In particular, if we put $\phi_1 - \phi_2 = 2\pi k/N$ radians, k an integer, the cosine term written above will also average to zero (for arbitrary ψ) over all values of $(n + m)$, but only for the cases where $n - m = 0$. For any other value of spatial lag, the third term will not be eliminated.

We thus find that, in general, the presence of specular path causes the observed process and the elemental outputs of the array to be nonstationary. Correspondingly, the correlation matrix $\mathbf{R}$ of this process is, in general, non-Toeplitz. Note, however, that this matrix is always Hermitian, that is, $R(n, m) = R^*(m, n)$.

4.4 THE FOURIER METHOD

Consider the spatial series $x(1), x(2), \ldots, x(N)$ generated in the manner described in Section 4.2. By analogy with the *Fourier analysis* of a time series, we define the *angle spectrum* of this spatial series as follows:

$$X(\phi) = \sum_{n=1}^{N} x(n) \exp\left[-j\left(n - \frac{N+1}{2}\right)\phi\right] \tag{4.30}$$

The squared magnitude, $|X(\phi)|^2$, is termed the *periodogram.* The electrical phase angle ϕ is related linearly to the vector wavenumber $\mathbf{v}$ [see Eqs. (4.2) and (4.5)] and the expectation of the periodogram equals the *wavenumber power spectrum.* Thus the periodogram of Eq. (4.30) provides the basis of a method for estimating the wavenumber power spectrum of the energy sources responsible for illuminating the array.

Except for the multiplying factor $\exp[j(N-1)\phi/2]$, which has unit magnitude and constant phase $(N-1)\phi/2$, we may realize Eq. (4.30) by using the structure of Fig. 4.5. It consists of a set of *phase shifters,* connected to the elemental outputs, and a *summer.*

Given the angle spectrum $X(\phi)$, we may estimate the directions of the signal sources by computing the *spectral peaks,* that is, the values of ϕ for which the magnitude of the angle spectrum $X(\phi)$ attains its maxima.

For the special case of a single source, the direction of which is defined by the electrical phase angle ϕ_m, we have (in the absence of noise)

$$s(n, m) = a_m \exp\left[j\left(n - \frac{N+1}{2}\right)\phi_m\right], \qquad n = 1, 2, \ldots, N \tag{4.31}$$

Hence, substituting Eq. (4.31) into (4.30), with $x(n)$ and $X(\phi)$ replaced by $s(n, m)$ and

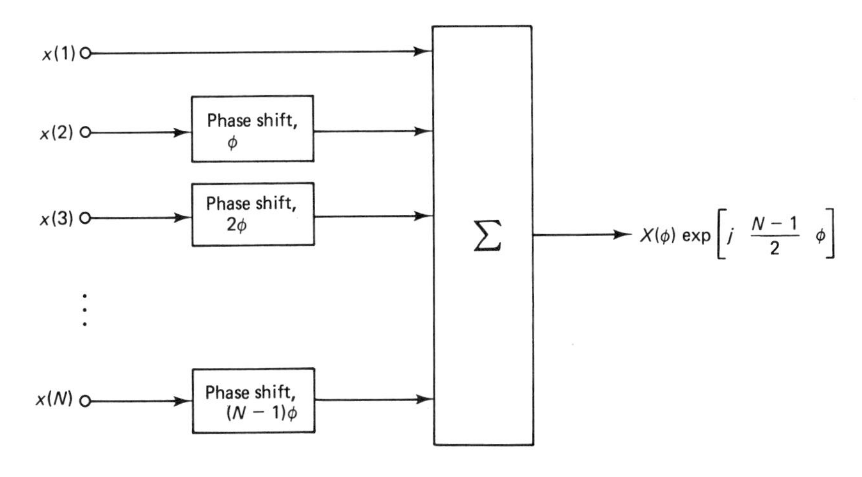

Figure 4.5 Network for computing the angle spectrum $X(\phi)$.

$S_m(\phi)$, respectively, we find that the angle spectrum of this source equals

$$S_m(\phi) = a_m \sum_{n=1}^{N} \exp\left[-j\left(n - \frac{N+1}{2}\right)(\phi - \phi_m)\right] \tag{4.32}$$

This represents the sum of a geometric series with a first term equal to $a_m \exp[-j(1-N)(\phi - \phi_m)/2]$, a common ratio equal to $\exp[-j(\phi - \phi_m)]$, and a total number of terms equal to N. Thus, using the formula for the sum of a geometric series, we get

$$S_m(\phi) = a_m \frac{\sin[N(\phi - \phi_m)/2]}{\sin[(\phi - \phi_m)/2]}$$

$$= a_m W(\phi - \phi_m) \tag{4.33}$$

where

$$W(\phi) = \frac{\sin(N\phi/2)}{\sin(\phi/2)} \tag{4.34}$$

The function $W(\phi)$ represents the familiar *radiation pattern* of a uniformly illuminated linear array. Figure 4.6 shows a plot of $W(\phi)$ versus the electrical phase angle ϕ. The pattern exhibits a *main lobe* or *beam* of finite width, and several *sidelobes*. The first null of $W(\phi)$ occurs at the angle

$$\frac{N\phi_B}{2} = \pi$$

or

$$\phi_B = \frac{2\pi}{N} \tag{4.35}$$

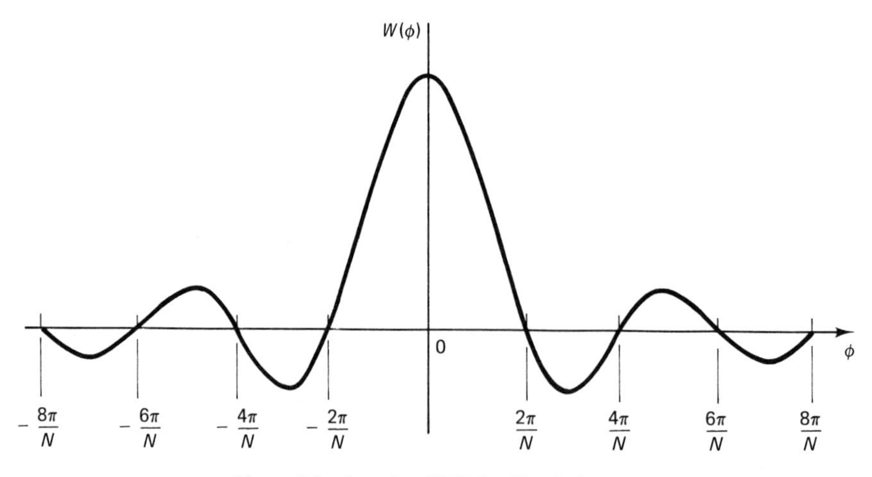

Figure 4.6 Function $W(\phi)$ for $N = 8$ elements.

We refer to ϕ_B, the angular separation between the peak of the main lobe and the first null in the response of Fig. 4.6, as a *standard beamwidth.* Denoting a standard beamwidth by BW, we find from Eq. (4.35) that one BW equals $2\pi/N$ radians.

The response of Fig. 4.6 illustrates some of the problems associated with spatial processing. We see that when a source having a well-defined direction is viewed through a linear array antenna of *finite length* or *aperture*, it appears to be coming from a dominant but diffused direction (represented by the main lobe) as well as from false directions due to the sidelobes. The sidelobes are due to the equal weighting assumed for the output of each element in the array. If the weights are tapered, we are able to reduce the amplitude of the sidelobes. However, this improvement is achieved at the expense of an increase in the width of the main lobe. Indeed, there exists a variety of weighting functions that can be used to realize this exchange (Collins and Zucker [14], and Harris [15]).

For the case when two or more sources exist at different points in space, we may use the *principle of superposition* to extend the result of Eq. (4.33). Thus, for M sources coming from directions $\theta_1, \theta_2, \ldots, \theta_M$ or, correspondingly, electrical phase angles $\phi_1, \phi_2, \ldots, \phi_M$, we may express the angle spectrum of the complete field as

$$\begin{aligned} S(\phi) &= \sum_{m=1}^{M} a_m \frac{\sin[N(\phi - \phi_m)/2]}{\sin[(\phi - \phi_m)/2]} \\ &= \sum_{m=1}^{M} a_m W(\phi - \phi_m) \end{aligned} \tag{4.36}$$

When the sources are separated from each other by two or more standard beamwidths, the magnitude of the angle spectrum $S(\phi)$ exhibits a corresponding number of distinct peaks, with each peak being located at or close to a point determined by the direction of the corresponding source. We say that the sources are *resolved* when the dominant peak locations coincide with the directions of the individual sources. When the sources are separated by an amount lying between one and two standard beamwidths, distinct peaks (corresponding to the directions of the sources) may or may not appear in the angle spectrum $S(\phi)$, depending on the values of the pertinent complex amplitudes. Ordinarily, we find that the Fourier method fails to resolve two sources that are separated by less than one standard beamwidth.

We may now state the advantages and limitations of the classical Fourier method. Its advantages are: (1) it is *nonparametric* in the sense that it does not rely on the knowledge of a model of the input process, (2) it is *robust* in that it is relatively insensitive to parameter changes, and (3) it is simple to implement. Its only limitation is that we have to increase the *aperture* of the array in order to improve the resolution capability of the method. This solution, however, is of limited practical value, as it requires increasing the physical size of the array.

This limitation of the Fourier method may be overcome by using *parametric* methods that require modeling the input process by a suitable set of parameters. Several such procedures are considered in the rest of this chapter.

4.5 MAXIMUM-LIKELIHOOD ESTIMATION

The first parametric procedure that we wish to describe for estimating the directions of the incident plane waves is based on a classical method known as the *maximum-likelihood estimation* (Van Trees [16]).

Let the spatial samples $x(1), x(2), \ldots, x(N)$ constitute the elements of an $N \times 1$ observed signal vector $\mathbf{x}$, and let ϕ be the value of a nonrandom real parameter of interest contained in this vector. We use $f(\mathbf{x} \mid \phi)$ to denote the *joint probability density function* of the elements of the vector $\mathbf{x}$, given that the parameter of interest has the value ϕ. The function $f(\mathbf{x} \mid \phi)$, viewed as a function of ϕ, may be regarded as the *likelihood function*. Frequently, we work with the logarithm, $\ln f(\mathbf{x} \mid \phi)$, and call it the *log-likelihood function*. Let $\hat{\phi}(\mathbf{x})$ denote the *maximum-likelihood* (ML) estimate of the parameter of interest. We define $\hat{\phi}(\mathbf{x})$ as that value of ϕ at which the likelihood function $f(\mathbf{x} \mid \phi)$ is a maximum. The physical significance of this method is that we are, in effect, choosing the value of the parameter of interest which most likely caused the given observed signal vector $\mathbf{x}$ to occur.

An important property of the maximum-likelihood estimate $\hat{\phi}(\mathbf{x})$, or simply $\hat{\phi}$, is that if it is *unbiased* (i.e., its expected value is equal to the true value of the parameter of interest), then it will satisfy the *Cramér–Rao bound* with an equality [16]. The Cramér–Rao bound defines the bound on the variance of any estimate of the real parameter under study.

As an application of the maximum-likelihood estimation, we consider first the special case of a symmetric multipath environment. Later, we consider the general case of an arbitrary number of plane waves incident upon the array.

4.6 THE MAXIMUM-LIKELIHOOD RECEIVER FOR SYMMETRIC MULTIPATH

Consider an ideal linear array that consists of N identical antenna elements, illuminated by the signal configuration depicted in Fig. 4.2. We assume that the array antenna is positioned at right angles to the reflecting surface, so that the returns from a distant target and its image impinge on the array in a symmetric manner with respect to the normal to the antenna. To emphasize the symmetric nature of the configuration shown in Fig. 4.2, we use the index n to define the position of the antenna element measured from the center of the array. Thus, for the case when the number of elements N in the array is odd, the index n takes on the integer values $-(N-1)/2, \ldots, -1, 0, 1, \ldots, (N-1)/2$. On the other hand, when the number of elements N is even, the index n takes on the values $-(N-1)/2, \ldots, -1/2, 1/2, \ldots, (N-1)/2$ (i.e., odd integer multiples of one-half). In both cases, however, $n = 0$ corresponds to the center of the array. When N is odd, the center element lies at this point; when N is even, $n = 0$ represents a fictitious point.

We may simplify the analysis by putting the electrical phase angle $\phi_1 = \phi$ for the direct component, and the complex amplitude $a_2 = a_1 \rho \exp(j\psi)$ and the electrical phase angle $\phi_2 = -\phi$ for the specular component.

Thus, adapting Eq. (4.21) to this situation, we may express the complex amplitude of the observed signal at the nth element of the array as follows:

$$\begin{aligned} x(n) &= x_c(n) + jx_s(n) \\ &= a_1 \exp(jn\phi) + \rho a_1 \exp[j(\psi - n\phi)] + w(n) \\ n &= -(N-1)/2, \ldots, (N-1)/2 \end{aligned} \tag{4.37}$$

where ρ is the surface reflection coefficient and ψ is the phase difference between the direct and specular components at the center of the array.

The *in-phase component* $x_c(n)$ equals the real part of the complex amplitude $x(n)$; hence we may write

$$x_c(n) = u_1 \cos(n\phi) + u_2 \sin(n\phi) + w_c(n) \tag{4.38}$$

where $w_c(n)$ is the in-phase component of $w(n)$, and

$$u_1 = a_1(1 + \rho \cos \psi) \tag{4.39}$$

$$u_2 = a_1 \rho \sin \psi \tag{4.40}$$

The *quadrature component* $x_s(n)$ equals the imaginary part of the complex amplitude $x(n)$; hence we may write

$$x_s(n) = v_1 \cos(n\phi) + v_2 \sin(n\phi) + w_s(n) \tag{4.41}$$

where $w_s(n)$ is the quadrature component of $w(n)$, and

$$v_1 = a_1 \rho \sin \psi \tag{4.42}$$

$$v_2 = a_1(1 - \rho \cos \psi) \tag{4.43}$$

Consider first the processing that must be done on the in-phase component $x_c(n)$ in Eq. (4.38). With the element index n as defined above, we find that $\cos(n\phi)$ and $\sin(n\phi)$ are orthogonal over the interval considered for all values of ϕ and N; that is,

$$\sum_n \cos(n\phi) \sin(n\phi) = 0 \tag{4.44}$$

However, in general, the sequences $\{\cos(n\phi)\}$ and $\{\sin(n\phi)\}$ do not have equal energy. The in-phase noise component $w_c(n)$ is a Gaussian random variable of zero mean and variance σ^2 for each n. Also, $w_c(n)$ and $w_c(m)$ are uncorrelated for $n \neq m$. We now define the column vector $\mathbf{w}_c$ whose elements are

$$w_c(n) = x_c(n) - u_1 \cos(n\phi) - u_2 \sin(n\phi) \tag{4.45}$$

The conditional probability density function (PDF) of $x_c(n)$ over the complete range of n, given u_1, u_2, and ϕ, equals

$$f(x_c \mid u_1, u_2, \phi) = (2\pi\sigma^2)^{-N/2} \exp\left(-\frac{1}{2\sigma^2} \mathbf{w}_c^T \mathbf{w}_c\right) \tag{4.46}$$

which is the likelihood function for this particular estimation problem. By substituting Eq. (4.45) in (4.46), we obtain

$$f(x_c \mid u_1, u_2, \phi) = (2\pi\sigma^2)^{-N/2} \exp\left\{-\frac{1}{2\sigma^2}\sum_n [x_c(n) - u_1 \cos(n\phi) - u_2 \sin(n\phi)]^2\right\} \tag{4.47}$$

We now consider writing the above in terms of *unit-energy basis signals.* We define

$$\begin{aligned} C_1 &= (2\pi\sigma^2)^{-N/2} \\ s_1(n, \phi) &= \frac{\cos(n\phi)}{\left[\sum_n \cos^2(n\phi)\right]^{1/2}} \\ s_2(n, \phi) &= \frac{\sin(n\phi)}{\left[\sum_n \sin^2(n\phi)\right]^{1/2}} \\ q_1 &= u_1\left[\sum_n \cos^2(n\phi)\right]^{1/2} \\ q_2 &= u_2\left[\sum_n \cos^2(n\phi)\right]^{1/2} \end{aligned} \tag{4.48}$$

Then, substituting these definitions in Eq. (4.47) and expanding, we may write

$$\begin{aligned} f(x_c \mid q_1, q_2, \phi) = C_1 &\exp\left[-\frac{1}{2\sigma^2}\sum_n x_c^2(n)\right] \\ &\times \exp\left\{-\frac{1}{2\sigma^2}\left[q_1^2 - 2q_1 \sum_n x_c(n) s_1(n, \phi)\right]\right\} \\ &\times \exp\left\{-\frac{1}{2\sigma^2}\left[q_2^2 - 2q_2 \sum_n x_c(n) s_2(n, \phi)\right]\right\} \end{aligned} \tag{4.49}$$

where we have made use of the orthonormality of $s_1(\cdot)$ and $s_2(\cdot)$.

We next define two new quantities $L_1(\phi)$ and $L_2(\phi)$, respectively, as

$$\begin{aligned} L_1(\phi) &= \sum_n x_c(n) s_1(n, \phi) \\ L_2(\phi) &= \sum_n x_c(n) s_2(n, \phi) \end{aligned} \tag{4.50}$$

and so rewrite Eq. (4.49) as

$$f(x_c \mid q_1, q_2, \phi) = C_1 \exp\left[-\frac{1}{2\sigma^2}\sum_n x_c^2(n)\right]$$

$$\times \exp\left\{-\frac{1}{2\sigma^2}\left[q_1^2 - 2q_1 L_1(\phi)\right]\right\} \exp\left\{-\frac{1}{2\sigma^2}\left[q_2^2 - 2q_2 L_2(\phi)\right]\right\} \quad (4.51)$$

The three-dimensional maximization of Eq. (4.51) with respect to q_1, q_2, and ϕ leads to ML estimates of these parameters. We note, however, that the only parameter of interest to us in Eq. (4.51) is ϕ; the remaining two parameters, q_1 and q_2, are unwanted parameters. It is therefore our intent to form an estimator which is independent of both q_1 and q_2. This may be accomplished by forming the *marginal density function* with respect to ϕ only, with the aid of *Bayes' rule*. The general definition of the marginal density function $f(x_c, \phi)$ is

$$f(x_c, \phi) = \int_{q_1}\int_{q_2} f(x_c, \phi, q_1, q_2)\, dq_1\, dq_2 \quad (4.52)$$

Using Bayes' rule, we have

$$f(x_c, \phi) = f(\phi \mid x_c) f(x_c)$$

and

$$f(x_c, \phi, q_1, q_2) = f(x_c \mid \phi, q_1, q_2) f(\phi, q_1, q_2)$$

Substituting the above two relations into Eq. (4.52), we obtain

$$f(\phi \mid x_c) = \frac{1}{f(x_c)} \int_{q_1}\int_{q_2} f(x_c \mid q_1, q_2, \phi) f(\phi, q_1, q_2)\, dq_1\, dq_2 \quad (4.53)$$

The expression on the left-hand side of Eq. (4.53) is called the *a posteriori probability density function.* We wish to find the value of ϕ for which this distribution is a maximum. The distribution $f(\phi, q_1, q_2)$ is called the *prior probability density function*; it describes the joint probability of the parameters ϕ, q_1, and q_2 before the data to be analyzed are gathered. The $f(x_c)$ is independent of any of the parameters of interest; hence it plays the role of a normalizing constant such that the left-hand side of Eq. (4.53) integrates to unity over the entire range of ϕ.

The integration in Eq. (4.53) may be performed by first setting the prior probability density function $f(\phi, q_1, q_2)$ equal to a constant over the range of all the parameters, and then completing the square in the exponents of Eq. (4.51) and substituting the result into Eq. (4.53). After lumping the constants $C_1/f(x_c)$ and the prior probability density function into a new constant C_2, we obtain

$$f(\phi \mid x_c) = C_2 \int_{q_1} \int_{q_2} \exp\left[-\frac{1}{2\sigma^2} \sum_n x_c^2(n)\right]$$
$$\times \exp\left\{-\frac{1}{2\sigma^2}[q_1 - L_1(\phi)]^2\right\} \exp\left\{-\frac{1}{2\sigma^2}[q_2 - L_2(\phi)]^2\right\}$$
$$\times \exp\left\{\frac{1}{2\sigma^2}[L_1^2(\phi) + L_2^2(\phi)]\right\} dq_1\, dq_2 \tag{4.54}$$

The integration with respect to q_1 and q_2 may be accomplished by realizing that the second and third exponential factors in the integrand of Eq. (4.54) describe Gaussian probability density functions (except for constant scaling factors) in q_1 and q_2, respectively. Therefore, they integrate to a constant. We note that since the object of this endeavor is only to maximize the distribution $f(\phi \mid x_c)$ with respect to ϕ, the actual value of the multiplicative constant is of no interest. Therefore, we may rewrite the a posteriori probability density function $f(\phi \mid x_c)$ as

$$f(\phi \mid x_c) = C_3 \exp\left\{\frac{1}{2\sigma^2}\left[L_c^2(\phi) - \sum_n x_c^2(n)\right]\right\} \tag{4.55}$$

where C_3 is a new constant, and

$$L_c^2(\phi) = \frac{\left[\sum_n x_c(n) \cos(n\phi)\right]^2}{\sum_n \cos^2(n\phi)} + \frac{\left[\sum_n x_c(n) \sin(n\phi)\right]^2}{\sum_n \sin^2(n\phi)} \tag{4.56}$$

Maximization of $L_c^2(\phi)$ in Eq. (4.56) with respect to ϕ maximizes the a posteriori density function $f(\phi \mid x_c)$ and gives the desired estimate $\hat{\phi}$ of ϕ. This value is called the *maximum-a posteriori (MAP) estimate* of ϕ. Because of the fact that the prior probability density function $f(q_1, q_2, \phi)$ in Eq. (4.53) was made equal to a constant. the MAP and the ML estimates of ϕ are equivalent for this case.

Following a procedure similar to that described above, we find that maximization of the a posteriori PDF of ϕ given the quadrature component $x_s(n)$ of the observed signal $x(n, t)$ is equivalent to maximizing a second function defined by

$$L_s^2(\phi) = \frac{\left[\sum_n x_s(n) \cos(n\phi)\right]^2}{\sum_n \cos^2(n\phi)} + \frac{\left[\sum_n x_s(n) \sin(n\phi)\right]^2}{\sum_n \sin^2(n\phi)} \tag{4.57}$$

which has the same structure as Eq. (4.56), except that it is now fed by $x_s(n)$ instead of $x_c(n)$.

The receiver structure for $L_c^2(\phi)$ and $L_s^2(\phi)$, for the case of equal noise powers at the elements of the array, is shown in Fig. 4.7. Note that, for a given value of ϕ, the two branches operating on $\cos(n\phi)$ and $\sin(n\phi)$ in Fig. 4.7, corresponding to the

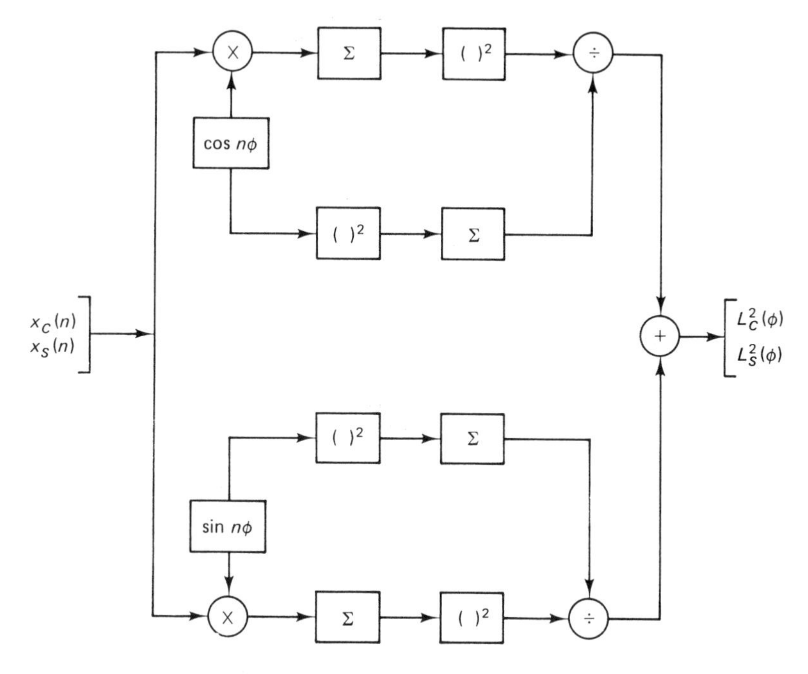

Figure 4.7 Maximum-likelihood receiver for symmetric multipath.

denominator terms in the right-hand side of Eq. (4.56) or (4.57), represent the equalization that must be applied to the two principal branches of the receiver operating on $x_c(n)$ or $x_s(n)$ in order to correct for the fact that cos $(n\phi)$ and sin $(n\phi)$ have unequal energy over the spatial interval spanned by the array.

Since the terms $w_c(n)$ and $w_s(n)$ in Eqs. (4.56) and (4.57), respectively, are Gaussian distributed and uncorrelated, the corresponding likelihood functions $f(x_c \mid u_1, u_2, \phi)$ and $f(x_s \mid v_1, v_2, \phi)$ are statistically independent. Therefore, the maximum likelihood estimate of ϕ, given both $x_c(n)$ and $x_s(n)$, is obtained by maximizing, with respect to ϕ, the sum of $L_c^2(\phi)$ and $L_s^2(\phi)$. Thus the objective function of interest is $L(\phi)$, defined by

$$L(\phi) = L_c^2(\phi) + L_s^2(\phi) \tag{4.58}$$

where $L_c^2(\phi)$ and $L_s^2(\phi)$ are themselves defined in Eqs. (4.56) and (4.57).

Maximization of the Objective Function $L(\phi)$

Equation (4.58) must be maximized with respect to ϕ to give the desired estimate of both electrical angles corresponding to the direct and specular waves incident on the array. This maximization is with respect to a single variable, ϕ, a fact that allows for a simple implementation of this ML estimator. We now discuss a simple efficient algorithm, called the *golden section search* (Bandler [17]), which may be employed

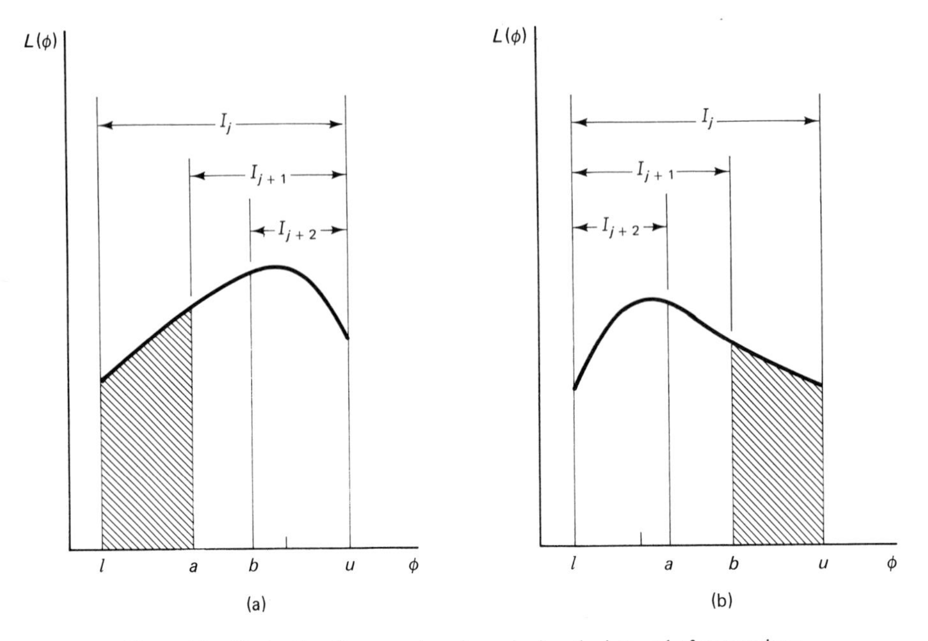

Figure 4.8 Illustrating the procedure for reducing the interval of uncertainty.

to effect this required maximization. We assume an interval of uncertainty in ϕ for which $L(\phi)$ is known to be unimodal. The object of the algorithm is to sequentially reduce the width of the pertinent interval until the value of ϕ is known within an acceptably narrow range.

At the beginning of the jth iteration, the interval of uncertainty I_j is given by (see Fig. 4.8)

$$I_j = u - l \tag{4.59}$$

Note that in Fig. 4.8 we have

$$l < a < b < u \tag{4.60}$$

After evaluating $L(a)$ and $L(b)$, one of two conclusions can be drawn:

1. If $L(a) < L(b)$, the maximum lies in $[a, u]$ and $I_{j+1} = u - a$, as in Fig. 4.8(a).
2. If $L(a) > L(b)$, the maximum lies in $[l, b]$ and $I_{j+1} = b - l$, as in Fig. 4.8(b).

Whatever the outcome of this comparison, we want

$$I_{j+1} = u - a = b - l \tag{4.61}$$

which is achieved by symmetrical placement of a and b on $[l, u]$. To increase efficiency, we would like to reuse one of the points relating to I_{j+1} in the new interval I_{j+2}. This means that a new point will be located at the positions of the crosshatch on the ϕ-axis (Fig. 4.8) in the evaluation of the new interval I_{j+2}.

Because we do not know the outcome of this decision in advance, we must have

$$I_{j+2} = u - b = a - l \tag{4.62}$$

Combining Eqs. (4.59) to (4.62), we obtain

$$I_j = I_{j+1} + I_{j+2} \tag{4.63}$$

The interval of uncertainty is to be reduced by a constant factor τ after each iteration, which implies that

$$\frac{I_j}{I_{j+1}} = \frac{I_{j+1}}{I_{j+2}} = \tau \tag{4.64}$$

Combining Eqs. (4.63) and (4.64), we obtain

$$\tau^2 = \tau + 1 \tag{4.65}$$

the solution of relevance being $\tau \simeq 1.618034$. The division of a line according to this ratio is called the *golden section* of a line.

At the jth iteration of this method, we thus have

$$\phi_a^j = \frac{1}{\tau^2} I_j + \phi_l^j$$

and

$$\phi_b^j = \frac{1}{\tau} I_j + \phi_l^j, \qquad j = 1, 2, 3, \ldots \tag{4.66}$$

where ϕ_a^j, ϕ_b^j, and ϕ_l^j are the values of ϕ at the jth iteration, corresponding to the positions a, b, and l, respectively.

Note that each iteration, except the first, involves only one function evaluation owing to the symmetry of the positioning of the points a and b in the interval. Depending on the outcome of the jth iteration, the appropriate conditions are set for the $(j + 1)$th iteration, and the procedure repeats.

One of the features of this scheme is that, for a desired accuracy in ϕ and for a given initial interval of uncertainty width, the number of iterations required is fixed in advance. After k iterations, we have

$$\frac{I_1}{I_k} = \tau^{k-1} \tag{4.67}$$

Therefore, for a desired accuracy of $\Delta\phi$, the number of iterations k should be chosen so that

$$\tau^{k-2} < \frac{\phi_u^1 - \phi_l^1}{\Delta\phi} \leq \tau^{k-1} \tag{4.68}$$

For example, if the initial interval is chosen to be 1.0 BW, and it is desired to obtain an accuracy of 0.001 BW, then 16 iterations are necessary.

It is informative to determine an explicit expression for the objective function $L(\phi)$ for the ideal case of infinite *signal-to-noise ratio* (SNR). Thus, if in Eqs. (4.38) and (4.41), we put both noise terms $w_c(n)$ and $w_s(n)$ equal to zero and also assign some fixed value, say ϕ_T, to the "true" angle of target arrival, we obtain

$$x_c(n) = u_1 \cos(n\phi_T) + u_2 \sin(n\phi_T) \tag{4.69}$$

and

$$x_s(n) = v_1 \cos(n\phi_T) + v_2 \sin(n\phi_T) \tag{4.70}$$

Then, substituting Eq. (4.69) into (4.56), and Eq. (4.70) into (4.57), and using these results in Eq. (4.58), we find that the explicit dependence of the objective function on the "test" angle ϕ takes on the following form (normalized with respect to the amplitude a_1):

$$L(\phi) = \frac{[S_1^2(\phi) + S_2^2(\phi)][NA - BS_3(\phi)] + 2S_1(\phi)S_2(\phi)[NB - AS_3(\phi)]}{N^2 - S_3^2(\phi)} \tag{4.71}$$

where

$$\begin{aligned} S_1(\phi) &= \frac{\sin\{N(\phi - \phi_T)/2\}}{\sin\{(\phi - \phi_T)/2\}} \\ S_2(\phi) &= \frac{\sin\{N(\phi + \phi_T)/2\}}{\sin\{(\phi + \phi_T)/2\}} \\ S_3(\phi) &= \frac{\sin(N\phi)}{\sin\phi} \\ A &= 1 + \rho^2 \\ B &= 2\rho\cos\psi \end{aligned} \tag{4.72}$$

Figure 4.9 shows plots of this objective function for the true electrical phase ϕ_T equal to 0.125, 0.25, 0.5, and 1 BW when the number of elements $N = 21$ and the reflection coefficient $\rho = 0.9$. The position of the vertical arrow indicates the true target angle ϕ_T. There are several interesting considerations which arise from this set of curves. First, when $\phi_T \leq 1$ BW, all curves are unimodal over the range $0 \leq \phi \leq 1$ BW. Therefore, the restriction that the respective objective function must be unimodal in order to apply the golden section search imposes no difficulty in using the method in this situation. Also, when $\phi_T \leq 1$ BW, we note that the maxima are relatively well defined for $\psi = 0°$, but become progressively more dispersed as ψ approaches 180°. Thus we expect that the confounding effects of noise will be greater for the case of $\psi = 180°$ than when $\psi = 0°$. We shall indeed find this to be true later.

We note that, for a fixed value of ψ, the maxima become more distinct as ϕ_T increases (up to a value of 1.0 BW). Therefore, we should again anticipate better estimation behavior in the presence of noise as ϕ_T increases up to this value. This statement will also be verified later.

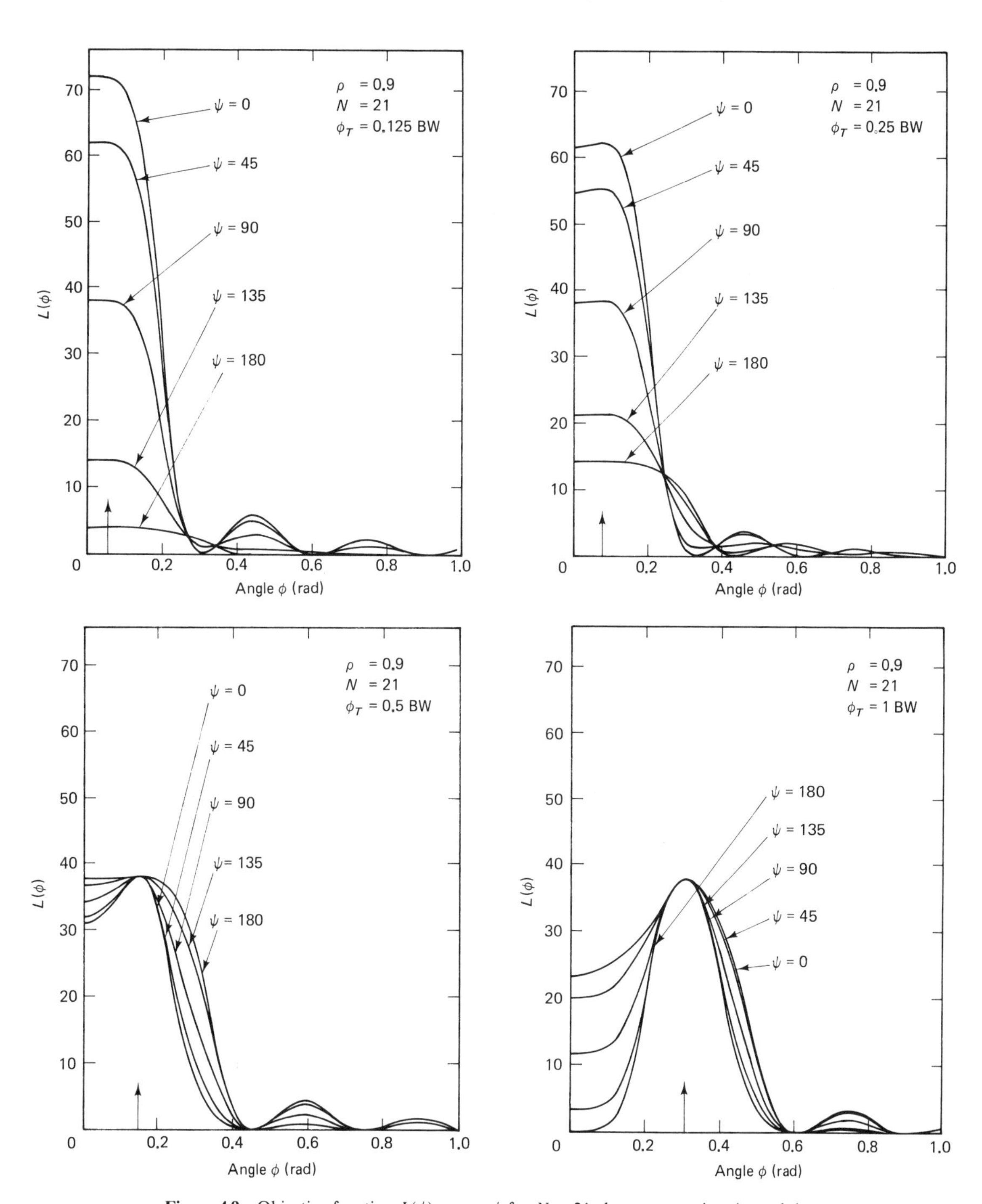

Figure 4.9 Objective function $L(\phi)$ versus ϕ for $N = 21$ elements, varying ϕ_T and ψ.

We note that, for many cases, the curves in Fig. 4.9 are nonsymmetrical about their maxima. They generally fall off much more quickly on the high side of the maximum than on the low side. This nonsymmetric behavior is a consequence of the nonlinear characteristics of the estimator and, as such, generally leads to biased results.

The curves of Fig. 4.9 are actually plots which are proportional to the log-posterior probability density function of ϕ. With this in mind, it is easy to determine that the confidence interval for ϕ will extend farther to the left than to the right for most of the cases in the figure. Consequently, the estimates will generally tend to be biased low, a fact which, except for a few exceptions, will be verified later as a result of computer simulations.

Theoretical Variance of the ML Receiver

We now develop the form of the theoretical variance of the ML receiver shown in Fig. 4.7. The technique used will be to treat $L(\phi)$ as if it were linear in ϕ, which of course is true only over a small range of ϕ. However, within this small range, the estimator may be shown to be linear and unbiased; hence the Cramér–Rao bound applies and the conditions necessary for the equality are satisfied (Haykin and Reilly [7]). Therefore, the variance of the estimate $\hat{\phi}$ may be determined by evaluating the pertinent value of the bound. We first write

$$f(x_c, x_s \mid \phi) = C_4 \exp\left\{\frac{1}{2\sigma^2}\left[L(\phi) - X_{c,s}\right]\right\} \tag{4.73}$$

where C_4 is a constant, $L(\phi)$ is defined by Eq. (4.58), and

$$X_{c,s} = \sum_n x_c^2(n) + x_s^2(n)$$

The form of Cramér–Rao bound of interest to us is given by [16]

$$\text{var}\,[\hat{\phi}] = E[(\hat{\phi} - \phi)^2] \geq \left\{-E\left[\frac{\partial^2 \ln f(x_c, x_s \mid \phi)}{\partial \phi^2}\right]\right\}^{-1} \tag{4.74}$$

Hence, substituting Eq. (4.73) into (4.74), we obtain the desired result:

$$\text{var}\,[\hat{\phi}] = -2\sigma^2\left\{E\left[\frac{\partial^2 L(\phi)}{\partial \phi^2}\right]\right\}^{-1} \tag{4.75}$$

The expression for the variance of $\hat{\phi}$ in Eq. (4.75) is applicable only to the case when the SNR is high. We also note that, at high SNR, the ML estimator is essentially unbiased. Thus, when the SNR is high, we may use Eqs. (4.71) and (4.75) to find the explicit dependence of the variance of the estimate $\hat{\phi}$ on the true value of the angle ϕ, the reflection coefficient ρ, and the phase difference ψ.

Plots of var $[\hat{\phi}]$ computed in this way, are shown in Fig. 4.10 when $N = 21$ elements. For the curves in this figure, 0 dB on the ordinate corresponds to 1 (degree)^2. We see that the variance depends very strongly on the parameter ϕ_T (the

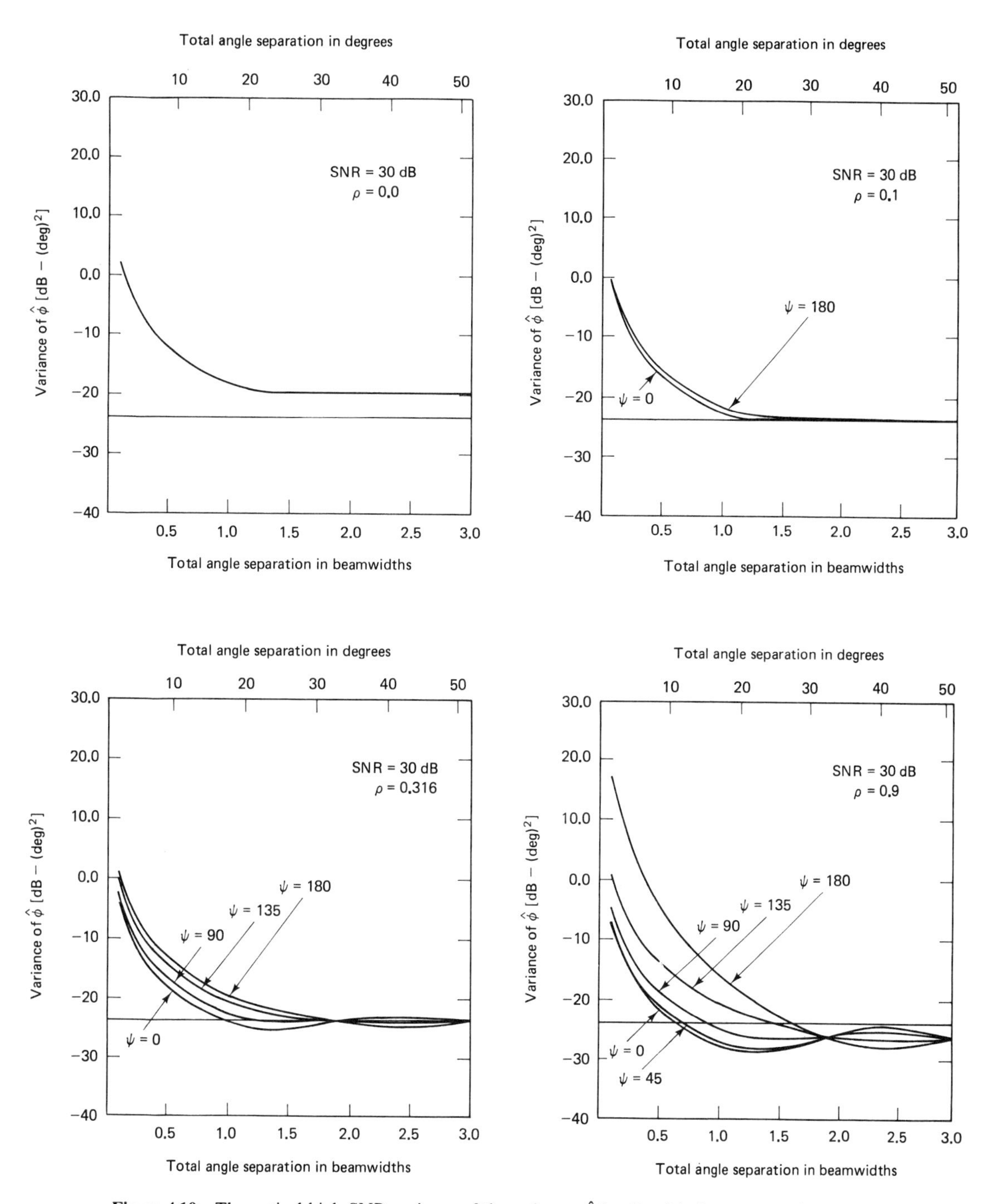

Figure 4.10 Theoretical high-SNR variance of the estimate $\hat{\phi}$ for $N = 21$ elements, varying ψ and ρ.

true angle of arrival) and on the parameters ψ and ρ. The dependence on ψ is not surprising; if $\psi = 180°$, the two incident signals will be largely canceled out at the center of the array, thereby decreasing the effective SNR. Also, as the separation between the two incidence angles goes to zero (for $\psi = 180°$), the total power received by the array decreases, because the two signal components cancel over a larger span of the array. This explains why the variances increase very sharply for $\psi = 180°$ as ϕ_T approaches zero. This effect is more pronounced for larger values of ρ because the signal cancellation in this case is more complete.

On the other hand, when $\psi = 0°$, we see that the variance is much lower than in the previous case for $\phi_T < 0.9$ BW. This is because the power incident on the array from the specular component adds to the direct component power and increases the overall SNR. We also note that the effect of variation in ψ becomes less pronounced as ρ decreases. This is because the strength of the specular component diminishes in this case, thereby causing the total received signal to become less sensitive to changes in the parameters of the specular component. When $\rho = 0$, the variance becomes independent of ψ.

The solid horizontal line in all the curves in Fig. 4.10 shows the equivalent variance of $\hat{\phi}$ when it is known that only one plane wave is incident on the array. Thus Fig. 4.10 shows the effect of the multipath component on the elevation-angle estimate.

The manner in which the variance of $\hat{\phi}$ varies with the number of elements N is shown in Fig. 4.11 for $\rho = 0.9$ and several different values of ϕ_T. The abscissa is plotted on a log scale, and it may be observed that doubling the number of elements decreases the variance by 9 dB in all situations when N is higher than about 5. Thus we see that a "3-dB increase" in N improves the variances by 9 dB, which implies that the variance of $\hat{\phi}$ varies as N^3. Therefore, increasing the number of elements in percentage terms buys a considerable margin in increased performance.

Computer Simulation Results

We now compare the theoretical results as previously discussed with those obtained by computer simulation. Simulations were run for parameter values corresponding to those used in Fig. 4.10 and the results are shown in Fig. 4.12. The variances obtained by simulation are shown as a dot, and the theoretical variances are shown superimposed as dashed lines. The simulations were executed at an SNR of 30 dB for $N = 21$ elements; these values were sufficiently high for the resulting simulation variances to agree with the high-SNR theoretical variance as given by Eq. (4.75). The simulation variances shown are the result of 99 independent trials. We see excellent agreement between the predicted theoretical variances and those obtained by simulation.

We now examine the dependence of the variance of the estimate $\hat{\phi}$ on the SNR. The results are shown in Fig. 4.13 for $N = 21$ elements. At sufficiently high SNR, for any value of the parameter ϕ_T, the variance is a linear function of the elemental SNR; however, for low SNR, the variance is no longer linear and asymp-

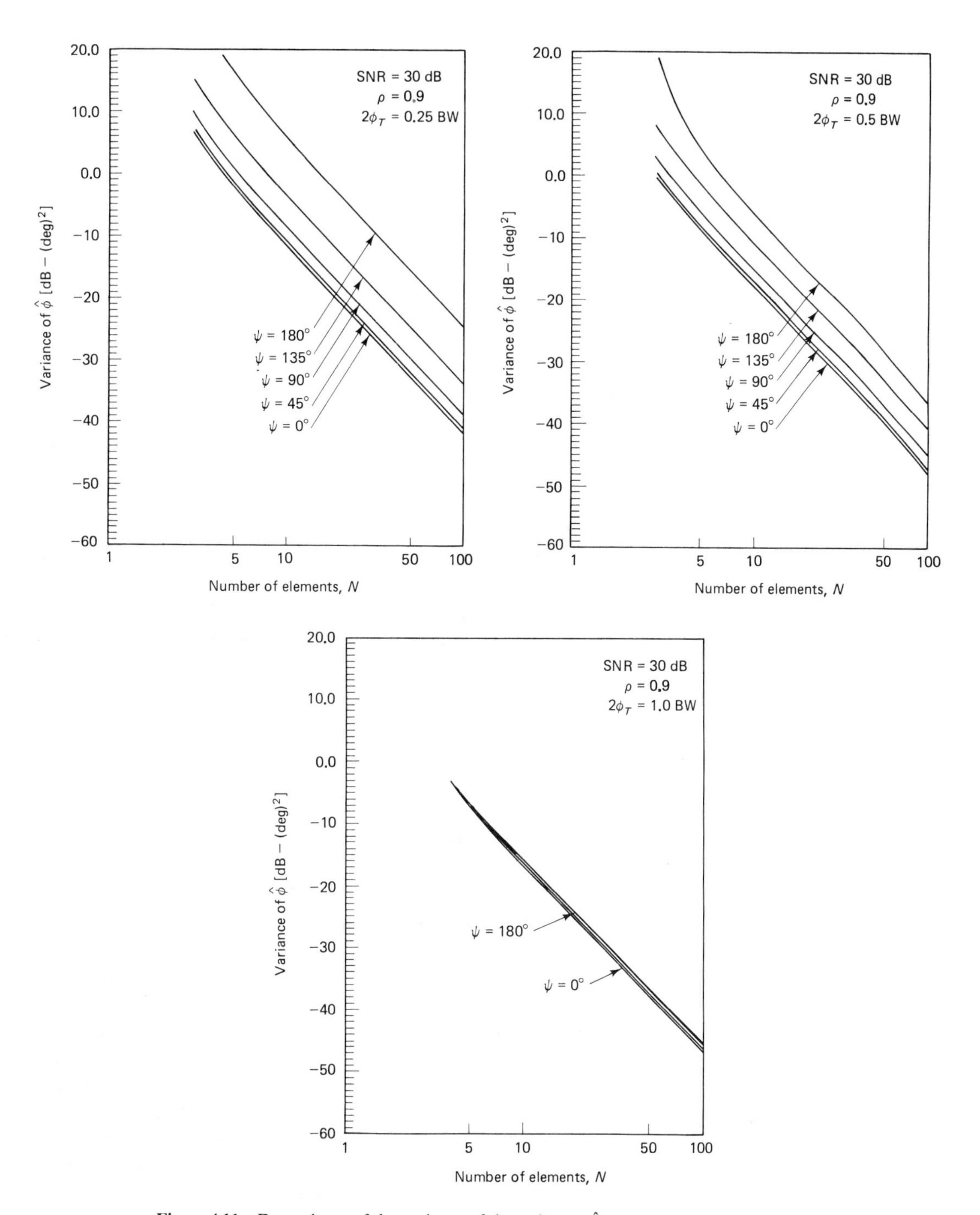

Figure 4.11 Dependence of the variance of the estimate $\hat{\phi}$ on number of elements N.

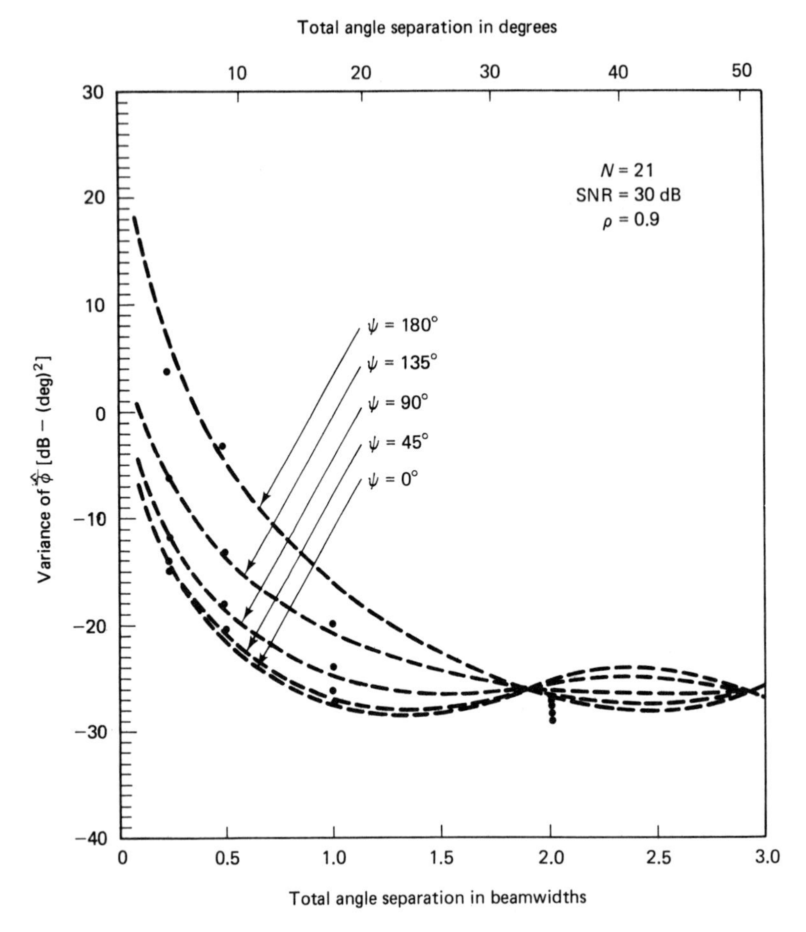

Figure 4.12 Variance of estimate $\hat{\phi}$ obtained by computer simulation for $N = 21$ elements and varying ψ.

totically approaches an upper limit as illustrated in Fig. 4.13. It is interesting to note that the signal-to-noise ratio *threshold*, SNR_T, where the variance departs from linearity, is about the same value for which the corresponding curve showing the mean of the estimate $\hat{\phi}$ departs from its true value (Haykin and Reilly [7]).

Mention should be made of the fact that White [18] has developed an estimator which is somewhat similar in nature to the symmetric ML estimator developed in this chapter. His approach is simply to take the model form of the elemental excitations as given by Eqs. (4.38) and (4.41) and form the resulting PDF of the data given all the parameter values, which is also the joint likelihood function of all the parameters. He then devises an analog computer structure, which involves nulling out an error voltage in a loop, to solve for the maximum of the likelihood function with respect to all five parameter values. His structure is quite complex, involving

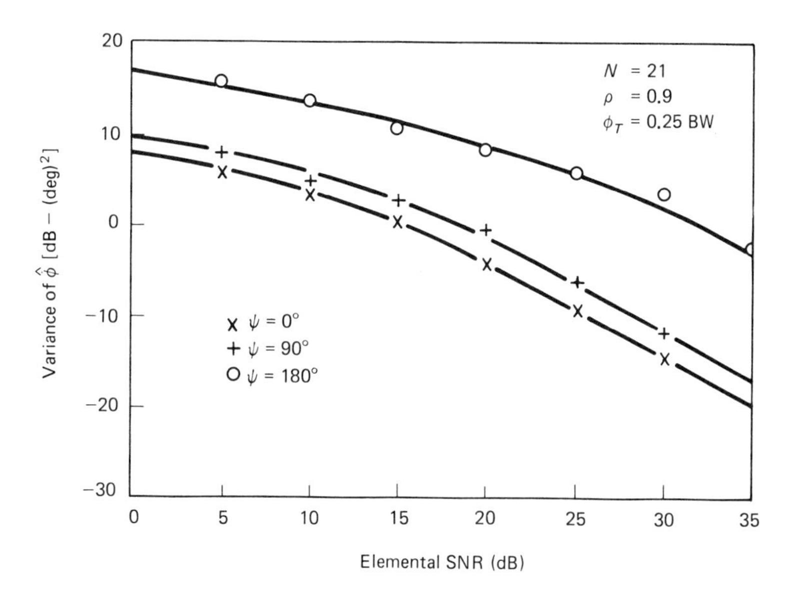

Figure 4.13 Variance of estimate $\hat{\phi}$ versus elemental SNR, as obtained by computer simulation, for $N = 21$ elements, $\rho = 0.9$, $\phi = 0.25$ BW.

$2N$ programmable phase shifters, two beamforming networks, a multitude of programmable attenuators, adders and multipliers, as well as a computer that must calculate the value of a variety of functions.

The symmetric ML estimator structure which has been developed here is inherently much simpler than White's processor because the estimator is a function solely of the one parameter of interest; all the other undesired parameters have been integrated out. Indeed, the symmetric ML estimator is simple enough for implementation on a microprocessor. The theoretical performances of both the symmetric ML and the White estimator are identical.

Temporal Averaging

In a practical situation where a number of snapshots of the environment are available, there is the question of how to utilize all the available data so as to produce the best overall direction estimator. Here we describe a method for doing this, based on maximum-likelihood principles (Reilly [19]).

In a typical radar system, the pulse width is on the order of 1 μs, and the pulse repetition frequency is about 1 to 10 kHz. The receiver bandwidth typically equals the reciprocal of the pulse width, which implies a bandwidth on the order of 1 MHz. Based on these parameter values, it is justifiable to assume that the noise present at the receiver output is uncorrelated from pulse to pulse. With the assumption of Gaussian noise, the noise samples are then statistically independent.

We now evaluate the likelihood function of ϕ based on data from each element of the array over a total of K snapshots in time. We assume that only one snapshot is gathered per transmitted pulse. Then, assuming a uniform prior distribution, we may express the likelihood function of ϕ, based on spatial data from snapshot k, as follows:

$$\begin{aligned} f(\mathbf{x}(k) \mid \phi) &= \frac{f(\mathbf{x}(k))}{f(\phi)} f(\phi \mid \mathbf{x}(k)) \\ &= C_4 f(\mathbf{x}(k)) \exp \left\{ \frac{1}{2\sigma^2} \left[L_k(\phi) - \sum_n x_c^2(n, k) - \sum_n x_s^2(n, k) \right] \right\} \end{aligned} \tag{4.76}$$

where C_4 is a constant, $x_c(n, k)$ and $x_s(n, k)$ are the in-phase and quadrature components of the observed data at element n and for snapshot k, and

$$L_{ck}^2(\phi) = \frac{\left[\sum_n x_c(n, k) \cos(n\phi) \right]^2}{\sum_n \cos^2(n\phi)} + \frac{\left[\sum_n x_c(n, k) \sin(n\phi) \right]^2}{\sum_n \sin^2(n\phi)}$$

$$L_{sk}^2(\phi) = \frac{\left[\sum_n x_s(n, k) \cos(n\phi) \right]^2}{\sum_n \cos^2(n\phi)} + \frac{\left[\sum_n x_s(n, k) \sin(n\phi) \right]^2}{\sum_n \sin^2(n\phi)}$$

$$L_k(\phi) = L_{ck}^2(\phi) + L_{sk}^2(\phi) \tag{4.77}$$

Let the $NK \times 1$ vector $\mathbf{X}$ denote all the available data over the complete ranges of n and k. Then, using the statistical independence of noise samples between snapshots, we may express the corresponding likelihood function of ϕ as

$$\begin{aligned} \mathbf{f}(\mathbf{X} \mid \phi) &= \prod_{k=1}^{K} f(\mathbf{x}(k) \mid \phi) \\ &= C_4^K \prod_{k=1}^{K} f(\mathbf{x}(k)) \exp \left\{ \frac{1}{2\sigma^2} \sum_{k=1}^{K} \left[L_k(\phi) - \sum_n x_c^2(n, k) - \sum_n x_s^2(n, k) \right] \right\} \end{aligned} \tag{4.78}$$

From Eq. (4.78), it is obvious that in order to maximize the likelihood function $f(\mathbf{x} \mid \phi)$, we must maximize the new objective function

$$Q(\phi) = \sum_{k=1}^{K} L_k(\phi) \tag{4.79}$$

where $L_k(\phi)$ is defined by Eq. (4.77). The value $\hat{\phi}$ of ϕ that maximizes $Q(\phi)$ is the maximum-likelihood estimate of ϕ over all available snapshots.

Note that the objective function $Q(\phi)$ is very similar to the single snapshot form used previously. The estimate $\hat{\phi}$ is obtained by finding that value of ϕ which maximizes the sum of all the functions formed from each snapshot of data.

In Fig. 4.14 we show the variances of $\hat{\phi}$ versus the number of snapshots used

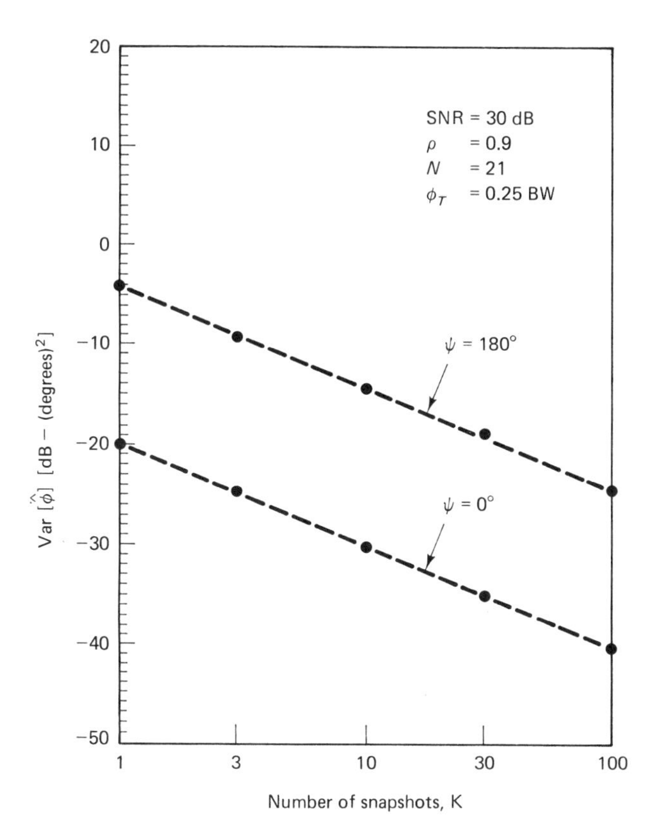

Figure 4.14 Performance of temporal smoothing technique showing variance of estimate $\hat{\phi}$ versus number of snapshots.

for two different values of ψ, 0° and 180°. The other parameters used in the computer simulation were: $\phi_T = 0.25$ BW, $\rho = 0.9$, $N = 21$, and SNR $= 30$ dB. We note from the figure that the estimator variance varies inversely with K. For example, if 100 snapshots are used, the variance improves by 20 dB over that obtained with a single snapshot. It should, however, be emphasized that this improvement in receiver performance is obtained on the assumption that the direction of the target remains essentially unchanged during the period of time over which the set of $K = 100$ snapshots is collected.

4.7 THE MAXIMUM-LIKELIHOOD RECEIVER: GENERAL CASE

In certain low-angle tracking-radar problems, where the range of the target is sufficient for the curvature of the earth to become significant, the symmetric specular multipath model used in the preceding section is no longer valid. Also, the situation is exacerbated by the phenomenon of atmospheric bending [1], where the electromagnetic radiation from the radar propagates along a path which bends toward the ground. This effect is caused by refraction due to a variation in atmospheric density with altitude. In tracking applications where these geometries are prevalent, it is necessary to develop a more general form of the maximum-likelihood estimator. In any event, it is highly desirable to develop such a generalized estimator so that we may deal with the problem of a number of plane waves arriving at the array from arbitrary directions.

Consider then the baseband form of the in phase elemental excitations $x_c(n)$ which result from M plane waves incident on the array from arbitrary electrical angles $\phi_1, \phi_2, \ldots, \phi_M$. The angle ϕ_m is related to the actual angle of arrival of the mth incident wave, θ_m, by Eq. (4.5). These excitations at element n of the array may be expressed in terms of $\phi_1, \phi_2, \ldots, \phi_M$ by

$$x_c(n) = \sum_{m=1}^{M} u_m \cos(n\phi_m) + \sum_{m=1}^{M} v_m \sin(n\phi_m) + w_c(n),$$

$$n = -(N-1)/2, \ldots, (N-1)/2 \tag{4.80}$$

where $u_m = a_m \cos \psi_m$ (4.81)

$v_m = -a_m \sin \psi_m$ (4.82)

a_m = complex amplitude of the mth plane wave

ψ_m = initial phase of the mth plane wave

The noise $w_c(n)$ is assumed to be Gaussian of zero mean and variance σ^2, for all n.

Let us now consider the following $2M \times N$ signal matrix:

$$\mathbf{C} = \begin{bmatrix} \cos[-\phi_1(N-1)/2] & \cdots & \cos\phi_1 & \cdots & \cos[\phi_1(N-1)/2] \\ \sin[-\phi_1(N-1)/2] & \cdots & \sin\phi_1 & \cdots & \sin[\phi_1(N-1)/2] \\ \cos[-\phi_2(N-1)/2] & \cdots & \cos\phi_2 & \cdots & \cos[\phi_2(N-1)/2] \\ \sin[-\phi_2(N-1)/2] & \cdots & \sin\phi_2 & \cdots & \sin[\phi_2(N-1)/2] \\ \vdots & & & & \\ \cos[-\phi_M(N-1)/2] & \cdots & \cos\phi_M & \cdots & \cos[\phi_M(N-1)/2] \\ \sin[-\phi_M(N-1)/2] & \cdots & \sin\phi_M & \cdots & \sin[\phi_M(N-1)/2] \end{bmatrix} \tag{4.83}$$

In other words, the matrix **C** is composed of a cos $(\cdot)$ and a sin $(\cdot)$ function defined over the complete range of n for each $\phi_m, m = 1, \ldots, M$.

We now define the signal vector $\{\mathbf{s}_k(n)\}$, $k = 1, \ldots, 2M$, which is obtained from the matrix **C** by performing a *Gram–Schmidt orthonormalization procedure* [12] on each row of **C**. We note that the Gram–Schmidt orthonormalization procedure introduces a linear transformation in the original signal space **C**, which results in a set of $2M$ orthonormal basis vectors $\mathbf{s}_k$. Provided that the row vectors of **C** are linearly independent over the corresponding ranges of n and m, the number of orthonormal basis vectors will equal the number of rows in the original matrix (i.e., $2M$). We note that the $\mathbf{s}_k$ are still functions of the ϕ_m.

We may therefore rewrite Eq. (4.80) as

$$x_c(n) = \sum_{k=1}^{2M} q_k s_k(n) + w_c(n) \tag{4.84}$$

where the q_k are the projections of the $u_m \cos(n\phi_m)$ or the $v_m \sin(n\phi_m)$ onto the corresponding $\mathbf{s}_k$. The resulting likelihood function then becomes

$$f(\mathbf{x}_c \mid \mathbf{q}, \boldsymbol{\phi}) = K_1 \exp\left\{-\frac{1}{2\sigma^2}\sum_n \left[x_c(n) - \sum_{k=1}^{2M} q_k s_k(n)\right]^2\right\} \tag{4.85}$$

where $\mathbf{x}_c$, $\boldsymbol{\phi}$, and $\mathbf{q}$ are the vectors of all $x_c(n)$, ϕ_k, and q_k, respectively, and K_1 is a constant.

We note that (by definition of orthonormal basis functions)

$$\sum_n s_k(n)s_j(n) = \delta_{kj} \tag{4.86}$$

where δ_{kj} is the Kronecker delta. Hence we may write

$$\begin{aligned}\sum_n \left[\sum_{k=1}^{2M} q_k s_k(n)\right]^2 &= \sum_{k=1}^{2M}\sum_{j=1}^{2M} q_k q_j \sum_n s_k(n)s_j(n) \\ &= \sum_{k=1}^{2M}\sum_{j=1}^{2M} q_k q_j \delta_{kj} \\ &= \sum_{k=1}^{2M} q_k^2\end{aligned} \tag{4.87}$$

Define the quantity $L_{ck}(\boldsymbol{\phi})$ as

$$L_{ck}(\boldsymbol{\phi}) = \sum_n x_c(n)s_k(n), \qquad k = 1, 2, \ldots, 2M \tag{4.88}$$

Then, expanding the squared term in the exponent of Eq. (4.85) and substituting Eqs. (4.87) and (4.88) in the expansion, we get

$$\begin{aligned}f(\mathbf{x}_c \mid \mathbf{q}, \boldsymbol{\phi}) = K_1 &\exp\left\{-\frac{1}{2\sigma^2}\left[\sum_n x_c^2(n) - \sum_k L_{ck}^2(\boldsymbol{\phi})\right]\right\} \\ &\times \exp\left\{-\frac{1}{2\sigma^2}\sum_k [q_k - L_{ck}(\boldsymbol{\phi})]^2\right\}\end{aligned} \tag{4.89}$$

The above is now in a convenient form for removing the dependence on q_k. By analogy with the application of Bayes' rule as in Section 4.6, we may express the a posteriori joint PDF of the vector $\boldsymbol{\phi}$, given the in-phase data vector $\mathbf{x}_c$, as follows:

$$f(\boldsymbol{\phi} \mid \mathbf{x}_c) = \frac{1}{f(\mathbf{x}_c)}\int_{q_1} \cdots \int_{q_{2M}} f(\mathbf{x}_c \mid \mathbf{q}, \boldsymbol{\phi}) f(\boldsymbol{\phi}, q_1, \ldots, q_{2M})\, dq_1 \cdots dq_{2M} \tag{4.90}$$

Here again, we assume a uniform prior $f(\boldsymbol{\phi}, \mathbf{q})$. Accordingly, after substituting Eq. (4.89) into (4.90), and lumping the three constants K_1, $1/f(x_c)$ and the constant of the prior PDF together into a new constant K_2, we get

$$f(\boldsymbol{\phi} \mid \mathbf{x}_c) = K_2 \exp\left\{-\frac{1}{2\sigma^2}\left[\sum_n x_c(n) - \sum_k L_{ck}^2(\boldsymbol{\phi})\right]\right\}$$

$$\times \int_{q_1} \cdots \int_{q_{2M}} \exp\left\{-\frac{1}{2\sigma^2}\sum_k [q_k - L_{ck}(\boldsymbol{\phi})]^2 \, dq_1 \cdots dq_{2M}\right\} \tag{4.91}$$

Now we note that the q_k are dependent on the a_k, the ψ_k, and the ϕ_k. On the other hand, $L_{ck}(\boldsymbol{\phi})$ is a function of only the observed data and the vector $\boldsymbol{\phi}$. In integrating out the dependence on the q_k, as shown in Eq. (4.91), our aim is basically to remove the dependence on the unwanted parameters, the a_k and the ψ_k. Thus the integrand in Eq. (4.91), except for the scaling factor $(2\pi\sigma^2)^{-M}$, represents the joint PDF of a multivariate Gaussian distribution of individual means $L_{ck}(\boldsymbol{\phi})$, $k = 1, 2, \ldots, 2M$, and a common variance σ^2. Since by definition the total volume under a joint PDF equals 1, it follows that the $2M$-fold integral in Eq. (4.91) equals $(2\pi\sigma^2)^M$, which is independent of $\boldsymbol{\phi}$. Hence, by combining this constant together with K_2 to form a new constant K_3, we may simplify Eq. (4.91) as follows:

$$f(\boldsymbol{\phi} \mid \mathbf{x}_c) = K_3 \exp\left\{-\frac{1}{2\sigma^2}\left[\sum_n x_c^2(n) - \sum_k L_{ck}^2(\boldsymbol{\phi})\right]\right\} \tag{4.92}$$

Following a procedure similar to that described above, the a posteriori PDF of $\boldsymbol{\phi}$, given the quadrature data vector $\mathbf{x}_s$, may be determined as

$$f(\boldsymbol{\phi} \mid \mathbf{x}_s) = K_4 \exp\left\{-\frac{1}{2\sigma^2}\left[\sum_n x_s^2(n) - \sum_k L_{sk}^2(\boldsymbol{\phi})\right]\right\} \tag{4.93}$$

where K_4 is a constant, and

$$L_{sk}(\boldsymbol{\phi}) = \sum_n x_s(n) s_k(n), \qquad k = 1, 2, \ldots, M \tag{4.94}$$

Since the inphase and quadrature data are statistically independent, it follows that the *a posteriori* PDF of $\boldsymbol{\phi}$, given $\mathbf{x}_c$ and $\mathbf{x}_s$, equals

$$f(\boldsymbol{\phi} \mid \mathbf{x}_c, \mathbf{x}_s) = K_5 \exp\left\{-\frac{1}{2\sigma^2}\left[\sum_n (x_c^2(n) + x_s^2(n)) - \sum_k (L_{ck}^2(\boldsymbol{\phi}) + L_{sk}^2(\boldsymbol{\phi}))\right]\right\} \tag{4.95}$$

where the constant $K_5 = K_3 K_4$. This PDF is maximized by maximizing the expression $L_A(\boldsymbol{\phi})$ given by

$$L_A(\boldsymbol{\phi}) = \sum_k [L_{ck}^2(\boldsymbol{\phi}) + L_{sk}^2(\boldsymbol{\phi})] \tag{4.96}$$

with respect to the ϕ_k, $k = 1, \ldots, M$. The desired joint maximum a posteriori (MAP) estimate $\hat{\boldsymbol{\phi}}$ of the vector $\boldsymbol{\phi}$ is that which maximizes the expression in Eq. (4.96). A receiver structure based on the use of Eq. (4.96) is shown in Fig. 4.15.

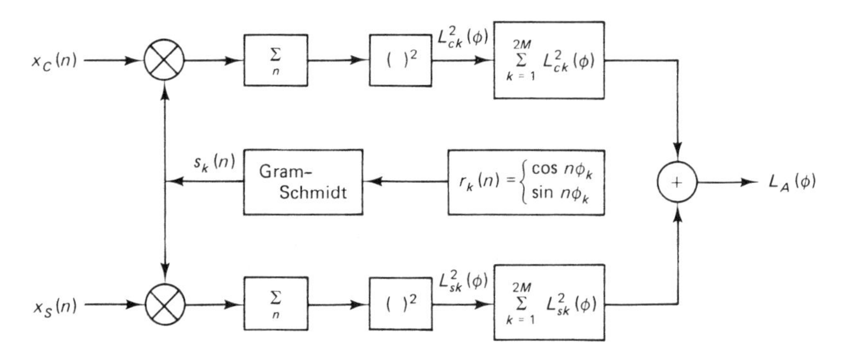

Figure 4.15 Generalized maximum-likelihood receiver.

We note that the estimation procedure presented here requires the estimation of only one parameter per plane wave incident on the array; that is, if M waves are impinging on the array, only an M-dimensional nonlinear maximization is required. We also note that we could have formed the likelihood function directly from Eq. (4.80) and found an ML estimate of the ϕ_m, u_m, and v_m, for $m = 1, \ldots, M$, by maximizing the corresponding likelihood function. This procedure would have required a $3M$-dimensional maximization. Since the order of the computation complexity varies as the dimensionality cubed, the proposed method is in the order of being 27 times more efficient. Therefore, the effort involved is integrating out all the undesired parameters in the model of Eq. (4.80) is expended to produce a more computationally efficient estimator.

Maximization of the Likelihood Function

In a specular multipath environment, only two plane waves are incident on the array. Hence, in the application of this ML estimator, a maximization must be performed in two variables. Unfortunately, the algorithms in existence for multidimensional optimization are considerably more complex than the golden section search applicable to the symmetric ML receiver, as discussed in Section 4.6.

The most widely used multidimensional optimization algorithms are the *conjugate-gradient methods* developed by Fletcher and Powell [20]. These algorithms are well documented from the statistical viewpoint by Bard [21] and from the numerical viewpoint by Bandler [17]. An efficient, well-debugged software version of this algorithm is available in [22].

Figure 4.16 shows a flow-graph implementation of the ML estimator for the general case of M incident plane waves. Note that the number of incident plane waves M is assumed known, whereas in practical applications, this may not be the case. Reilly and Haykin [8] describe a test, based on the *F-statistic*, to determine whether or not the value of M used in the computation is high enough. Wax and Kailath [23] describe another procedure, based on the *Akaike information criterion*, for estimating the value of M.

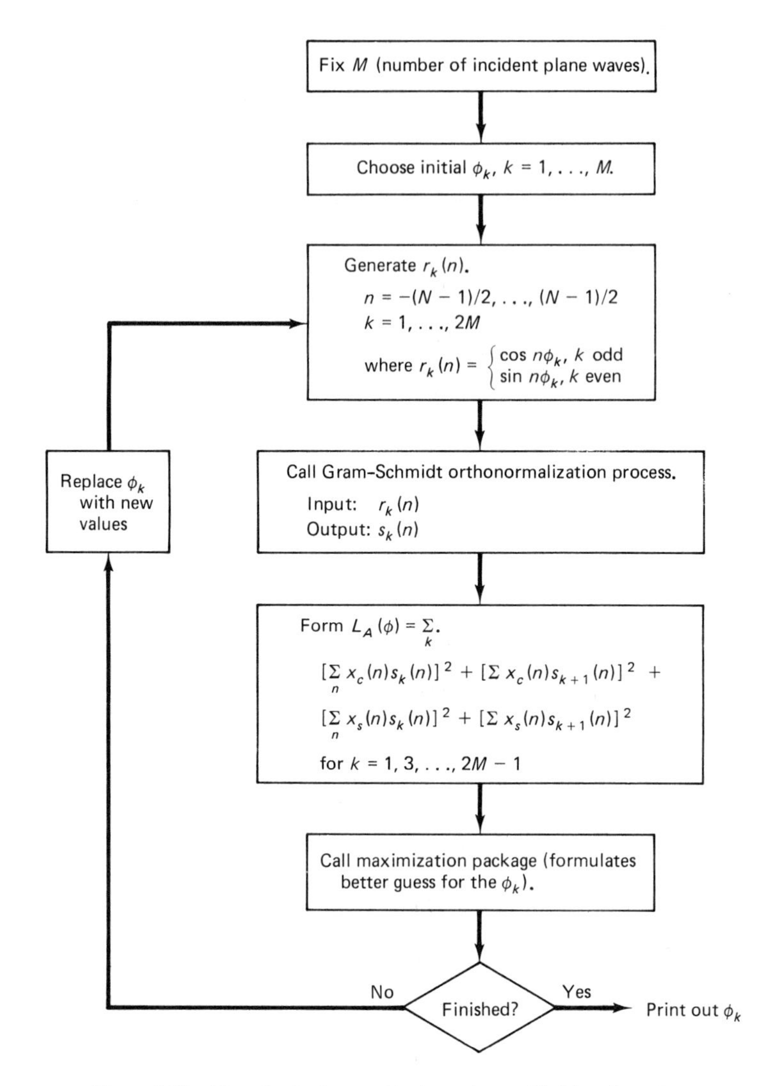

Figure 4.16 Flowchart of generalized maximum-likelihood receiver.

The first step in the computation is to form the $2M \times N$ matrix $\mathbf{C}$ of values of cos $(n\phi_k)$ and sin $(n\phi_k)$ using initial estimates $\hat{\boldsymbol{\phi}}$. For each ϕ_k, two rows of $\mathbf{C}$ are formed: one for cos $(n\phi_k)$ and one for sin $(n\phi_k)$. The Gram–Schmidt orthonormalization procedure is then applied to the matrix $\mathbf{C}$. The output is another $2M$ by N matrix $\mathbf{S}(\cdot)$, the rows of which all have unit energy and are all mutually orthogonal.

Each row of the matrix $\mathbf{S}(\cdot)$ is next correlated with the in-phase data vector $\mathbf{x}_c(n)$ and the quadrature data vector $\mathbf{x}_s(n)$ over the range of n. The sum of the squares of all $2 \times 2M$ correlations is then determined, and this sum determines the objective function used in the maximization routine. The process iterates, each time with a better guess of $\hat{\boldsymbol{\phi}}$, until the estimates have sufficiently converged.

Results. The high-SNR theoretical variances for this estimator may be determined by exactly the same method that was used in Section 4.6. As an example, consider the case of $N = 21$ elements and $M = 2$ plane waves. One incident plane wave is fixed at an actual direction corresponding to the electrical phase angle $\phi_1 = 26.4°$ ($\simeq 1.5$ BW for a 21-element array) below the normal. Figure 4.17 shows the theoretical variance of the angle separation between the two incident plane waves. The variance is calibrated in decibels–(degrees)^2, and the angle separation in BW. The SNR value used in this instance was 30 dB; the SNR refers to the equivalent free-space SNR.

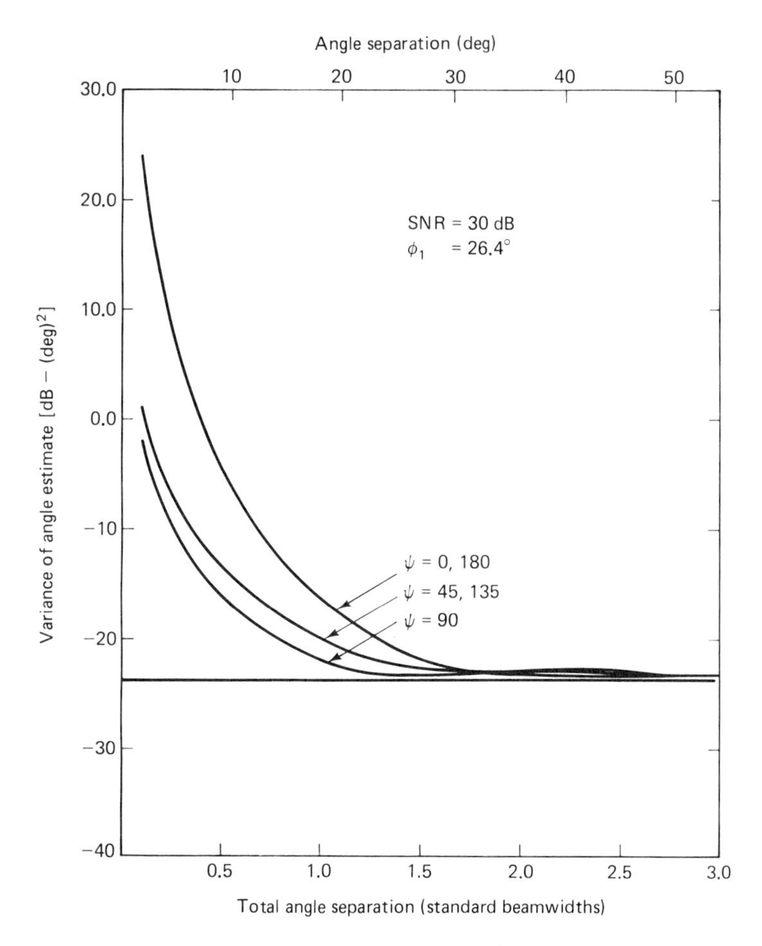

Figure 4.17 Theoretical high-SNR variance estimate $\hat{\phi}$ for generalized ML receiver for $N = 21$ elements.

There are some interesting facts which are evident from Fig. 4.17. The first is that, whereas the equivalent set of curves for the symmetric ML estimator shows that the variances are best for $\psi = 0°$ and degrade steadily as $\psi \rightarrow 180°$, the variances shown in Fig. 4.17 are best for $\psi = 90°$. Also, the behavior is symmetric in ψ, with the variances degrading as ψ approaches either 0 or 180°. We note that, when $\psi = 90°$, the two incident signal components are spatially uncorrelated over the span of the array for any incident direction angle. We also note from Fig. 4.17 that the best variances attainable with the asymmetric estimator are about 5 dB worse than the best attainable with the symmetric estimator. This is because the extra knowledge gleaned in the model by assuming symmetric incidence angles is not available in the asymmetric realization. The curves of Fig. 4.17 display a poorer variance at low-angle separations (e.g., 0.1 BW), but are not necessarily uniformly worse.

The horizontal line in Fig. 4.17 indicates the variance attainable with an estimator which knows that only one plane wave is incident on the array. Thus we see that, for this asymmetric estimator, the presence of more than one incident plane wave always degrades the performance of the estimator. This is in contrast to the symmetric structure, which in some cases exhibits a performance that improves in the presence of an additional incident wave.

Since this asymmetric ML technique is a multivariate estimator, there is a different variance associated with each ϕ_j, $j = 1, \ldots, M$. The curves in Fig. 4.17 correspond to the case where $M = 2$ and $\rho = 0.9$ and show only the results having the better variance; that is, they show the variances associated with the stronger (direct) signal. The variances associated with the reflected signal (specular component) are about 1 dB worse uniformly than those shown in Fig. 4.17.

Three-dimensional plots of the objective function $L_A(\boldsymbol{\phi})$ as defined by Eq. (4.96) against ϕ_1 and ϕ_2 are shown in Fig. 4.18 for an angle separation of 0.5 BW. The number of elements $N = 21$, and the true value of electrical angle ϕ_1 is 90°. Since the estimator is now multivariate, multidimensional surfaces are required to plot the objective function $L_A(\boldsymbol{\phi})$. Note that an equivalue contour on these plots corresponds to a contour of equal a posteriori probability of the joint estimate $\hat{\boldsymbol{\phi}}$. This may be seen as follows. Suppose that we wish to establish a set of points $\{\phi_1, \ldots, \phi_M\}$ all with the same a posteriori probability. We then find that these points define equivalue contours of the objective function $L_A(\boldsymbol{\phi})$. It therefore follows that all points on the interior of a particular contour define a confidence region on the estimate $\hat{\boldsymbol{\phi}}$, the level of significance being related to the value associated with the contour.

Note that Fig. 4.18 shows multiple maxima. This is not an undesirable situation, since the second maximum corresponds to the point where the pair (ϕ_1, ϕ_2) is reversed in order. Hence the plot of Fig. 4.18 is saying that (1.62, 1.57) is almost as good an estimate of $\boldsymbol{\phi}$ as is (1.57, 1.62), which, of course, is true.

Numerical problems may sometimes be encountered in the optimization routine described above, because the surfaces are poorly conditioned. The maxima in Fig. 4.18 are quite broad and hence computer truncation errors may create numerical problems. Also, we observe that many saddle points and ridges exist, the pres-

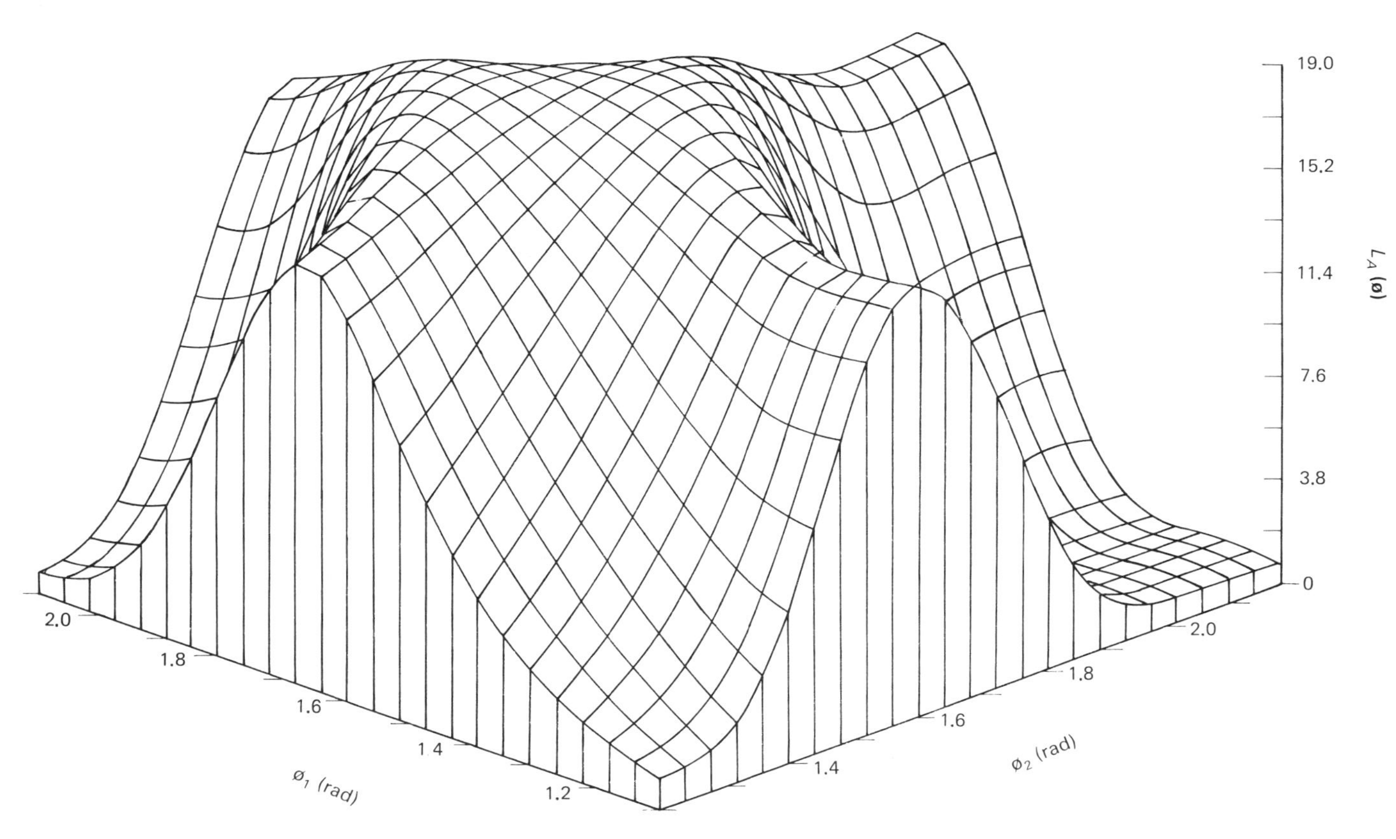

Figure 4.18 Objective function $L_A(\boldsymbol{\phi})$ of the generalized maximum-likelihood receiver for $\phi_1 = 90^\circ$, angle separation = 0.5 BW, $\rho = 0.9$, $N = 21$ elements.

ence of which slows down the rate of convergence of the algorithm. If these problems are found to arise in practice, it would be necessary to use higher-precision arithmetic.

4.8 AN ADAPTIVE ANTENNA WITH A CALIBRATION CURVE: SYMMETRIC MULTIPATH

We now describe a second technique, based on the use of an *adaptive antenna*, that is well suited for estimating the direction of a target for the case of symmetric multipath (Haykin and Kesler [10], Kesler [24]). In effect, we are assuming a multipath environment identical to that used for deriving the symmetric maximum-likelihood receiver of Section 4.6.

A block diagram of the system is shown in Fig. 4.19. It consists of a vertical linear array of N elements, a beamforming network, and an adaptive interference canceler. Three beams are used as inputs to the adaptive canceler, as described below:

1. A *reference beam*, pointing along the horizontal direction
2. Two *auxiliary beams*, steered at angles $+\theta_B$ and $-\theta_B$ with respect to the normal to the array

The reference beam, denoted by X_R, enters the adaptive canceler as the *desired response*. The auxiliary beams, denoted by X_1 and X_2, are summed and multiplied with an *adjustable weight*, w. The resulting output is then subtracted from the

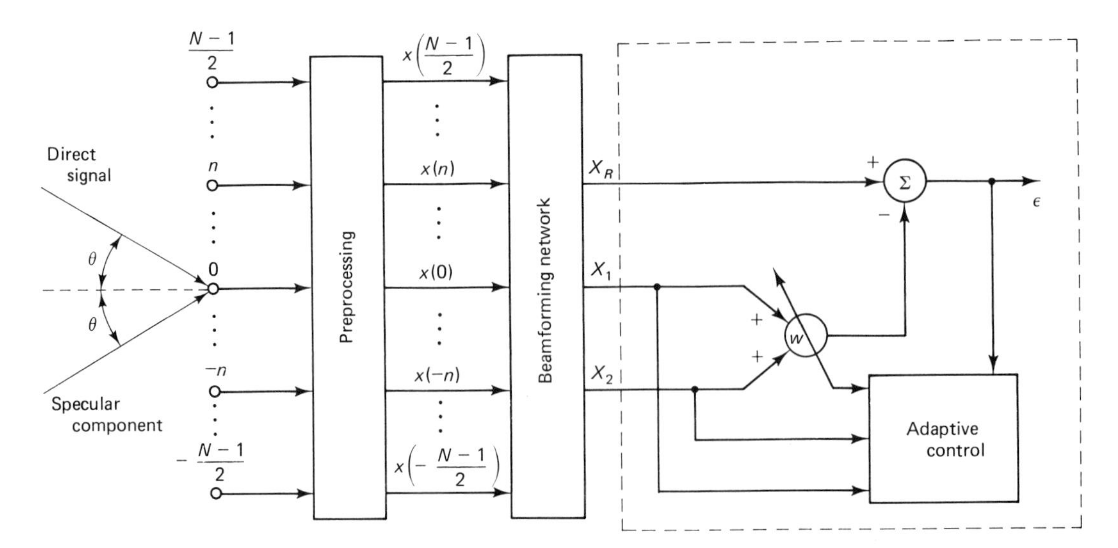

Figure 4.19 Block diagram of the adaptive antenna.

reference beam to produce an *error*, ϵ. The weight w is adjusted by means of an *adaptive algorithm* so as to minimize the mean-square value of the error. By so doing, two *nulls* are automatically placed in the reference beam along the directions of the direct and specular components of the received signal. This, in turn, minimizes the total average power in the reference beam. For the adaptive algorithm to adjust the weight, we may, for example, use the complex form of the *least-mean-square (LMS) algorithm* (Widrow et al. [25]).

The automatic nulling of the direct and specular components of the received signal is the result of a basic property of adaptive antennas. Specifically, when an adaptive antenna operates in an environment with two symmetrically positioned coherent sources, both sources are nulled out when steady-state conditions are reached.

The scheme above is similar to the *double-nulling* technique described by White [18]. There is, however, one major difference between White's double-nulling technique and that described here. Specifically, by supplying the adaptive antenna of Fig. 4.19 with a calibration curve, the resulting system is capable of estimating the angle of arrival of the direct path, as described below.

The Optimum Weight

For the special case of symmetric multipath, the observed signal at the output of element n in the array is given by Eq. (4.37), reproduced here for convenience.

$$x(n) = a_1 \exp(jn\phi) + \rho a_1 \exp[j(\psi - n\phi)] + w(n) \tag{4.97}$$

The reference beam at the output of the beamforming network is defined by

$$X_R = \sum_{n=-(N-1)/2}^{(N-1)/2} b_n x(n) \tag{4.98}$$

where $\{b_n\}$ is a set of real weights applied to the individual elemental outputs. The two auxiliary beams are defined by

$$X_1 = \sum_{n=-(N-1)/2}^{(N-1)/2} b_n \exp(jn\phi_B) x(n) \tag{4.99}$$

and

$$X_2 = \sum_{n=-(N-1)/2}^{(N-1)/2} b_n \exp(-jn\phi_B) x(n) \tag{4.100}$$

The electrical phase angle ϕ_B is related to the direction of the first auxiliary beam, θ_B, by

$$\phi_B = \frac{2\pi d}{\lambda} \sin \theta_B \tag{4.101}$$

The sum of the two auxiliary beams equals

$$\begin{aligned} X_A &= X_1 + X_2 \\ &= \sum_{n=-(N-1)/2}^{(N-1)/2} 2b_n \cos(n\phi_B)x(n) \end{aligned} \tag{4.102}$$

The error, at the system output, is defined by

$$\epsilon = X_R - wX_A \tag{4.103}$$

where, in general, the weight w is complex valued. The mean-square value of the error equals

$$\begin{aligned} \xi &= E[|\epsilon|^2] \\ &= E[|X_R|^2] - wE[X_R^* X_A] - w^* E[X_R X_A^*] + |w|^2 E[|X_R|^2] \end{aligned} \tag{4.104}$$

Differentiating the objective function ξ with respect to the real and imaginary parts of the weight w, and setting both results equal to zero, we may determine the real and imaginary parts of the *optimum weight*, respectively. Denoting this optimum weight by w_{opt}, we may thus write

$$w_{\text{opt}} = \frac{E[X_R X_A^*]}{E[|X_A|^2]} \tag{4.105}$$

Using Eqs. (4.98) and (4.102), we have

$$\begin{aligned} E[X_R X_A^*] &= E\left[\sum_{n=-(N-1)/2}^{(N-1)/2} b_n x(n) \sum_{m=-(N-1)/2}^{(N-1)/2} 2b_m \cos(m\phi_B)x^*(m)\right] \\ &= \sum_{n=-(N-1)/2}^{(N-1)/2} \sum_{m=-(N-1)/2}^{(N-1)/2} 2b_n b_m \cos(m\phi_B) E[x(n)x^*(m)] \\ &= \sum_{n=-(N-1)/2}^{(N-1)/2} \sum_{m=-(N-1)/2}^{(N-1)/2} 2b_n b_m \cos(m\phi_B) R(n, m) \end{aligned} \tag{4.106}$$

where $R(n, m) = E[x(n)x^*(m)]$. Putting $\phi_1 = -\phi_2 = \phi$ in Eq. (4.97), we get

$$\begin{aligned} \frac{R(n, m)}{E[|a_1|^2]} &= \exp[j(n-m)\phi] + \rho^2 \exp[-j(n-m)\phi] \\ &\quad + 2\rho \cos[(n+m)\phi - \psi] + \frac{1}{\text{SNR}} \delta_{nm} \end{aligned} \tag{4.107}$$

We assume that (1) the number of elements in the array is odd, so that $(N-1)/2$ has an integer value, and (2) the weighting function used in the beamforming network is symmetrical, so that we may write $b_{-n} = b_n$. Then, substituting Eq.

(4.107) into (4.106), and performing some algebraic manipulations, we get

$$\frac{E[X_R X_A^*]}{E[|a_1|^2]} = 2b_0^2\left(C + \frac{1}{\text{SNR}}\right) + 4b_0 C \sum_{k=1}^{(N-1)/2} b_k \cos(k\phi)$$
$$+ 4b_0 C \sum_{k=1}^{(N-1)/2} b_k \cos(k\phi)\cos(k\phi_B)$$
$$+ 8C \sum_{k=1}^{(N-1)/2} \sum_{l=1}^{(N-1)/2} b_k b_l \cos(k\phi)\cos(l\phi)\cos(l\phi_B)$$
$$+ \frac{4}{\text{SNR}} \sum_{k=1}^{(N-1)/2} b_k^2 \cos(k\phi_B) \tag{4.108}$$

where

$$C = 1 + 2\rho\cos\psi + \rho^2 \tag{4.109}$$

Similarly, we may show that

$$\frac{E[|X_A|^2]}{E[|a_1|^2]} = 4b_0^2\left(C + \frac{1}{\text{SNR}}\right) + 16b_0 C \sum_{k=1}^{(N-1)/2} b_k \cos(k\phi)\cos(k\phi_B)$$
$$+ 16C\left[\sum_{k=1}^{(N-1)/2} b_k \cos(k\phi)\cos(k\phi_B)\right]^2$$
$$+ \frac{8}{\text{SNR}} \sum_{k=1}^{(N-1)/2} [b_k \cos(k\phi_B)]^2 \tag{4.110}$$

Examining Eq. (4.108), we see that the whole expression is real. Accordingly, the optimum weight w_{opt} is a real quantity.

Six factors affect the value of the optimum weight. They are:

1. The angle of arrival, θ
2. The auxiliary beam position, θ_B
3. The signal-to-noise ratio, SNR
4. The phase difference ψ, measured at the center of the array
5. The surface reflection coefficient, ρ
6. The weighting function of the beamforming network, $\{b_n\}$

Calibration Curve

When the input signal-to-noise is infinitely large, 1/SNR is zero, and as a consequence, the constant C cancels out when we form the ratio of Eq. (4.105). Let

$$w_{\text{cal}} = w_{\text{opt}}\Big|_{\text{SNR}=\infty} \tag{4.111}$$

Then, from Eqs. (4.108) and (4.110) we deduce that (after cancelling common factors)

$$w_{\text{cal}} = \frac{b_0 + 2 \sum_{k=1}^{(N-1)/2} b_k \cos(k\phi)}{2b_0 + 4 \sum_{k=1}^{(N-1)/2} b_k \cos(k\phi) \cos(k\phi_B)} \tag{4.112}$$

This formula is independent of SNR, ρ, and ψ; that is, it does not depend on the unknown parameters associated with the multipath environment in which the array operates. It depends only on the weighting function $\{b_n\}$ used in the beamforming network and the auxiliary beam position ϕ_B. Indeed, these are system parameters under the designers' control, and as such they may be chosen to optimize the performance of the system.

The curve obtained by plotting w_{cal} versus the electrical phase angle ϕ may be viewed as a *calibration curve.* Specifically, given the value of the optimum weight w_{opt} that minimizes the mean-square value of the error at the system output, we may put $w_{\text{opt}} \simeq w_{\text{cal}}$, and thereby use the calibration curve to estimate the electrical angle ϕ and, hence, the direction of the target. Clearly, there is an estimation error incurred in using such a procedure. This error increases as the angle of arrival or the SNR decreases.

Haykin and Kesler [10] have shown that the use of a *binomially weighted array* yields the best overall performance. For example, Fig. 4.20 shows the calibration curve for such a system when $\phi_B = 1$ standard beamwidth. The dashed curve in Fig. 20 represents the calibration curve obtained by using Eq. (4.112). The solid curve in this figure represents the curve obtained by plotting w_{opt} versus ϕ for the

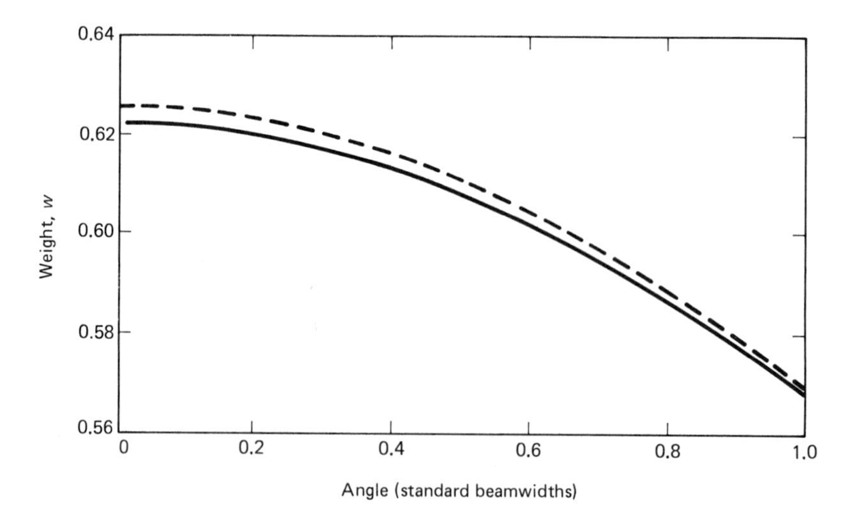

Figure 4.20 Optimum weight versus angle for varying SNR. Solid line: SNR = 0 dB; dashed line represents calibration curve for binomial weighting, $\phi_B = 1$ standard beamwidth.

case when the signal-to-noise ratio SNR = 0 dB. The difference between the two curves is noticeable only for very small angles of arrival.

4.9 EXPERIMENTAL RESULTS

In this section we present some experimental results obtained by testing the symmetric maximum-likelihood receiver of Section 4.6 and the adaptive antenna scheme of Section 4.8 with real radar data [11, 26]. The data were obtained by means of a one-dimensional sampled aperture radar facility with a seven-element receiving array. Each element consists of a horn antenna and a total receiver with its output heterodyned down to baseband. The radar data were recorded along the Ottawa river, using a boat equipped with a radar beacon. The height of the beacon was 8 m

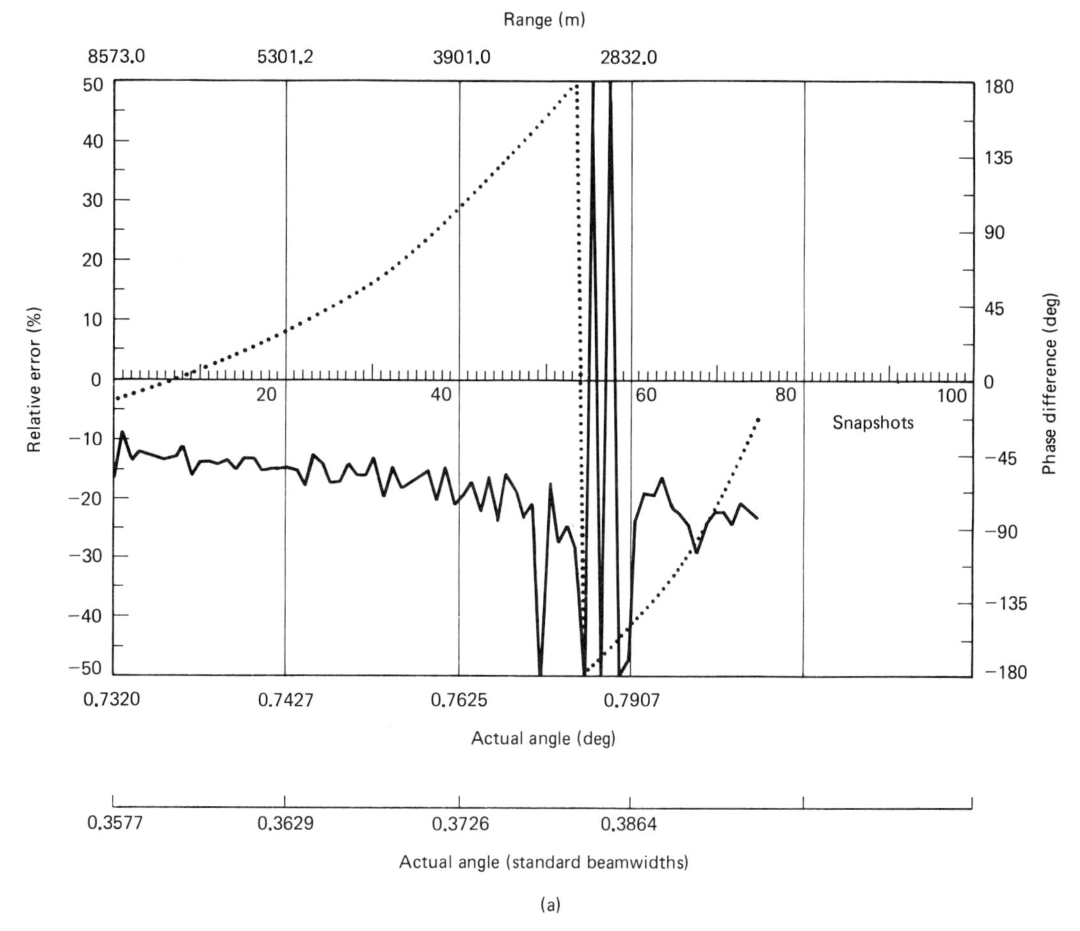

Figure 4.21 Measured relative error (solid line) and calculated phase difference ψ (dotted line) for: (a) symmetric ML receiver; (b) adaptive antenna.

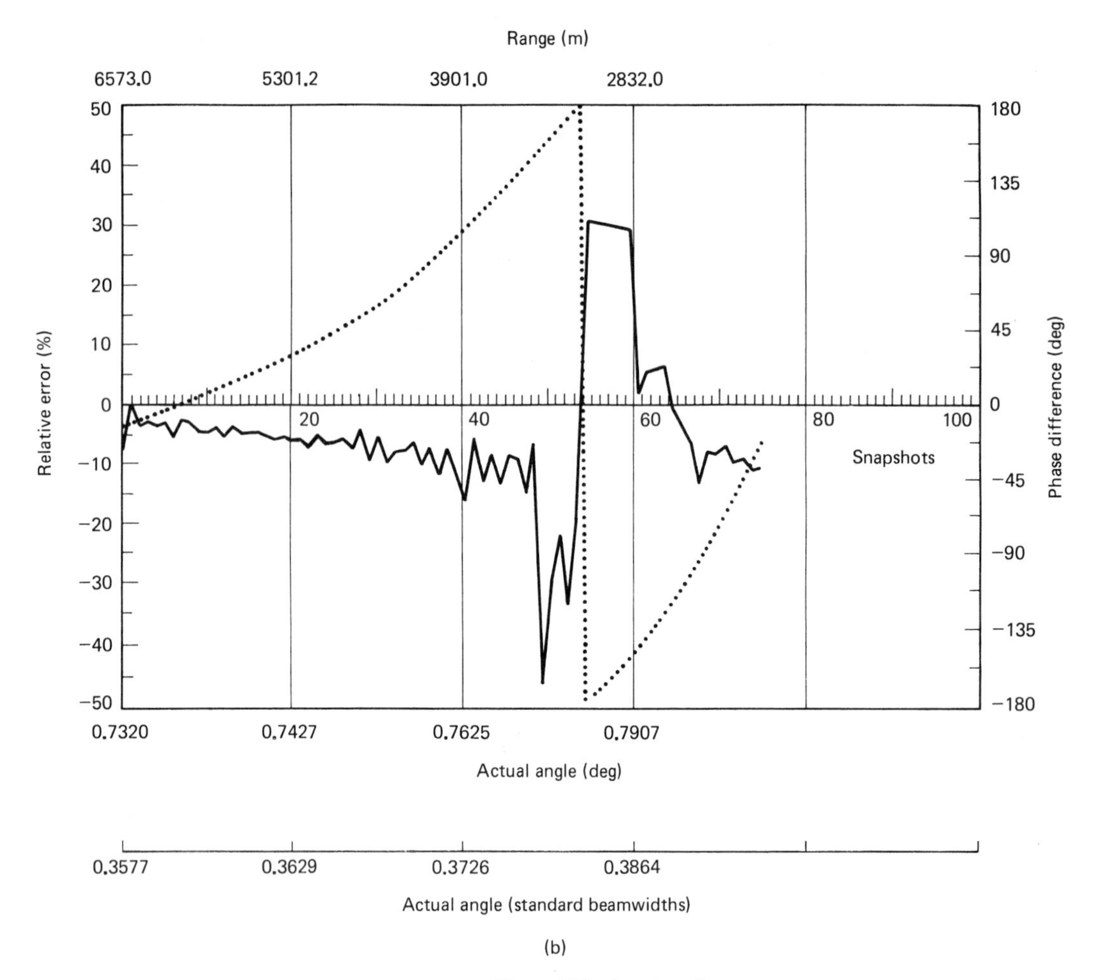

(b)

Figure 4.21 *(continued)*

and that of the radar 3.65 m. The angular separation of the beacon and its image varied from 1.6° down to 0.83° as the range varied from 577 to 1106 m, so that their separation was always less than the standard beamwidth of the receiving linear array (which equals 1.8°).

In Fig. 4.21 we show the results obtained by using the data generated with this facility to test the symmetric maximum-likelihood receiver and the adaptive antenna scheme of Sections 4.6 and 4.8, respectively. These two processors were tested using six files of multipath data, with each file containing a total of 75 to 100 snapshots. In Fig. 4.21(a) and (b) we have plotted the percentage estimation error versus the actual angle of arrival (expressed in degrees or standard beamwidths) for these two receivers, respectively. In the figure we have also included a plot of the phase difference, ψ, between the direct and specular components, measured at the center of the array.

For each snapshot, the actual angle of arrival, θ, and the phase difference ψ were calculated from the geometry of the environment prevailing at the time when the snapshot was taken.

In Fig. 4.22 we have plotted the estimate of the angle of arrival versus the actual value of this angle. Part (a) of the figure refers to the maximum-likelihood receiver, while part (b) refers to the adaptive antenna scheme. In both parts of the figure, each point refers to the result obtained from one snapshot of data.

Based on the results presented in Figs. 4.21 and 4.22 and the analysis performed on five other data files, we may draw the following conclusions:

1. When the conditions are right, both estimators are capable of resolving a target at small angles of arrival, by using one snapshot of data, even when the angle of arrival is a small fraction of a beamwidth.
2. The estimator using an adaptive antenna appears to have, consistently, a slightly better performance than the maximum-likelihood receiver. This may be due to the ability of the adaptive antenna to respond to deviations from the idealized symmetric multipath model of Fig. 4.2 somewhat more effectively than the maximum-likelihood reciever.

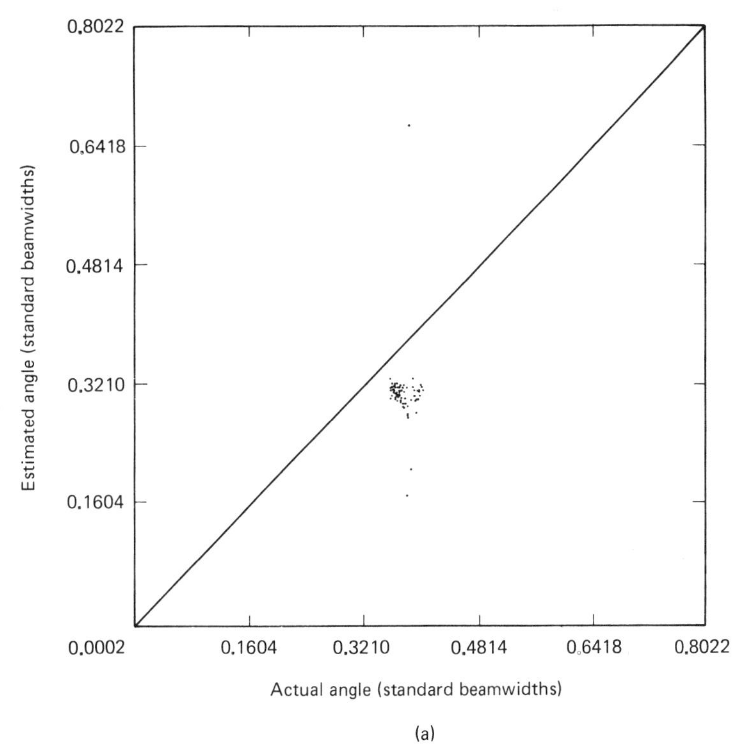

Figure 4.22 Estimated angle versus actual angle, using 50 snapshots, with one point for each snapshot for: (a) symmetric ML receiver; (b) adaptive antenna.

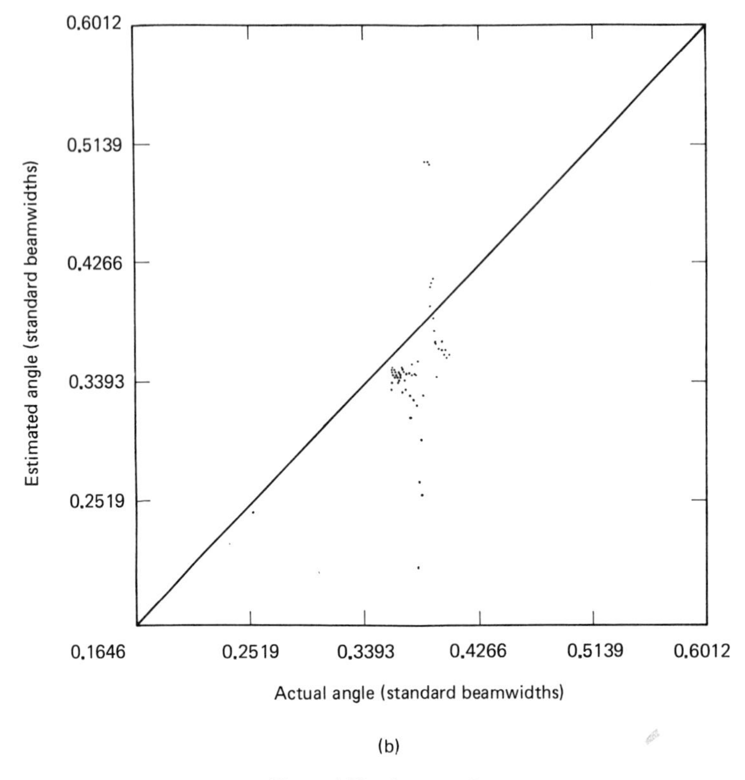

(b)

Figure 4.22 *(continued)*

3. As expected, both estimators fail when the phase difference ψ between the direct and specular components of the received signal approaches 180°. The breakdown of the adaptive antenna scheme, however, appears to be more graceful than that of the maximum-likelihood receiver.

4.10 PARAMETRIC SPECTRAL ESTIMATION METHODS

The Fourier method, discussed in Section 4.4, offers the basis of a nonparametric approach for estimating the wavenumber power spectrum of a set of plane waves incident on a linear array. By locating the spectral peaks, we are able to estimate the directions of the incident plane waves. When the aperture of the array, measured in wavelengths, is large, the Fourier method is quite effective. When, however, the aperture of the array is relatively small, the Fourier method suffers from a limited resolution capability. We may overcome this limitation by using the maximum-likelihood receiver or the adaptive antenna scheme, as described in the previous sections. In particular, for the special case of a symmetric multipath environment, we may use the maximum-likelihood receiver of Fig. 4.7 or the adaptive antenna scheme of Fig. 4.19 supplied with a calibration curve, both of which are simple to implement. For the more general case of two or more plane waves arriving at the array from arbitrary directions, we may use the maximum-likelihood receiver of Fig. 4.15. However, the practical limitation of the latter receiver is the fact that there is no simple routine for maximizing the resultant multidimensional likelihood function.

There is, therefore, special interest in the use of modern *parametric spectral estimation techniques* that are *moldel-based.* For an array of fixed aperture, we are usually able to realize with these techniques a spatial resolution capability far in excess of that achievable with the classical Fourier method. Although these high-resolution spectral estimation techniques are usually much more sophisticated than the Fourier method, nevertheless, with the continuing improvement in the cost and flexibility of digital signal processing hardware, their implementation has become a practical reality.

An important feature of parametric spectral estimation techniques is their data adaptivity. That is, the parameters involved in the spectral estimation vary in accordance with the incoming data. This is unlike the Fourier method, where the design of the processor is fixed ahead of time.

A myriad of parametric spectral estimation techniques have been developed during the last decade or so. For more details, the interested reader is referred to Kay and Marple [27], Johnson [28], Haykin [29], Childers [30], and special issues on spectral estimation [31–35] that are devoted to the subject. Also, Reilly [19], and Evans et al. [5] present detailed discussions of many of these techniques, supported by computer simulation and experimental results, as applied to the problem of estimating the angle(s) of arrival in a radar environment. In the remainder of this chapter, however, we will be mainly concerned with the use of the linear prediction method and related issues. In a section toward the end of the chapter, we summarize the important features of some of the other methods.

4.11 LINEAR PREDICTION

Suppose that we are given the complex-valued spatial samples $x(n-1)$, $x(n-2), \ldots, x(n-L)$ as observed at the outputs of array elements $n-1$, $n-2, \ldots, n-L$, respectively. We assume that $L < N$, where N is the total number of elements in the array. Let $\hat{x}(n)$ denote the *linear prediction* of the value of the spatial sample $x(n)$ at the output of element n, given this set of data. We define $\hat{x}(n)$ as a linear combination of $x(n-1), x(n-2), \ldots, x(n-L)$, as shown by

$$\hat{x}(n) = \sum_{l=1}^{L} h(l)x(n-l), \qquad L < n \leq N \tag{4.113}$$

where $h(1), h(2), \ldots, h(L)$ are the *prediction filter coefficients.* We refer to L, the number of spatial samples used in the prediction, as the *order* of the prediction filter. Figure 4.23 shows the signal-flow graph representation of Eq. (4.113). We define the

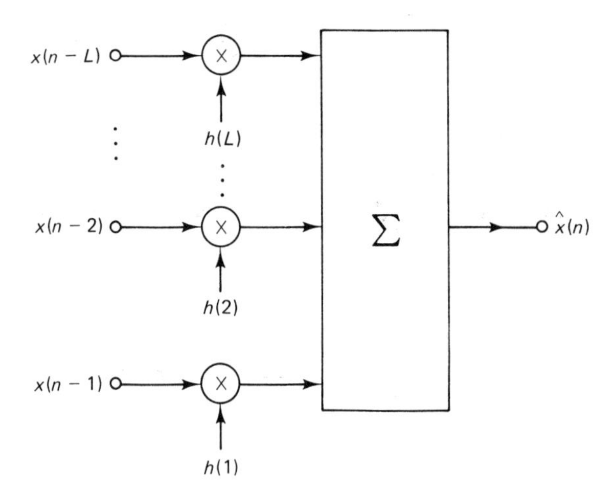

Figure 4.23 Linear prediction filter.

prediction error or *residual*, $f(L, n)$, as the difference between the actual value $x(n)$ of the spatial sample derived from element n and the predicted value $\hat{x}(n)$. The dependence of the prediction error $f(L, n)$ on the filter order L is shown explicitly for reasons that will become apparent later. We may thus write

$$\begin{aligned} f(L, n) &= x(n) - \hat{x}(n) \\ &= x(n) - \sum_{l=1}^{L} h(l)x(n-l) \end{aligned} \tag{4.114}$$

Define a new set of filter coefficients related to the prediction filter coefficients as follows:

$$g(l) = \begin{cases} 1, & l = 0 \\ -h(l), & l = 1, 2, \ldots, L \end{cases} \tag{4.115}$$

Then we may rewrite Eq. (4.114) in the form

$$f(L, n) = \sum_{l=0}^{L} g(l)x(n-l), \qquad L < n \leq N \tag{4.116}$$

We see that the new filter operates on the input set of spatial samples $x(n)$, $x(n-1), \ldots, x(n-L)$ to produce an output equal to the prediction error $f(L, n)$, as illustrated in Fig. 4.24. We refer to the spatial filter defined by the set of coefficients $\{g(l)\}$, $l = 0, 1, \ldots, L$, as a *prediction-error filter* of order L. Figure 4.25 shows, in block-diagrammatic form, the relation between the prediction filter and the prediction-error filter. Note that, although according to our convention, the prediction-error filter has the same order as the prediction filter, nevertheless, the length of the prediction-error filter exceeds that of the prediction filter by one spatial unit.

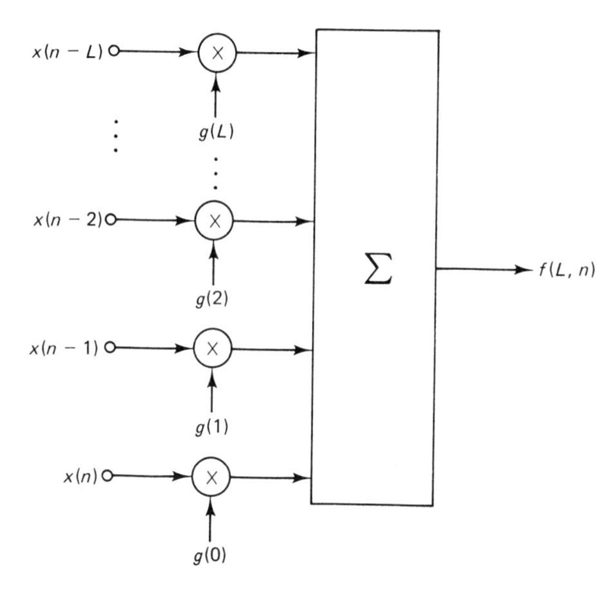

Figure 4.24 Prediction-error filter.

Define an $(L + 1) \times 1$ *prediction-error filter vector*

$$\mathbf{g} = \begin{bmatrix} g(0) \\ g(1) \\ \vdots \\ g(L) \end{bmatrix} \tag{4.117}$$

and a corresponding $(L + 1) \times 1$ *input vector*

$$\mathbf{x}(n) = \begin{bmatrix} x(n) \\ x(n-1) \\ \vdots \\ x(n-L) \end{bmatrix} \tag{4.118}$$

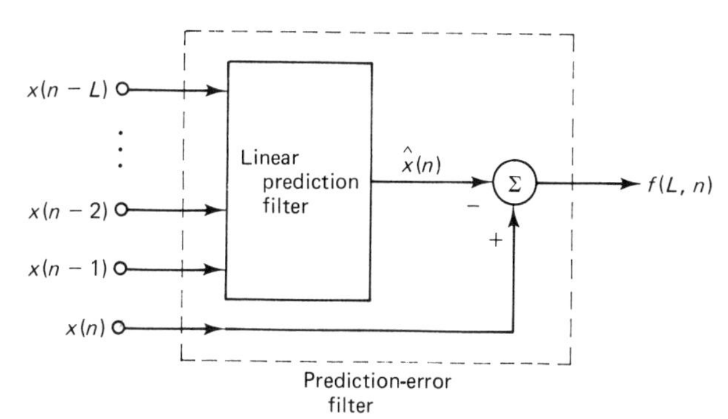

Figure 4.25 Illustrating the relationship between the linear prediction filter and the prediction-error filter.

Then we may rewrite Eq. (4.116) in the compact form

$$f(L, n) = \mathbf{g}^T\mathbf{x}(n) \tag{4.119}$$

or, equivalently,

$$f(L, n) = \mathbf{x}^T(n)\mathbf{g} \tag{4.120}$$

The mean-square value of the forward prediction error equals

$$\begin{aligned}\epsilon_f &= E[|f(L, n)|^2] \\ &= E[f^*(L, n)f(L, n)] \\ &= E[\mathbf{g}^H\mathbf{x}^*(n)\mathbf{x}^T(n)\mathbf{g}] \\ &= g^H E[\mathbf{x}^*(n)\mathbf{x}^T(n)]\mathbf{g} \\ &= g^H\mathbf{R}\mathbf{g}\end{aligned} \tag{4.121}$$

where the superscript H denotes *Hermitian* transposition (combining the operations of complex conjugation and ordinary transposition) and the $(L + 1) \times (L + 1)$ matrix $\mathbf{R}$ is the correlation matrix of the input signal, defined by

$$\mathbf{R} = E[\mathbf{x}^*(n)\mathbf{x}^T(n)]$$

We assume that the input vector $\mathbf{x}(n)$ is spatially stationary, so that the matrix $\mathbf{R}$ is Toeplitz. We wish to choose the prediction-error filter vector $\mathbf{g}$ so as to minimize the mean-square value ϵ_f, subject to the constraint that the element $g(0)$ of the vector $\mathbf{g}$ equals 1. This constraint may be expressed as

$$\boldsymbol{\delta}^T\mathbf{g} = \mathbf{g}^T\boldsymbol{\delta} = 1 \tag{4.122}$$

where $\boldsymbol{\delta}$ is an $(L + 1) \times 1$ vector defined by

$$\boldsymbol{\delta} = \begin{bmatrix} 1 \\ 0 \\ \vdots \\ 0 \end{bmatrix} \tag{4.123}$$

In general, the vector $\mathbf{g}$ is complex-valued. We may therefore formulate the problem as one of finding the prediction-error filter vector $\mathbf{g}$ for which the objective function

$$J = \mathbf{g}^H\mathbf{R}\mathbf{g} - \lambda\boldsymbol{\delta}^T\mathbf{g} - \lambda^*\mathbf{g}^H\boldsymbol{\delta} \tag{4.124}$$

is minimized. The λ is the *Lagrange multiplier* to be determined. We may rewrite Eq. (4.124) in the equivalent form

$$J = (\mathbf{g}^H - \lambda^*\boldsymbol{\delta}^T\mathbf{R}^{-1})\mathbf{R}(\mathbf{g} - \lambda\mathbf{R}^{-1}\boldsymbol{\delta}) - |\lambda|^2\boldsymbol{\delta}^T\mathbf{R}^{-1}\boldsymbol{\delta} \tag{4.125}$$

Since the correlation matrix $\mathbf{R}$ is nonegative definite, we find immediately from Eq. (4.125) that J attains its minimum value when

$$\mathbf{g} = \lambda\mathbf{R}^{-1}\boldsymbol{\delta} \tag{4.126}$$

Using Eqs. (4.121), (4.122), and (4.126), we find that

$$\lambda = \epsilon_{f,\ \min} = \frac{1}{\boldsymbol{\delta}^T \mathbf{R}^{-1} \boldsymbol{\delta}} \tag{4.127}$$

where $\epsilon_{f,\min}$ is the minimum mean-square value of the prediction error $f(L, n)$. Accordingly, we may rewrite Eq. (4.126) in the form

$$\mathbf{R}\mathbf{g} = \epsilon_{f,\min} \boldsymbol{\delta} \tag{4.128}$$

or in the expanded form

$$\sum_{l=0}^{L} g(l)R(k-l) = \begin{cases} \epsilon_{f,\min}, & k = 0 \\ 0, & k = 1, 2, \ldots, L \end{cases} \tag{4.129}$$

We refer to the system of $(L + 1)$ simultaneous equations (4.129) as the *augmented normal equations.* Note that we have assumed that the input signal is spatially stationary.

The prediction process described above is said to be in the *forward* direction. We may also perform a prediction in the *backward* direction.* In this second case, we use the set of L spatial samples $x(n - L + 1), \ldots, x(n - 1), x(n)$ to make a prediction of $x(n - L)$. Denoting the result of this prediction by $\hat{x}(n - L)$, we define the *backward prediction* error as

$$b(L, n) = x(n - L) - \hat{x}(n - L) \tag{4.130}$$

In a corresponding way to the forward prediction-error filter, the backward prediction-error filter operates on the input set of spatial samples $x(n - L)$, $x(n - L + 1), \ldots, x(n - 1), x(n)$ to produce an output signal equal to the backward prediction error $b(L, n)$.

Define an $(L + 1) \times 1$ *backward prediction-error filter vector*

$$\mathbf{g}_b = \begin{bmatrix} g_b(0) \\ \vdots \\ g_b(L-1) \\ g_b(L) \end{bmatrix} \tag{4.131}$$

Then we may express the backward prediction error

$$b(L, n) = \mathbf{g}_b^T \mathbf{x}(n) \tag{4.132}$$

or, equivalently,

$$b(L, n) = \mathbf{x}^T(n)\mathbf{g}_b \tag{4.133}$$

where the input vector $\mathbf{x}(n)$ is as defined in Eq. (4.118).

*In array processing, forward and backward predictions are also referred to as *left* and *right spatial predictions* [48]. We have adopted the same terminology as that used in time-series analysis.

Here again we wish to find the backward prediction-error filter vector $\mathbf{g}_b$ that will minimize the mean-square value of the backward prediction error:

$$\epsilon_b = E[|b(L, n)|^2] \tag{4.134}$$

subject to the constraint that the last element, $g_b(L)$, in the vector $\mathbf{g}_b$ equals 1. Following a procedure similar to that described for minimizing the forward prediction-error filter, we find that the minimum mean-square value of the backward prediction error, $\epsilon_{b,\min}$, is attained when

$$\mathbf{g}_b = \epsilon_{b,\min} \mathbf{R}^{-1} \boldsymbol{\delta}_b \tag{4.135}$$

where $\mathbf{R}$ is the correlation matrix of the observed signal and $\boldsymbol{\delta}_b$ is an $(L+1) \times 1$ vector defined by

$$\boldsymbol{\delta}_b = \begin{bmatrix} 0 \\ 0 \\ \vdots \\ 0 \\ 1 \end{bmatrix} \tag{4.136}$$

In effect, the vector $\boldsymbol{\delta}_b$ is obtained by arranging the elements of the vector $\boldsymbol{\delta}$ in reverse order. We may rewrite Eq. (4.135) in the expanded form

$$\sum_{l=0}^{L} g_b(l) R(k-l) = \begin{cases} 0, & k = 0, 1, \ldots, L-1 \\ \epsilon_{b,\min} & k = L \end{cases} \tag{4.137}$$

where $R(k-l)$ is the (k, l)th element of $\mathbf{R}$.

Replacing the index l by $L-l$ and the index k by $L-k$ in Eq. (4.129), and using the Hermitian property of the correlation matrix, we get

$$\sum_{l=0}^{L} g^*(L-l) R(k-l) = \begin{cases} 0, & k = 0, 1, \ldots, L-1 \\ \epsilon_{f,\min} & k = L \end{cases} \tag{4.138}$$

Now we may compare Eqs. (4.137) and (4.138), pertaining to backward and forward predictions, respectively, and thus make the following observations when the input signal is spatially stationary:

1. The minimum mean-square values of the forward and backward prediction errors are equal; that is,

$$\epsilon_{f,\min} = \epsilon_{b,\min} \tag{4.139}$$

2. The coefficients of the forward and backward prediction-error filters are related by

$$g_b(l) = g^*(L-l), \qquad l = 0, 1, \ldots, L \tag{4.140}$$

That is, we may obtain the backward prediction-error filter by complex-conjugating the forward prediction-error filter coefficients and arranging them in reverse order, as in Fig. 4.26. Then, by feeding this filter with the input samples $x(n)$,

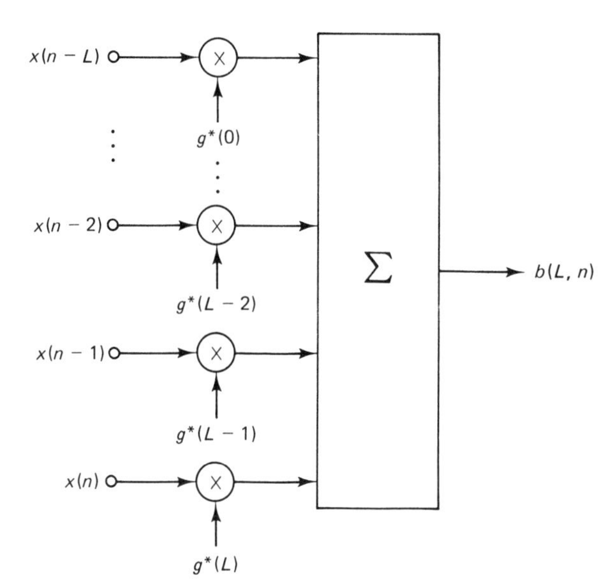

Figure 4.26 Illustrating the procedure for computing the backward prediction error.

$x(n-1), \ldots, x(n-L)$, in their regular order, we obtain the backward prediction error:

$$b(L, n) = \sum_{l=0}^{L} g^*(L-l)x(n-l) \tag{4.141}$$

Note that we may also express the backward prediction error $b(L, n)$ in the equivalent form

$$b(L, n) = \sum_{l=0}^{L} g^*(l)x(n-L+l) \tag{4.142}$$

4.12 AUTOREGRESSIVE/MAXIMUM-ENTROPY SPECTRAL ANALYSIS

We say that the spatial samples $x(n), x(n-1), \ldots, x(n-L)$ are samples of an *autoregressive* (*AR*) *process* if they satisfy the following difference equation:

$$x(n) = -\sum_{l=1}^{L} a(l)x(n-l) + w(n) \tag{4.143}$$

where $a(1), a(2), \ldots, a(L)$ are constant parameters, and $w(n)$ is the sample of a zero-mean white noise process. We refer to L as the *order* of the AR process. According to Eq. (4.143), we may generate the AR process $\{x(n)\}$ by applying the white noise process $\{w(n)\}$ to an *all-pole filter*, as shown in Fig. 4.27. In this figure, z^{-1} repre-

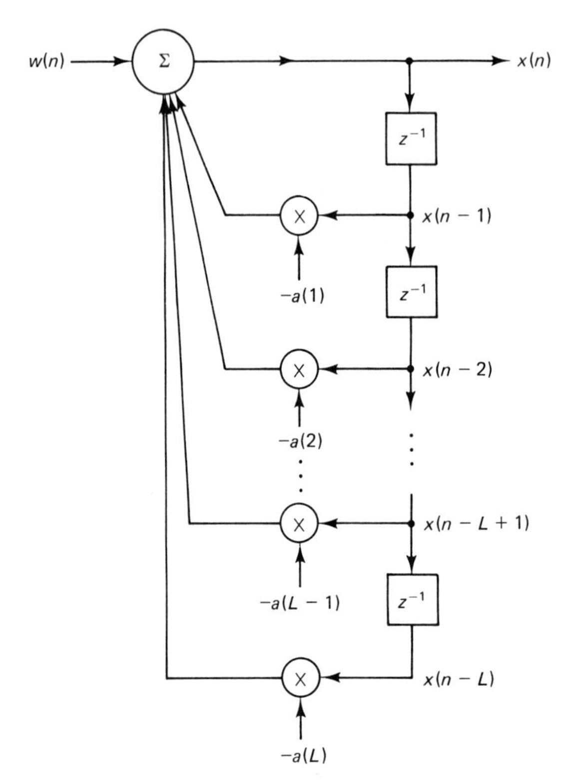

Figure 4.27 All-pole filter for generating an AR process.

sents a *unit spatial-delay operator*, with

$$z^{-1}[x(n)] = x(n-1) \tag{4.144}$$

We may also view z as a *complex variable* with a real as well as an imaginary part. For points on the *unit circle*, as in Fig. 4.28, we have z equal to $\exp(j\phi)$, where ϕ is the electrical phase angle.

The transfer function of the all-pole filter of Fig. 4.27 is defined by

$$H(z) = \frac{1}{1 + \sum_{l=1}^{L} a(l)z^{-l}} \tag{4.145}$$

Let σ_w^2 denote the variance of sample $w(n)$ of the white noise process. Then from Fig. 4.27 it follows that $S(\phi)$, the power spectrum of the AR process at the output of the all-pole filter, is equal to the power spectrum of the filter input multiplied by the squared magnitude of the transfer function $H(z)$, evaluated at $z = \exp(j\phi)$. We may thus write

$$S(\phi) = \frac{\sigma_w^2}{\left|1 + \sum_{l=1}^{L} a(l) \exp(-jl\phi)\right|^2} \tag{4.146}$$

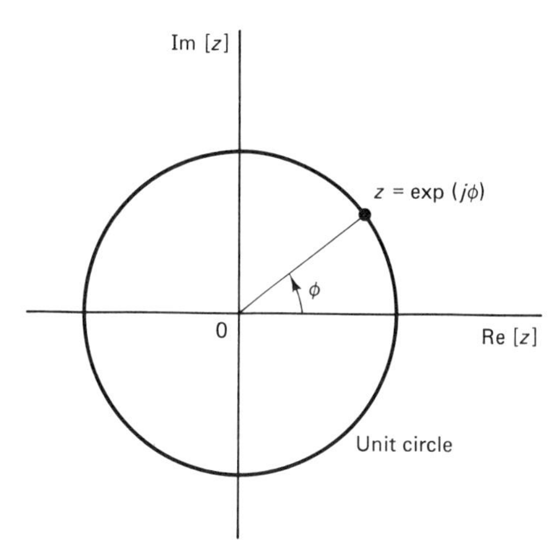

Figure 4.28 z-plane.

Suppose now that we apply the AR process $\{x(n)\}$ of order L to a prediction-error filter that is also of order L. When the average power of the prediction error is minimized, we find that the transfer function of the prediction-error filter takes on a value equal to the reciprocal of the transfer function $H(z)$ in Eq. (4.145), with the prediction-error filter coefficients equal to the respective AR parameters. When this happens, the resultant prediction error takes on the form of a white noise process with a constant power spectrum. It is for this reason that the prediction-error filter is also referred to as a *whitening filter*.

For the all-pole filter of Fig. 4.27 to be stable, we have to ensure that all the poles of the transfer function $H(z)$ lie inside the unit circle in the z-plane. Correspondingly, this requires that all the zeros of the prediction-error filter lie inside the unit circle in the z-plane. A filter whose zeros are all restricted to lie inside the unit circle in the z-plane is referred to as a *minimum-phase filter* (Oppenheim and Schafer [36]). We may sum up these observations by stating that if the prediction-error filter input is an AR process of the same order as the filter itself, the resultant prediction error is a white noise process, and knowledge of the prediction-error filter parameters provides a means of estimating the power spectrum of the input signal.

It is of interest to note that the result of Eq. (4.146) may also be derived by considering the *entropy* of the process $\{x(n)\}$ as the quantity of interest. In particular, it may be shown that by maximizing the entropy of the process $\{x(n)\}$, subject to the constraint that the autocorrelation function of the process has a set of known values, we get an estimate of the power spectrum of the process $\{x(n)\}$ that is identical to the AR spectrum of Eq. (4.146). This result, known as the *maximum-entropy spectral estimate*, was first derived by Burg [37, 38]. For the derivation of this result, see also Ulrych and Bishop [39], Haykin and Kesler [40], and McDonough [41].

To emphasize the fact that the use of an autoregressive model and the maximum-entropy method yield identical spectral estimates, we refer to the formula of Eq. (4.146) as the *autoregressive/maximum entropy* or *AR/ME spectral estimate*. It is important to realize that this result only holds in the case of a linear array. In the multidimensional case (as in a planar array, for example) we find that the true maximum-entropy spectral estimate is distinctly different from the spectrum derived by the use of an AR model (McLellan [42]).

To use the AR/ME spectral estimate of Eq. (4.146), we need a procedure for computing the AR parameters $a(1), a(2), \ldots, a(L)$, or, equivalently, the prediction-error filter coefficients $g(1), g(2), \ldots, g(L)$. There are several procedures available for doing this computation (Ulrych and Ooe [43]). We will only consider two procedures: (1) the Burg technique, and (2) the forward–backward linear prediction method and its modified forms.

4.13 THE BURG TECHNIQUE

Consider a prediction-error filter of order L, driven by a spatially stationary process represented by the samples $x(n), x(n-1), \ldots, x(n-L)$. When this filter is operated in the forward direction, we have the system of $(L+1)$ simultaneous equations (4.129), which we may rewrite in expanded matrix form as follows:

$$\begin{bmatrix} R(0) & R(-1) & \cdots & R(-L) \\ R(1) & R(0) & \cdots & R(-L+1) \\ \vdots & \vdots & & \vdots \\ R(L) & R(L-1) & \cdots & R(0) \end{bmatrix} \begin{bmatrix} g(L, 0) \\ g(L, 1) \\ \vdots \\ g(L, L) \end{bmatrix} = \begin{bmatrix} \epsilon_{\min}(L) \\ 0 \\ \vdots \\ 0 \end{bmatrix} \tag{4.147}$$

where $\epsilon_{\min}(L)$ is the mean-square value of the forward prediction error or that of the backward prediction error, and $g(L, 0), g(L, 1), \ldots, g(L, L)$ are the forward prediction-error filter coefficients for a filter order L. We have shown the dependence of these parameters on the filter order L for reasons that will become apparent shortly. When the prediction-error filter is operated in the backward direction, we have the corresponding system of $(L+1)$ simultaneous equations [see Eqs. (4.137) and (4.140)]:

$$\begin{bmatrix} R(0) & R(-1) & \cdots & R(-L) \\ R(1) & R(0) & \cdots & R(-L+1) \\ \vdots & \vdots & & \vdots \\ R(L) & R(L-1) & \cdots & R(0) \end{bmatrix} \begin{bmatrix} g^*(L, L) \\ g^*(L, L-1) \\ \vdots \\ g^*(L, 0) \end{bmatrix} = \begin{bmatrix} 0 \\ 0 \\ \vdots \\ \epsilon_{\min}(L) \end{bmatrix} \tag{4.148}$$

We wish to combine these two systems of $(L+1)$ simultaneous equations into a single system of $(L+2)$ simultaneous equations. The motivation for doing this is to develop recursive relations that would enable us to compute the parameters of a prediction-error filter of order $L+1$, given the parameters of a corresponding filter of order L. We note that the correlation matrix in Eqs. (4.147) and (4.148) has exactly the same form. We further note that this correlation matrix is Toeplitz, which is another way of saying that the input process is spatially stationary. Thus, combining Eqs. (4.147) and (4.148), we may write

$$\begin{bmatrix} R(0) & R(-1) & \cdots & R(-L) & R(-L-1) \\ R(1) & R(0) & \cdots & R(-L+1) & R(-L) \\ \vdots & \vdots & & \vdots & \vdots \\ R(L) & R(L-1) & \cdots & R(0) & R(-1) \\ R(L+1) & R(L) & \cdots & R(1) & R(0) \end{bmatrix} \cdot \begin{bmatrix} g(L, 0) \\ g(L, 1) + \Gamma_{L+1} g^*(L, L) \\ \vdots \\ g(L, L) + \Gamma_{L+1} g^*(L, 1) \\ \Gamma_{L+1} g^*(L, 0) \end{bmatrix} = \begin{bmatrix} \epsilon_{\min}(L) + \Gamma_{L+1}\Delta_{L+1} \\ 0 \\ \vdots \\ 0 \\ \Delta^*_{L+1} + \Gamma_{L+1}\epsilon_{\min}(L) \end{bmatrix} \quad (4.149)$$

where Γ_{L+1} and Δ_{L+1} are constants to be determined. This new system of equations is obtained by multiplying both sides of Eq. (4.148) by the constant Γ_{L+1}, and then adding the result of Eq. (4.147) in such a way that we may increase the order of the correlation matrix from $L + 1$ to $L + 2$ and at the same time maintain the Toeplitz character of the expanded correlation matrix. In order to achieve this objective, we have to choose Δ_{L+1} so as to satisfy the condition

$$\Delta_{L+1} = \sum_{l=0}^{L} R(-l-1) g^*(L, L-l) \quad (4.150)$$

Note that since the correlation matrix is Hermitian, we may equivalently write

$$\begin{aligned} \Delta^*_{L+1} &= \sum_{l=0}^{L} R(l+1) g(L, L-l) \\ &= \sum_{l=0}^{L} R(L+1-l) g(L, l) \end{aligned}$$

We may view the system of $(L + 2)$ simultaneous equations (4.149) as the augmented normal equations of a prediction-error filter of order $L + 1$. But, for such a filter operated in the forward direction, we have

$$\begin{bmatrix} R(0) & R(-1) & \cdots & R(-L) & R(-L-1) \\ R(1) & R(0) & \cdots & R(-L+1) & R(-L) \\ \vdots & \vdots & & \vdots & \vdots \\ R(L) & R(L-1) & \cdots & R(0) & R(-1) \\ R(L+1) & R(L) & \cdots & R(1) & R(0) \end{bmatrix} \cdot \begin{bmatrix} g(L+1, 0) \\ g(L+1, 1) \\ \vdots \\ g(L+1, L) \\ g(L+1, L+1) \end{bmatrix} = \begin{bmatrix} \epsilon_{\min}(L+1) \\ 0 \\ \vdots \\ 0 \\ 0 \end{bmatrix} \quad (4.151)$$

Therefore, comparing the two systems of $(L + 2)$ simultaneous equations (4.149) and (4.151), we deduce the following three relations:

$$g(L+1, l) = g(L, l) + \Gamma_{L+1} g^*(L, L+1-l), \qquad l = 0, 1, \ldots, L \tag{4.152}$$

$$\epsilon_{\min}(L+1) = \epsilon_{\min}(L) + \Gamma_{L+1}\Delta_{L+1} \tag{4.153}$$

and

$$0 = \Delta^*_{L+1} + \Gamma_{L+1}\epsilon_{\min}(L) \tag{4.154}$$

Eliminating Δ_{L+1} between Eqs. (4.153) and (4.154), we get

$$\epsilon_{\min}(L+1) = \epsilon_{\min}(L)(1 - |\Gamma_{L+1}|^2) \tag{4.155}$$

We note from Eq. (4.152) that

$$g(L+1, l) = \begin{cases} 1, & l = 0 \\ \Gamma_{L+1}, & l = L+1 \\ 0, & l > L+1 \end{cases} \tag{4.156}$$

The coefficient $g(L + 1, l)$ is zero for $l > L + 1$ because such a coefficient is nonexistent for a prediction-error filter of order $L + 1$. Note also that Γ_{L+1} equals the last coefficient, $g(L + 1, L + 1)$, in a prediction-error filter of order $L + 1$. The recursive relation (4.155) has a mathematical form that is analogous to the transmission of power through a terminated two-port device. For this reason we find that Γ_{L+1} is referred to as the *reflection coefficient.** Equation (4.152) states that, given the set of prediction-error filter coefficients $g(L, 0), g(L, 1), \ldots, g(L, L)$ for filter order L, and the reflection coefficient $\Gamma_{L+1} = g(L + 1, L + 1)$, we may compute the remaining coefficients $g(L + 1, 0), g(L + 1, 1), \ldots, g(L + 1, L)$ for filter order $L + 1$. This recursive relation is known as the *Levinson* recursion [44].†

From Eq. (4.155) we see that when $|\Gamma_l| < 1$ for $l = 1, 2, \ldots, L$, the mean-square value of the prediction error decreases with increasing filter order. Furthermore, by restricting the reflection coefficients in this manner, it may be shown that the resulting prediction-error filter is minimum phase [Burg [38], Haykin and Kesler [40]).

As a consequence of the Levinson recursion, we may write

$$\begin{aligned} f(l, n) &= f(l-1, n) + \Gamma_l b(l-1, n-1) \\ b(l, n) &= b(l-1, n-1) + \Gamma^*_l f(l-1, n) \end{aligned} \tag{4.157}$$

where the forward prediction error $f(l - 1, n)$ and spatially delayed backward pre-

*The parameter Γ_{L+1} is also referred to in the statistical literature as a *partial correlation (PARCOR) coefficient*.

†The recursive relation of Eq. (4.152) was first discovered by Levinson in 1947. It was rediscovered at a later date by Dubin [45]. In recognition of this, the relation is sometimes referred to as the *Levinson–Durbin recursion*.

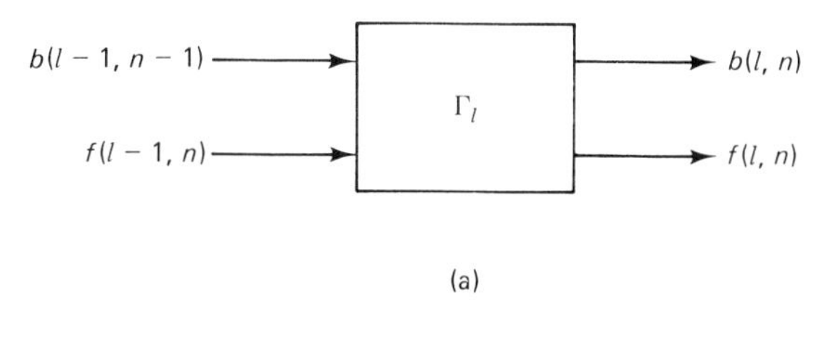

(a)

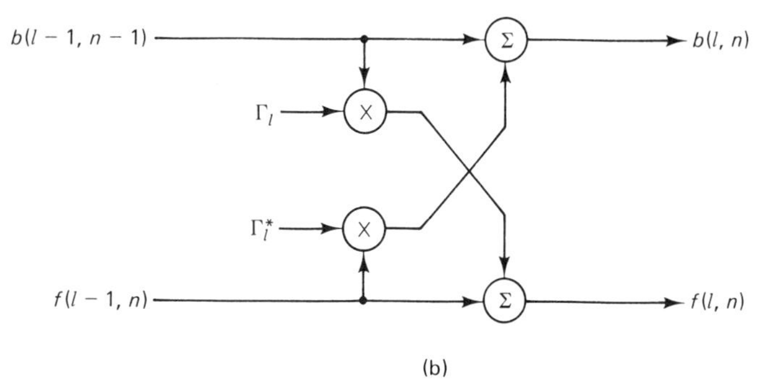

(b)

Figure 4.29 (a) Black box for representing Eqs. (4.157); (b) details of the black box.

diction error $b(l-1, n-1)$ pertain to filter order $l-1$, and the forward prediction error $f(l, n)$ and the backward prediction error $b(l, n)$ refer to filter order l. We may represent Eqs. (4.157) in block-diagrammatic form as shown in Fig. 4.29(a). The characterization of the black box labeled Γ_l is determined completely by specifying the reflection coefficient Γ_l. The detailed structure of the black box is shown in Fig. 4.29(b). Note that this structure exhibits conjugate symmetry in the sense that the coefficient required in Eqs. (4.157) to determine the backward prediction error $b(l, n)$ is the complex conjugate of the coefficient required to determine the forward reflection coefficient. This property is a direct consequence of the assumed Toeplitz character of the correlation matrix $\mathbf{R}$.

Given the spatial samples $x(n)$, $x(n-1)$, ..., $x(n-L)$, and using the *initial conditions*

$$f(0, n) = b(0, n) = x(n)$$

we may use Eqs. (4.157) to compute the sequence of forward prediction errors $f(1, n), f(2, n), \ldots, f(L, n)$ and the corresponding sequence of backward prediction errors $b(1, n), b(2, n), \ldots, b(L, n)$ for filter order L. The operations involved in this computation are illustrated by the signal flow graph shown in Fig. 4.30 for the case of $L = 3$. This spatial filter is known as a *lattice filter* because the basic building block used for its construction has the appearance of a lattice. Note that the number of stages in the lattice structure of Fig. 4.30 is equal to the order of the prediction filter, with stage 1 corresponding to Γ_1, stage 2 to Γ_2, etc.

The lattice filter structure of Fig. 4.30 offers the following interesting features:

1. For a stationary process represented by the spatial samples $x(n)$, $x(n-1), \ldots, x(n-L)$, the corresponding backward prediction errors $b(0, n)$, $b(1, n), \ldots, b(L, n)$ are *orthogonal* to each other, as shown by

$$E[b(k, n)b^*(l, n)] = \begin{cases} \epsilon_{\min}(l), & k = l \\ 0, & k \neq l \end{cases} \tag{4.158}$$

   Based on this property, we say that the individual stages of the lattice structure are *decoupled* from each other. Accordingly, we may replace the spatially correlated set of samples $x(n), x(n-1), \ldots, x(n-L)$ by a corresponding set of uncorrelated backward prediction errors $b(L, 0), b(L, 1), \ldots, b(L, L)$ *with no loss of information.* We may view this transformation as a form of Gram–Schmidt orthogonalization applied to the input process.*
2. In using the lattice structure to compute the forward prediction error $f(L, n)$ and the backward prediction error $b(L, n)$ for filter order L, we get useful by-products in the form of lower-order prediction errors, namely, $f(L-1, n), \ldots, f(1, n)$ and $b(L-1, n), \ldots, b(1, n)$.
3. Computationally speaking, it is relatively easy to increase the filter order from $L-1$ to L. The signal-flow graph of Fig. 4.30 depicts the operations involved in computing the forward prediction error $f(L, n)$ and the backward prediction error $b(L, n)$ for filter order $L = 3$. The operations involved in computing $f(L-1, n)$ and $b(L-1, n)$ for filter order $L-1 = 2$ are enclosed inside the dashed rectangle in Fig. 4.30. We thus see that all the computations performed for filter order $L-1 = 2$ are retained when we increase the filter order to $L = 3$. To go from filter order $L-1 = 2$ to $L = 3$ we simply have to compute a new reflection coefficient Γ_3, and use it to expand the scope of the computation as depicted in Fig. 4.30.

In the *Burg technique* we use the lattice structure to minimize an objective function defined as the sum of the mean-square values of the forward and backward prediction errors, as shown by

$$J(l) = E[|f(l, n)|^2 + |b(l, n)|^2] \tag{4.159}$$

Then, substituting Eqs. (4.157) into (4.159), differentiating $J(l)$ with respect to Γ_l (assumed to be complex valued), and setting the result equal to zero, we find that the optimum value of Γ_l is defined by [40, 47]

$$\Gamma_{\text{opt}, l} = -\frac{2E[f(l-1, n)b^*(l-1, n-1)]}{E[|f(l-1, n)|^2 + |b(l-1, n-1)|^2]} \tag{4.160}$$

*By using the sequence of backward prediction errors $b(L, 0), b(L, 1), \ldots, b(L, L)$ as the input to an adaptive filter rather than the original input $x(n), x(n-1), \ldots, x(N-L)$ it is possible to improve the convergence properties of an adaptive antenna used to suppress the effect of interferences impinging on the array from unknown directions [Griffiths (46)].

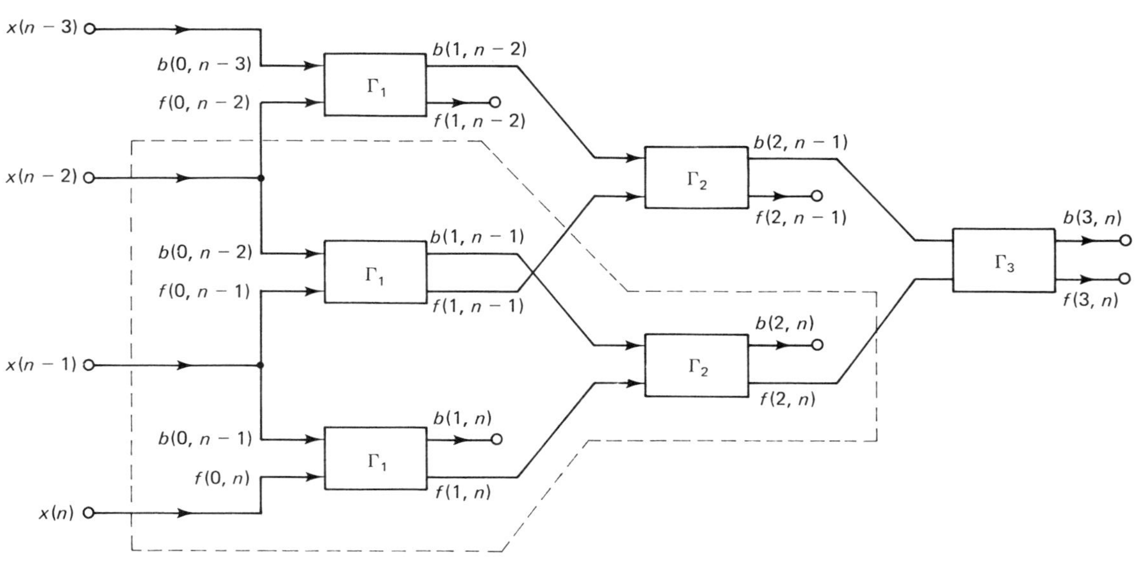

Figure 4.30 Spatial lattice filter of order $L = 3$. The portion of the filter enclosed inside the dashed lines corresponds to a spatial filter of order $L - 1 = 2$.

This result is known as the *Burg formula* [Burg [37]. A useful feature of this formula is the fact that it guarantees that $|\Gamma_{\text{opt},\,l}| \le 1$ for all prediction filter order l.

Using spatial averages as estimates of the expectations in the numerator and denominator of the right-hand side of Eq. (4.160), we may express the estimate of this optimum reflection coefficient as follows:

$$\hat{\Gamma}_{\text{opt},\,l} = -\frac{2\sum_{n=l+1}^{N} f(l-1,n)b^*(l-1,n-1)}{\sum_{n=l+1}^{N} |f(l-1,n)|^2 + |b(l-1,n-1)|^2}, \qquad l = 1, 2, \ldots, L \qquad (4.161)$$

where N is the length of the array and L is the desired prediction filter order.

The computation proceeds as follows:

1. Start the computation with the initial conditions:
$$f(0,n) = b(0,n) = x(n), \qquad n = 1, 2, \ldots, N$$
and
$$\epsilon_{\min}(0) = \frac{1}{N}\sum_{n=1}^{N} |x(n)|^2$$
2. Use Eq. (4.161) with $l = 1$, and thereby compute the reflection coefficient estimate $\hat{\Gamma}_{\text{opt},\,1}$.
3. Use Eqs. (4.157) with $l = 1$ to compute the forward prediction error $f(1, n)$ and backward prediction error $b(1, n)$ for filter order $l = 1$.
4. Use Eq. (4.155) to compute the mean-square value of the prediction error, $\epsilon_{\min}(1)$, for filter order $L = 1$.
5. Increment l by one, and go back to step 2, and repeat the procedure.
6. Stop the computation when the desired filter order is reached.

The procedure described above works well for spatially stationary processes. However, in general, it breaks down for spatially nonstationary processes. Sherman and Durrani [48] describe a lattice-based spatial filter that is suitable for stationary or nonstationary processes.

Experimental Results

One target. We first present results of an experiment [49] for the case when there is only one source of energy illuminating the array. The objective here was to utilize the spatial series obtained from the array in order to estimate the direction of the source, which is indicated by the peak of the AR/ME power spectrum of the spatial series.

For the source, a transmitting antenna was placed at one end of an anechoic chamber. The receiving array was a linear vertical array composed of eight equally

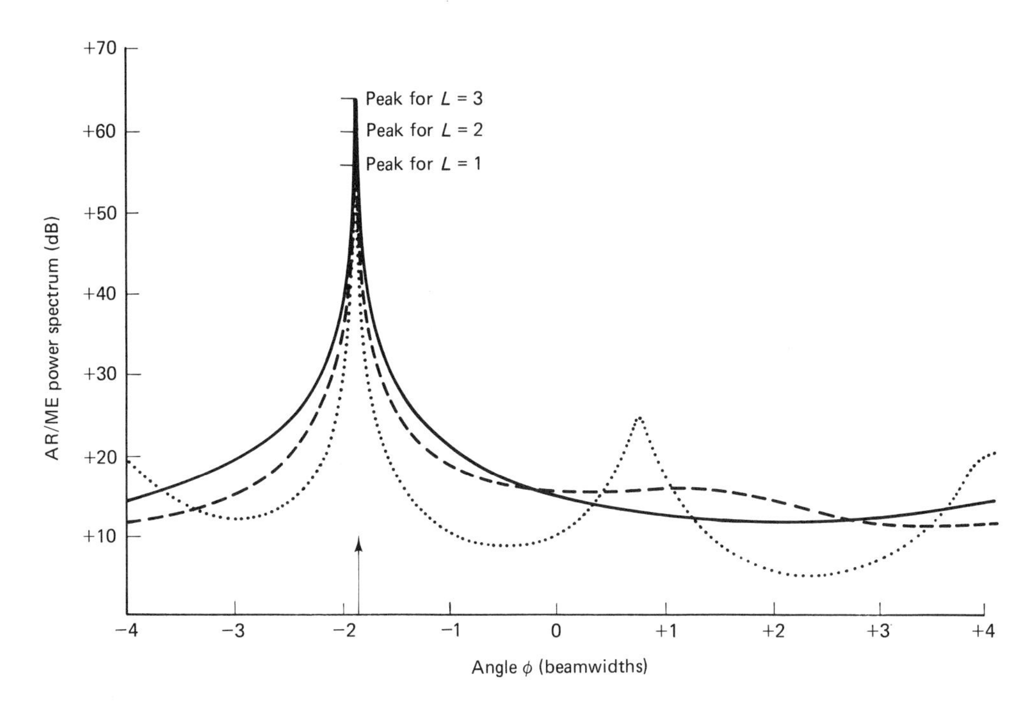

Figure 4.31 AR/ME power spectra for varying prediction filter order L. Solid line: $L = 1$; dashed line: $L = 2$; dotted line: $L = 3$.

spaced horn antennas, placed at the other end of the chamber. The signal from each element of this array was heterodyned down to baseband by means of in-phase and quadrature mixing. The frequency of operation was 9.79 GHz, and the physical separation of the elements was such that one standard beamwidth of the array was equal to 4.27°. The elemental signal-to-noise ratio was estimated at 18 dB.

Figure 4.31 shows how the AR/ME power spectrum of the resultant complex-valued spatial series changes with prediction filter order L for the case when the source was positioned to produce an electrical phase angle $\phi = -1.87$ BW. We see from the figure that (1) the location of the spectral peak coincides very closely with the actual direction of the source, (2) the resolution of this estimator increases with increasing filter order L (within limits), and (3) spurious spectral peaks appear as L is increased. In general, there may be as many as L peaks in the spectrum.

Coherent sources. When the sources of energy illuminating the array are coherent, the Burg technique may fail to operate satisfactorily. Such a situation arises in a multipath environment with a specular component that is correlated with the direct component (Reilly and Haykin [50], and Reilly [19]).

We now describe the results of a second experiment in which one transmitting antenna, simulating the target or source of the direct signal component, was placed at one end of an anechoic chamber. In close proximity to this antenna, a second transmitting antenna was placed, simulating the image target or source of the specular signal component. The feed of this second antenna was designed to introduce a variable attenuation and phase difference, both measured with respect to the feed of the first transmitting antenna. The magnitude of the attenuation corresponds to the value of the reflection coefficient, ρ, of the surface being modeled. The receiving array was a linear vertical array composed of eight equally spaced horn antennas, located at the other end of the chamber. Here again, the signal from each element of the receiving array was heterodyned down to baseband by means of in-phase and quadratic mixing. The frequency of operation was 9.79 GHz. The angular separation between the direct and specular components was 0.5 BW. The prediction order L was 3. Three curves are shown in Fig. 4.32 for $\psi = 0°$, $90°$, and $180°$, where ψ is the phase difference between these two components measured at the center of the array.

In Section 4.3 it was shown that the process is spatially nonstationary unless the phase difference ψ is an odd multiple of 90°. We therefore expect that in the general case of specular multipath, the AR/ME power spectrum computed by using the Burg technique would be misleading. This assertion is firmly validated by the curves shown in Fig. 4.32. We see that when $\psi = 0°$, the two signal components coalesce into a single peak. When $\psi = 180°$, the two spectral peaks are well resolved, but their positions bear no resemblance to the true angular positions of the direct and specular components. It is only when $\psi = 90°$ that the Burg technique

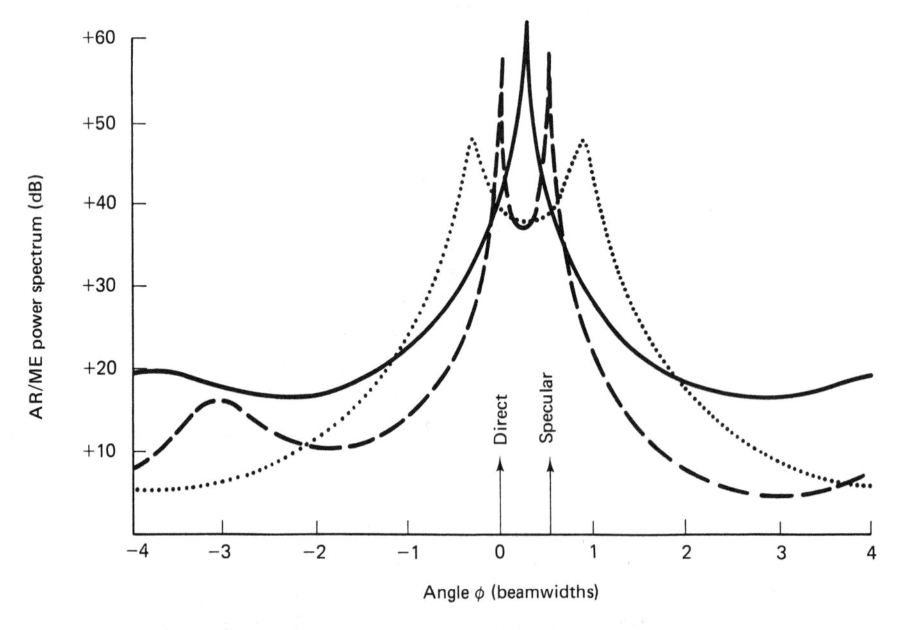

Figure 4.32 AR/ME power spectra for specular multipath for angle separation of 0.5 BW and varying phase difference ψ. Solid line: $\psi = 0°$; dashed line: $\psi = 90°$; dotted line: $\psi = 180°$.

yields useful results. The SNR corresponding to the curve for $\psi = 90°$ was about 28 dB.

The angular separation between the direct and specular paths in Fig. 4.32 equals 0.5 BW. For this separation, when $\psi = 0°$ and the reflection coefficient ρ is close to one, the array antenna output in effect consists of the sampled version of *one quarter* of a cosine wave *symmetrically* positioned with respect to the midpoint between the direct and specular paths, due to the signals received along these two paths. The additive noise has the further effect of making the resultant signal appear to the array as if there were a single target located midway between the direct and specular paths. On the other hand, when $\psi = 180°$ the array antenna output, in effect, consists of the sampled version of *one quarter* of a sine wave *antisymmetrically* positioned with respect to the midpoint between the two paths. In this case the additive noise has the further effect of making the resultant signal appear to the array as if there were two targets separated by an angle larger than its true value.

In Fig. 4.33 we have plotted the AR/ME power spectrum (again based on a prediction filter order $L = 3$), corresponding to an angular separation between the direct and specular components equal to 1.1 BW. Other parameters of the system were left unchanged. The two curves shown in Fig. 4.33 represent the extremes with respect to variation in ψ. In this case we find that the direct and specular components are almost uncorrelated, with the result that the peaks of the AR/ME power

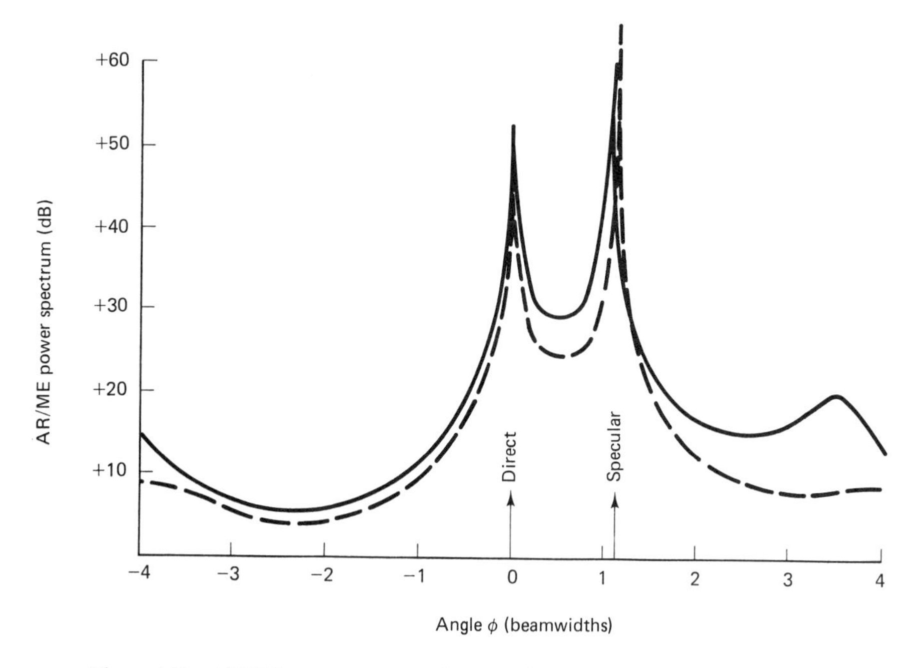

Figure 4.33 AR/ME power spectrum for specular multipath for angle separation of 1.1 BW and two extreme values of phase difference ψ.

spectrum provide reasonably accurate estimates of the actual directions of these two components for any ψ.

Herring [51] has shown theoretically that the Burg technique will yield accurate unbiased spectral estimates for the case when the two incident signal components are separated by $N/(N-1)$ beamwidths, for arbitrary phase difference ψ. This is true despite the fact that the process is nonstationary for $\psi \neq 90°$. With the number of antenna elements in the array, $N = 8$, the experimental curves shown in Fig. 4.33 support this finding, where the signal separation is closely equal to $N/(N-1) = 8/7 = 1.14$ beamwidths. The curves indicate that, in this case, the peaks of the AR/ME power spectrum obtained by using the Burg technique are reasonably accurate descriptions of the true situation for any value of phase difference ψ.

The conclusion, based on these results, is that, in general, the Burg technique is of little use for estimating the direction of a target in the presence of specular multipath, when the direct and specular paths lie inside a beamwidth.

4.14 THE FORWARD–BACKWARD LINEAR PREDICTION METHOD

We now describe a second method based on the use of *least squares* for estimating the AR parameters or, equivalently, the forward prediction-error filter coefficients for a given sequence of complex-valued spatial samples. In this method we recognize from the outset that only one finite-length sequence of samples is available for analysis. The quantity to be minimized is the prediction error energy defined as the sum of the squared forward prediction errors and the squared backward prediction errors. The desired solution is obtained by minimizing this energy with respect to *all* the AR parameters simultaneously for a specified prediction filter order. Although this method does not offer the flexibility enjoyed by the Burg technique, nevertheless it works well with stationary or nonstationary processes.

The method is referred to in the literature as the *forward–backward linear prediction (FBLP) method* or the *least squares method*; we will use the former terminology. It was apparently developed independently by Ulrych and Clayton [52] and Nuttal [53].

We define the prediction-error energy as

$$\epsilon = \sum_{n=L+1}^{N} |f(L, n)|^2 + |b(L, n)|^2 \tag{4.162}$$

where $f(L, n)$ is the forward prediction error and $b(L, n)$ is the backward prediction error for a prediction filter order L, and N is the array length. We express the forward prediction error in matrix form as

$$\begin{aligned} f(L, n) &= \mathbf{x}^T(n)\mathbf{g} \\ &= \mathbf{g}^T\mathbf{x}(n) \end{aligned} \tag{4.163}$$

where $\mathbf{g}$ is the $(L + 1) \times 1$ forward prediction-error filter vector:

$$\mathbf{g} = \begin{bmatrix} 1 \\ g(L, 1) \\ \vdots \\ g(L, L) \end{bmatrix} \tag{4.164}$$

and $\mathbf{x}(n)$ is the $(L + 1) \times 1$ input vector:

$$\mathbf{x}(n) = \begin{bmatrix} x(n) \\ x(n - 1) \\ \vdots \\ x(n - L) \end{bmatrix} \tag{4.165}$$

We express the corresponding backward prediction error in matrix form as

$$\begin{aligned} b(L, n) &= \mathbf{x}_b^T(n)\mathbf{g}^* \\ &= \mathbf{g}^H\mathbf{x}_b(n) \end{aligned} \tag{4.166}$$

The $(L + 1) \times 1$ vector $\mathbf{x}_b(n)$ contains the same elements as the input vector $\mathbf{x}(n)$, but in reverse order, as shown by

$$\mathbf{x}_b(n) = \begin{bmatrix} x(n - L) \\ x(n - L + 1) \\ \vdots \\ x(n) \end{bmatrix} \tag{4.167}$$

Substituting Eqs. (4.163) and (4.167) into (4.162), we get

$$\epsilon = \sum_{n=L+1}^{N} [\mathbf{g}^H\mathbf{x}^*(n)\mathbf{x}^T(n)\mathbf{g} + \mathbf{g}^H\mathbf{x}_b(n)\mathbf{x}_b^H(n)\mathbf{g}] \tag{4.168}$$

Define the $(L + 1) \times (L + 1)$ *deterministic correlation matrix*:

$$\mathbf{C}' = \sum_{n=L+1}^{N} \mathbf{x}^*(n)\mathbf{x}^T(n) + \mathbf{x}_b(n)\mathbf{x}_b^H(n) \tag{4.169}$$

where we have used a prime in the symbol for the correlation matrix to distinguish it from the corresponding $L \times L$ correlation matrix $\mathbf{C}$ to be defined later. Note that the correlation matrix $\mathbf{C}'$ is Hermitian; in general, however, it is non-Toeplitz. To distinguish this deterministic correlation matrix from the ensemble-averaged correlation matrix $\mathbf{R}$ considered earlier, we have purposely used a different symbol for it.

We may thus rewrite Eq. (4.162) in the form

$$\epsilon = \mathbf{g}^H\mathbf{C}'\mathbf{g} \tag{4.170}$$

The requirement is to find the prediction-error filter vector $\mathbf{g}$ that minimizes the prediction error energy ϵ subject to the constraint that

$$\mathbf{g}^T\boldsymbol{\delta} = \boldsymbol{\delta}^T\mathbf{g} = 1 \tag{4.171}$$

where $\boldsymbol{\delta}$ is the $(L + 1) \times 1$ vector defined in Eq. (4.123). To carry out this constrained optimization, we define the objective function

$$J = \mathbf{g}^H\mathbf{C}'\mathbf{g} - \lambda\delta^T\mathbf{g} - \lambda^*\mathbf{g}^H\boldsymbol{\delta} \tag{4.172}$$

where λ is the Lagrange multiplier. Following a procedure similar to that described in Section 4.11, we find that the optimum prediction-error filter vector that minimizes the objective function J is defined by

$$\mathbf{C}'\mathbf{g} = \epsilon_{\min}\,\boldsymbol{\delta} \tag{4.173}$$

where $\epsilon_{\min}$ is the minimum value of the prediction-error energy.

Computer Simulation Results

We now present computer simulation results using artificial data based on the symmetric specular multipath model of Eq. (4.21) for selected values of ρ, ϕ_1, ϕ_2, ψ, and SNR (Reilly [19]). Independent, zero-mean, Gaussian-distributed noise samples of the proper variance were added to the spatial samples representing two plane waves so as to produce test data with the desired SNR. Figure 4.34 shows the AR/ME power spectrum computed using the FBLP method for the following parameter values:

number of elements, $N = 21$
electrical angle of direct path, $\phi_1 = 0.25$ BW
electrical angle of specular path, $\phi_2 = -0.25$ BW
reflection coefficient, $\rho = 0.9$
phase difference at center of array, $\psi = 0°, 90°, 180°$
elemental signal-to-noise ratio, SNR $= 30$ dB

The results of Fig. 4.34 show that (1) the AR/ME power spectrum computed by using the FBLP method does resolve the direct and specular paths reasonably well, even though they lie within a fraction of beamwidth; and (2) the spectrum does not vary significantly with the phase difference ψ for the SNR used in the test.

Figure 4.35 shows the variations of the AR/ME power spectrum with the prediction-filter order L for values inside the range defined by $M \leq L \leq N/2$, where M is the number of incident plane waves. We see that for this range of values the spectral peak positions (corresponding to the direct and specular paths) are relatively insensitive to the value of L, and that only the background noise spectrum changes significantly with the filter order. This, of course, is true only if the signal-to-noise ratio is high, which is indeed the case in Fig. 4.35.

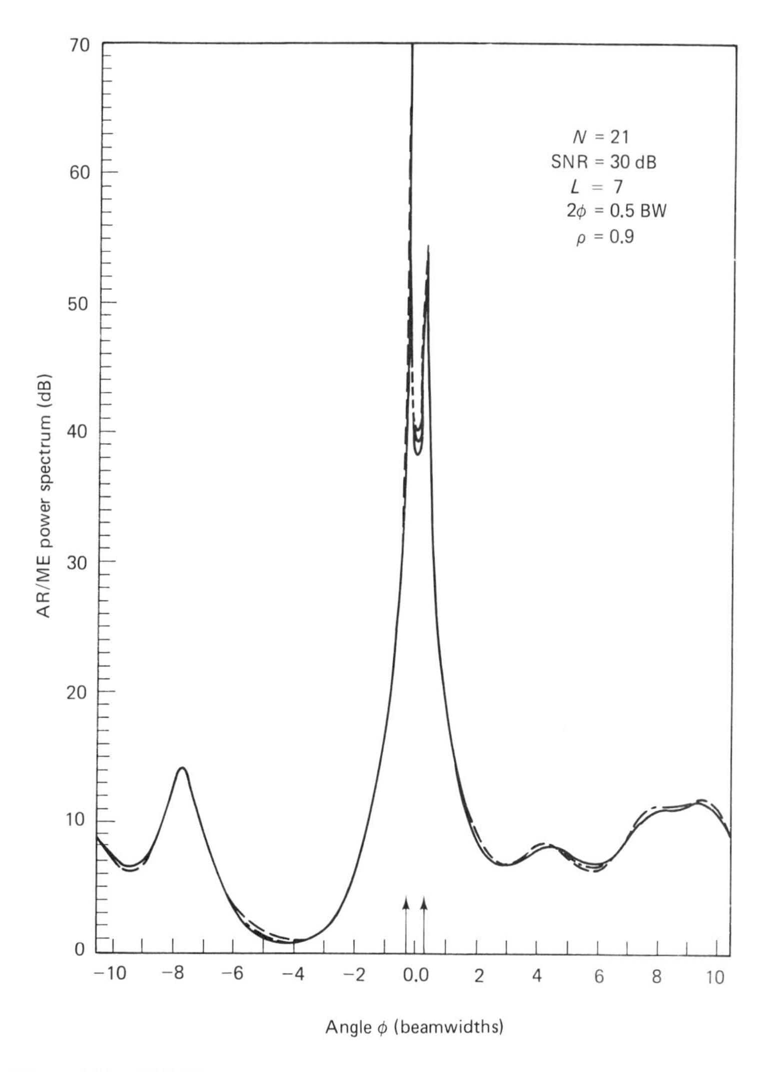

Figure 4.34 AR/ME power spectra obtained by using the FBLP method for specular multipath with an angle separation of 0.5 BW between the direct and specular paths. The figure represents the superposition of 3 curves for $\psi = 0°, 90°, 180°$.

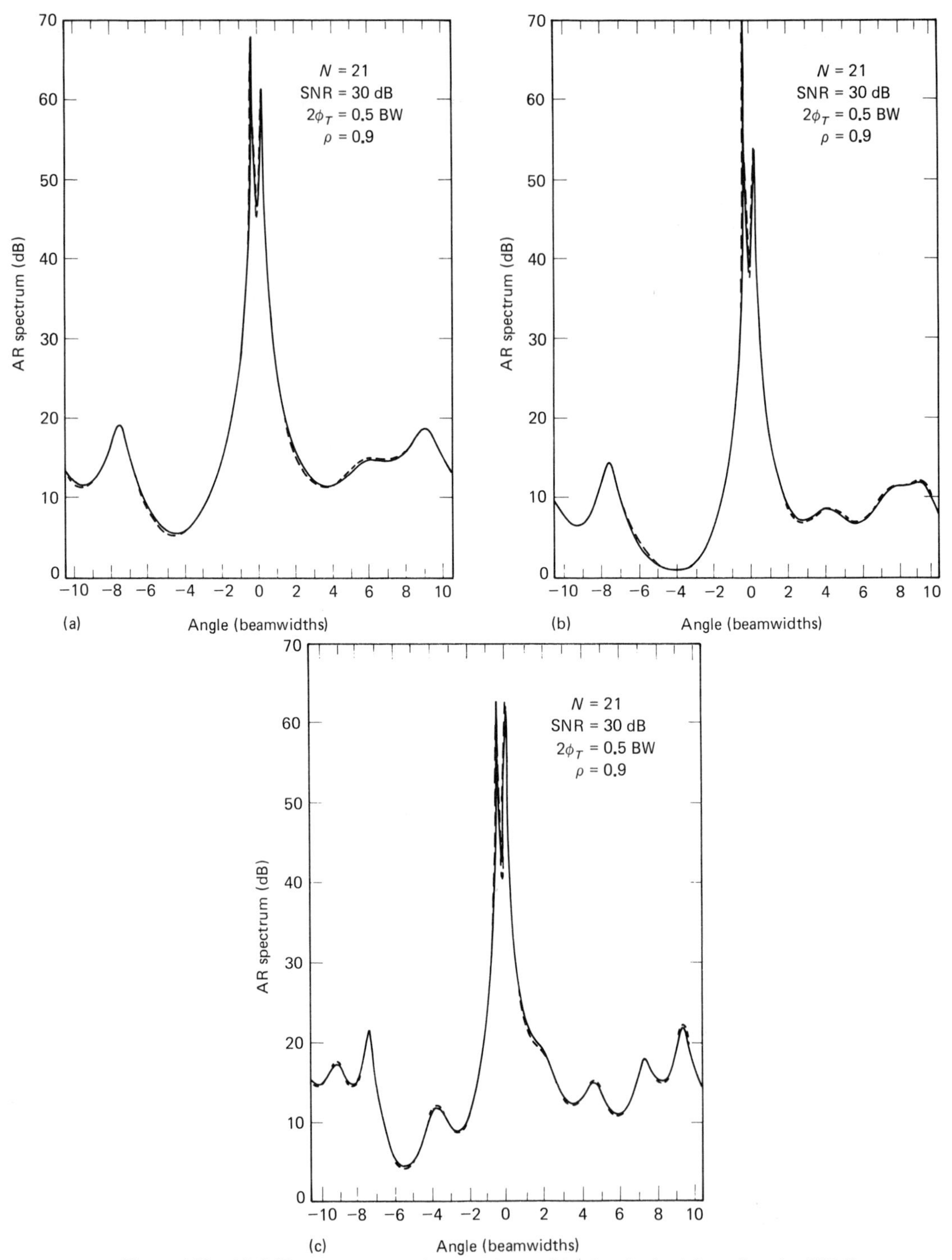

Figure 4.35 AR/ME power spectra for specular multipath, obtained by using the FBLP method, for varying prediction-filter order. The figure represents the superposition of 3 curves for $\phi = 0^\circ, 90^\circ, 180^\circ$. (a) $L = 5$; (b) $L = 7$; (c) $L = 9$.

4.15 LIMITATIONS OF THE FBLP METHOD

Tufts and Kumaresan [54, 55], using computer simulation, have compared the sample variances of angular frequency estimates obtained by the FBLP method with the the Cramér–Rao bound. They report two serious shortcomings of the FBLP method:

1. At high SNR, the FBLP method is a few decibels poorer than the Cramér–Rao bound, which is attained by the classical maximum-likelihood estimation procedure.
2. The FBLP method exhibits a *threshold* that appears to be reached at relatively high SNR values, above which the performance of the method deteriorates rapidly compared to the Cramér–Rao bound. This threshold effect is caused by one of two factors, depending on the prediction-filter order. When the prediction-filter order L is a large fraction of the array length N, spurious spectral peaks are induced by the presence of additive noise in the data; the threshold effect is caused by the spurious spectral peaks overpowering the effect of the signal. On the other hand, when the prediction-filter order L is small, the threshold effect is caused by the merging of closely spaced spectral peaks.

To overcome the aforementioned problems associated with the FBLP method, Tufts and Kumaresan [54, 55] have developed a two-step procedure for modifying the method. The result is impressive in that the performance of the *modified FBLP method* is close to the Cramér–Rao bound, even for closely spaced angles of arrival at lower values of SNR than would be possible with other linear prediction methods.

Before presenting details of the modified FBLP method as developed by Kumaresan and Tufts, we will digress by first developing an eigenvector representation of the prediction filter and then studying the properties of this representation for the special case of noiseless data. In so doing, we develop valuable insight into the ways in which the conventional FBLP method may be modified to make it a more useful signal processing tool.

4.16 EIGENVECTOR REPRESENTATION OF PREDICTION FILTER

Equation (4.173) defines the $(L+1)$ by 1 prediction-error-filter vector $\mathbf{g}$. We now wish to determine the corresponding expression for the $L \times 1$ prediction-filter vector $\mathbf{h}$. We first partition the $(L+1) \times (L+1)$ correlation matrix $\mathbf{C}'$ as follows:

$$\mathbf{C}' = \begin{bmatrix} \zeta & \mathbf{r}^H \\ \mathbf{r} & \mathbf{C} \end{bmatrix} \tag{4.174}$$

The ζ is a scalar defined by

$$\zeta = \sum_{n=L+1}^{N} |x(n)|^2 + |x(n-L)|^2 \tag{4.175}$$

The cross-correlation vector $\mathbf{r}$ is an $L \times 1$ vector with its ith element defined by

$$r(i) = \sum_{n=L+1}^{N} x(n)x^*(n-i) + x^*(n-L)x(n-L+i), \qquad i = 1, 2, \ldots, L \tag{4.176}$$

The correlation matrix $\mathbf{C}$ is an $L \times L$ matrix with its (t, u)th element defined by

$$c(t, u) = \sum_{n=L+1}^{N} x^*(n-t)x(n-u) + x(n-L+t)x^*(n-L+u),$$

$$t, u = 1, 2, \ldots, L \tag{4.177}$$

Then, substituting Eq. (4.174) into (4.173), and expressing the prediction-error filter $\mathbf{g}$ in terms of the prediction-filter vector $\mathbf{h}$, we may write

$$\begin{bmatrix} \zeta & \mathbf{r}^H \\ \mathbf{r} & \mathbf{C} \end{bmatrix} \begin{bmatrix} 1 \\ -\mathbf{h} \end{bmatrix} = \epsilon_{\min} \boldsymbol{\delta} \tag{4.178}$$

Since the first element of the vector $\boldsymbol{\delta}$ equals 1 and its remaining L elements equal zero, we deduce from Eq. (4.178) the following two relations:

$$\mathbf{Ch} = \mathbf{r} \tag{4.179}$$

and

$$\epsilon_{\min} = \zeta - \mathbf{r}^H\mathbf{h} \tag{4.180}$$

Define the $2(N - L) \times L$ *data matrix*

$$\mathbf{A} = \begin{bmatrix} x(L) & x(L-1) & \cdots & x(1) \\ x(L+1) & x(L) & \cdots & x(2) \\ \vdots & \vdots & & \vdots \\ x(N-1) & x(N-2) & \cdots & x(N-L) \\ x^*(2) & x^*(3) & \cdots & x^*(L+1) \\ x^*(3) & x^*(4) & \cdots & x^*(L+2) \\ \vdots & \vdots & & \vdots \\ x^*(N-L+1) & x^*(N-L+2) & \cdots & x^*(N) \end{bmatrix} \tag{4.181}$$

Then we may express the deterministic correlation matrix $\mathbf{C}$ in terms of the data matrix $\mathbf{A}$ as follows:

$$\mathbf{C} = \mathbf{A}^H\mathbf{A} \tag{4.182}$$

It is also useful to define the $2(N-L+1) \times 1$ *data vector*:

$$\mathbf{b} = \begin{bmatrix} x(L+1) \\ x(L+2) \\ \vdots \\ x(N) \\ x^*(1) \\ x^*(2) \\ \vdots \\ x^*(N-L) \end{bmatrix} \tag{4.183}$$

We may then express the input–output relation of the prediction filter in matrix form as follows:

$$\mathbf{Ah} = \mathbf{b} \tag{4.184}$$

The matrix product $\mathbf{Ah}$ equals a $2(N-L+1) \times 1$ vector whose ith element equals either the forward prediction of the input sample $x(i+L)$, given the input samples $x(i+L-1), x(i+L-2), \ldots, x(i)$, for $1 \le i \le N-L$, or it equals the backward prediction of the complex-conjugated input sample $x^*(i-N+L)$, given the complex-conjugated input samples $x^*(i+1-N+L)$, $x^*(i+2-N+L)$, $\ldots$, $x^*(i-N+2L)$, for $N-L+1 \le i \le 2N-2L$.

The cross-correlation vector $\mathbf{r}$ is defined in terms of the data matrix $\mathbf{A}$ by

$$\mathbf{r} = \mathbf{A}^H\mathbf{b} \tag{4.185}$$

Of all the vectors $\mathbf{h}$ for which the prediction-error energy is minimum, the prediction filter vector of interest is the one that has *minimum norm* in the Euclidean sense. This particular value, known as the minimum norm solution, is given in terms of the data matrix $\mathbf{A}$ by [59, 60]

$$\mathbf{h} = \mathbf{A}^{\#}\mathbf{b}$$

where $\mathbf{A}^{\#}$ is the *pseudo-inverse* or *Moore–Penrose generalized inverse* of the data matrix $\mathbf{A}$. If $2(N-L) > L$, we have more equations than unknowns, and the system of equations described in matrix form by (4.184) is *overdetermined*. In this case, the data matrix $\mathbf{A}$ has rank L, and its pseudo-inverse is defined by

$$\mathbf{A}^{\#} = [\mathbf{A}^H\mathbf{A}]^{-1}\mathbf{A}^H$$

If, on the other hand, $2(N-L) < L$, we have more unknowns than equations, and the system of equations in (4.184) is *underdetermined*. In this second case, the data matrix $\mathbf{A}$ has rank $2(N-L)$, and its pseudo-inverse is defined by

$$\mathbf{A}^{\#} = \mathbf{A}^H[\mathbf{A}\mathbf{A}^H]^{-1}$$

Equivalently, we may express the minimum norm solution in terms of the correlation matrix $\mathbf{C}$ as

$$\mathbf{h} = \mathbf{C}^{\#}\mathbf{r} \tag{4.186}$$

where $\mathbf{C}^{\#}$ is the pseudo-inverse of $\mathbf{C}$. When $2N - L > L$, the pseudo-inverse $\mathbf{C}^{\#}$ equals the ordinary inverse $\mathbf{C}^{-1}$.

Let $\lambda_1, \lambda_2, \ldots, \lambda_L$ denote the eigenvalues of the correlation matrix $\mathbf{C}$, and let $\mathbf{u}_1, \mathbf{u}_2, \ldots \mathbf{u}_L$ denote the associated eigenvectors, respectively. Since the correlation matrix $\mathbf{C}$ is nonnegative definite, all the eigenvalues are real and nonnegative. We assume that the eigenvectors are all normalized to have unit length. Since the correlation matrix $\mathbf{C}$ is also Hermitian, the eigenvectors are orthogonal to each other. Accordingly, we have

$$\mathbf{u}_i^H \mathbf{u}_k = \begin{cases} 1, & i = k \\ 0, & i \neq k \end{cases}$$

We may express the pseudo-inverse of the correlation matrix $\mathbf{C}$ in terms of its eigenvalues and eigenvectors as follows [60]

$$\mathbf{C}^{\#} = \sum_{k=1}^{W} \frac{1}{\lambda_k} \mathbf{u}_k \mathbf{u}_k^H \tag{4.187}$$

where W is the rank of data matrix $\mathbf{A}$. Hence, substituting Eq. (4.187) in (4.186), we get

$$\mathbf{h} = \sum_{k=1}^{W} \frac{\mathbf{u}_k}{\lambda_k} (\mathbf{u}_k^H \mathbf{r}) \tag{4.188}$$

We usually find that the *signal subspace* eigenvectors of the correlation matrix $\mathbf{C}$, that is, those associated with the large and generally well-separated (in magnitude) eigenvalues $\lambda_1, \ldots, \lambda_M$, are relatively insensitive to noise. On the other hand, the *noise subspace* eigenvectors of $\mathbf{C}$, that is, those associated with the remaining L–M eigenvalues are highly sensitive to noise in that they can change their directions abruptly. These noise subspace eigenvectors can make significant contributions to the prediction filter vector $\mathbf{h}$, as they are amplified by the reciprocal of their usually small eigenvalues. This, in turn, gives rise to spurious peaks in the AR/ME spectrum.

4.17 SPECIAL CASE: NOISELESS DATA

For noiseless data, the FBLP method reduces to *Prony's method* [57, 58] for the following special set of conditions:

1. The prediction-filter order L equals the number of complex exponentials, M, contained in the input signal (i.e., the number of plane waves incident on the array equals L).
2. The number of elements in the array, N, equals $2L$.
3. The prediction filter is used in the forward direction only; that is, we use only the first $N - L$ equations in the system of $2(N - L)$ simultaneous equations described by Eq. (4.184).

Under these conditions, we may write

$$\begin{bmatrix} x(L) & x(L-1) & \cdots & x(1) \\ x(L+1) & x(L) & \cdots & x(2) \\ \vdots & \vdots & & \vdots \\ x(2L-1) & x(2L-2) & \cdots & x(L) \end{bmatrix} \begin{bmatrix} h(1) \\ h(2) \\ \vdots \\ h(L) \end{bmatrix} = \begin{bmatrix} x(L+1) \\ x(L+2) \\ \vdots \\ x(2L) \end{bmatrix} \tag{4.189}$$

The $L \times L$ matrix on the left-hand side of Eq. (4.189) is both square and of full rank. Accordingly, we can directly solve this system of simultaneous equations for the prediction-filter coefficients $h(1), h(2), \ldots, h(L)$. The L roots of the transfer function

$$H(z) = 1 + \sum_{k=1}^{L} h(k)z^{-k}$$

can now be determined. For the case of noiseless data, the angular positions of the L plane waves incident on the array can thus be determined exactly. However, a serious limitation of Prony's method is that it is very sensitive to the presence of noise in the observed signal even though it may be relatively small [58].

For the case when the prediction filter order, L, is greater than the number of incident plane waves, M, and the number of elements in the array, N, is greater than $2M$, the system of $2(N - L)$ simultaneous equations defined in matrix form by (4.184) can be *underdetermined* or *overdetermined*, depending on the value of L.

Let the matrix $\mathbf{A}_0$ denote the noiseless version of the data matrix in Eq. (4.181). When the observed signal is noiseless, it is easy to verify that the largest size of nonzero determinant that can be formed from matrix $\mathbf{A}_0$ is M. Hence the matrix $\mathbf{A}_0$ has rank M. Let $\mathbf{C}_0$ denote the $L \times L$ correlation matrix of the noiseless observed signal. Since $\mathbf{C}_0 = \mathbf{A}_0^H \mathbf{A}_0$ is an $L \times L$ matrix, it follows that with $L > M$ the matrix $\mathbf{C}_0$ is singular. In such a case, we express the minimum norm value of the prediction filter vector as

$$\mathbf{h} = \mathbf{C}_0^{\#} \mathbf{r}_0 \tag{4.190}$$

where $\mathbf{C}_0^{\#}$ is the *pseudo-inverse* of the correlation matrix $\mathbf{C}_0$, and $\mathbf{r}_0$ is the noiseless version of the $L \times 1$ vector $\mathbf{r}$.

Since the matrix $\mathbf{A}_0$ has rank M, it follows that $L - M$ of the eigenvalues of the correlation matrix $\mathbf{C}_0$ are zero, and only M of its eigenvalues are nonzero. Hence, the pseudo-inverse of $\mathbf{C}_0$ is defined by [60]

$$\mathbf{C}_0^{\#} = \sum_{k=1}^{M} \frac{1}{\lambda_{0k}} \mathbf{u}_{0k} \mathbf{u}_{0k}^H \tag{4.191}$$

where λ_{0k}, $k = 1, 2, \ldots, M$, are the nonzero eigenvalues of $\mathbf{C}_0$, and $\mathbf{u}_{0k}$, $k = 1, 2, \ldots, M$, are the corresponding eigenvectors. Thus, substituting Eq. (4.191) in (4.190), we get

$$\mathbf{h} = \sum_{k=1}^{M} \frac{\mathbf{u}_{0k}}{\lambda_k} (\mathbf{u}_{0k}^H \mathbf{r}_0) \tag{4.192}$$

Comparing this result with that of Eq. (4.188), we see that for noiseless data the prediction filter vector consists of a linear combination of the M *principal eigenvectors* of the correlation matrix $\mathbf{C}_0$.

This solution for a noiseless observed signal has a number of important properties, as described below (Tufts and Kumaresan [55]):

Property 1. *The $(L + 1) \times 1$ prediction-error-filter vector, $\mathbf{g}$, is an eigenvector of the $(L + 1) \times (L + 1)$ correlation matrix $\mathbf{C}'_0$, and the corresponding eigenvalue is zero.*

For a noiseless observed signal, the prediction error is zero. Hence, putting $\epsilon_{\min} = 0$ in Eq. (4.173) and using $\mathbf{C}'_0$ for the pertinent correlation matrix, we get

$$\mathbf{C}'_0 \mathbf{g} = \mathbf{0} \tag{4.193}$$

where $\mathbf{0}$ is an $(L + 1) \times 1$ null vector. We note that the correlation matrix $\mathbf{C}'_0$ has only M nonzero eigenvalues and $L + 1 - M$ zero eigenvalues. Thus Eq. (4.193) is an eigen-equation corresponding to a zero eigenvalue, and the prediction-error-filter vector $\mathbf{g}$ is a corresponding eigenvector.

Property 2. *The transfer function of the prediction-error filter has M zeros on the unit circle in the z-plane at angular locations corresponding to the electrical phase angles of the M incident plane waves, provided that the prediction-filter order L satisfies the condition*

$$M \leq L \leq N - M/2$$

where N is the number of elements in the array.

To prove this property, we first show that the prediction-error filter-vector $\mathbf{g}$ is orthogonal to a set of $(L + 1) \times 1$ *sinusoidal vectors*, $\mathbf{e}_m$, defined by

$$\mathbf{e}_m = \begin{bmatrix} 1 \\ \exp(j\phi_m) \\ \exp(j2\phi_m) \\ \vdots \\ \exp(jL\phi_m) \end{bmatrix}, \qquad m = 1, 2, \ldots, M \tag{4.194}$$

Putting the noise $w(n) = 0$ in Eq. (4.10), we find that the noiseless observed signal is defined by

$$x(n) = \sum_{m=1}^{M} a_m \exp(jn\phi_m), \qquad n = 1, 2, \ldots, N \tag{4.195}$$

where we have redefined the complex amplitude a_m such that $a_m \exp(j\phi_m)$ now represents the received signal at element 1 of the array due to the mth plane wave.

The (t, u)th element of the $(L + 1) \times (L + 1)$ correlation matrix $\mathbf{C}'_0$ is defined by [see Eq. (4.169)]

$$C'_0(t, u) = \sum_{n=L+1}^{N} [x^*(n - t)x(n - u) + x(n - L + t)x^*(n - L + u)],$$

$$t, u = 0, 1, \ldots, L \qquad (4.196)$$

Substituting Eq. (4.195) into (4.196), and rearranging the summations, we get (after some manipulation)

$$C'_0(t, u) = \sum_{m=1}^{M} \sum_{l=1}^{M} X_{ml} \exp [j(t\phi_m - u\phi_l)] \qquad (4.197)$$

where X_{ml} is defined by

$$X_{ml} = \sum_{n=L+1}^{N} a_m^* a_l \exp [jn(\phi_l - \phi_m)] + a_m a_l^* \exp [j(n - L)(\phi_m - \phi_l)] \qquad (4.198)$$

Define the $2(N - L) \times 1$ vectors:

$$\mathbf{f}_m = \begin{bmatrix} a_m \exp [j\phi_m(L + 1)] \\ a_m \exp [j\phi_m(L + 2)] \\ \vdots \\ a_m \exp (j\phi_m N) \\ a_m^* \exp (-j\phi_m) \\ a_m^* \exp (-j2\phi_m) \\ \vdots \\ a_m^* \exp [-j\phi_m(N - L)] \end{bmatrix}, \qquad m = 1, 2, \ldots, M \qquad (4.199)$$

Then we may express X_{ml} as the inner product of the vectors $\mathbf{f}_m$ and $\mathbf{f}_l$, as shown by

$$X_{ml} = \mathbf{f}_m^H \mathbf{f}_l \qquad (4.200)$$

We may now write the $(L + 1) \times (L + 1)$ correlation matrix $\mathbf{C}'_0$ as a product of three matrices, as shown by

$$\mathbf{C}'_0 = \mathbf{SPS}^H \qquad (4.201)$$

where $\mathbf{S}$ is an $(L + 1) \times M$ sinusoidal matrix defined by

$$\mathbf{S} = [\mathbf{e}_1, \mathbf{e}_2, \ldots, \mathbf{e}_M]$$

$$= \begin{bmatrix} 1 & 1 & \cdots & 1 \\ \exp (j\phi_1) & \exp (j\phi_2) & \cdots & \exp (j\phi_M) \\ \exp (j2\phi_1) & \exp (j2\phi_2) & \cdots & \exp (j2\phi_M) \\ \vdots & \vdots & & \vdots \\ \exp (jL\phi_1) & \exp (jL\phi_2) & & \exp (jL\phi_M) \end{bmatrix} \qquad (4.202)$$

and $\mathbf{P}$ is an $M \times M$ matrix defined by

$$\mathbf{P} = \begin{bmatrix} X_{11} & X_{12} & \cdots & X_{1M} \\ X_{21} & X_{22} & \cdots & X_{2M} \\ \vdots & \vdots & & \vdots \\ X_{M1} & X_{M2} & \cdots & X_{MM} \end{bmatrix} \tag{4.203}$$

The square matrix $\mathbf{P}$ shows the interaction between the M plane waves incident on a linear array of finite aperture. It has the following properties:

1. The matrix $\mathbf{P}$ is Hermitian (i.e., $\mathbf{P}^H = \mathbf{P}$).
2. The mth element on the main diagonal of the matrix $\mathbf{P}$ equals

$$X_{mm} = 2(N - L)|a_m|^2 \tag{4.204}$$

where a_m is the complex amplitude of the mth plane wave received by the array.

3. Each element of the matrix $\mathbf{P}$ is an inner product of two vectors in the set $\{\mathbf{f}_m\}$, $m = 1, 2, \ldots, M$. Such a matrix is called a *Gramian* matrix [61].
4. From Eq. (4.199) we see that the vectors $\mathbf{f}_m$, $m = 1, 2, \ldots, M$, are linearly independent; hence, the matrix $\mathbf{P}$ is nonsingular.

Since the prediction-error-filter vector $\mathbf{g}$ is an eigenvector of the correlation matrix $\mathbf{C}'_0$, corresponding to an eigvenvalue of zero, we may write

$$\mathbf{g}^H \mathbf{C}'_0 \mathbf{g} = 0 \tag{4.205}$$

Substituting Eq. (4.201) into (4.205), we get

$$\mathbf{g}^H \mathbf{S} \mathbf{P} \mathbf{S}^H \mathbf{g} = 0 \tag{4.206}$$

Since the matrix $\mathbf{P}$ is nonsingular, this condition can be satisfied if and only if

$$\mathbf{S}^H \mathbf{g} = \mathbf{0} \tag{4.207}$$

where $\mathbf{0}$ is an $M \times 1$ null vector. Equivalently, we may use the first line of Eq. (4.202) to write that the prediction-error-filter vector $\mathbf{g}$ is orthogonal to the set of $(L + 1) \times 1$ sinusoidal vectors $\{\mathbf{e}_m\}$, $m = 1, 2, \ldots, M$, as shown by

$$\begin{aligned} \mathbf{e}_m^H \mathbf{g} &= 1 + \sum_{k=1}^{L} g(k) \exp(-jk\phi_m), \qquad m = 1, 2, \ldots, M \\ &= 0 \end{aligned} \tag{4.208}$$

This result is recognized as the transfer function of the prediction-error filter, $H(z)$, evaluated on the unit circle at $z = \exp(j\phi_m)$, for $m = 1, 2, \ldots, M$. We have thus proved that the transfer function of the prediction-error filter has M zeros on the unit circle in the z-plane, which determine the angular positions of the M incident plane waves.

We may now find the range of values of the prediction-filter order L for which this property is true. First, we observe that the minimum value of L is M, the number of incident plane waves. For a fixed number of elements in the array, N, as the prediction-filter order L is increased up to the point when $2(N - L) < M$, we find that the $2(N - L) \times 1$ vectors $\mathbf{f}_m$, $m = 1, 2, \ldots, M$ cannot be linearly independent, and consequently the matrix $\mathbf{P}$ is singular. Then Eq. (4.208) will no longer be true, in general. The maximum value of L is therefore $N - M/2$. Thus, the prediction-filter order L should satisfy the condition

$$M \leq L \leq N - \frac{M}{2} \tag{4.209}$$

Property 3. *For a prediction-filter order L satisfying the condition $M \leq L \leq N - M/2$, the $L - M$ extraneous zeros of the transfer function of the prediction-error filter, $H(z)$, are uniformly distributed in angle around the inside of the unit circle in the z-plane.*

For a proof of this property, the reader is referred to [62].

For the special case of noiseless data with $L = M$, the $(L + 1) \times (L + 1)$ correlation matrix $\mathbf{C}_0'$ has a single zero eigenvalue, and the prediction-error filter vector $\mathbf{g}$ is in a unique position in that it equals the eigenvector associated with this zero eigenvalue. When $M \leq L \leq N - M/2$, the correlation matrix $\mathbf{C}_0'$ has $L + 1 - M$ zero eigenvalues, with the result that the *null space* of $\mathbf{C}_0'$ has more than one eigenvector. However, the prediction-error filter vector $\mathbf{g}$ is once again unique in that (1) it corresponds to a zero eigenvalue in accordance with Property 1, and (2) the corresponding prediction filter vector $\mathbf{h}$ is defined by the minimum norm solution given in Eq. (4.190); that is, the vector $\mathbf{h}$ has the smallest length possible in the Euclidean sense.

4.18 THE MODIFIED FBLP METHOD

We now describe the two-step procedure of Tufts and Kumaresan [55] for modifying the FBLP method. The motivation here is to overcome the problems associated with the conventional FBLP method, which were discussed in Section 4.15. First, we find a square matrix $\hat{\mathbf{C}}$ of specified rank to replace the correlation matrix $\mathbf{C}$, because we want a better estimate of the signal correlation matrix. In this connection we make use of a theorem due to Eckart and Young [63], which we shall state later. In so doing we achieve an effective increase in SNR. Second, we increase the order of the prediction filter to values beyond those which would be appropriate for the conventional FBLP method, so as to improve the resolution capability of the prediction filter. We are unable to do this in the conventional FBLP method because of the instabilities introduced into the prediction-error filter. We shall now explain these two steps.

Step 1: Finding an estimate of the Signal Correlation Matrix. From Section 4.17 we recall that in the ideal case of noiseless data, the rank of the correlation matrix $\mathbf{C}_0$ equals the number of incident plane waves, M, even though the dimension of this matrix, L, may be higher than M. At low SNR, it is useful to try to return to this ideal situation. The means to do this is provided by the *Eckart–Young theorem* [63], which provides a method for best fit of signal correlation matrix $\hat{\mathbf{C}}$ of lower rank to the given signal-plus-noise correlation matrix $\mathbf{C}$ of full rank. It is fortunate that this improvement is not very sensitive to exact knowledge of the number of incident plane waves, M. If M is not known, we may try different values of M. A version of the Eckart–Young theorem can be stated as follows:

> Let $\mathbf{C}$ be an $L \times L$ matrix of rank K, which has complex-valued elements. Let $\mathbf{S}$ be the set of all $L \times L$ matrices of rank $M < K$. Then for all matrices $\mathbf{B}$ in $\mathbf{S}$, we have*
>
> $$\| \mathbf{C} - \hat{\mathbf{C}} \| \leq \| \mathbf{C} - \mathbf{B} \| \tag{4.210}$$
>
> where $\hat{\mathbf{C}}$ is an $L \times L$ matrix involving the M signal subspace eigenvalues and eigenvectors of $\mathbf{C}$, as shown by
>
> $$\hat{\mathbf{C}} = \sum_{k=1}^{M} \lambda_k \mathbf{u}_k \mathbf{u}_k^H \tag{4.211}$$

In order that the matrix $\hat{\mathbf{C}}$, as defined above, be a useful estimate, the largest M eigenvalues of $\mathbf{C}$ and the corresponding eigenvectors should be the perturbed versions of the M nonzero eigenvalues and the corresponding eigenvectors of $\mathbf{C}_0$. Indeed, there is experimental evidence [55] to suggest that, for high values of SNR, the M *principal* eigenvalues and eigenvectors of the $L \times L$ correlation matrix $\mathbf{C}$ for signal plus noise remain "close" to their noiseless case by using the signal correlation matrix estimate $\hat{\mathbf{C}}$ in place of $\mathbf{C}$.

Thus, in terms of the eigen-decomposition of $\hat{\mathbf{C}}$ given in Eq. (4.211), we may express the corresponding prediction filter vector as

$$\hat{\mathbf{h}} = \sum_{k=1}^{M} \frac{\mathbf{u}_k}{\lambda_k} (\mathbf{u}_k^H \mathbf{r}) \tag{4.212}$$

Note that the $L - M$ noise subspace eigenvectors have dropped out in the above formula.

We may therefore sum up Step 1 by stating that it involves partitioning the space spanned by the eigenvectors of the correlation matrix $\mathbf{C}$ into a *signal subspace* and a *noise subspace*. Let $\lambda_1, \lambda_2, \ldots, \lambda_L$ denote the eigenvalues of $\mathbf{C}$, and let $\mathbf{u}_1, \mathbf{u}_2, \ldots, \mathbf{u}_L$ denote the associated eigenvectors, respectively. We assume that $\lambda_1 \geq \lambda_2 \geq \ldots \lambda_M \geq \ldots \geq \lambda_L$. The principal eigenvectors $\mathbf{u}_1, \mathbf{u}_2, \ldots, \mathbf{u}_M$ are said to span the signal subspace; these eigenvectors define the improved estimate of the prediction filter vector $\mathbf{h}$, as in Eq. (4.212). The remaining eigenvectors $\mathbf{u}_{M+1}, \ldots, \mathbf{u}_L$ are said to span the noise subspace; these eigenvectors are simply ignored.

*For a square matrix $\mathbf{X}$, we define the matrix norm $\| \mathbf{X} \|^2$ as the *trace* of the matrix product $\mathbf{X}^H\mathbf{X}$.

Step 2: Increasing the Order of Prediction Filter. Intuitively speaking, the order L should be as large as possible in order to have a large aperture for the prediction filter. However, the use of large values of L causes serious problems in the conventional FBLP method, as discussed in Section 4.15. For a practical maximum value of L, Ulrych and Clayton [52], and Lang and McLellan [64] use the value $N/2$. For best performance of the FBLP method, Lang and McLellan [64] suggest the value of $N/3$ for L.

Once the undesirable effects of the noise subspace eigenvectors have been removed, as described in Step 1, we may increase the prediction filter order L beyond these limits. This increase in L helps to improve the resolution capability of the prediction filter since it represents an increase in the aperture or degrees of freedom available in the design of the filter.

However, even in using the modified FBLP method there is a practical limit to how large a value we may use for the prediction filter order L. At very large values of L, the effective SNR improvement decreases, with the result that the resolution capability of the FBLP method decreases despite the increased aperture. This decrease in effective SNR improvement is due to the fact that, with increased L, fewer product terms [equal to $2(N - L)$ in number] are averaged in computing the elements of the correlation matrix $\mathbf{C}$. This, in turn, results in larger perturbations of the eigenvalues λ_k and eigenvectors $\mathbf{u}_k$. We thus find that, on the one hand, increasing the value of L provides the potential for higher resolution, and yet, on the other hand, it also increases the fluctuations in the matrix $\mathbf{C}$, and hence in the estimate $\hat{\mathbf{C}}$ which is formed with λ_k and $\mathbf{u}_k$ with $k = 1, 2, \ldots, L$. Thus we need to use a compromise value of filter order L, balancing the two effects of resolution and stability. Tufts and Kumaresan [55] have experimentally determined the compromise value of $3N/4$ for L.

The Kumaresan–Prony Case

As explained in Section 4.17, the maximum permissible value of L equals $N - M/2$, beyond which the angular positions of the M incident plane waves cannot be found from the zeros of the prediction-error filter even in the case of noiseless data. At $L = N - M/2$, we have a set of only $2(N - L) = M$ simultaneous equations. Correspondingly, the matrix $\mathbf{A}$ is of order $M \times L$. The correlation matrix $\mathbf{C}$ itself is of rank M. Hence $\hat{\mathbf{C}} = \mathbf{C}$. Thus, when $L = N - M/2$, the first step in the improvement procedure is redundant. In particular, since $\mathbf{C}$ has only M nonzero eigenvalues, the noise eigenvectors are automatically eliminated, thereby removing the associated ill effects. This special case is known as the *Kumaresan–Prony case* [55].

The use of the Kumaresan-Prony method offers the following advantages:

1. For a prescribed number of plane waves M incident on the array, it enjoys the advantage of the maximum possible aperture $N - M/2$ for the prediction filter. For example, in the idealized specular model with two paths, we have $M = 2$ and the prediction-filter order $L = N - 1$, where N is the number of elements in the array.

2. It has a computational advantage in that the prediction-filter vector **h** can be computed directly from the prediction equations: $\mathbf{Ah} = \mathbf{b}$ by using the appropriate pseudo-inverse of **A** as shown by

$$\mathbf{h} = \mathbf{A}^H[\mathbf{A}\mathbf{A}^H]^{-1}\mathbf{b} \tag{4.213}$$

No eigenvalue and eigenvector decomposition are involved in this computation. We only have to invert an $M \times M$ matrix, $\mathbf{AA}^H$, where M is usually quite small. For example, in the aforementioned specular model with $M = 2$, the computation of **h** involves the inversion of a 2×2 matrix, which is trivial. For a complex environment involving a large number M of plane waves, Marple [65] describes a fast algorithm for computing the vector **h**.

However, these advantages of the Kumaresan-Prony method are realized at the cost of some stability in **C**, since the least possible number of terms, $2(N - L) = M$, are averaged for each element of **C**. Accordingly, the performance of the Kumaresan-Prony method is usually inferior to that of the modified FBLP method, for which the prediction-filter order is set equal to its optimum value (i.e., $3N/4$).

4.19 COMPUTER SIMULATION RESULTS

In this section we present computer simulation results for: (a) the noiseless case, (b) a comparison of the performance of the FBLP method with that of the modified FBLP method, (c) an assessment of the sensitivity of the modified FBLP method to variations in the phase difference ψ between the direct and specular paths at the center of the array, signal-to-noise ratio (SNR), surface reflection coefficient and number of elements, and (d) the Kumaresan-Prony case. The data used in the simulation were based on a symmetric specular multipath model described by the equation:

$$x(n) = \exp(jn\phi) + \rho \exp[j(\psi - n\phi)] + w(n)$$
$$n = -(N-1)/2, \ldots, (N-1)/2$$

where the direct component (with electrical angle ϕ) has been normalized to have unit amplitude. Correspondingly, the specular component (with electrical angle $-\phi$) has a normalized amplitude equal to the surface reflection coefficient ρ. The angle ψ represents the phase difference between these two components, measured at the center of the array. The noise samples $w(n)$ are independent complex Gaussian random variables with zero mean and variance determined by the prescribed SNR. As before, the SNR is measured in free space (i.e., with the specular component reduced to zero).

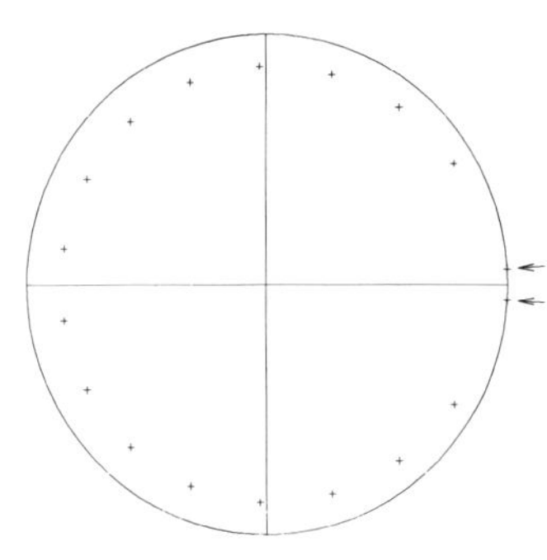

Figure 4.36 z-plane locations for the prediction-error filter zeros for the noiseless case with $N = 25$, $L = 18$, and $2\phi = 0.5$ BW. Sinusoidal locations are shown by arrows.

(a) Noiseless Case

Figure 4.36 shows the zero locations of the prediction-error filter in the z-plane for the special case of a noiseless observed signal (i.e., infinite SNR) and the following parameters:

number of elements, $N = 25$
prediction-filter order, $L = 18$
angular separation, $2\phi = 0.5$ BW

We see that, as expected, there are two *signal zeros* located at angles $\pm\phi$ on the unit circle, and the remaining 16 *extraneous zeros* are distributed uniformly inside the unit circle. The true angular locations of the two incident components are shown by arrows, a practice that is followed in all subsequent results.

(b) Comparison of the FBLP and Modified FBLP Methods

Figure 4.37 shows the locations of the prediction-error filter zeros in the z-plane for both the FBLP and modified FBLP methods for the following parameters:

number of elements, $N = 25$
prediction-filter order, $L = 18$
angular separation, $2\phi = 0.5$ BW
reflection coefficient, $\rho = 0.9$
phase difference, $\psi = 90°$
signal-to-noise ratio, SNR $= 10$ dB

Part (a) of the figure shows the results obtained by using the conventional FBLP method, and Part (b) shows the corresponding results obtained by using the modified FBLP method. Both parts of the figure include results of 50 independent trials.

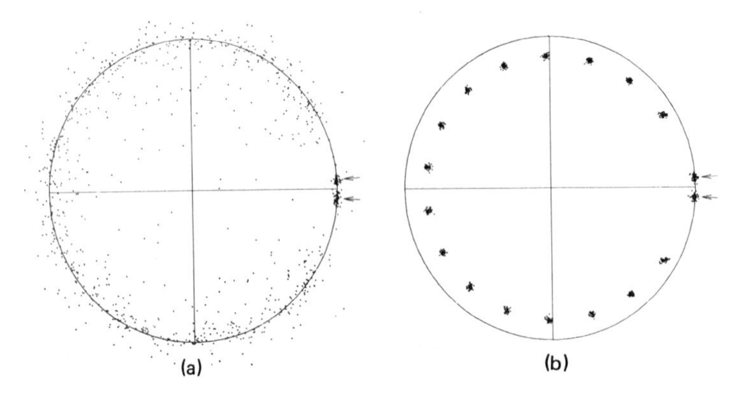

Figure 4.37 Comparison of (a) FBLP method, and (b) modified FBLP method (50 independent trials of both methods) for $N = 25$, $L = 18$, $2\phi = 0.5$ BW, $\rho = 0.9$, $\psi = 90°$, and SNR = 10 dB.

From this figure two points are immediately apparent:

1. In the conventional FBLP method,* the presence of noise eigenvectors causes considerable fluctuations in the locations of the prediction-error filter zeros, even though the direct and specular signals components are orthogonal to each other (for $\psi = 90°$).
2. In the modified FBLP method, the spreads of the noise-zero clusters are considerably reduced, with the result that the composite zero-pattern for the 50 independent trials closely resembles the zero-pattern for the corresponding idealized noiseless case. This means that the chance of spurious frequency estimates is considerably reduced.

(c) Sensitivity of the Modified FBLP Method to Parameter Variations

For this evaluation, the prediction-filter order L was set close to its optimum value, that is, $3N/4$. Thus, with the number of elements $N = 25$, we have $L = 18$. This value of L is used in Figs. 4.38–4.40, which shows the z-plane locations of the prediction-error filter zeros for 50 independent trials of the modified FBLP method, with the angular separation between the direct and specular paths corresponding to $2\phi = 0.5$ BW. The values of the remaining parameters were as described below.

Figure 4.38 shows the results obtained for the reflection coefficient $\rho = 0.85$, SNR = 10 dB, and for varying phase difference ψ between the direct and specular paths at the center of the array. Parts (a)–(h) of the figure correspond to $\psi = 0°$,

*We may have been somewhat unfair to the conventional FBLP method in Fig. 4.37. In the normal application of the method, the value used for L (typically $N/3$) is much smaller than that used in this example. Nevertheless, the observation made here still applies.

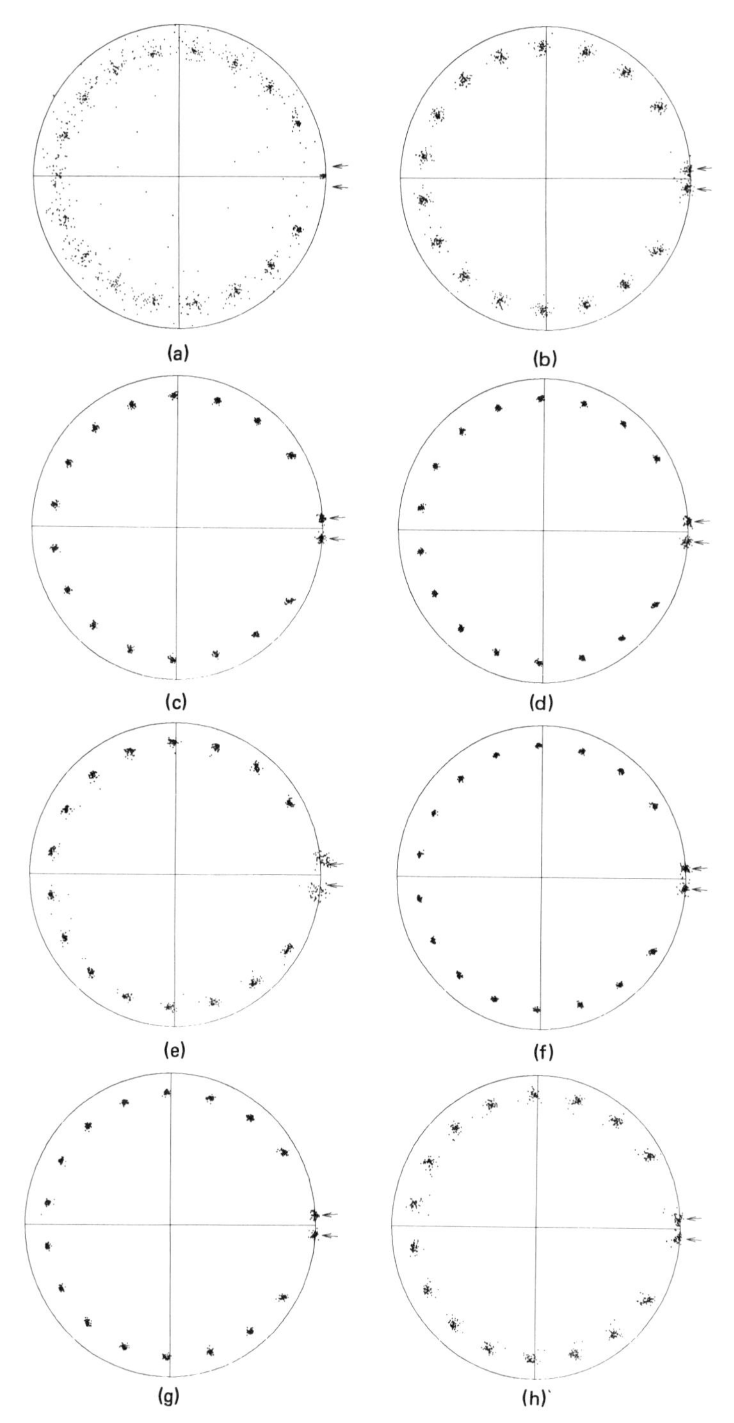

Figure 4.38 Results of 50 independent trials of modified FBLP method for $N = 25$, $L = 18$, $2\phi = 0.5$ BW, $\rho = 0.85$, SNR = 10 dB. (a) $\psi = 0°$; (b) $\psi = 45°$; (c) $\psi = 90°$; (d) $\psi = 135°$; (e) $\psi = 180°$; (f) $\psi = -135°$; (g) $\psi = -90°$; (h) $\psi = -45°$.

$\pm 45°$, $\pm 90°$, $\pm 135°$, $180°$. The following points are apparent from this figure:

1. The phase difference $\psi = 0°$, shown in Part (a) of the figure, appears to have the most degrading effect on the performance of the modified FBLP method. In this case the method fails by responding to the combined action of the direct and specular paths as if there were only a single target located midway between the true angular positions of the two paths (i.e., normal to the array in the example being considered here).
2. When $\psi = 180°$, as shown in Part (e) of the figure, the centroids of the two signal-zero clusters are noticeably displaced from the true angular locations of the direct and specular paths, with the result that the estimate obtained for the angular separation between the two paths is greater than the true value of 2ϕ.
3. When $\psi = \pm 45°$, as in Parts (b) and (h) of the figure, the noise-zero clusters develop large spreads, but nevertheless the centroids of the signal-zero clusters are close to the true angular locations of the direct and specular paths.
4. The best performance is achieved when the phase difference $\psi = \pm 90°$, as in Parts (c) and (g) of the figure, for which the direct and specular signal components are orthogonal to each other.

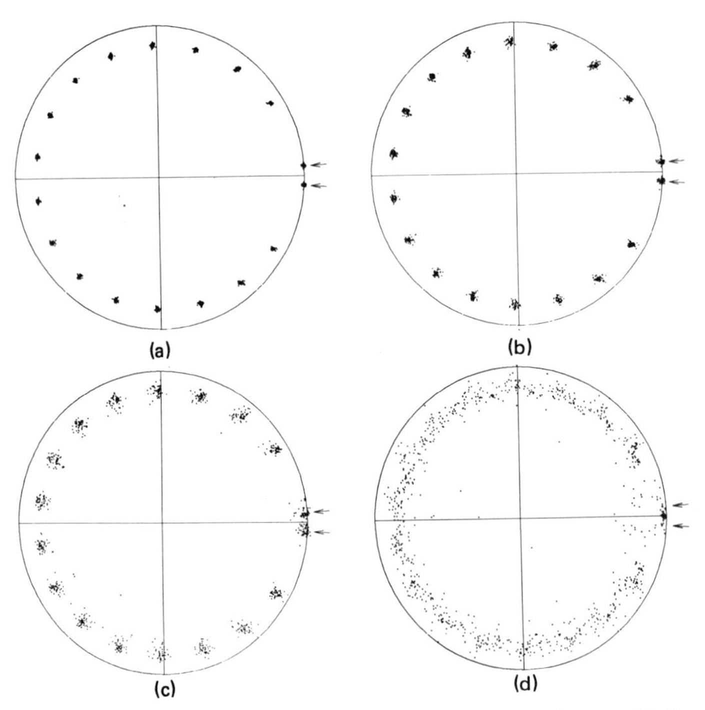

Figure 4.39 Results of 50 independent trials of FBLP method for $N = 25$, $L = 18$, $2\phi = 0.5$ BW, $\rho = 0.85$, $\psi = 0°$. (a) SNR = 30 dB; (b) SNR = 25 dB; (c) SNR = 20 dB; (d) SNR = 15 dB.

Figure 4.39 shows the results obtained for the reflection coefficient $\rho = 0.85$, phase difference $\psi = 0°$, and for incremental changes of 5 dB in the signal-to-noise ratio in the range 15–30 dB. These results show that, for the above set of parameters, the modified FBLP method is able to resolve the direct and specular paths accurately [as in Parts (a) and (b)] at signal-to-noise ratios down to 25 dB, and with some loss of accuracy at 20 dB as in Part (c). At a SNR of 15 dB, as in Part (d), the method breaks down.

Figure 4.40 shows the results obtained for the phase difference $\psi = 180°$, SNR = 10 dB, and three different values of the surface reflection coefficient ρ, namely, 0.5, 0.7, and 0.9. We see that as ρ becomes smaller (i.e., the specular component becomes weaker), three things happen: (1) the signal-zero cluster corresponding to the specular path becomes increasingly more diffuse, (2) this cluster tends to move slightly inward, and (3) the spreads of the noise-zero clusters increase. For other values of ψ (e.g. 90°), the movement of the cluster pertaining to the specular path is less pronounced for decreasing ρ.

Figure 4.41 shows the results obtained for an array with $N = 8$ elements, prediction-filter order $L = 6$, SNR = 30 dB, reflection coefficient $\rho = 0.9$, and three different values of the phase difference ψ, namely, 0°, 90°, and 180°. Here again we

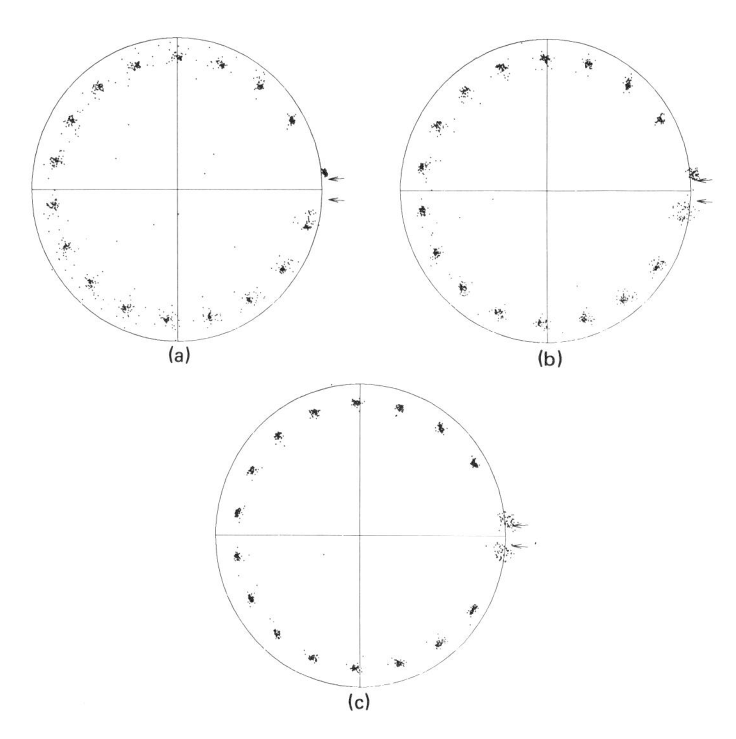

Figure 4.40 Results of 50 independent trials of FBLP method for $N = 25$, $L = 18$, $2\phi = 0.5$ BW, $\psi = 180°$, SNR = 10 dB. (a) $\rho = 0.5$; (b) $\rho = 0.7$; (c) $\rho = 0.9$.

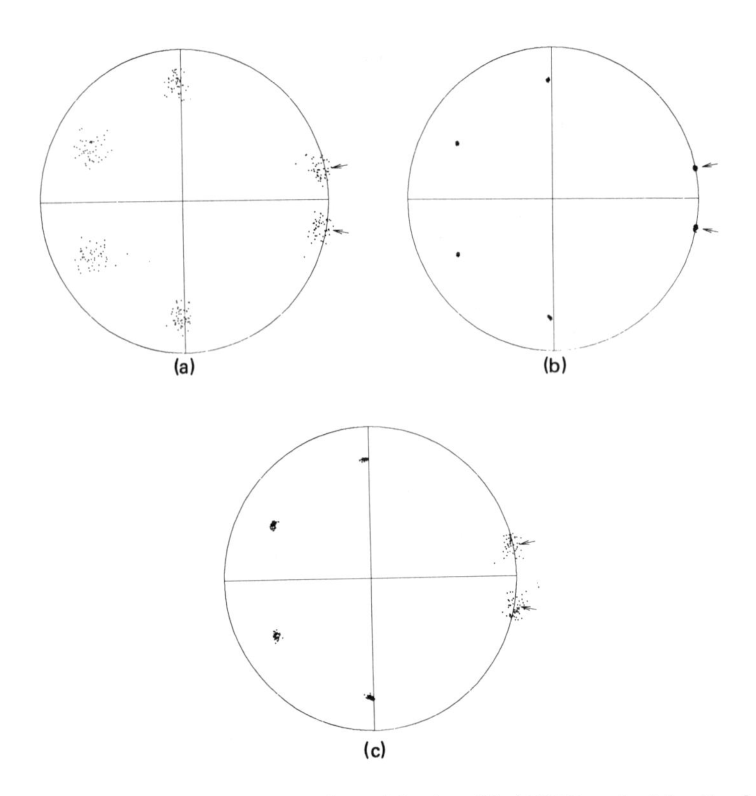

Figure 4.41 Results of 50 independent trials of modified FBLP method for $N = 8$, $L = 6$, $2\phi = 0.5$ BW, $\rho = 0.9$, SNR = 30 dB. (a) $\psi = 0°$; (b) $\psi = 90°$; (c) $\psi = 180°$.

have chosen the order L so that it is optimum for the value of N used here. Also, the angular separation between the direct and specular paths corresponds to $2\phi = 0.5$ BW. Comparing Figs. 4.41 (a) and 4.39 (a), both of which correspond essentially to the same set of environmental conditions, except for a change in the number of elements, we see that reducing the number of array elements has the effect of increasing the spreads of both the signal-zero and noise-zero clusters. This, in turn, results in an increase in the variance of the estimate obtained for the angular separation between the two paths. Also, comparing the three plots in Fig. 4.41 that pertain to three different values of ψ, we see that the method is indeed able to resolve the direct and specular paths, even for the two critical values of ψ, 0° and 180°. The variances of the estimates obtained at these two values of ψ are roughly the same, but significantly larger than the variance of the estimate obtained at $\psi = 90°$, as judged by the spreads of the respective signal-zero clusters. It is also of interest to note that the spread of the noise-zero clusters is relatively small for $\psi = 180°$, compared to the situation for $\psi = 0°$, and that it is smallest for $\psi = 90°$.

A comparison of the results given here for the modified FBLP method and those for the Burg technique considered in Section 4.13 reveals two interesting

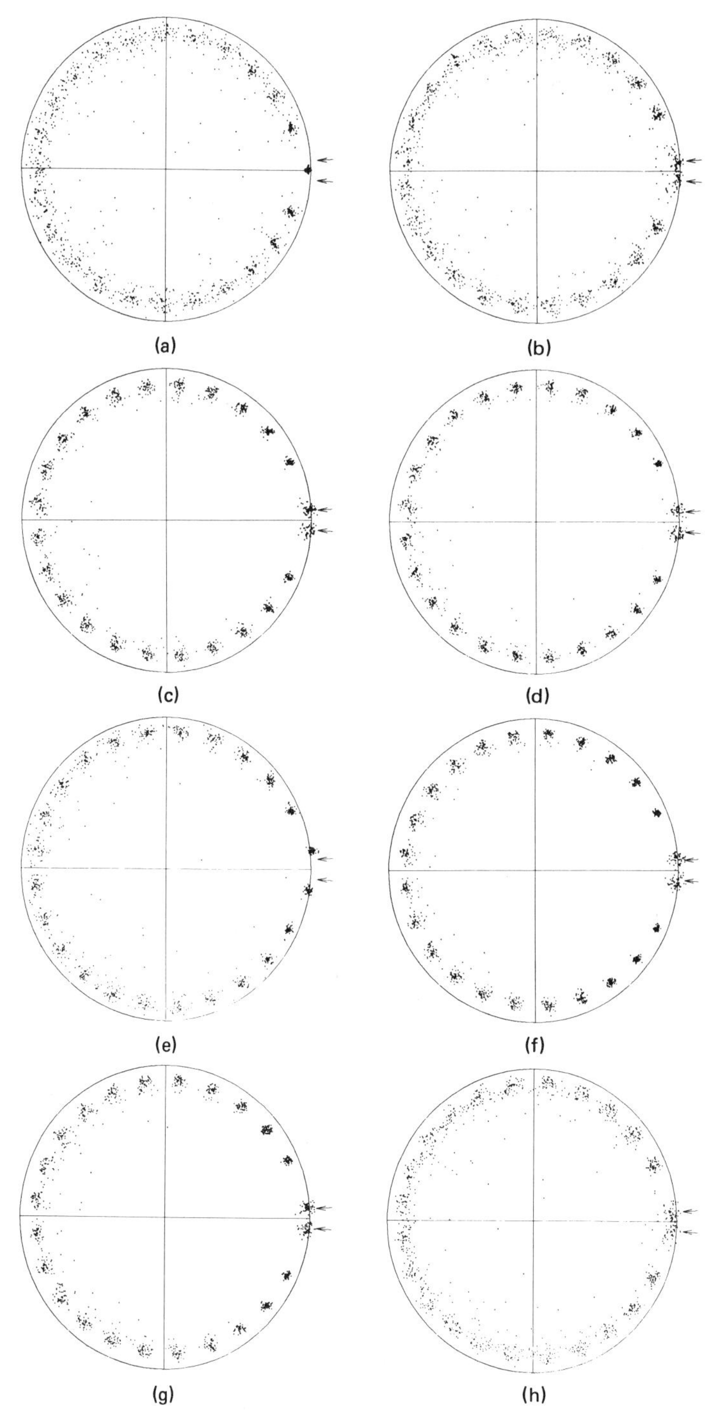

Figure 4.42 Results of 50 independent trials of the Kumaresan-Prony case for $N = 25$, $L = 24$, $2\phi = 0.5$ BW, $\rho = 0.85$, SNR = 10 dB. (a) $\psi = 0°$; (b) $\psi = 45°$; (c) $\psi = 90°$; (d) $\psi = 135°$; (e) $\psi = 180°$; (f) $\psi = -135°$; (g) $\psi = -90°$; (h) $\psi = -45°$.

points:

1. The phase difference ψ affects the estimates of the angular separation between the direct and specular paths obtained by using the modified FBLP and Burg methods in a similar way, as evidenced by comparing the implications of Figs. 4.38 and 4.32.
2. The degrading effect of the critical values of $\psi = 0°, 180°$ on the estimate of the angular separation between the direct and specular paths becomes visible in the Burg technique at a larger signal-to-noise ratio than in the modified FBLP method. This is evidenced by comparing the implications of Figs. 4.41 and 4.32, which pertain essentailly to the same set of parameters.

We may now summarize the results of Figs. 4.38 to 4.41 by stating that (1) the modified FBLP method can resolve the direct and specular paths even when the phase difference ψ assumes either one of its critical values (i.e., 0° and 180°) provided the signal-to-noise ratio is high enough, and (2) the patterns obtained for the signal-zero and noise-zero clusters reveal useful information about the multipath environment itself.

(d) The Kumaresan-Prony Case

Finally, in Fig. 4.42 we show the results obtained for the Kumaresan-Prony case for the following parameters:

number of elements, $N = 25$
angular separation, $2\phi = 0.5$ BW
signal-to-noise ratio, SNR = 10 dB
reflection coefficient, $\rho = 0.85$

With the number of plane waves M equal to 2 (represented by the direct and specular paths), the prediction-filter order $L = N - M/2 = 24$. In Fig. 4.42, the phase difference ψ is given the values 0°, $\pm 45°$, $\pm 90°$, $\pm 135°$, 180. Comparing the results of this figure with the corresponding ones shown in Fig. 4.38, it is clear that for the same environmental conditions, the Kumaresan-Prony method is more sensitive to variations in the phase difference ψ than the modified FBLP method.

In conclusion, it should be pointed out that although in this example only a symmetric multipath model was considered, nevertheless the modified FBLP method and the Kumaresan–Prony method (and for that matter, the conventional FBLP method) do work with a nonsymmetric specuiar situation just as well.

4.20 OPERATION IN A DIFFUSE MULTIPATH ENVIRONMENT

All the results presented above on the FBLP method and its different forms have been based on a purely specular model, for which the array antenna output consists of the sampled version of two complex exponentials plus additive white noise. In this idealized situation, the Kumaresan–Prony method or the modified FBLP method works effectively, provided the signal–to–noise ratio is high enough. However, as mentioned in Section 4.3, in a real–life situation the multipath phenomenon may also give rise to the presence of a diffuse component in the received signal. In such a situation the array antenna output consists of the sampled version of one or two complex exponentials plus a colored noise process. Marple [65] has tested the performance of the Kumaresan–Prony method under conditions that are representative of a diffuse multipath environment. The two significant points emerging from Marple's results, rephrased to correspond to a diffuse multipath environment, are as follows:

1. The Kumaresan–Prony method is a poor estimator of the wavenumber spectrum of the spatially diffuse component.
2. The accuracy of the estimate of the angular position of the direct path is unaffected by the presence of the diffuse component.

Although Marple only considers the Kumaresan–Prony method, nevertheless it would seem that the modified FBLP method behaves in a similar fashion when applied to a diffuse multipath environment.

We may therefore state that if the primary requirement is to estimate reliably the angle of arrival of a target in the presence of specular and/or diffuse multipath, then the modified FBLP method or the Kumaresan–Prony method should be capable of providing such an estimate, with the proviso, of course, that the signal–to–noise ratio is above the pertinent threshold. Indeed, it is quite likely that the reliability of both estimators may improve in the presence of a spatially diffuse component, because it is not correlated with the useful signal component received directly from the target. If, on the other hand, the requirement also calls for an assessment of the prevalent multipath conditions by measuring the wavenumber spectrum, then some other method should be used.

4.21 OTHER METHODS

In this chapter we have considered only two parametric procedures, the Burg technique, and the forward–backward linear prediction method (and its modified forms), for estimating the power spectrum of a set of plane waves arriving at a linear array from different directions. There are, however, many other parametric procedures that may also be used for this purpose. Mention should be made of the *maximum-likelihood method* (MLM) of spectral analysis, which was originally formulated by Capon for seismic array processing under conditions of directional interferences [66]. This method should not be confused with the classical maximum-likelihood estimation procedure described in Sections 4.6 and 4.7. The MLM may be represented in terms of a minimum-variance unbiased spectral estimate of the complex signal from a given angle, while rejecting all other spectral components of the input in an optimum manner. It is well recognized that the resolution capability of the MLM is inferior to the AR/ME method [67, 68]. This issue was discussed in some detail in Section 2.8. The main virtue of the MLM is that it gives an indication of the relative source powers with any degree of accuracy.

Schmidt [69] describes a parametric method for computing the angle spectrum $S(\phi)$, which also relies on an eigenvalue analysis of the correlation matrix $\mathbf{C}'$. This method, however, differs from the modified FBLP method in that it utilizes the noise subspace eigenvectors.

In yet another approach, the sum of one or more plane waves in the presence of additive white noise is modeled by an *autoregressive–moving average* (*ARMA*) *process*, as shown by [70]

$$x(n) = -\sum_{k=1}^{L} a_k x(n-k) + \sum_{k=1}^{K} b_k w(n-k) + w(n) \tag{4.214}$$

where $\{w(n)\}$ is a zero-mean uncorrelated sequence of samples with variance σ_w^2, and the a_k and b_k are constant coefficients. Indeed, it may be shown that the model of Eq. (4.214) remains appropriate even when diffuse multipath is present. A useful feature of the ARMA model of Eq. (4.214) is that it contains the pure AR model as a special case, which is obtained by putting $b_k = 0$ for $k = 1, 2, \ldots, K$. However, the determination of the ARMA model parameters in Eq. (4.214) poses analytic difficulties. Box and Jenkins [71], Kaveh [72], and others have developed procedures that may be used to estimate ARMA model parameters. For a discussion of these ARMA modeling techniques and other spectral estimation techniques as applied to the problem of estimating angles of arrival in a radar environment, the reader is referred to the report by Evans et al. [5].

4.22 SUMMARY

In this chapter we have considered the use of a linear array of uniformly spaced antenna elements to solve two basic radar problems:

1. *The direct estimation of angle(s) of arrival(s) of incident plane waves:* In the special case of a symmetric multipath environment, where the direct and specular components arrive at the array in a symmetric manner with respect to the normal to the array, we may use the symmetric maximum-likelihood receiver or the adaptive antenna (based on adaptive cancellation) supplied with a calibration curve. Both schemes are relatively simple to implement. In the more general case, where we have two or more plane waves arriving at the array from arbitrary directions, we may use the generalized form of the maximum-likelihood receiver whose derivation is based on the Gram–Schmidt orthogonalization procedure. The only limitation of this latter receiver is that there is no simple procedure presently available for computing the maxima of the pertinent objective function.

2. *Power spectrum estimation:* We described the use of linear prediction as the basis of a method for estimating the autoregressive/maximum-entropy power spectrum of a set of plane waves (arriving at the array from arbitrary directions) as a

function of the wavenumber or the electrical phase angle ϕ. By computing the spectral peaks (i.e., the values of ϕ at which the power spectrum attains its maxima) we may indirectly estimate the angles of arrival of the incident plane waves. To compute the autoregressive/maximum-entropy power spectrum, we require a knowledge of the pertinent set of autoregressive or prediction-error-filter coefficients. For the case of noncoherent sources, we may use the Burg technique based on a spatial lattice filter. A more general method is the forward–backward linear prediction or least-squares method, which works for noncoherent or coherent sources, provided the signal-to-noise ratio is high enough. Finally, we described a two-step modification of this second method (due to Tufts and Kumaresan) that makes linears prediction perform almost as well as the classical maximum-likelihood estimation procedure.

ACKNOWLEDGMENTS

The author of this chapter is grateful to Dr. J. P. Reilly and Dr. J. Kesler for their contributions to the material described in this chapter, and to Mr. W. Stehwien for computing the results given in Section 4.19. He is indebted to Mr. D. Mabey and Dr. J. Litva, Communications Research Centre, Ottawa, for many stimulating discussions on the multipath problem in low–angle tracking radar. He is grateful to the Natural Sciences and Engineering Research Council, Ottawa, and the Department of Communications, Ottawa, for supporting the research reported in this Chapter. He is also grateful to the other researchers whose contributions have been acknowledged in the list of references.

REFERENCES

1 M. I. Skolnik, *Introduction to Radar Systems*, McGraw-Hill, New York, 1962.

2 S. Haykin, ed., *Array Processing: Applications to Radar*, Dowden, Hutchinson, & Ross, Stroudsburg, Pa., 1980.

3 R. A. Monzingo and T. W. Miller, *Introduction to Adaptive Arrays*, Wiley-Interscience, New York, 1980.

4 J. E. Hudson, *Adaptive Array Principles*, Peter Peregrinus, Stevenage, U.K., 1981.

5 J. E. Evans, J. R. Johnson, and D. F. Sun, "Application of Advanced Signal Processing Techniques to Angle of Arrival Estimation in ATC Navigation and Surveillance Systems," Massachusetts Institute of Technology, Lincoln Laboratory, Technical Report 582, June 23, 1982.

6 S. Haykin, J. P. Reilly, and D. P. Taylor, "New Realization of Maximum-Likelihood Receiver for Low-Angle Tracking Radar," Electron. Lett., Vol. 16, pp. 288–289, 1980.

7 S. Haykin and J. P. Reilly, "Maximum-Likelihood Receiver for Low-Angle Tracking Radar, Part 1: The Symmetric Case," Proc. IEE, Vol. 129, Part F, pp. 261–272, 1982.

8 J. P. Reilly and S. Haykin, "Maximum-Likelihood Receiver for Low-Angle Tracking Radar, Part 2: The Nonsymmetric Case," Proc. IEE, Vol. 129, Part F, pp. 331–340, 1982.

9 J. Kesler and S. Haykin, "A New Adaptive Antenna for Elevation Angle Estimation in the Presence of Multipath," IEEE AP-S Int. Symp., Vol. I, Quebec, June 2–6, 1980, pp. 130–133.

10 S. Haykin and J. Kesler, "Adaptive Canceller for Elevation Angle Estimation in the Presence of Multipath," Proc. IEE, Vol. 30, Part F, pp. 303–308, June 1983.

11 S. Haykin, J. Kesler, and J. Litva, "Evaluation of Angle of Arrival Estimators Using Real Multipath Data," Proc. IEEE ICASSP '83, Boston, Apr. 14–16, 1983.

12 S. Haykin, *Communication Systems*, 2nd ed., Wiley, New York, 1983.

13 D. K. Barton, "Low-Angle Radar Tracking," Proc. IEEE, Vol. 62, pp. 687–704, June 1974.

14 R. E. Collins and F. J. Zucker, *Antenna Theory*, Part I, McGraw-Hill, New York, 1969.

15 F. J. Harris, "On the Use of Windows for Harmonic Analysis with the Discrete Fourier Transformation," Proc. IEEE, Vol. 66, pp. 51–83, 1978.

16 H. L. Van Trees, *Detection, Estimation, and Modulation Theory*, Part I, Wiley, New York, 1968.

17 J. W. Bandler, "Computer-Aided Circuit Optimization," in *Modern Filter Theory and Design*, ed. G. C. Temes and S. K. Mitra, Wiley, New York, 1973.

18 W. D. White, "Low-Angle Radar Tracking in the Presence of Multipath," IEEE Trans. Aerosp. Electron. Syst., Vol. AES-10, pp. 835–852, 1974.

19 J. P. Reilly, "Nonlinear Array Processing Techniques with Applications to Correlated Multipath," Ph.D. thesis, McMaster University, Hamilton, Ont., 1981.

20 R. Fletcher and M. J. D. Powell, "A Rapidly Convergent Descent Method for Minimization," Comput. J., Vol. 6, pp. 163–168, 1963.

21 Y. Bard, *Nonlinear Parameter Estimation*, Academic Press, New York, 1974.

22 J. W. Bandler and N. K. Sinha, "FLOPTV. A Program for Accelerated Least-*p*th Algorithm," SOC Report, Faculty of Engineering, McMaster University, Hamilton, Ont., Feb. 1980.

23 M. Wax and T. Kailath, "Determining the Number of Sources Impinging on a Passive Array by Akaike's Information Criterion," Information Systems Laboratory, Stanford University, 1983.

24 J. Kesler, "Adaptive Interference Cancelling in Multiple Beam Antennas with Applications to Multipath," Ph.D. thesis, McMaster University, Hamilton, Ont., Mar. 1981.

25 B. Widrow, J. McCool, and M. Ball, "The Complex LMS Algorithm," Proc. IEEE, Vol. 63, pp. 719–720, Apr. 1975.

26 S. Haykin and J. Kesler, "Performance Evaluation of Two Angles of Target Arrival Estimation Schemes Using Real Multipath Data," CRL Integral Report 107, Communications Research Laboratory, McMaster University, Hamilton, Ont., Dec. 1982.

27 S. M. Kay and S. L. Marple, Jr., "Spectrum Analysis—A Modern Perspective," Proc. IEEE, Vol. 69, pp. 1380–1419, 1981.

28 D. H. Johnson, "The Application of Spectral Estimation Methods to Bearing Estimation Problems," Proc. IEEE, Vol. 70, pp. 1018–1028, Sept. 1982.

29 S. Haykin, ed., *Nonlinear Methods of Spectral Analysis*, [2nd ed.] Springer-Verlag, New York, 1983.

30 D. G. Childers, *Modern Spectrum Analysis*, IEEE Press, New York, 1978.

31 Special Issue on Spectral Estimation, Proc. IEEE, Sept. 1982.

32 Proc. First IEEE ASSP Workshop on Spectral Estimation, McMaster University, Hamilton, Ont., Aug. 1981.

33 Proc. RADC Spectrum Estimation Workshop, Rome Air Development Center, Rome, N.Y., Oct. 1978.

34 Proc. RADC Spectrum Estimation Workshop, Rome Air Development Center, Rome, N.Y., Oct. 1979.

35 Proceedings IEE Conf. Spectral Anal. Use Underwater Acoust., London, 1982.

36 A. V. Oppenheim and R. W. Schafer, *Digital Signal Processing*, Prentice-Hall, Englewood Cliffs, N.J., 1975.

37 J. P. Burg, "Maximum Entropy Spectral Analysis," Proc. 37th Mtg. Soc. Explor. Geophysicists, 1967.

38 J. P. Burg, "Maximum Entropy Spectral Analysis," Ph.D. dissertation, Stanford University, 1975.

39 T. J. Ulrych and T. N. Bishop, "Maximum Entropy Spectral Analysis and Autoregressive Decomposition," Rev. Geophys. Space Phys., Vol. 13, pp. 183–200, 1975.

40 S. Haykin and S. Kesler, "Prediction-Error Filtering and Maximum-Entropy Spectral Estimation," in [29].

41 R. N. McDonough, "Application of the Maximum-Likelihood and the Maximum-Entropy Method to Array Processing," in [29].

42 J. H. McLellan, "Multidimensional Spectral Estimation," Proc. IEEE, Vol. 70, pp. 1029–1039, Sept. 1982.

43 T. J. Ulrych and M. Ooe, "Autoregressive and Mixed Autoregressive Moving Average models and Spectra," in [29].

44 N. Levinson, "The Wiener RMS (Root Mean Square) Error Criteria in Filter Design and Prediction," J. Math. Phys., Vol. 25, pp. 261–278, 1947.

45 J. Durbin, "The Fitting of Time-Series Models," Rev. Int. Stat. Inst., Vol. 28, pp. 233–243, 1960.

46 L. J. Griffiths, "Adaptive Structures for Multiple-Input Noise Cancelling Applications," Proc. IEEE ICASSP '79, Washington, D.C., Apr. 2–4, 1979, pp. 925–928.

47 S. Haykin and S. Kesler, "The Complex Form of the Maximum Entropy Method for Spectral Estimation," Proc. IEEE, Vol. 64, pp. 822–823, 1976.

48 K. C. Sherman and T. S. Durrani, "A Triangular Adaptive Lattice Filter for Spatial Filtering," Proc. IEEE ICASSP '83, Boston, Apr. 14–16, 1983.

49 J. P. Reilly and S. Haykin, "Direction Finding in Array Processing Using MEM Spectral Estimators," Internal Report 93, Communications Research Laboratory, McMaster University, Hamilton, Ont., Feb. 1981.

50 J. P. Reilly and S. Haykin, "An Experimental Study of the MEM Applied to Array Processing in the Presence of Multipath," Proc. IEEE ICASSP '80, Denver, Colo., Apr. 1980, pp. 120–123.

51 R. W. Herring, "The Cause of Line Splitting in Burg Maximum-Entropy Spectral Analysis," IEEE Trans. Acoustics, Speech, and Signal Processing, vol. ASSP-28, pp. 692–701, [Dec.] 1980.

52 T. J. Ulrych and R. W. Clayton, "Time Series Modelling and Maximum Entropy," J. Phys. Earth Planet. Inter., Vol. 12, pp. 188–200, 1976.

53 A. H. Nuttal, "Spectral Analysis of a Univariate Process with Bad Data Points via Maximum Entropy and Linear Predictive Techniques," in Naval Underwater System Center (NUSC)

Scientific and Engineering Studies, Spectral Estimation, NUSC, New London, Conn., Mar. 1976.

54 R. Kumaresan and D. W. Tufts, "Singular Value Decomposition and Spectral Analysis," Proc. First IEEE ASSP Workshop on Spectral Estimation, McMaster University, Hamilton, Ont., Aug. 1981.

55 D. W. Tufts and R. Kumaresan, "Estimation of Frequencies of Multiple Sinusoids: Making Linear Prediction Perform Like Maximum Likelihood," Proc. IEEE, Vol. 70, pp. 975–989, Sept. 1982.

56 G. Hadley, *Linear Algebra*, Addison-Wesley, Reading, Mass., 1961.

57 R. Prony, "Essai expérimental et analytique, etc.," L'Ecole Polytechnique, Paris, Vol. 1, pp. 24–26, 1795.

58 F. B. Hildebrand, *Introduction to Numerical Analysis*, McGraw-Hill, New York, 1956.

59 C. L. Lawson and R. J. Hanson, *Solving Least Square Problems*, Prentice-Hall, Englewood Cliffs, N.J., 1974.

60 G. W. Stewart, *Introduction to Matrix Computations*, Prentice-Hall, Englewood Cliffs, N.J., 1973.

61 S. A. Tretter, *Introduction to Discrete Time Signal Processing*, Wiley, New York, 1976.

62 R. Kumaresan, "Estimating the Parameters of Exponentially Damped/Undamped Sinusoidal Signals in Noise," Ph.D. dissertation, University of Rhode Island, Aug. 1982.

63 C. Eckart and G. Young, "The Approximation of a Matrix by Another of Lower Rank," Psychometrika, Vol. 1, pp. 211–218, 1931.

64 S. W. Lang and J. H. McClellan, "Frequency Estimation with Maximum Entropy Spectral Estimators," IEEE Trans. Acoust. Speech Signal Process., Vol. ASSP-28, pp. 716–724, 1980.

65 S. L. Marple, Jr., "A Fast Computational Algorithm for and Performance of the Kumaresan-Prony method of Spectrum Analysis," Proc. IEEE ICASSP '83, Boston, April 14–16, 1983.

66 J. Capon, "High-Resolution Frequency Wavenumber Spectrum Analysis," Proc. IEEE, Vol. 57, pp. 1408–1418, 1969.

67 R. T. Lacoss, "Data Adaptive Spectral Analysis Methods," Geophysics, Vol. 36, pp. 661–675, Aug. 1971.

68 J. P. Burg, "The Relationship between Maximum Entropy Spectra and Maximum Likelihood Spectra," Geophysics, Vol. 37, pp. 375–376, 1972.

69 R. O. Schmidt, "A Signal Subspace Approach to Multiple Emitter Location and Spectral Estimation," Ph.D. dissertation, Stanford University, 1981.

70 S. Haykin and J. P. Reilly, "Mixed Autoregressive-Moving Average Modelling of the Response of a Linear Array Antenna to Incident Plane Waves," Proc. IEEE, Vol. 68, pp. 622–623, May 1980.

71 G. E. P. Box and G. M. Jenkins, *Time Series Analysis, Forecasting and Control*, Holden-Day, San Francisco, 1970.

72 M. Kaveh, "High Resolution Spectral Estimation for Noisy Signals," IEEE Trans. Acoust. Speech Signal Process., Vol. ASSP-28, pp. 753–755, 1980.

5

J. L. YEN
Department of Electrical Engineering
University of Toronto, Ontario

Image Reconstruction in Synthesis Radio Telescope Arrays

5.1 INTRODUCTION

In the past few decades astronomy has witnessed a rapid succession of major discoveries. The traditional optical window for earthly observation of heavenly bodies, first opened up when Galileo turned his telescope to the sky, has been rapidly supplemented by looking through the radio and infrared windows from the earth and space observations ranging from infrared through visible and ultraviolet to x-ray and γ-ray. The vastly expanding horizons of observations have led to the discoveries of pulsars, neutron stars, black holes, complex molecules, and quasars, objects of enormous power in great distances away. We are presently at a stage where such questions as how stars are formed and how the universe began and evolved can be discussed with increasingly more observational evidences. All these have come about because technological advances brought along a continuing series of new instruments. After the initial discovery of different celestial objects using new instruments, the challenge turned to understanding the astrophysical processes responsible for the radiation from such objects. For this, images and their variation with time and frequency need to be obtained. Furthermore, survey and classification of objects according to their characteristics are required. Finally, in combining observations from x-ray through visible and infrared to radio spectrum, proper interpretation of the nature of the objects can proceed.

As seen from the earth, radio emission from celestial sources appears either as distributed radiation from diffused media or from isolated sources of finite angular extent. Depending on the radiation mechanism and the state of the emitting region, the radio emission shows broad continuum spectral features, narrow emission, or absorption line structures.* Radiation from intersteller media in the Milky Way at decimeter wavelength or longer appears over large portion or even the entire sky. On the other hand, distant galaxies or quasars often appear as discrete sources despite their large linear dimension, because of their great distances from earth. Some objects with main features extending over 30 arc min have important compact components on the order of 1 milli-arc second. Radio telescopes of different resolution are therefore required. Because radio waves are roughly 1 million times longer than light waves, radio images from single telescopes are severely limited by the constraint of manageable size. As a result, single telescopes are only suitable for low-resolution study of nearby objects. To observe the brightness and spectral density distributions of more distant sources, telescope arrays of great resolving power and sensitivity are required. Because of the incoherent nature of celestial radio emission, the direction from which a wavefront comes from can be obtained by measuring the cross-correlation or the mutual coherence of signals received by two antennas at different locations. By combining the cross-correlation from antenna pairs forming *interferometers with baselines* distributed over a region, the image of an isolated object can be reconstructed. Furthermore, because of the stationary statistical nature of celestial radiation, it is not necessary that all the interferometers be present simultaneously. By moving telescopes and by making use of earth's rotation with respect to the sky, large baseline regions can be synthesized using a small number of discrete antenna elements.

Since its introduction by Ryle [4], the *earth rotation synthesis telescope* has become the principal radio astronomical instrument for the study of distant objects ranging from 1° in angular size down to less than 1 milli-arc second. Arrays of tens of elements extending from hundreds of meters to nearly the diameter of the earth with dimension up to hundreds of millions of wavelength are used in the investigation of an enormous variety of celestial objects. The signal from each element is down-converted to a lower-frequency band by mixing, and then cross-correlated as the earth rotates. For baselines up to few tens of kilometers, such as the Cambridge 5-km telescope [5], the Westerbork synthesis radio telescope [6], and the very large array (VLA) of the National Radio Astronomy Observatory [7] in New Mexico, both the signal and the local oscillator for each element are distributed via transmission lines. For baselines up to a few hundred kilometers [8], terrestrial microwave links are employed. When the distances span thousands of kilometers or across the ocean as in *very long baseline interferometer* (VLBI) arrays, independent local oscillators at each element are used to convert the signal to baseband and then recorded on magnetic tape for subsequent cross-correlation processing [9, 10]. With the advent of communication satellites, satellite channels have been used both for signal

*For an introduction to radio astronomy and radio telescopes, refer to standard texts [1–3].

transmission and local oscillator distribution [11]. For even greater resolution, earth–space or space–space instruments have been proposed and attempted. In addition to having great angular resolution, the synthesis telescope is also an instrument of very high sensitivity. This is because the array observes the entire source as the earth rotates, and the resulting long integration time allows detection of minute signals as low as 10^{-8} to 10^{-6} times the system noise of an individual array element.

In a rotation synthesis radio telescope, all the elements of the array track the radio source under study. Cross-correlations between every $N(N-1)/2$ interferometer pair of the N array elements are determined. As the earth rotates, the interferometer baselines scan over curves in a volume as viewed from the radio source. According to *Van Cittert–Zernike theorem* in the theory of partial coherence, the mutual spectral density over a plane is the Fourier transform of the brightness spectral density distribution of the source [12]. The synthesis telescope samples mutual coherence along discrete curves. From these samples an *image* or *map* of a continuum radio source can be reconstructed. The principles of the instrument can be found in previous reviews [13–16]. If the radio source is a molecular line source, the cross-spectral density for every baseline may be derived by using the Wiener–Khinchin theory, from which an image for each frequency component is derived. It is not uncommon to construct continuum images with 10^6 to 10^7 picture elements, and for line sources the number is multiplied by the number of frequency channels required, which may vary from 10^2 to 10^3.

Although complete information on temporal coherence is easy to obtain, spatial coherence is only sparsely sampled by the sweeping baselines. As a result, a major problem in image reconstruction is that of incomplete data. Various image reconstruction methods based on the assumption of isolated sources such as CLEAN [17] have been extensively used to remove large sidelobes produced by incomplete sampling. The measurement accuracy is further degraded by propagation and instrument defects. The ionosphere and the troposphere over each array element are different and may vary with time, thereby introducing uncertainties in phase. Incoherency of local oscillators, in particular in VLBI arrays, further degrades the phase. Path attenuation due to rain, cloud, pointing error, and gain fluctuation produce amplitude errors. Despite elaborate calibration procedures, observed data are often unavoidably laden with errors. As a result, sophisticated and extensive processing is necessary to reconstruct meaningful images of radio sources from the incomplete, imprecise, and noisy array data. Approaches that have been used include model fitting, closure phase and amplitude, and maximum entropy [18–20].

The principal aim of this chapter is to introduce signal processing methods for image reconstruction of unpolarized or partially polarized continuum and spectral line radio sources. The data base consists of incomplete samples of mutual coherence or cross-spectral density obtained by rotation synthesis radio telescope arrays in the presence of gain and phase uncertainties as well as additive noise. Emphasis will be placed on resolution, ambiguity, and dynamic range of the reconstructed maps. For completeness, a description of the wave fields emitted by radio sources, the measurement of mutual coherence by a two-element interferometer, and the spatial sampling of an array by earth rotation are also included.

5.2 WAVE FIELDS OF RADIO SOURCES

Radio emission from celestial sources as seen on earth can be decomposed into plane waves from different directions, and because of their great distances, the earth is in the far field. The waves are random in nature, having either broad continuum spectral behavior or narrow line features with various Doppler shifts, depending on the radiation mechanism. In most cases, the radiation is ergodic so that all observable quadratic ensemble averages can be replaced by time averages. Radiation from different elements of a source are, in general, statistically independent or mutually incoherent. Sometimes the waves are partially polarized because of the presence of magnetic field in the emitting region and the intervening media. The characteristics of random wave fields emitted by celestial sources are described by the *theory of partial coherence* [12]. In this theory, the distribution of sources of radiation in space, time, and polarization are related to the observable mutual coherence or cross-correlation of the wave field near the earth in space, time, and polarization. All radio astronomical observations and image reconstruction are based on these relations. In this section we introduce the essentials of partial coherence for unpolarized plane wave fields, which will serve as the basis for the main discussions of the chapter.

Consider a polarization component of the plane-wave field emitted by an unpolarized source element in the direction of the unit vector $\mathbf{n}$ as in Fig. 5.1. It is convenient to express the field at $\mathbf{r}$ as the real part of an *analytic signal.** Let this analytic signal be denoted by $e(t, \mathbf{r}, \mathbf{n})$. The wave fields at different positions $\mathbf{r}_1$ and $\mathbf{r}_2$ are related by the delay of propagation:

$$e(t, \mathbf{r}_1, \mathbf{n}) = e\left(t + \frac{\mathbf{n} \cdot (\mathbf{r}_1 - \mathbf{r}_2)}{c}, \mathbf{r}_2, \mathbf{n}\right) \tag{5.1}$$

Let the Fourier transform of the real field $e_r(t, \mathbf{r}, \mathbf{n})$, truncated to a finite time $|t| \leq T$, be

$$\tfrac{1}{2}\epsilon_T(f, \mathbf{r}, \mathbf{n}) = \int_{-T}^{T} e_r(t, \mathbf{r}, \mathbf{n}) \exp(-j2\pi ft)\, dt$$

The factor $\frac{1}{2}$ is introduced so that the complex analytic field $e(f, \mathbf{r}, \mathbf{n})$ has the transform $\epsilon_T(f, \mathbf{r}, \mathbf{n})$, $f \geq 0$, and zero, $f < 0$. The normalization of the field is chosen such that the brightness spectral density distribution of the source element is the

*An analytic signal is complex-valued, with its real and imaginary parts forming a Hilbert transform pair. An important property of the analytic signal is that its Fourier transform vanishes for negative frequencies.

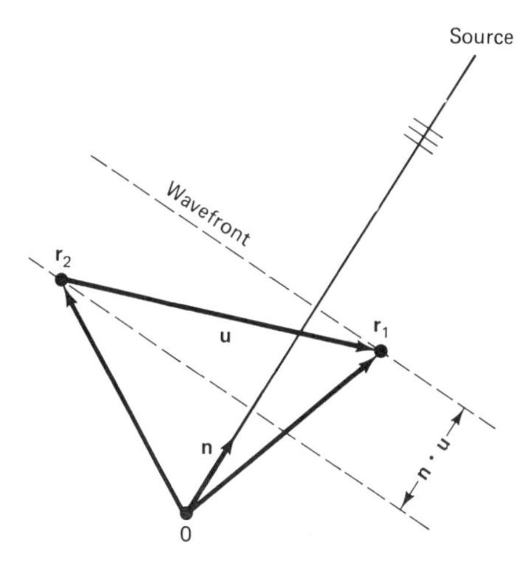

Figure 5.1 Wavefront and propagation delay.

average

$$I(f, \mathbf{n}) = \begin{cases} 2 \lim_{T \to \infty} \dfrac{1}{2T} \epsilon_T(f, \mathbf{r}, \mathbf{n}) \epsilon_T^*(f, \mathbf{r}, \mathbf{n}), & f > 0 \\ 0, & f < 0 \end{cases}$$

It is positive real and is independent of $\mathbf{r}$ because for a plane-wave field a change in position only introduces a time delay. The factor of 2 accounts for radiation in the orthogonal polarization. The source autocorrelation or self-coherence is the average

$$I(\tau, \mathbf{n}) = 2E[e(t + \tau, \mathbf{r}, \mathbf{n}) e^*(t, \mathbf{r}, \mathbf{n})] \tag{5.2}$$

where E is the expectation operator. According to the Wiener–Khinchin theorem, the self-coherence and spectral density for a stationary random process form a Fourier transform pair, as shown by

$$I(\tau, \mathbf{n}) = \int_0^\infty I(f, \mathbf{n}) \exp(j2\pi f\tau)\, df \tag{5.3}$$

The real and imaginary parts of $I(\tau, \mathbf{n})$ are related by the Hilbert transform. The total brightness of the source element is $I(\tau, \mathbf{n})$ for $\tau = 0$.

For an extended source the wave field is a superposition of plane waves from different directions; thus

$$e(t, \mathbf{r}) = \int e(t, \mathbf{r}, \mathbf{n})\, d\mathbf{n}$$

The basic observable is the mutual coherence or cross-correlation of the fields at $\mathbf{r}_j$ and $\mathbf{r}_k$; that is,

$$E[e(t + \tau, r_j) e^*(t, \mathbf{r}_k)] = E\left[\int e(t + \tau, \mathbf{r}_j, \mathbf{n})\, d\mathbf{n} \int e^*(t, \mathbf{r}_k, \mathbf{n}')\, d\mathbf{n}'\right]$$

Since waves from different source elements are statistically independent, the average becomes

$$\int E[e(t+\tau, \mathbf{r}_j, \mathbf{n})e^*(t, \mathbf{r}_k, \mathbf{n})]\, d\mathbf{n} = \frac{1}{2}\int I\left(\tau + \frac{\mathbf{n}\cdot\mathbf{u}}{c}, \mathbf{n}\right) d\mathbf{n}$$
$$= \Gamma(\tau, \mathbf{u}) \tag{5.4}$$

In deriving the relation above we have made use of Eq. (5.2), the propagation delay between $\mathbf{r}_j$ and $\mathbf{r}_k$ in Eq. (5.1), and we have introduced the baseline $\mathbf{u} = \mathbf{r}_j - \mathbf{r}_k$. The fact that the mutual coherence depends only on the baseline implies that the wave field is stationary in space as well as in time. For a source element in direction $\mathbf{n}$ the contribution to $\Gamma(\tau, \mathbf{u})$ is a replica of the source self-coherence $I(\tau, \mathbf{n})$ with appropriate shift, scale conversion, and magnification. On a fixed baseline $\mathbf{u}$, a shift in τ by the propagation delay $\mathbf{n} \cdot \mathbf{u}/c$ is introduced. For a fixed τ and for $\mathbf{u}$ along the source element direction $\mathbf{n}$, the variation of $\Gamma(\tau, \mathbf{u})$ is again a replica of $I(\tau, \mathbf{n})$ except for a scale conversion from time to distance. As the baseline rotates away from $\mathbf{n}$, a

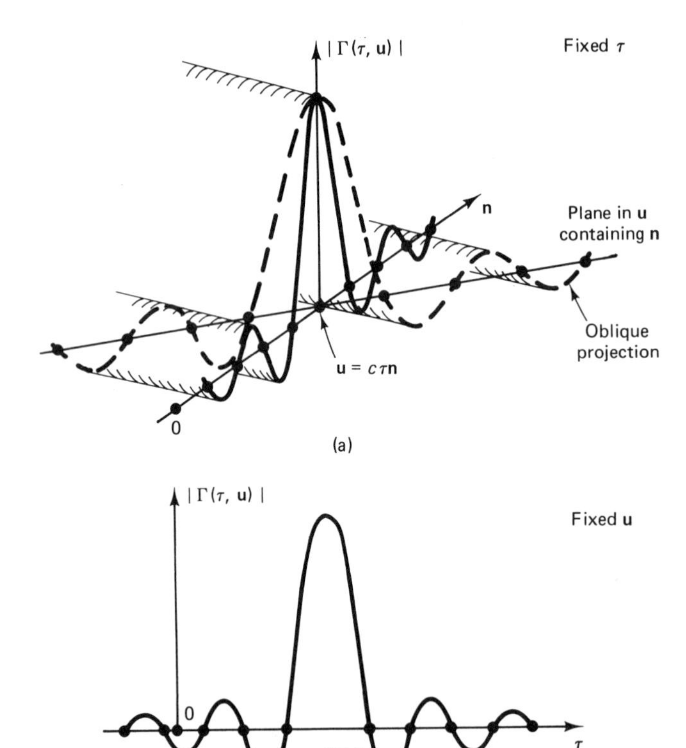

Figure 5.2 Mutual coherence $\Gamma(\tau, \mathbf{u})$ due to a source element.

magnification due to oblique projection is introduced. When the baseline is perpendicular to the source element direction, the magnification becomes infinite because now $\mathbf{u}$ is parallel to plane wavefronts from $\mathbf{n}$ and $\Gamma(\tau, \mathbf{u})$ becomes independent of baseline length. Figure 5.2 shows an example of $|\Gamma(\tau, \mathbf{u})|$ as a function of τ and $\mathbf{u}$ for a source element characterized by

$$I(f, \mathbf{n}) = \begin{cases} \delta(\mathbf{n}), & f_0 - \dfrac{\Delta f}{2} < f < f_0 + \dfrac{\Delta f}{2} \\ 0, & \text{otherwise} \end{cases}$$

$$I(\tau, \mathbf{n}) = \exp(j2\pi f_0 \tau) \frac{\sin \pi \Delta f \tau}{\pi \tau} \delta(\mathbf{n})$$

The cross-spectral density of the wave field at $\mathbf{r}_j$ and $\mathbf{r}_k$ is the average

$$\Gamma(f, \mathbf{u}) = \lim_{T \to \infty} \frac{1}{2T} \epsilon_T(f, \mathbf{r}_j + \mathbf{u}) \epsilon_T^*(f, \mathbf{r}_j)$$

where $\epsilon_T(f, \mathbf{r}), f \geq 0$, is the transform of the truncated version of the real field $e_r(t, \mathbf{r})$ in interval $2T$. Again, from the Wiener–Khinchin theorem, we have

$$\Gamma(\tau, \mathbf{u}) = \int_0^{\infty} \Gamma(f, \mathbf{u}) \exp(j2\pi f \tau)\, df \tag{5.5}$$

On using the Fourier transform relation of Eq. (5.3), the mutual coherence $\Gamma(\tau, \mathbf{u})$ and the mutual spectral density $\Gamma(f, \mathbf{u})$ can be expressed in terms of the source spectral density as follows, respectively:

$$\Gamma(\tau, \mathbf{u}) = \frac{1}{2} \iint_0^{\infty} I(f, \mathbf{n}) \exp\left[j2\pi f \left(\tau + \frac{\mathbf{n} \cdot \mathbf{u}}{c} \right) \right] df\, d\mathbf{n} \tag{5.6}$$

$$\Gamma(f, \mathbf{u}) = \frac{1}{2} \int I(f, \mathbf{n}) \exp\left(j2\pi f \frac{\mathbf{n} \cdot \mathbf{u}}{c} \right) d\mathbf{n} \tag{5.7}$$

This pair of relations constitutes the well-known *Van Cittert–Zernike theorem.* The corresponding pair of expressions in terms of source self-coherence is given by Eq. (5.4) and

$$\Gamma(f, \mathbf{u}) = \frac{1}{2} \int_{-\infty}^{\infty} \int I\left(\tau + \frac{\mathbf{n} \cdot \mathbf{u}}{c}, \mathbf{n} \right) \exp(-j2\pi f \tau)\, d\mathbf{n}\, d\tau \tag{5.8}$$

The mapping of radio sources by synthesis radio telescopes makes use of measured or estimated mutual coherence of the wave field $\Gamma(\tau, \mathbf{u})$ or mutual spectral density $\Gamma(f, \mathbf{u})$ over an entire baseline plane $\mathbf{u}$. By Fourier inversion of the Van Cittert–Zernike theorem, namely, Eqs. (5.6) and (5.7), the source spectral density distribution becomes

$$I(f, \mathbf{n}) = 2 \int_{\mathbf{u}\text{ plane}} \Gamma(f, \mathbf{u}) \exp\left(-j2\pi f \frac{\mathbf{n}\cdot\mathbf{u}}{c}\right) fd\,\frac{\mathbf{u}}{c} \tag{5.9}$$

$$= 2 \int_{\mathbf{u}\text{ plane}} \int_{-\infty}^{\infty} \Gamma(\tau, \mathbf{u}) \exp\left[-j2\pi f\left(\tau + \frac{\mathbf{n}\cdot\mathbf{u}}{c}\right)\right] df\tau\, d\frac{\mathbf{u}}{c} \tag{5.10}$$

Alternatively, the source self coherence can be expressed as

$$I(\tau, \mathbf{n}) = 2 \int_{0}^{\infty} \int_{\mathbf{u}\text{ plane}} \Gamma(f, \mathbf{u}) \exp\left[j2\pi f\left(\tau - \frac{\mathbf{n}\cdot\mathbf{u}}{c}\right)\right] fd\,\frac{\mathbf{u}}{c}\,df \tag{5.11}$$

$$= \frac{1}{j\pi} \int_{\mathbf{u}\text{ plane}} \frac{d}{d\tau}\Gamma\left(\tau - \frac{\mathbf{n}\cdot\mathbf{u}}{c}, \mathbf{u}\right) d\,\frac{\mathbf{u}}{c} \tag{5.12}$$

Although the entire visible sky contributes to the wave field, only sources within the field of view selected by the array element beam need be considered. Furthermore, regardless of whatever spectral features that the source may have, only signals within the receiver bandwidth contribute to the measurement. In most synthesis telescope arrays the bandwidth of the receiver is small compared with the band center frequency. Image reconstruction is therefore commonly based on the source spectral density using Eqs. (5.9) and (5.10). Only in certain special instruments is wide bandwidth used either directly [21, 22] or in synthesized form [23]. A broadband instrument is best described by Eq. (5.12); however, it has yet to be used for image reconstruction.

5.3 MEASUREMENT OF MUTUAL COHERENCE AND MUTUAL SPECTRAL DENSITY

In Section 5.2 we discussed the relations between the mutual coherence and the mutual spectral density of a random wave field, and between the source self-coherence and spectral density distributions. In a rotation synthesis array the measurement of mutual coherence and mutual spectral density is complicated by many factors, such as the motion of the array elements; the pattern, polarization, and frequency response of the array elements; the receiver transfer function and system noise; and sampling and signal processing used to obtain estimates of $\Gamma(\tau, \mathbf{n})$ and $\Gamma(f, \mathbf{u})$. In this section we describe these measurements in a two-element interferometer.

To observe an extended radio source located near a reference direction $\mathbf{n}_0$, all elements of a synthesis array track the reference position as the earth rotates, or in the case of a space-borne element, as the spacecraft traverses its orbit. A plane wavefront from source element $\mathbf{n}_0$ will arrive at different array elements at different times. To designate wavefronts from $\mathbf{n}_0$ regardless of element location, let us introduce a hypothetical clock attached to the source, and disseminate time by sending out timing pulses. Arriving at an array element j, these timing pulses are used to set a wavefront clock t_w in addition to its own station clock t_j. Regardless of the location of the array elements, when their wavefront clocks are the same they see the same wavefront, that is,

$$e[t_w, \mathbf{r}_j(t_w), \mathbf{n}_0] = e[t_w, \mathbf{r}_k(t_w), \mathbf{n}_0]$$

For convenience, the epoch of the wavefront clock is set equal to the station clock when that wavefront reaches the position origin. Due to propagation delay, the wavefront clock of element j is related to the station clock t_j and position $\mathbf{r}_j(t)$ according to

$$t_w = t_j + \frac{\mathbf{n}_0 \cdot \mathbf{r}_j(t_j)}{c} \tag{5.13}$$

Since each element has two clocks, all quantities at that element can be expressed in either time t_j or t_w.

For source elements in the neighborhood of $\mathbf{n}_0$ such that $\mathbf{n} = \mathbf{n}_0 + \mathbf{s}$, a residual delay as shown in Fig. 5.3 need be added. We may thus write

$$e(t_j, \mathbf{r}_j(t_j), \mathbf{n}) = e\left(t_w + \frac{\mathbf{s} \cdot \mathbf{r}(t_w)}{c}, \mathbf{u}, \mathbf{s}\right)$$

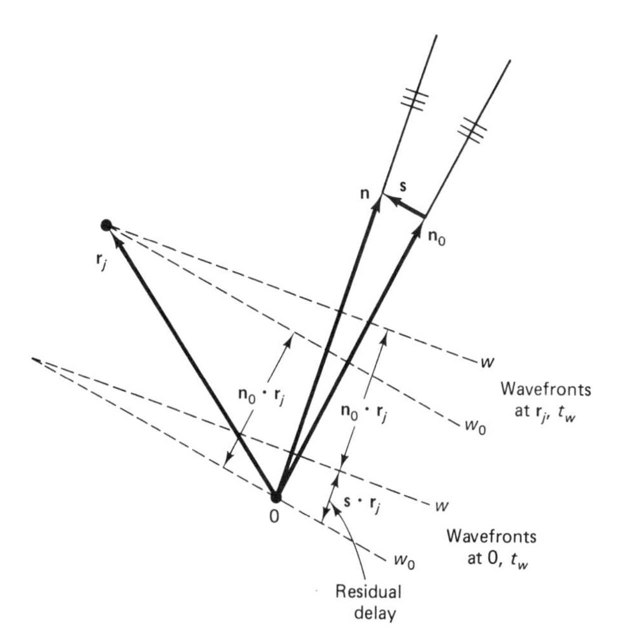

Figure 5.3 Reference direction and residual delay.

The mutual coherence of Eq. (5.4) becomes

$$\Gamma(\tau_w, \mathbf{u}) = \int E[e(t_w + \tau_w, \mathbf{r}_j(t_w + \tau_w), \mathbf{s})e^*(t_w, \mathbf{r}_k(t_w), \mathbf{s})\, d\mathbf{s}$$

$$= \int I\left(\tau_w + \frac{\mathbf{s} \cdot \mathbf{u}}{c}, \mathbf{s}\right) d\mathbf{s}$$

The baseline $\mathbf{u} = \mathbf{r}_j(t_w + \tau_w) - \mathbf{r}_k(t_w)$ is a varying quantity determined by the element motions. Similarly, all equations (5.5) to (5.12) remain valid if τ, f, and $\mathbf{n}$ are replaced by τ_w, f_w, and $\mathbf{s}$, respectively.

Although the mutual coherence or spectral density is best measured in terms of wavefront clock t_w as described above, the behavior of antenna elements and the associated receivers is usually expressed in terms of frequency responses with respect to station clock t_j. Within a short time interval where the motion of an antenna element can be considered as linear, the relation between wavefront clock t_w and station clock t_j as given in Eq. (5.13) is approximated by

$$t_w = \frac{dt_w}{dt_j} t_j + \beta_j$$

where dt_w/dt_j and β_j are slowly varying. A transfer function $H_j(f)$ with respect to station clock t_j is therefore transformed to $H_j[(dt_w/dt_j)f_w]$ when referred to the wavefront clock t_w. However, for earth-fixed arrays, $|(d/dt)\mathbf{r}_j(t_j)|/c < 1.4 \times 10^{-6}$; $H_j[(dt_w/dt_j)f_w]$ can therefore be approximated by $H_j(f_w)$. Let $\epsilon(f_w, \mathbf{r}_j, \mathbf{s})$ be the spectrum of the analytic field $e(t_w, \mathbf{r}_j, \mathbf{s})$ at $\mathbf{r}_j$ and let $l_j(f_w, \mathbf{s})$ be the field pattern of the jth antenna for the polarization under consideration with beam center pointing toward $\mathbf{n}_0$. The voltage received by the jth element has the spectrum

$$E_j(f_w) = \int l_j(f_w, \mathbf{s})\epsilon(f_w, \mathbf{r}_j, \mathbf{s})\, d\mathbf{s}$$

In each receiver, as shown in Fig. 5.4, the signal is amplified, down-converted in frequency, and filtered. Let the local oscillator be based on the wavefront clock with a residual phase ϕ_j in the form $\cos(2\pi f_{0_w} t_w - \phi_j)$. The output spectrum at the IF

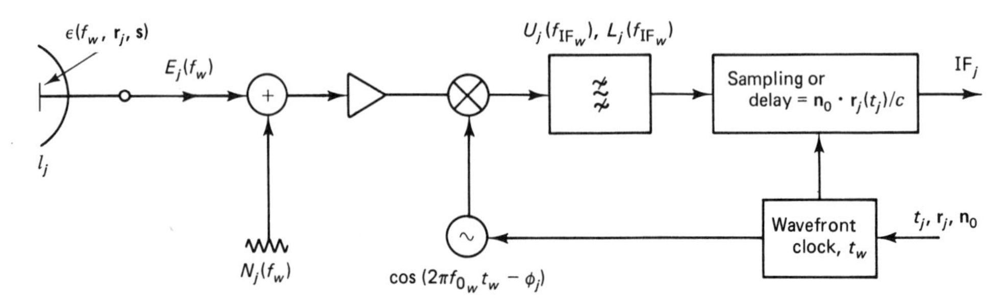

Figure 5.4 Receiver block diagram.

frequency f_{IF_w} can thus be written as

$$Z_j(f_{\mathrm{IF}_w}) = U_j(f_{\mathrm{IF}_w})[E_j(f_{0_w} + f_{\mathrm{IF}_w}) + N_j(f_{0_w} + f_{\mathrm{IF}_w})] \exp(j\phi_j) + L_j(f_{\mathrm{IF}_w})[E_j^*(f_{0_w} - f_{\mathrm{IF}_w}) + N_j^*(f_{0_w} - f_{\mathrm{IF}})] \exp(-j\phi_j)$$

Here $U_j(f_{\mathrm{IF}_w})$ and $L_j(f_{\mathrm{IF}_w})$ are the equivalent transfer functions for the upper and lower sidebands, and $N_j(f_w)$ represents noise generated in the receiver as referred to the input. Since all signals of interest are expressed in their analytic form, their spectra vanish for negative frequencies.

The implementation of wavefront clock t_w at each element is usually carried out in two steps, as shown in Fig. 5.4. The local oscillator phase $2\pi f_{0_w} t_w - \phi_j$ when derived from the station clock t_j requires a continuously varying frequency computed from the element motion and source position according to Eq. (5.13) so as to track the interference "fringe." After mixing, amplification, and filtering, an appropriate delay in the form of a variable delay line or variable sampling rate is inserted to track the delay according to the wavefront clock. In a connected synthesis array, the signals $Z_j(f_{\mathrm{IF}_w})$ from all elements are transmitted to a central processor for further real-time processing. In a VLBI array, each signal is either recorded on tape and then transported to a processor or transmitted by a communication link for further processing. In the former case, fringe and delay tracking can be performed during processing.

A two-element interferometer using array elements j and k estimates the cross-spectral density $S_{jk}(f_{\mathrm{IF}_w})$ and cross-correlation $R_{jk}(\tau_w)$ of the outputs $Z_j(f_{\mathrm{IF}_w})$ and $Z_k(f_{\mathrm{IF}_w})$. We may thus write

$$S_{jk}(f_{\mathrm{IF}_w}) = E[Z_j(f_{\mathrm{IF}_w})Z_k^*(f_{\mathrm{IF}_w})] = S_{kj}^*(f_{\mathrm{IF}_w}) \tag{5.14}$$

$$R_{jk}(\tau_w) = \int_0^\infty S_{jk}(f_{\mathrm{IF}_w}) \exp(j2\pi f_{\mathrm{IF}_w}\tau_w)\, df_{\mathrm{IF}_w} \tag{5.15}$$

Because of the incoherent nature of radio sources and the fact that noises in the two telescopes are neither correlated with the signal nor with each other, the cross-spectral density becomes

$$\begin{aligned} S_{jk}(f_{\mathrm{IF}_w}) &= \tfrac{1}{2} \exp[j(\phi_j - \phi_k)] U_j(f_{\mathrm{IF}_w}) U_k^*(f_{\mathrm{IF}_w}) \int l_j(f_{0_w} + f_{\mathrm{IF}_w}, \mathbf{s}) \\ &\quad \times l_k^*(f_{0_w} + f_{\mathrm{IF}_w}, \mathbf{s}) I(f_{0_w} + f_{\mathrm{IF}_w}, \mathbf{s}) \exp\left[j2\pi(f_{0_w} + f_{\mathrm{IF}_w}) \frac{\mathbf{s}\cdot\mathbf{u}}{c}\right] d\mathbf{s} \\ &\quad + \tfrac{1}{2} \exp[-j(\phi_j - \phi_k)] L_j(f_{\mathrm{IF}_w}) L_k^*(f_{\mathrm{IF}_w}) \int l_j^*(f_{0_w} - f_{\mathrm{IF}_w}, \mathbf{s}) l_k^*(f_{0_w} - f_{\mathrm{IF}_w}, \mathbf{s}) \\ &\quad \times I^*(f_{0_w} - f_{\mathrm{IF}_w}, \mathbf{s}) \exp\left[j2\pi(f_{0_w} - f_{\mathrm{IF}_w}) \frac{\mathbf{s}\cdot\mathbf{u}}{c}\right] d\mathbf{s} \end{aligned} \tag{5.16}$$

For simplicity we confine further discussions to upper sideband systems. In an ideal situation all elements have identical transfer functions, phase residuals, patterns, and have zero pointing error. Under these assumptions, Eq. (5.16) becomes

$$S_{jk}(f_{\mathrm{IF}_w}) = \tfrac{1}{2}|U(f_{\mathrm{IF}_w})|^2 \int |l(f_{0_w} + f_{\mathrm{IF}_w}, \mathbf{s})|^2 \times I(f_{0_w} + f_{\mathrm{IF}_w}, \mathbf{s}) \exp\left[j2\pi(f_{0_w} + f_{\mathrm{IF}_w}) \frac{\mathbf{s}\cdot\mathbf{u}}{c}\right] d\mathbf{s} \tag{5.17}$$

Comparing Eq. (5.17) with (5.7), we see that $S_{jk}(f_{\mathrm{IF}_w})$ is a frequency- and direction-weighted version of the mutual spectral density $\Gamma(f_{0_w} + f_{\mathrm{IF}_w}, \mathbf{s})$. On taking the Fourier transform,

$$R_{jk}(\tau_w) = \frac{1}{2}\int_0^\infty |U(f_{\mathrm{IF}_w})|^2 \int |l(f_{0_w} + f_{\mathrm{IF}_w}, \mathbf{s})|^2 \times I(f_{0_w} + f_{\mathrm{IF}_w}, \mathbf{s}) \exp\left\{j2\pi\left[(f_{0_w} + f_{\mathrm{IF}_w}) \frac{\mathbf{s}\cdot\mathbf{u}}{c} + f_{\mathrm{IF}_w}\tau_w\right]\right\} d\mathbf{s}\, df_{\mathrm{IF}_w} \tag{5.18}$$

Comparing Eq. (5.18) with (5.6), we see that $R_{jk}(\tau_w)$ is a weighted version of the mutual coherence demodulated by the local oscillator with the factor $\exp(-j2\pi f_{0_w}\tau_w)$. This demodulation results in a reduction of wavefront clock delay τ_w accuracy requirement from that of much less than $1/f_{0_w}$ to $1/\Delta f$, where Δf is the IF bandwidth. Detailed relations between frequency response and array performance can be found in [24].

For partially polarized waves, let two orthogonally polarized components of a plane wave at the position $\mathbf{r}$ with frequency f, coming from direction $\mathbf{n}$, be denoted by the row vector $[\epsilon_a(f, \mathbf{r}, \mathbf{n}), \epsilon_b(f, \mathbf{r}, \mathbf{n})]^T$, where the superscript T denotes transposition. The intensity of the source and its polarization is described by the Hermitian coherence matrix whose elements are quadratic ensemble averages defined by

$$[J(f, \mathbf{n})] = \begin{bmatrix} E[(\epsilon_a(f, \mathbf{r}, \mathbf{n})\epsilon_a^*(f, \mathbf{r}, \mathbf{n})] & E[\epsilon_a(f, \mathbf{r}, \mathbf{n})\epsilon_b^*(f, \mathbf{r}, \mathbf{n})] \\ E[\epsilon_b(f, \mathbf{r}, \mathbf{n})\epsilon_a^*(f, \mathbf{r}, \mathbf{n})] & E[\epsilon_b(f, \mathbf{r}, \mathbf{n})\epsilon_b^*(f, \mathbf{r}, \mathbf{n})] \end{bmatrix}$$

The brightness distribution of the source $I(f, \mathbf{n})$ is the *trace* of $[J(f, \mathbf{n})]$. To obtain the polarization image of a partially polarized source, each antenna must have two channels responding to two different polarizations. Instead of Eq. (5.15), the received voltage spectrum of the two channels of the jth array element can now be expressed as

$$\begin{bmatrix} E_{jA}(f_w) \\ E_{jB}(f_w) \end{bmatrix} = \int [l_j(\mathbf{f}, \mathbf{n})] \begin{bmatrix} \epsilon_a(f_w, \mathbf{r}_j, \mathbf{n}) \\ \epsilon_b(f_w, \mathbf{r}_j, \mathbf{n}) \end{bmatrix} d\mathbf{n}$$

where $[l_j(f, \mathbf{n})]$ is the 2×2 matrix field pattern of the jth element. Following the signal through a dual-channel receiver at each element, it is readily seen that in place of Eqs. (5.14) and (5.15), we have two matrices

$$[S_{jk}(f_{\mathrm{IF}_w})] = E[[Z_j(f_{\mathrm{IF}_w})][Z_k^*(f_{\mathrm{IF}_w})]^T]$$

$$[R_{jk}(\tau_w)] = \int_0^\infty [S_{jk}(f_{\mathrm{IF}_w})] \exp(j2\pi f_{\mathrm{IF}_w}\tau_w)\, df_{\mathrm{IF}_w}$$

The estimation of cross-correlation $R_{jk}(\tau_w)$ and cross-spectral density $S_{jk}(f_{\mathrm{IF}_w})$ from the receiver outputs $Z_j(f_{\mathrm{IF}_w})$ and $Z_k(f_{\mathrm{IF}_w})$ can be implemented either with analog correlators and filter banks or with digital processing. Except for millimeter wave observations, where large bandwidths preclude digital processing, the majority of present large synthesis telescopes makes use of digital correlation processing. Figure 5.5 shows a typical correlator for one baseline [25]. After sampling using the wavefront clock t_w for the two elements, the output of the jth element is fed into a shift register of appropriate length (short for continuum observations, long for spectral line work). The output of each stage of the shift register is correlated with the samples of the kth element, accumulated, and decimated by filtering to obtain an estimate of the cross-correlation with appropriate integration time. Since all signals are real, only the real part of $\hat{R}_{jk}(\tau_w)$ is estimated. The imaginary part of $\hat{R}_{jk}(\tau_w)$ is obtained by Hilbert-transforming the real part. From these, an estimate of the complex cross-spectral density $\hat{S}_{jk}(f_w)$ is obtained, normally by means of conventional spectral analysis using windowed fast Fourier transformation (FFT). Although

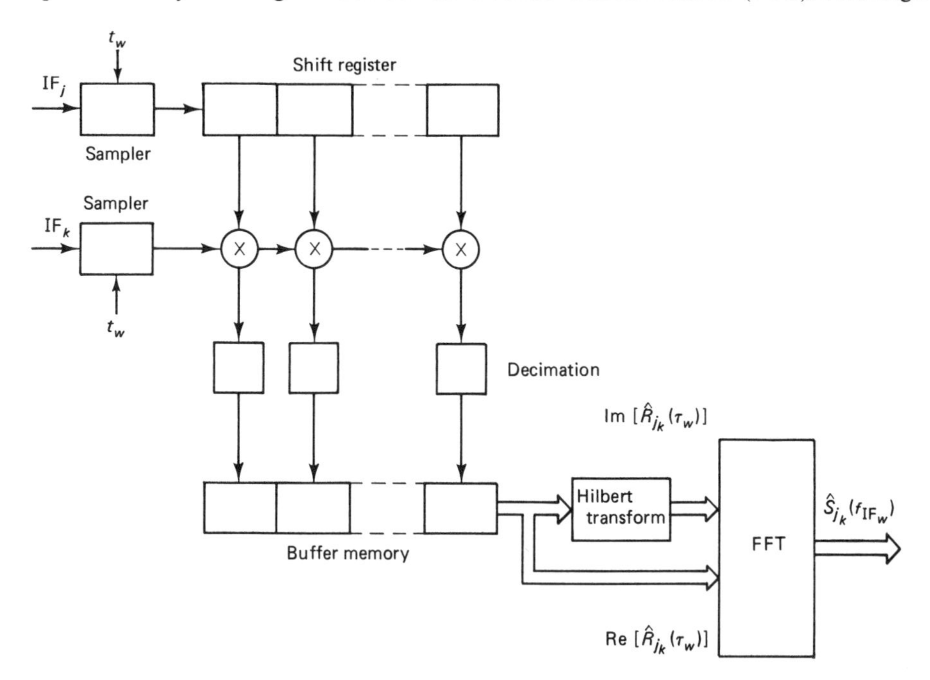

Figure 5.5 Digital correlator for one baseline.

maximum-entropy analysis has also been proposed [18], it has not been widely used because of the complexities of most spectral densities and the need for accurate line shape and amplitude determinations. For continuum observations, the spectrum is governed by the receiver transfer function and a knowledge of $\hat{R}_{jk}(0)$ is sufficient for image reconstruction. However, if the bandwidth $f_{\mathrm{IF}_{w0}} \pm \Delta f/2$, is too wide, integration of Eq. (5.18) with respect to f_{IF_w} first yields

$$R_{jk}(0) = \int I(f_{0_w} + f_{\mathrm{IF}_{w0}}, \mathbf{s}) \frac{\sin\left(\pi \Delta f \dfrac{\mathbf{s} \cdot \mathbf{u}}{c}\right)}{\pi \dfrac{\mathbf{s} \cdot \mathbf{u}}{c}} \exp\left[j2\pi(f_{0_w} + f_{\mathrm{IF}_{w0}}) \frac{\mathbf{s} \cdot \mathbf{u}}{c}\right] d\mathbf{s} \tag{5.19}$$

If the residual delay $\mathbf{s} \cdot \mathbf{u}/c$ from any part of the source to any part of the baseline sampled is small compared with $1/\Delta f$, the sinc function

$$\frac{\sin\left(\pi \Delta f \dfrac{\mathbf{s} \cdot \mathbf{u}}{c}\right)}{\pi \Delta f \dfrac{\mathbf{s} \cdot \mathbf{u}}{c}}$$

may be replaced by unity.

So far, we have not discussed the contributions of noise in the receivers. Observation of random noise of bandwidth Δf for Δt seconds collects $\Delta f \, \Delta t$ independent samples. Let the noise in element j be represented by its noise temperature T_j. The variance in the estimation of both the real and imaginary parts of $\hat{R}_{jk}(\tau_w)$ is

$$\sigma_{jk}^2 = \frac{T_j T_k}{\Delta f \, \Delta t}$$

In a spectral line system, the above is the variance in the estimation of the real and imaginary parts of $\hat{S}_{jk}(f_{\mathrm{IF}_w})$, with Δf taken to be the bandwidth per frequency channel. A unique character of radio astronomy is the extremely low signal received from extra terrestrial sources. Even with large radio telescopes and very low noise receivers, the signal-to-noise ratio usually lies in the range 10^{-6} to 10^{-3}. In fact, the fainter the sources the telescope can reach, the larger the number of sources can be seen. Estimation of cross-correlation and cross-spectral density of such weak signals requires a long integration time, and for continuum sources, a large total bandwidth. For a large synthesis array of many elements, the number of signals to be processed is very large. Because of the enormous number of signal samples involved, special processors using very large scale integrated (VLSI) circuits backed up by powerful computers are necessary. For arrays with 10 to 30 elements the number of baselines varies from 45 to 435. A typical continuum bandwidth of 50 MHz requires sampling at 10^8 samples per second. For a real correlator with eight delay lags and for four elements of polarized correlation matrix, the total number of multiplications for a 30-element array is 1.4×10^{12} per second. For spectral line observations with

5 MHz total bandwidth and 10^3 delay lags, the same array requires 1.7×10^{13} multiplications per second. Since it is impossible to perform multiplications with precision at such rates, it is customary to quantize the signal to two or a small number of levels.

The simplest quantization is to clip the signal so that only two levels remain. In a classical work, Van Vleck [26] showed that the cross-correlation of the clipped signal $R_{jk_c}(\tau_w)$ is related to the unclipped one according to

$$R_{jk}(\tau_w) = \sin\left[\frac{\pi}{2} R_{jk_c}(\tau_w)\right]$$

To obtain the proper scale, the relation above needs to be multiplied by the system temperatures. Because of its ease of implementation and immunity to receiver gain fluctuations, the one-bit correlator has been extensively used since its introduction [27]. One effect of clipping is to scatter some of the signal power into adjacent frequencies so as to increase the noise variance by a factor of $(\pi/2)^2$. To reduce the degradation, we may resort to oversampling or multibit quantization [28].

Another factor of practical interest is that the number of correlator banks required increases rapidly with array element number N according to $N(N-1)/2$. An alternative approach is to window the signal from each element and compute a short-time Fourier transform using a hardware FFT processor. For every pair of elements, one FFT processor is required to evaluate the complex spectra of the two real signals. As shown in Fig. 5.6, the power spectrum of each element and the complex cross-spectral density for each baseline can be computed using a single complex multiplier and accumulator [29]. The advent of VLSI has made high-speed FFT possible [30]. Furthermore, since the FFT processor output can be arranged in a single queue, high-density memories can be used as accumulators. Finally, because the processor can handle a relatively large number of bits, multibit quantization can be used. For a large spectral line synthesis array, this approach may be simpler than the correlation spectrometer.

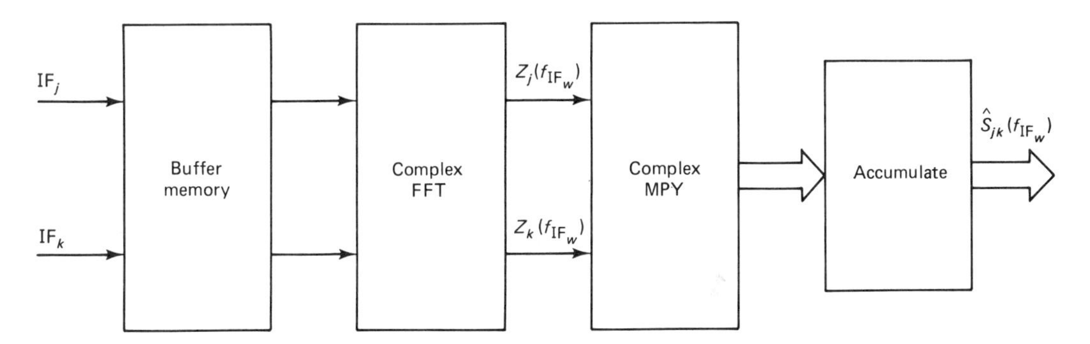

Figure 5.6 FFT cross-spectrum analyzer for one baseline.

5.4 ARRAY GEOMETRY AND ROTATIONAL SAMPLING OF BASELINE SPACE*

In a rotation synthesis radio telescope, the mutual coherence and mutual spectral density are estimated for all instantaneous baselines formed by the array. As the earth rotates, the baseline space is sampled by baseline tracks. Using an inversion procedure based on Eqs. (5.9) to (5.12), images of the source total intensity and spectral density can be reconstructed. The resolution of the images increases with the baseline region sampled. The finite spacing between baseline tracks causes aliasing of images, the field of view investigated must be kept small enough so that this does not occur. Proper sampling of the baseline space is therefore essential for image reconstruction.

Rotation synthesis arrays have been realized in many different forms, starting from simple linear arrays to those with cross-, T-, Y-, II-, or ring-shaped geometries in various parts of the world, as well as collections of existing telescopes over different continents without any particular order. The distribution of the elements is uniform in some arrays and nonuniform in others, some with fixed elements while others may have movable ones. The array dimension ranges from a few hundred meters to almost the diameter of the earth. In fact, with the advent of earth satellites, earth–space and space–space baselines have been proposed. The sampling of the baseline space can be redundant or nonredundant, depending on the array configuration. Adequate sampling of baseline space for image reconstruction depends also on the source direction. In particular, the projection of baseline tracks on the reference wavefront plane is of prime importance. The selection of an array configuration and the treatment of baseline space sampling by rotation are therefore complex problems dependent on many factors [31]. For earth-fixed VLBI arrays spanning the earth, the angular resolution is so great that the wobble of the earth's rotation axis, the tidal deformation of the solid earth, and the entire spectrum of tectronic motions of the earth's crust have to be taken into account. In fact, because of the great precision with which a VLBI measures positions in the sky relative to the array elements on earth, it has become a major instrument for the study of geodynamics [32].

Consider an array with elements at $\mathbf{r}_j(t)$ observing a source at reference postion $\mathbf{n}_0$. The estimates of $R_{jk}(\tau_w)$ and $S_{jk}(f_{\mathrm{IF}_w})$ measured in terms of wavefront clocks at each element depend on the element delay to position origin $\mathbf{n}_0 \cdot \mathbf{r}_j(t_j)/c$ and the residual delay across baseline $\mathbf{s} \cdot \mathbf{u}/c$. For an earth-fixed array, the element positions are given in a right-handed orthogonal coordinate system $(x, y, z)_e$, where z_e is from the geocenter to the pole of rotation and x_e is on the plane of the equator pointing toward Greenwich. The corresponding spherical components (r, ϕ, λ), where ϕ is the latitude and λ the east longitude, are also used. The position $\mathbf{n}$ of radio sources in the sky is defined in a sky-fixed coordinate system $(x, y, z)_s$ related to the earth-fixed coordinates by earth's rotation about z_e for an angle θ equal to the apparent sidereal time as shown in Fig. 5.7, with x_s along the intersection of the equator and

*This section was written with assistance from Y. L. Chow, University of Waterloo.

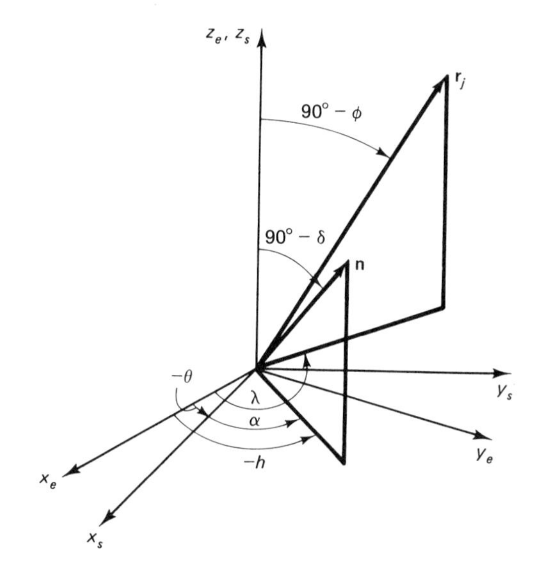

Figure 5.7 Earth-fixed and sky-fixed coordinates.

the plane of the ecliptic pointing toward the ascending node. The source postion $\mathbf{n}$ is given by the corresponding spherical coordinates in right ascention α and declination δ. If the geocenter is used as the position origin, the element delay from $\mathbf{r}_j$ to the origin along reference position $\mathbf{n}_0$ becomes

$$\frac{\mathbf{n}_0 \cdot \mathbf{r}_j(\theta)}{c} = \frac{r_j}{c} [\cos \phi_j \cos \delta_0 \cos (\theta - \alpha_0 + \lambda_j) + \sin \phi_j \sin \delta_0]$$

Here the delay is expressed in terms of the station clock sidereal time θ.

The baseline $\mathbf{u} = \mathbf{r}_j(t_w + \tau_w) - \mathbf{r}_k(t_w)$ can be approximated by $\mathbf{u} = \mathbf{r}_j(\theta) - \mathbf{r}_k(\theta)$ in most arrays except in some extreme VLBI situations. In 24 hours, each baseline sweeps over a circle in the sky on the plane $z_s = (z_j - z_k)_e$ with radius $\rho_s = [(x_j - x_k)_e^2 + (y_j - y_k)_e^2]^{1/2}$, as shown in Fig. 5.8. Reversing the roles of the two

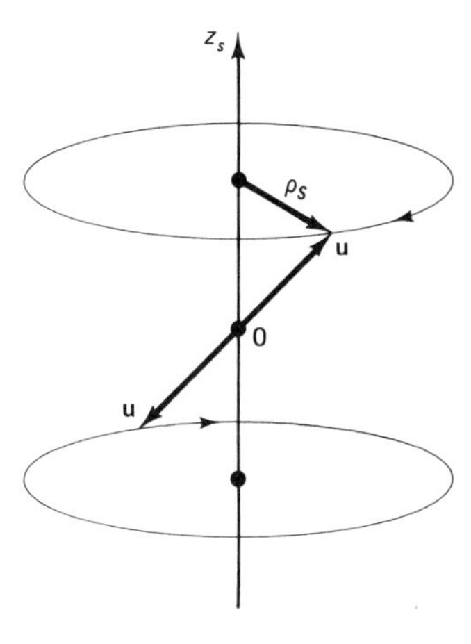

Figure 5.8 Sweeping of baseline space by earth rotation.

elements results in a baseline reflected through the geocenter. Due to earth blocking, sometimes only a part of the circular track is visible. To evaluate the residual delay $\mathbf{s} \cdot \mathbf{u}/c$, it is customary to introduce a third coordinate system (l, m, n) fixed to the reference source, with n along the reference direction $\mathbf{n}_0$, l on the equatorial plane pointing toward the east and m pointing north. As shown in Fig. 5.9, a source at (α, δ) has the (l, m, n) components

$$\begin{bmatrix} l \\ m \\ n-1 \end{bmatrix} = \begin{bmatrix} \cos\delta \sin(\alpha - \alpha_0) \\ -\sin\delta_0 \cos\delta \cos(\alpha - \alpha_0) + \cos\delta_0 \sin\delta \\ \cos\delta_0 \cos\delta \cos(\alpha - \alpha_0) + \sin\delta_0 \sin\delta - \end{bmatrix} \approx \begin{bmatrix} (\alpha - \alpha_0)\cos\delta_0 \\ (\delta - \delta_0) \\ 0 \end{bmatrix} \tag{5.20}$$

where the approximation is valid for small $\alpha - \alpha_0$ and $\delta - \delta_0$. Let the components of the baseline $\mathbf{u} = \mathbf{r}_j - \mathbf{r}_k$ along this coordinate system be (u, v, w). It is readily seen that, on introducing the hour angle $h_0 = \theta - \alpha_0$,

$$\begin{bmatrix} u \\ v \\ w \end{bmatrix} = \begin{bmatrix} -\sin h_0 & \cos h_0 & 0 \\ -\sin\delta_0 \cos h_0 & -\sin\delta_0 \sin h_0 & \cos\delta_0 \\ \cos\delta_0 \cos h_0 & \cos\delta_0 \sin h_0 & \sin\delta_0 \end{bmatrix} \begin{bmatrix} x_j - x_k \\ y_j - y_k \\ z_j - z_k \end{bmatrix}_e$$

The residual delay is therefore

$$\frac{\mathbf{s} \cdot \mathbf{u}}{c} = \frac{lu + mv + (n-1)w}{c}, \qquad n = \sqrt{1 - l^2 - m^2} \tag{5.21}$$

As seen from the reference direction $\mathbf{n}_0$, each baseline projects onto an ellipse in the (u, v)-plane as the earth rotates, with center at $(u, v) = (0, \cos\delta_0)$ and semiaxes ρ_s and $\rho_s \sin\delta_0$. For source features near $\mathbf{n}_0$, the n-component of $\mathbf{s}$ is nearly unity; information is to be found on baselines with different u, v components. Adequate sampling of mutual coherence $R_{jk}(\tau_w)$ and mutual spectral density $S_{jk}(f_{\mathrm{IF}_w})$ in the (u, v)-plane by a synthesis array therefore requires many instantaneous baselines sweeping the baseline space as the earth rotates.

A synthesis array of N elements has $N(N-1)/2$ baselines, each of which sweeps over a circular arc in the baseline space as the earth rotates. The distribution of baseline tracks depends on array configuration as well as its location on earth. The projections of the baseline tracks on the (u, v)-plane further depend on the

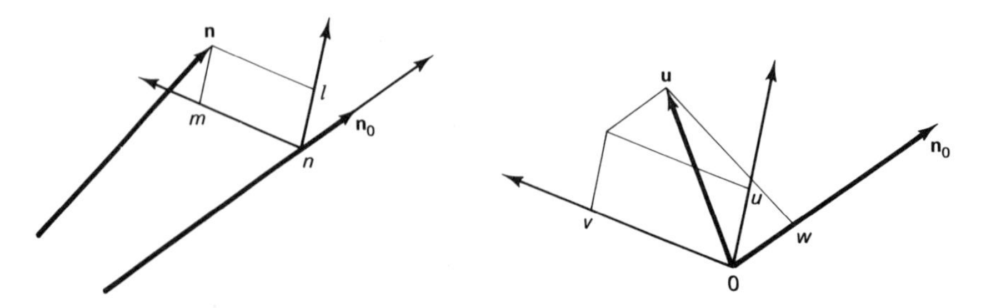

Figure 5.9 Reference source-fixed coordinates.

source declination and the tracking time. They are usually nonuniformly spaced, often with irregular boundaries and some times have holes. For some arrays the tracks for different pairs may coincide in the (u, v)-plane resulting in redundancy. The speed with which each track is scanned not only varies from track to track but also along each track, reaching a maximum at $u = 0$ and minimum at maximum $|u|$. This means the measurement will have different integration time and hence varying signal-to-noise ratio at different parts of the (u, v)-plane. In the following, we illustrate these properties by a few simple array configurations and consider the problems of designing array configurations with good (u, v)-plane sampling.

The simplest array is a uniformly spaced linear array of N elements, for which the number of different baselines is $N - 1$. However, since the total number of baselines is $N(N - 1)/2$, they are highly redundant. To reduce redundancy it is necessary to use nonuniform spacing [33]; for example, with spacings of 1, 1, 4, 3, a five-element array covers nine baselines from one to nine unit lengths with only one redundant spacing. As the earth rotates, the line containing the baselines sweeps over a biconical surface at uniform speed in a center-symmetric fashion, completely covering the surface in 24 hours. If the baseline is along the east-west direction, the biconical surface coalesces onto the equatorial plane in the baseline space, and 12 hours of tracking scans the entire plane because of symmetry about the origin. If the source is at $\delta = \pm 90°$, the biconical surface projects onto a circle in the equatorial baseline plane so that the (u, v) tracks are uniformly spaced circular arcs. At $\delta = 0°$ the biconical surface is seen edge on, so the (u, v) tracks degenerate into line segments spanning an X-shaped region. For an east-west array the region further coalesces into a single line. For in between declinations, the tracks are ellipses whose track length as limited by the horizon is dependent of the source declination δ and the element latitude ϕ. Tracks for negative declinations are the same as those of positive declinations except for shorter lengths due to earth blocking. In practice, the track length may be limited by considerations of the efficient use of these very large synthesis radio telescopes. Figure 5.10 illustrates most of the features of linear arrays described above.

A linear array parallel to the earth's axis has its baselines along the earth axis; hence such arrays do not scan the baseline space as the earth rotates. However, two such arrays parallel to each other, which can be easily implemented near the earth's equator, is a very useful configuration [34]. The instantaneous baselines lie on three straight lines, one along the earth's axis, the others parallel to it and displaced to each side by the spacing between the two linear arrays. As the earth rotates, the two nonaxial lines sweep over two symmetrical sections of circular cylinder in the baseline space which closes in 12 hours of tracking time. For $\delta = \pm 90°$ the cylinder projects onto a circle in the (u, v)-plane. This means that the array is not suitable for the study of polar sources. For $\delta = 0°$, however, the cylinder is projected onto a rectangle, and is therefore a most useful configuration. As long as the declination is not too high, the projection on the (u, v)-plane will have no hole in the center. Twelve-hour tracking, however, is necessary to avoid a hole along the v-axis. Figure 5.11 illustrates the main features of the parallel array.

Arrays having instantaneous baselines distributed over a plane instead of a

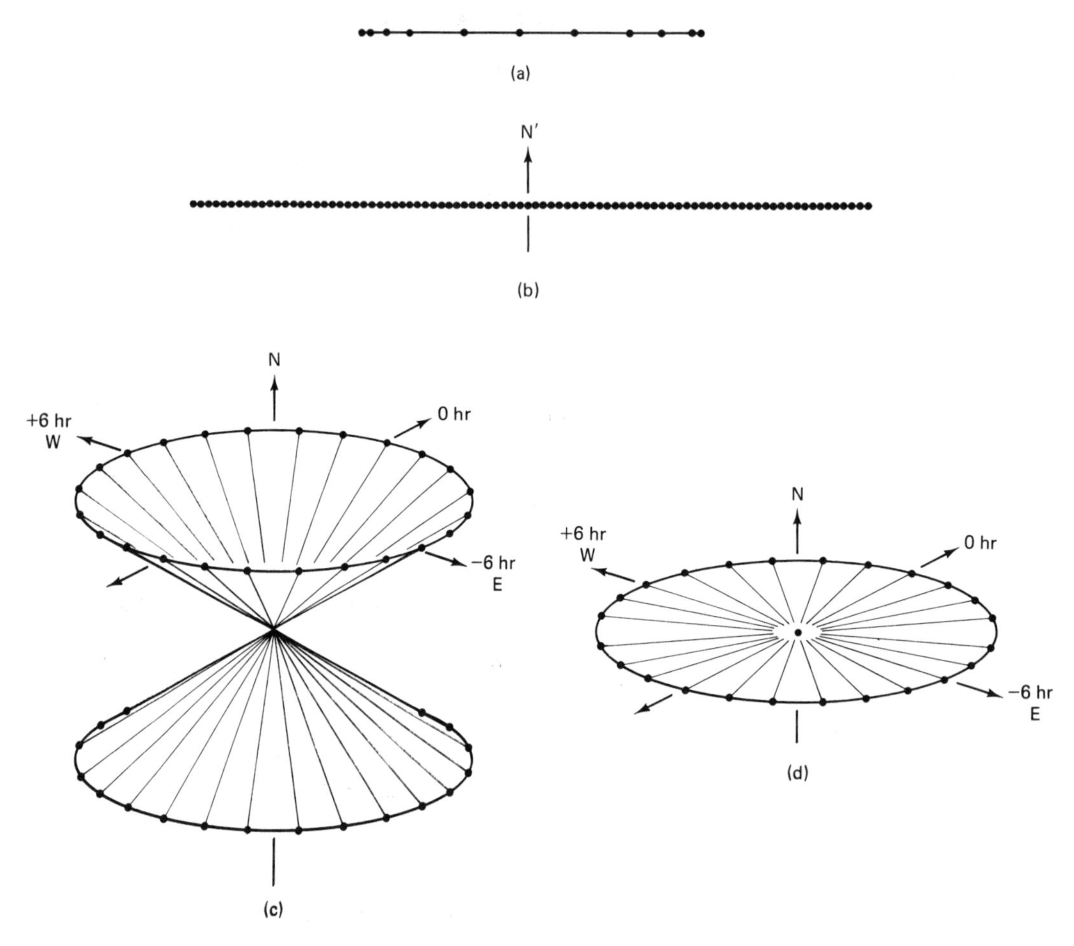

Figure 5.10 Linear array: (a) array configuration; (b) instantaneous baseline configuration; (c) baseline scan non-east-west array; (d) baseline scan east-west array.

line scan a volume instead of a surface as the earth rotates. As a result, the (u, v)-plane projections of the baseline tracks are less dependent on source declination. For example, the uniformly spaced ring array has baselines distributed fairly uniformly within a circle [35]. In particular, an odd-element array has no redundent baselines. In the ideal case of a ring array located at 0° latitude so that the earth's axis is parallel to the array plane, the baselines sweep over a complete spherical volume in 24 hours of tracking. A complete sphere has the advantage of being always projected onto a circle of the same radius on the (u, v)-plane for any source declination. As a result, the synthesized beamwidth remains constant, a desirable characteristic that no other array configuration can duplicate. If the ring is at latitude ϕ, the circular baseline plane sweeps out a portion of the sphere with polar caps of half-width ϕ missing. If the ring is on a latitude circle such as VLBI array,

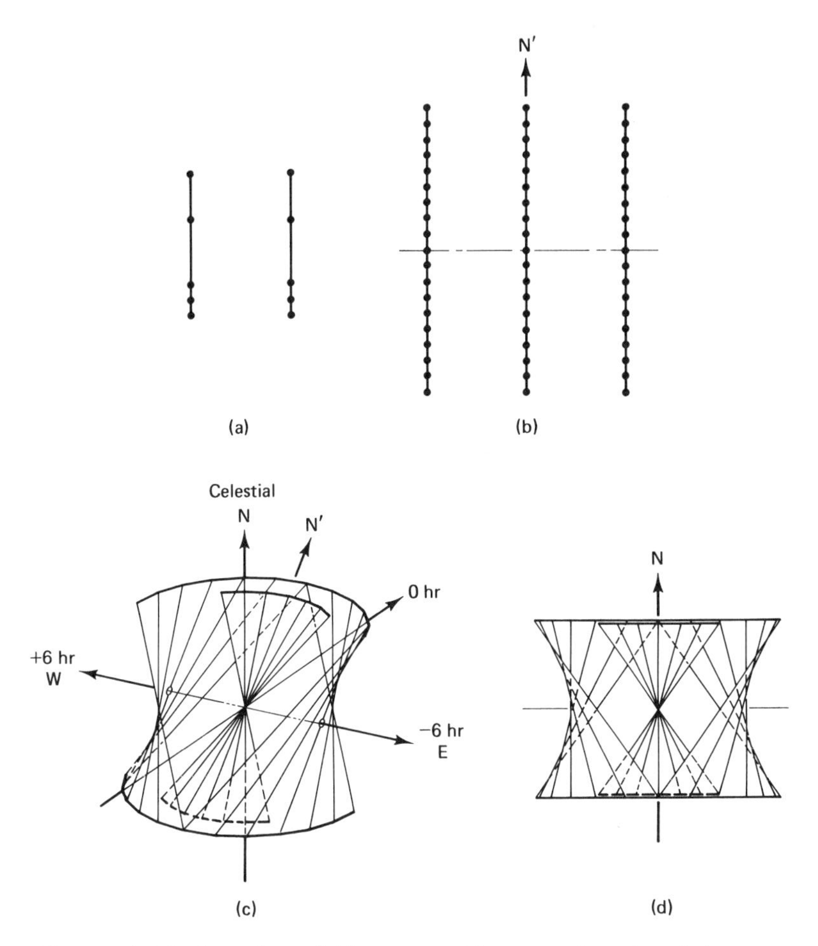

Figure 5.11 North-south parallel linear array: (a) array configuration; (b) instantaneous baseline configuration; (c) baseline scan; (d) (u, v)-plane projection, $\delta = 0$.

the baselines do not rotate out of the equatorial plane, and therefore have the same property as in an east-west linear array. Figure 5.12 illustrates the major properties of the ring array.

At moderate latitudes the ring array has the desirable property that when viewed from source positions other than near zero declination, the scanned baseline volume projects onto a circle in the (u, v)-plane. However, since the radius of the ring is a fixed dimension, the scanned baseline volume cannot be changed to accommodate variable resolution. To allow for baseline contraction and expansion, an array of open structure with instantaneous baselines approximately distributed within a circle is required. Among the different possible configurations, the Y array of Fig. 5.13 has an instantaneous baseline bounded roughly within a circle. With one of the three arms slightly tilted from the north-south direction, the baselines have

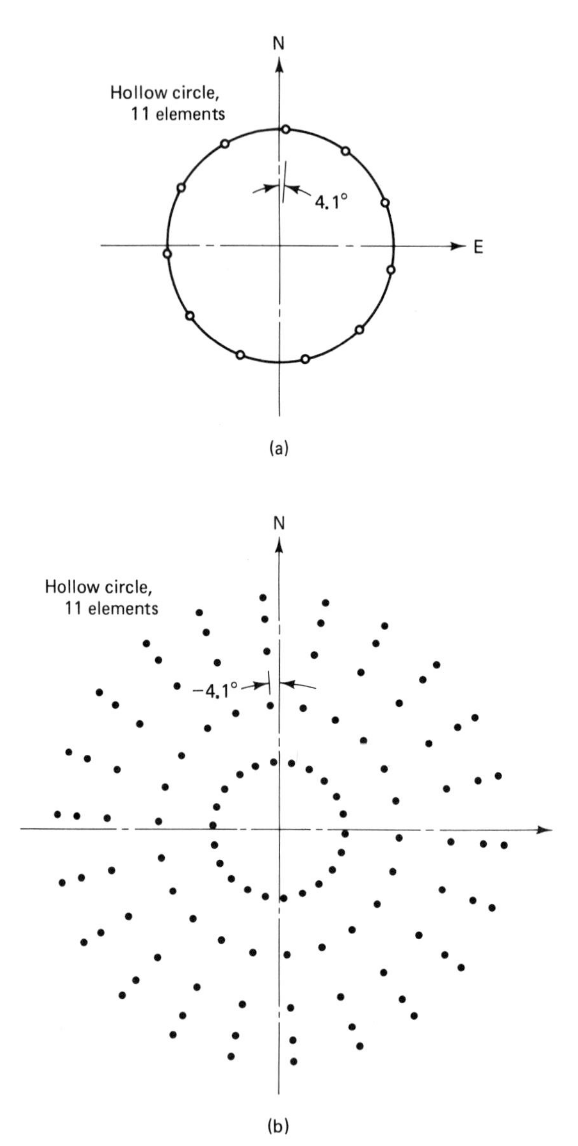

Figure 5.12 Circular array: (a) array configuration; (b) instantaneous baseline configuration.

little redundancy. Such an array would track nearly the same baseline volume as a ring. To complete the specification of the array, it remains to determine the element distribution on each arm such that the projected track density on the (u, v)-plane remains fairly uniform to allow uniform sampling of $R_{jk}(\tau_w)$ and $S_{jk}(f_{\mathrm{IF}_w})$. The projected baseline track density depends not only on array configuration and source declination but also on track length. With ± 6 hours of tracking time, the entire sphere minus the polar caps is almost covered, as shown in Fig. 5.14. The thickness of the volume as projected on a (u, v)-plane decreases with increasing distance from

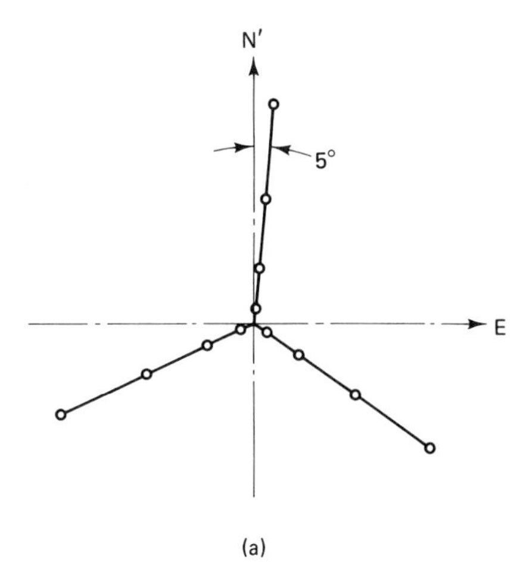

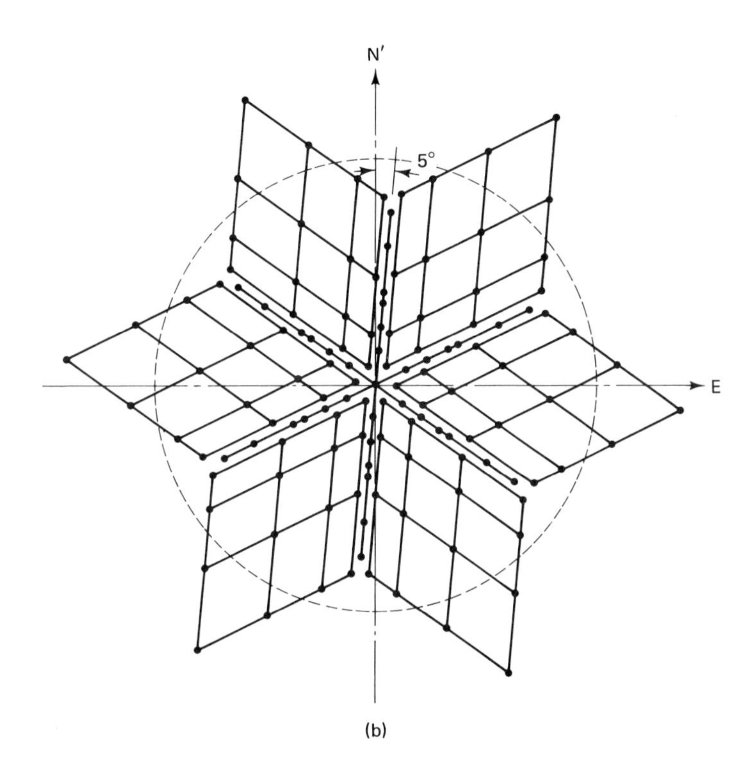

Figure 5.13 *Y* array: (a) array configuration; (b) instantaneous baseline configuration.

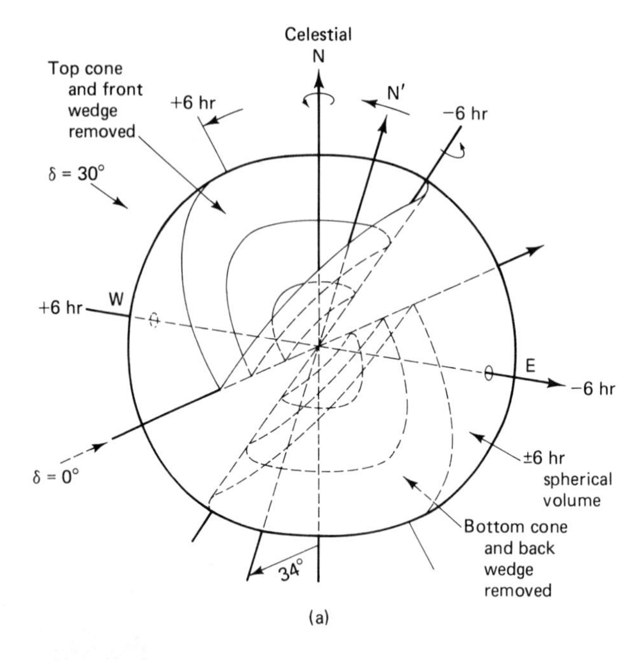

(a)

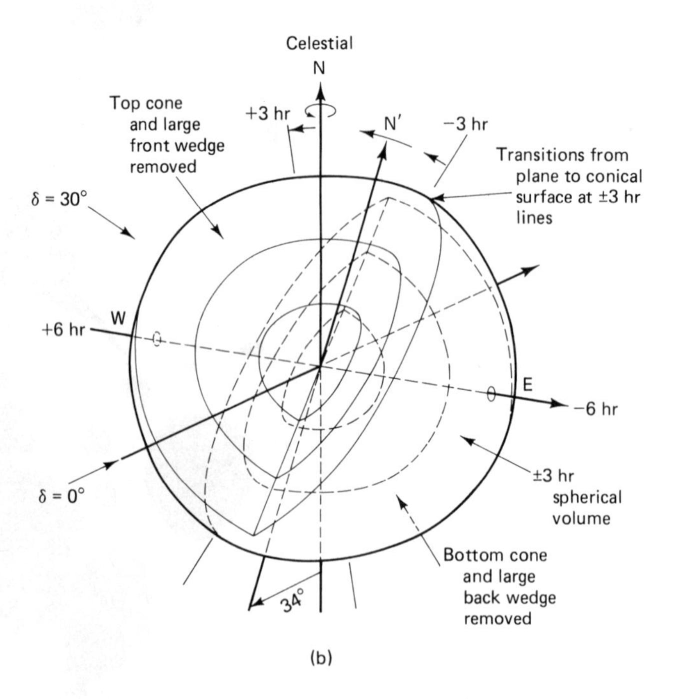

(b)

Figure 5.14 Baseline space sweeping by a plane circle: (a) ± 6 hours; (b) ± 3 hours.

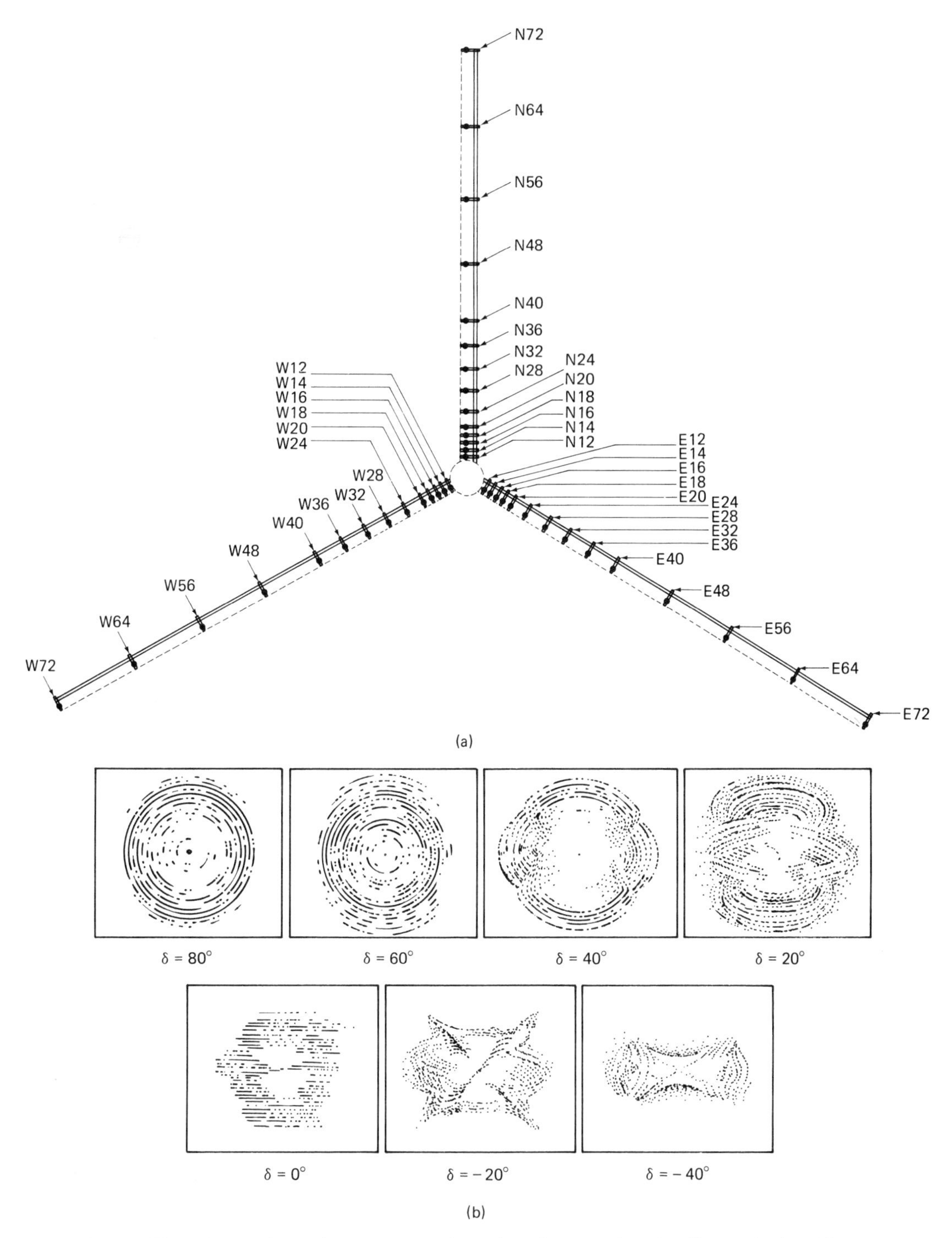

Figure 5.15 The VLA: (a) array configuration; (b) (u, v)-plane baseline tracks for different source declinations. [By permission of *Science* (vol. 216, p. 1279, June 18, 1982). From Hjellming and Bignell, "Radio Astronomy with Very Long Baseline Array."]

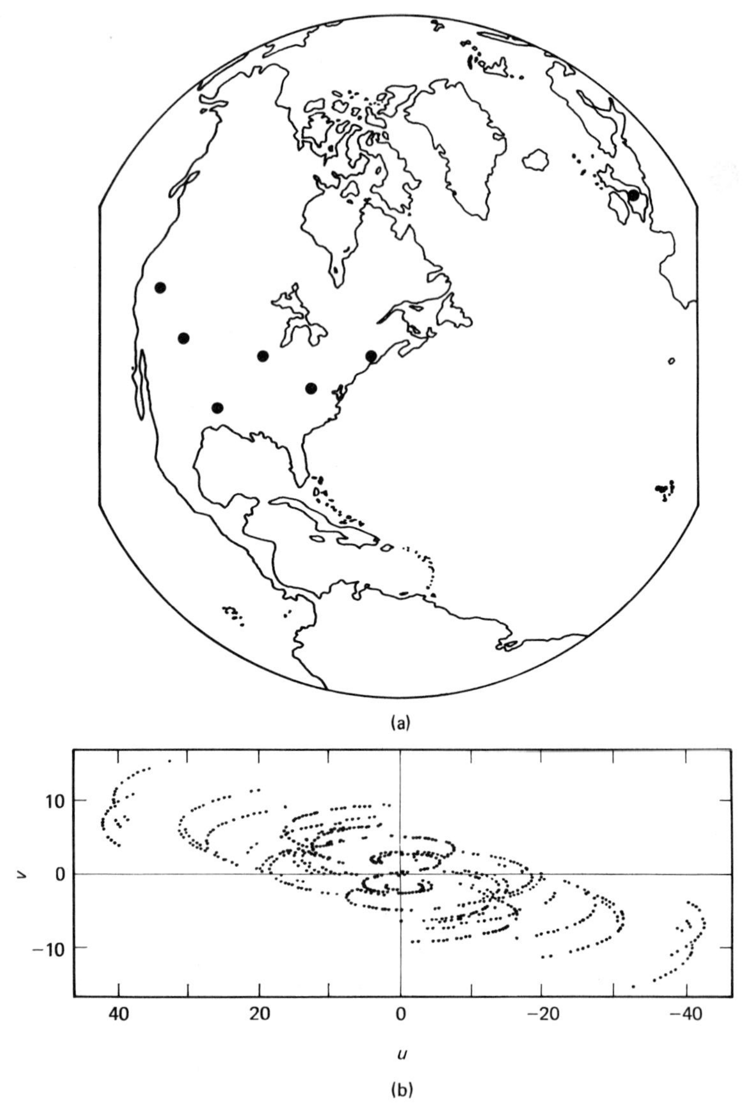

Figure 5.16 VLBI example: (a) array configuration; (b) (u, v)-plane baseline tracks. [By permission of the *Astrophysical Journal* (Vol. 263, p. 615, 1982). From Reid et al. "VLBI Observations of the Nucleus and Jet of M87."]

the origin. On the other hand, with ± 3 hours of tracking, a large central wedge is left out, as shown in the same figure. As a result, the thickness of the volume increases with increasing distance from the origin. To compensate for the linear increase of thickness from the center, the square of the element density along each arm should be inversely proportional to the radial distance. This means that the elements should be located in a square-law fashion along the arms [31]. However, to accommodate the decrease of thickness from origin for ± 6 hours of tracking, a reduction of the power law from 2 to 1.6 to 1.8 is an acceptable compromise.

Based on the configurations above, the geometry of the very large array (VLA) mentioned above is chosen. It is a Y array of 27 elements which can be positioned by means of railway tracks in four configurations of dimensions 35, 11, 3.15, and 1 km, with a scale factor of 3.285 between adjacent configurations. The elements on each arm are distributed according to the power-law relation $m^{1.716}$, where m is the element number. The power 1.716 is chosen so that element m on any configuration coincides with element $2m$ in the next smaller configuration because $2^{1.716} = 3.285$ is the scale factor. At the minimum observing wavelength of 1.3 cm, the maximum dimension of the array is 2.7×10^6 wavelengths. Figure 5.15 shows the configuration and some (u, v)-plane tracks of the VLA.

The highest-resolution astronomical studies are provided by VLBI, where telescopes distributed over the entire earth are used to form arrays with dimensions of thousands of kilometers up to 10^9 wavelengths. Quite often such arrays make use of existing large telescopes built for other purposes and tend to be distributed in irregular configurations. Figure 5.16 shows an example using telescopes located on two continents and the corresponding (u, v) coverage. For these arrays, the horizons of the extreme elements often limit the tracking times of the longer baselines to a few hours.

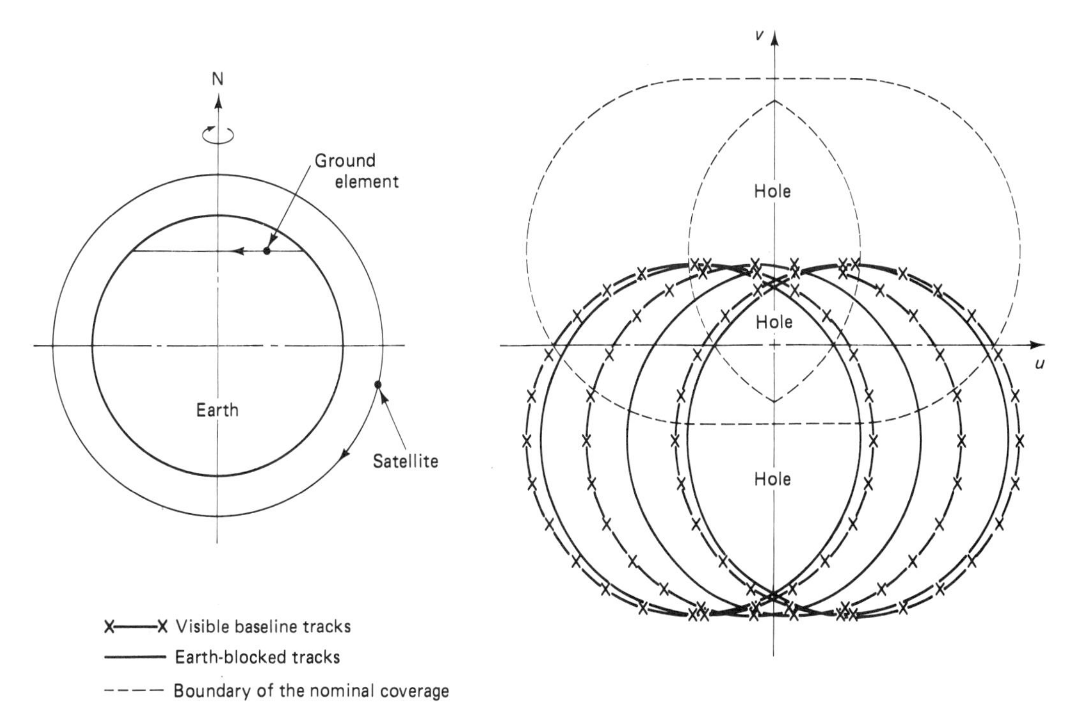

Figure 5.17 Satellite-to-earth (u, v)-plane baseline tracks.

As a final example, we consider an earth–space baseline. Unlike an earth-based array, the two elements of an earth–space interferometer need not have the same axes or rate of rotation. As a result, the sweeping of the baseline space is due to two rotations and the (u, v)-plane projection depends not only on the declination of the source but also on its right ascention. Much more considerations are therefore necessary before the selection of a satellite orbit and earth element location can be made. Figure 5.17 shows the (u, v)-plane coverage for a polar satellite orbit and a source declination of 0°. Note the presence of holes in the baseline tracks.

5.5 IMAGE RECONSTRUCTION BY FOURIER INVERSION

During an observation interval a synthesis array estimates the cross-correlation $\hat{R}_{jk}(\tau_w)$ or the cross-spectral density $\hat{S}_{jk}(f_{\mathrm{IF}_w})$ of each element pair j, k as a function of the instantaneous baseline $\mathbf{u} = \mathbf{r}_j - \mathbf{r}_k$. As the baselines are swept by earth's rotation, samples of $\hat{R}$ or $\hat{S}$ are obtained. The sample points can be either on a surface or in a volume as discussed in Section 5.4, and usually distributed nonuniformly within an irregular boundary. The estimated sample values are effected further by propagation through the changing trophosphere and ionosphere as well as system instabilities. The task is therefore to reconstruct an image such that when substituted into Eqs. (5.16) and (5.15), the resulting $\hat{R}_{jk}(\tau_w)$ or $\hat{S}_{jk}(f_{\mathrm{IF}_w})$ agrees with the measured samples to within the accuracy of the data. In this section we discuss image reconstruction by *Fourier inversion.* For simplicity, all elements of the array are assumed identical so that Eqs. (5.17) and (5.18) are valid. The beamwidths of the array elements are assumed to be wide enough so that the field pattern $l_j(f_w, \mathbf{s})$ can be replaced by a constant. For a spectral line source it is required to reconstruct an image for each frequency sample from the cross-spectral density $S_{jk}(f_{\mathrm{IF}_w})$ as given in Eq. (5.17). For a continuum source it is further assumed that the field of view is small enough so that the residual delay of Eq. (5.21) is everywhere negligible in comparison with the inverse bandwidth $1/\Delta f$ of observation. An image is to be reconstructed from the cross-correlation at zero delay $R_{jk}(0)$ as given in Eq. (5.19) with the sinc function omitted. For both cases we can write

$$V(u, v, w) = \int I(l, m) \exp\left\{ j2\pi \frac{f_w}{c} [lu + mv + (n-1)w] \right\} dl\, dm \tag{5.22}$$

Here we have replaced $\hat{R}_{jk}(0)$ and $\hat{S}_{jk}(f_{\mathrm{IF}_w})$ by $V(u, v, w)$, commonly known as the *complex visibility*. Note that u, v, w are the components of baseline $\mathbf{u} = \mathbf{r}_j - \mathbf{r}_k$ in the source-fixed coordinates, and the band-center frequency $f_{0_w} + f_{\mathrm{IF}_{w0}}$ is written as f_w. The visibility $V(u, v, w)$ is therefore the integral of the source spectral density $I(l, m)$ weighted by a phase shift due to the residual delay $\mathbf{s} \cdot \mathbf{u}/c$.

A simple and naive way of looking at image reconstruction proceeds as follows. Equation (5.22) states that for every baseline (u, v, w), each element of a source distribution $I(l, m)\, dl\, dm$ contributes to the complex visibility $V(u, v, w)$ by an amount proportional to its brightness $I(l, m)$ weighted by the residual delay phase

$\mathbf{s} \cdot \mathbf{u}/c$. If there were only one source element present, its intensity is proportional to the sum of the measured visibilities deweighted by the residual delay phase, regardless of where the samples in baseline space are taken. For a distributed source, the sum

$$I'(l, m) = \sum_{\text{samples}} V(u, v, w) \exp\left\{-j2\pi \frac{f_w}{c}[lu + mv + (n-1)w]\right\}$$

should resemble the source if the samples are well distributed. This result can, of course, be derived somewhat more rigorously and we proceed as follows.

Image reconstruction by Fourier inversion is most simple when the field of view is so small that the contribution to the residual delay by the baseline component along the reference direction [i.e., $(n-1)w$] can be neglected. In this case, Eq. (5.22) becomes

$$V(u, v) = \int I(l, m) \exp\left[j2\pi \frac{f_w}{c}(lu + mv)\right] dl\, dm \tag{5.23}$$

Since only the projection of baseline distribution on the (u, v)-plane normal to the reference direction is involved, we have omitted the dependence on w in the visibility. The reconstruction of an image, $I(l, m)$, from samples of visibility $V(u, v)$, with the two forming a *two-dimensional Fourier transform pair*, is the familiar problem of sampling of bandwidth-limited signals. Much is known about the reconstruction process if the sampling is periodic.

Unfortunately, the sampling of baseline space by a rotation synthesis array is nonuniform. Let i denote the instantaneous baseline number, h the sampling time, and let $(u, v) = (u_{hi}, v_{hi})$ be the position where a sample of visibility is taken. As discussed in Section 5.4, the baseline tracks of synthesis arrays are circles perpendicular to and centered on the earth's axis. The circles are often incomplete because of blocking by the earth. When sampled at equal time intervals, the shorter baselines close to the origin will always have denser sample points than the longer ones. The projection (u_{hi}, v_{hi}) of sample points with respect to a reference direction are therefore nonuniformly distributed with varying density inside an irregular boundary, all of which depend on the source declination. Even in the most regular configuration, namely, the east-west array with uniformly spaced baselines, the sample points are projected onto equispaced ellipses with density decreasing from the baseline origin. The well-known theory of reconstruction under periodic sampling no longer applies. If the sample-point distribution is fairly regularly distributed, a simple image reconstruction is the direct inversion of Eq. (5.23) on the sampled values

$$I'(l, m) = \sum_{h, i} W_{hi} V(u_{hi}, v_{hi}) \exp\left[-j2\pi \frac{f_w}{c}(lu_{hi} + mv_{hi})\right] \tag{5.24}$$

where W_{hi} is a set of weights. The visibility $V(u, v)$ at unsampled positions are set to zero. The reconstructed image is now a smeared version of the true image whose quality can be described by the image of a discrete source located at the reference direction. For such a source, the visibility $V(u, v)$ is a constant in the entire (u, v)-

plane, and the reconstructed image is defined by

$$B(l, m) = \sum_{h, i} W_{hi} \exp\left[-j2\pi \frac{f_w}{c}(lu_{hi} + mv_{hi})\right] \tag{5.25}$$

The quantity $B(l, m)$ is known as the *synthesized beam*, whose shape depends on the position of sample points (u_{hi}, v_{hi}) and the weights W_{hi}. Since $V(u, v)$ is the Fourier transform of the true image $I(l, m)$, the reconstructed image $I'(l, m)$ can be expressed as the convolution integral

$$I'(l, m) = \iint I(l', m')B(l - l', m - m')\, dl'\, dm' \tag{5.26}$$

This reconstruction is usually called the *principal solution* [36]. If $B(l, m)$ has a well-defined main beam with small sidelobes, $I'(l, m)$ is then a smoothed version resembling the true image $I(l, m)$. With poorly distributed sample points inside an irregular boundary having large holes, high sidelobes with peculiar structures occur both near and far from the main beam. Not only would the source in the desired field of view smear out into an irregular shape, sources outside the field may also appear inside to confuse the desired image. Figure 5.18 shows some examples of baseline track distributions and the synthesized beams. Further processing must therefore be performed to obtain a more reasonable map, as discussed below.

To reconstruct an image of N^2 *map points* or *pixels* (picture elements), the number of baseline plane samples available must be greater than N^2. To evaluate the direct Fourier transform of Eq. (5.24), we require at least N^4 complex multiplications. For N on the order of 10^3, this number becomes prohibitive. Means to reduce the computational burden must therefore be introduced. For an east-west array with uniformly spaced baselines, one possibility is to evaluate the transform for each sampling time h (i.e., over a radial line). Since the baselines are uniformly distributed, the fast Fourier transform algorithm can be used. This process is equivalent to mapping the source by fan beams in different orientations [37], involving the well-known *projection slice theorem.* However, even with this saving the computation is usually still excessive. Therefore, regardless of the array configuration, fast algorithms in evaluating the two-dimensional Fourier transform should be used. To accomplish this, sample values lying on a regular lattice on the (u, v)-plane are required. The observed values of $V(u, v)$ at nonuniformly distributed sample points must therefore first be weighted, interpolated, smoothed, and then resampled on the lattice or grid in a process commonly known as *gridding.*

A simple way to derive a set of samples on a regular lattice from the set of weighted nonuniform samples of observation $W_{hi} V(u_{hi}, v_{hi})$ is to convolve the discrete nonuniform samples with a *smoothing function* $c(u, v)$ and then resample at the desired lattice [38]. After convolution a continuous visibility is again obtained, as shown by

$$V'(u, v) = \sum_h \sum_i W_{hi} V(u_{hi}, v_{hi})c(u - u_{hi}, v - v_{hi}) \tag{5.27}$$

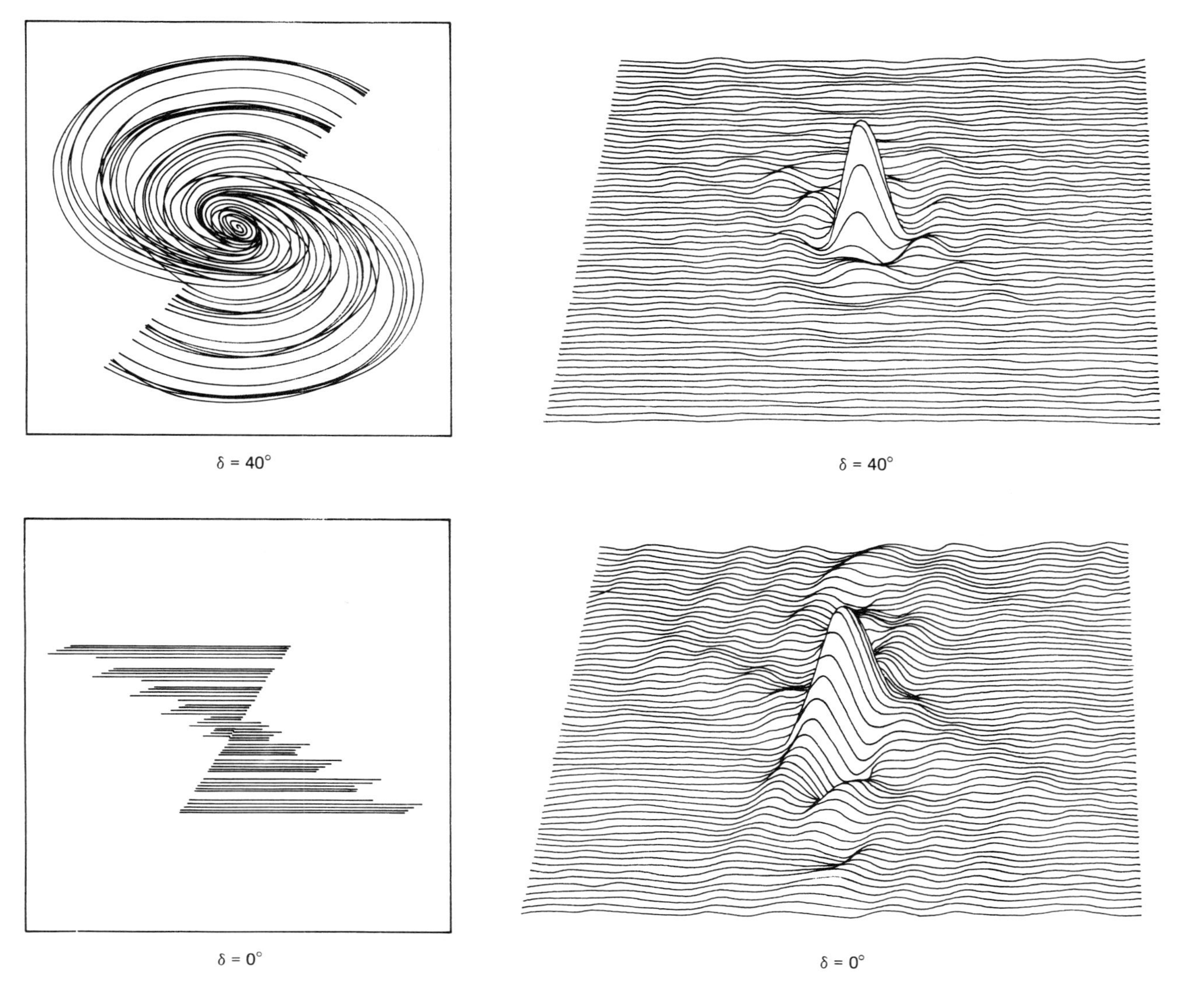

Figure 5.18 Baseline tracks (a) and synthesized beam (b) for different source declinations. [By permission of the National Radio Astronomy Observatory.]

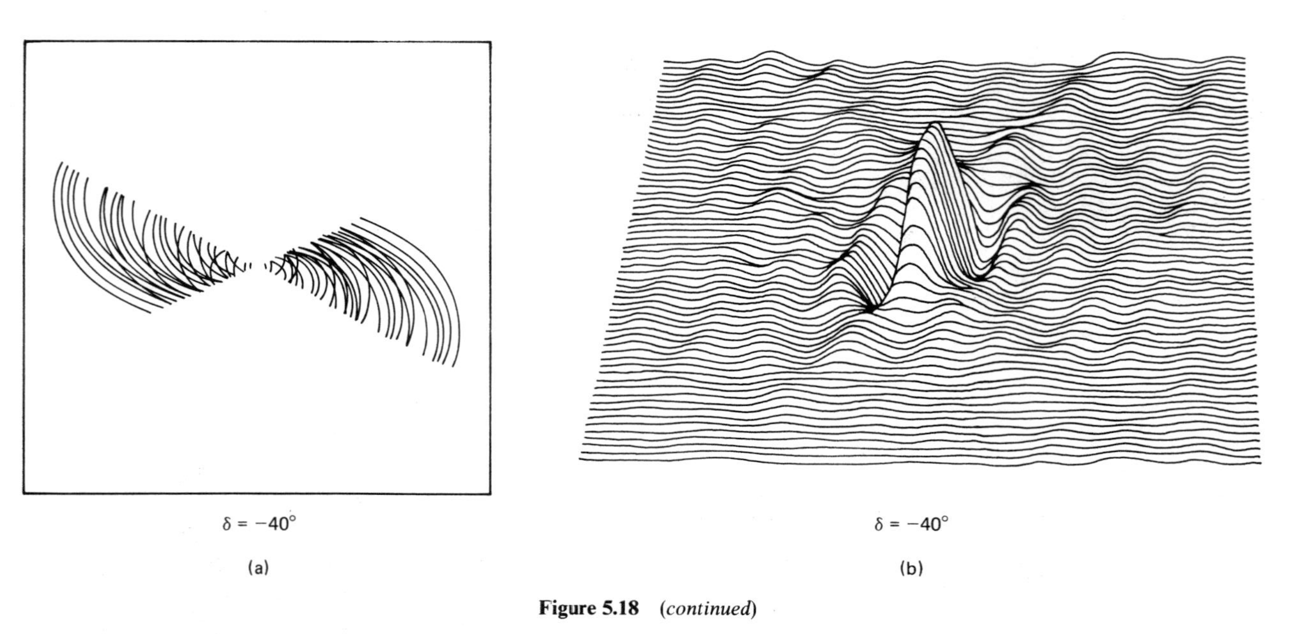

$\delta = -40°$

(a)

$\delta = -40°$

(b)

Figure 5.18 *(continued)*

Let $C(l, m)$ be the Fourier transform of the smoothing function $c(u, v)$; the Fourier transform of $V'(u, v)$ is then given by

$$I''(l, m) = I'(l, m)C(l, m)$$

In order that the gridding operation does not further distort the reconstructed image, $C(l, m)$ should be unity for l, m in the field of view. Furthermore, $C(l, m)$ should be zero outside the field of view so that resampling of $I''(l, m)$ on the regular lattice has nothing to aliase into the field of view. Unfortunately, these requirements imply a sinc-type smoothing function $c(u, v)$ which has a narrow main peak but remains significant over a large portion of the (u, v)-plane. For computational reasons, $c(u, v)$ must have a small finite support, otherwise the large number of multiplications required for convolution would defeat the purpose of gridding. The result is a smoothing function which deweights somewhat the edge of the field and allows a small amount of aliasing due to gridding. The deweighting can be corrected by weighting the reconstructed image in the field of view by $C(l, m)^{-1}$, but aliasing cannot be removed.

Let the resampling of $V'(u, v)$ of Eq. (5.27) on the baseline plane $\mathbf{u}$ be on a lattice characterized by the generators $\mathbf{a}$, $\mathbf{b}$ such that sample $V_{\mu\nu}$ is taken at $\mathbf{u}_{\mu\nu} = \mu\mathbf{a} + \nu\mathbf{b}$ with μ, ν being integers between 0 and $N - 1$. The simplest smoothing function $C(u, v)$ is a pillbox, as shown in Fig. 5.19, with unit value in a cell defined by the *Dirichlet region* [39] of the lattice, that is, a polygon consisting of all points nearer to the origin than to any other lattice point, and zero elsewhere. The convolution of Eq. (5.27) becomes simply equal to the summing of all samples within the cell. However, the transform of a pillbox is a sinc-type function in the image plane, which does not decrease rapidly, thereby introducing strong aliasing. Other smoothing functions are therefore preferred, such as the truncated Gaussian and some two-dimensional version of *prolate spheroidal wave functions*. The latter is capable of giving minimum fractional energy of $C(l, m)$ outside the field of view.

The weighting W_{hi} in Eq. (5.27) provides control of the image character. It is usually expressed as a product, $d_{hi} t_{hi}$, where d_{hi} corrects for the varying number of observed samples in each resampling cell and t_{hi} introduces a taper to reduce the side lobes. Two extremes of density corrections are often used. The first, known as *natural weighting*, weights all observed samples equally (i.e., $d_{hi} = 1$). In this case the weight of each resampled visibility is proportional to the number of observed visibilities contributing to that sample. Since the density of observed samples is always higher for short baselines, a broad plateau under the main beam is introduced. To remove the broad plateau, we may choose d_{hi} equal to the inverse of the number of observed samples contributing to the lattice point. This is known as *uniform weighting*, and the beam is less dependent on the observed sample point distribution. However, since not all observed visibilities are equally weighted, there will be a degradation in signal-to-noise ratio. If the observed samples are sparsely distributed as in a "snapshot" observation, the number of observed samples within a number of cells surrounding the lattice point can be used to determine d_{hi}, leading to a beam more dependent on the taper t_{hi}. As for the taper t_{hi}, a convenient choice would be a truncated Gaussian function. This, of course, reduces the sidelobe at the expense of a broader main beam. All these parameters must be tailored to the specific image to be reconstructed and the data set available.

To convert the direct Fourier transform of Eq. (5.24) into a discrete Fourier transform, we begin with the resampled visibility at the lattice points $\mathbf{u}_{\mu\nu} = \mu\mathbf{a} + \nu\mathbf{b}$ given by

$$V_{\mu\nu} = \sum W_{hi} V(u_{hi}, v_{hi}) C(u_{\mu\nu} - U_{hi}, V_{\mu\nu} - V_{hi})$$

where $u_{\mu\nu}$ and $v_{\mu\nu}$ are the u, v components of $\mathbf{u}_{\mu\nu}$. In the image plane, $\mathbf{s}$, we introduce a reciprocal lattice with generators $\mathbf{a}'$ and $\mathbf{b}'$ defined by

$$N \frac{f_w}{c} \mathbf{a}' \cdot \mathbf{a} = N \frac{f_w}{c} \mathbf{b}' \cdot \mathbf{b} = 1, \qquad \mathbf{a} \cdot \mathbf{b}' = \mathbf{a}' \cdot \mathbf{b} = 0$$

such that a point on the image plane can be expressed as $\mathbf{s} = \alpha\mathbf{a}' + \beta\mathbf{b}'$. The quantity N^2 is both the number of pixels in the field of view and the number of resampled visibilities. The direct Fourier transform of Eq. (5.23) becomes

$$I_{\alpha\beta} = \sum_{\mu,\,\nu=0}^{N-1} V_{\mu\nu} \exp\left[-\frac{j2\pi}{N}(\alpha\mu + \beta\nu)\right] \tag{5.28}$$

Here we have assumed the sample points to extend from $\mu, \nu = 0$ to $\mu, \nu = N - 1$. For integer α, β, Eq. (5.28) is a two-dimensional discrete Fourier transform. It is periodic in both α and β, with period N representing image-plane aliasing in replication on the lattice $N\mathbf{a}'$, $N\mathbf{b}'$. Furthermore, the phase factor is periodic in both μ and ν with period N, so that the domain of summation in μ, ν can be translated to a Dirichlet domain for the lattice $N\mathbf{a}$, $N\mathbf{b}$ about the baseline plane origin. Although the discrete Fourier transform over a rectangular lattice is commonly used, the less familiar hexagonal lattice [40] shown in Fig. 5.19 offers some advantages for nearly

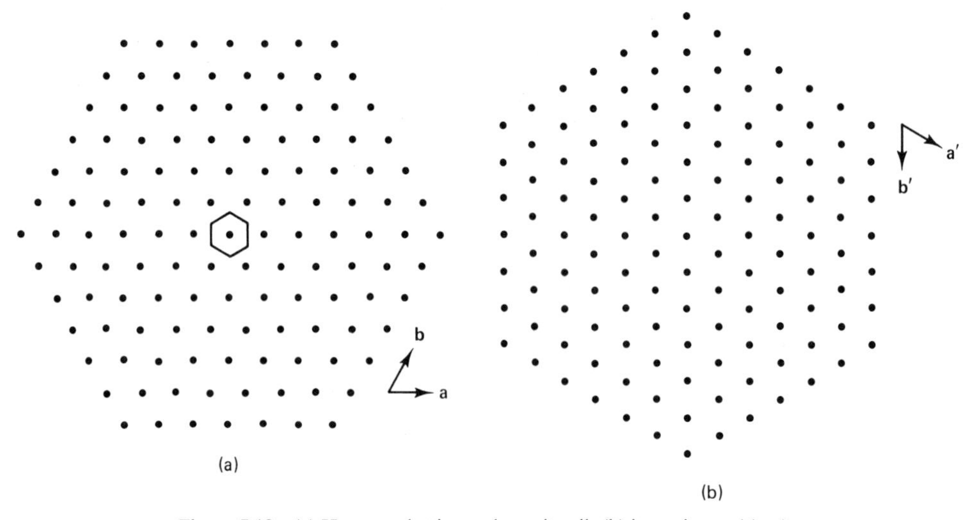

Figure 5.19 (a) Hexagon lattice and a unit cell; (b) its reciprocal lattice.

circular images. To evaluate a discrete Fourier transform of N^2 pixels, $2N^2 \log_r N$ multiplications are required, using an FFT algorithm with radix r. For large N, this represents a several-order-of-magnitude saving in computing effort.

We now turn to the problem of wide-field observations, where the contribution $(n - 1)w$ of the baseline component in the reference direction w to the residual delay $\mathbf{s} \cdot \mathbf{u}/c$ in Eq. (5.21) can no longer be neglected. In the case where the baselines are coplanar and passing through the origin, so that

$$w = au + bv$$

a simple transformation of the sky coordinates [15]

$$l' = l + a(n - 1), \qquad m' = m + a(n - 1)$$

can be used to eliminate the factor depending on w. This happens with an east-west array or for a snapshot observation of an array of arbitrary configuration covering an area where the earth's curvature can be ignored. On the sampling plane above, Eq. (5.21) becomes

$$V(u, v) = \int I(l, m) \exp\left\{j2\pi \frac{f_w}{c} [l + a(n - 1)]u + [m + b(n - 1)]v\right\} dl\, dm$$

Since u, v completely specifies the position of a sample point, the w coordinate of the sample point is omitted. On introducing the new variables l', m' and transforming the variables of integration from l, m to l', m', the relation above becomes a two-dimensional Fourier transform. Direct Fourier inversion on the sample values yields

$$I(l, m) = \left(1 - \frac{al + bm}{n}\right) \sum_{h, i} V(u_{hi}, v_{hi}) \exp\left\{-j2\pi \frac{f_w}{c} [l'u_{hi} + m'v_{hi}]\right\}$$

The first factor in the right-hand side comes from the transformation of integration. The direct Fourier transform of the observed visibilities therefore gives the image in a distorted coordinate system. Since $n - 1 \approx -\frac{1}{2}(l^2 + m^2)$, the distorted sky coordinates l', m' need only be introduced when the field of view is large enough, so that $(l^2 + m^2) f_w w/c$ is comparable to unity.

When the baseline samples do not lie on a plane, Eq. (5.21) can be expressed as a three-dimensional Fourier transform by introducing the dispersion relation for electromagnetic waves in free space, which is equivalent to $\mathbf{n}$ being a unit vector; we may thus write

$$V(u, v, w) = \int I(l, m)\, \delta(n - \sqrt{1 - l^2 - m^2}) \exp\left\{j2\pi \frac{f_w}{c} [lu + mv + (n - 1)w]\right\} dl\, dm\, dn$$

Direct Fourier inversion and then integration over n gives

$$I(l, m) = \iint V(u, v, w) \exp\left\{-j2\pi \frac{f_w}{c} [lu + mv + (n - 1)w]\right\} du\, dv\, dw\, dn$$

Since the integration in n is to extract the value at $n = \sqrt{1 - l^2 - m^2}$, the direct Fourier inversion from the observed samples of $V(u_{hi}, v_{hi}, w_{hi})$ can be taken as

$$I'(l, m) = \sum_{h,i} V(u_{hi}, v_{hi}, w_{hi}) \exp\left\{-j2\pi \frac{f_w}{c} [lu_{hi} + mv_{hi} + \sqrt{1 - l^2 - m^2}\, w_{hi}]\right\} \tag{5.29}$$

This result is anticipated from our previous simple physical reasoning. The presence of the w_{hi} term complicates the evaluation of the direct Fourier transform. It is necessary to evaluate the transform with respect to w_{hi} first before conversion into a two-dimensional discrete Fourier transform. A further interpretation of the problem can be found in [41].

For continuum observations it is often necessary to use large bandwidths to achieve high sensitivity. When the observation bandwidth is increased, the residual delay $\mathbf{s} \cdot \mathbf{u}/c$ from the edge of a wide field of view may become comparable to the inverse bandwidth $1/\Delta f$. This delay error introduces a decorrelation between signals received at the two ends of a baseline. The result is a deweighting of contributions to the observed cross-correlation $R_{jk}(0)$ from the outer region of the field as given by the sinc function of Eq. (5.19). The field of view is said to be greater than the delay beam width. When the broadband visibility is used in Eq. (5.24) to reconstruct an image, radial smearing will occur. To investigate this in more detail, we rewrite the broadband visibility of Eq. (5.19) in the form of Eq. (5.23), so that

$$V(u, v) = \int I(l, m) \frac{\sin (\pi \Delta f/c)(lu + mv)}{(\pi/c)(lu + mv)} \exp\left[j2\pi \frac{f_w}{c} (lu + mv)\right] dl\, dm \tag{5.30}$$

This relation is no longer a Fourier transform. However, an approximate image can be obtained by the inverse transform of $V(u, v)$ just as in the narrowband narrow-field case. If the visibility is sampled over the entire baseline plane, we have

$$I'(l, m) = \int V(u, v) \exp\left[-j2\pi \frac{f_w}{c} (lu + mv)\right] du\, dv$$

Using Eq. (5.30), this relation can be written in terms of a position-dependent synthesized beam according to

$$I'(l, m) = \int I(l', m')B(l, m; l', m)\, dl'\, dm'$$

where

$$B(l, m; l', m) = \int \frac{\sin [\pi(\Delta f/c)](l'u + m'v)}{(\pi/c)(l'u + m'v)} \exp\left\{-j2\pi \frac{f_w}{c} [(l' - l)u + (m' - m)v]\right\} du\, dv \tag{5.31}$$

is the synthesized beam with main beam direction located at (l', m'). In view of the radial symmetry of Eq. (5.31), the synthesized beam depends on the radial position only. Consider the beam located at $(l', m') = (l', 0)$. Then, Eq. (5.31) assumes the form

$$B(l, m; l', 0) = \int \frac{\sin [\pi(\Delta f/c)]l'u}{(\pi/c)l'u} \exp \left\{-j2\pi \frac{f_w}{c} [(l' - l)u - mv)]\right\} du\, dv$$

$$= \begin{cases} \frac{1}{l'} \delta(m), & \left(1 + \frac{\Delta f}{2f_w}\right)l' > l > \left(1 - \frac{\Delta f}{2f_w}\right)l' \\ 0 & \text{otherwise} \end{cases}$$

The beam is therefore smeared radially by an amount proportional to the distance from the center of field, with a peak value inversely proportional to the same distance. In interpreting images reconstructed from direct Fourier inversion of broadband data sample over a finite baseline plane, the effects of radial smearing must be recognized.

To avoid radial smearing, a different approach to image reconstruction from broadband observations is required. Instead of considering the spectral density $I(f, \mathbf{s})$, let us consider the source self-coherence $I(\tau, \mathbf{s})$ whose value at $\tau = 0$ gives the total brightness of the source. According to Eq. (5.12), $I(\tau, \mathbf{s})$ is given by the mutual coherence of the wave field $\Gamma(\tau, \mathbf{u})$ through a simple integral relation. If we assume the bandwidth of emission from the source to be that given by the receiver transfer function $U(f_{\mathrm{IF}_w})$ and that the antenna field pattern is wide, then, from Eqs. (5.18) and (5.6), the observed cross-correlation $R(\tau, \mathbf{u})$ for baseline $\mathbf{u}$ is related to the mutual coherence $\Gamma(\tau, \mathbf{u})$ by a frequency conversion with the local oscillator frequency f_{0_w}, as shown by

$$\Gamma(\tau, \mathbf{u}) = R(\tau, \mathbf{u}) \exp (j2\pi f_{0_w} \tau)$$

In terms of the sky components l, m and baseline components u, v, Eq. (5.12) becomes

$$I(\tau; l, m) = 2 \int \left[f_{0_w} R\left(\tau - \frac{lu + mv}{c}; u, v\right) + \frac{1}{j2\pi} \frac{d}{d\tau} R\left(\tau - \frac{lu + mv}{c}; u, v\right)\right] \times \left[\exp j2\pi f_{0_w}\left(\tau - \frac{lu + mv}{c}\right)\right] \frac{du\, dv}{c}$$

Using the observed samples $R(\tau; u_{hi}, v_{hi})$, the reconstructed total brightness of the source is therefore

$$I'(0; l, m) = 2 \sum_{h, i} \left[\frac{f_{0_w}}{c} R\left(-\frac{lu_{hi} + mv_{hi}}{c}; u_{hi}, v_{hi}\right) + \frac{1}{j2\pi c} \frac{d}{d\tau} R\left(-\frac{lu_{hi} + mv_{hi}}{c}; u_{hi}, v_{hi}\right)\right] \times \exp \left[-j2\pi\left(\frac{f_{0_w}}{c} (lu_{hi} + mv_{hi}\right)\right] \tag{5.32}$$

For every image point, the observed cross-correlation $R(\tau; u_{hi}, v_{hi})$ at discrete samples of τ must first be interpolated to correct for the residual delay $\tau = -(1/c)(lu_{hi} + mv_{hi})$ and then substituted into Eq. (5.32). However, since the residual delay τ varies slowly in the (u, v)-plane, each interpolation can be applied to a finite region in the (u, v)-plane. Excluding interpolation, the number of multiplications required to reconstruct an image from Eq. (5.32) is the same as the direct Fourier transform. However, fast algorithms cannot be directly applied here by gridding. Note that Eq. (5.32) is similar to Eq. (5.24) if the visibility is taken at a delay appropriate for both the image point and the baseline location. The $d/d\tau$ term in Eq. (5.32) partly accounts for the difference between the local oscillator frequency f_{0_w} in Eq. (5.32) and the band-center frequency f_w in Eq. (5.24). Equation (5.12) from which Eq. (5.32) is derived, is obtained from a two-dimensional Fourier transform relation. Equation (5.32) is therefore valid only when the w-component of the baseline can be neglected. However, because of the simple integrand of Eq. (5.12) it is reasonable to believe the direct reconstruction of Eq. (5.32) can be modified to take into account the w-component just as in Eq. (5.29). Thus, for an arbitrary bandwidth and field of view, an image can be reconstructed according to

$$I'(0; l, m) = 2 \sum_{h,i} \left[\frac{f_w}{c} R\left(-\frac{lu_{hi} + mv_{hi} + (n-1)w_{hi}}{c}; u_{hi}, v_{hi}, w_{hi} \right) + \frac{1}{-j2\pi c} \frac{d}{d\tau} R\left(-\frac{lu_{hi} + mv_{hi} + (n-1)w_{hi}}{c}; u_{hi}, v_{hi}, w_{hi} \right) \right] \times \exp\left\{ -j2\pi \frac{f_{0_w}}{c} (lu_{hi} + mv_{hi} + (n-1)w_{hi}) \right\}$$

From the discussions above, we see that direct reconstruction of an image from visibility data depends on many factors, some of which are image dependent, while many are instrument dependent. For this reason, detailed information about a particular instrument such as that for the VLA [42] is required before analysis can proceed.

5.6 IMAGE RESTORATION BY THE METHOD "CLEAN"

Direct Fourier inversion of observed visibility samples gives the principal solution of the image with all unsampled visibilities set to zero. The quality of the resulting image depends entirely on sampling in the baseline space. In many situations, the sampling is nonuniformly distributed within an irregular boundary, sometimes having large holes. The reconstructed principal solution image will show complex structures resulting from these irregularities. Figure 5.20 shows a poorly sampled (u, v)-plane and the synthesized beam. Clearly, the true image cannot be so complex that the visibility vanishes at all positions that happened to be not sampled by the observation. There must be image components invisible to the instrument with nonvanishing visibilities at the unsampled positions. Proper astrophysical interpre-

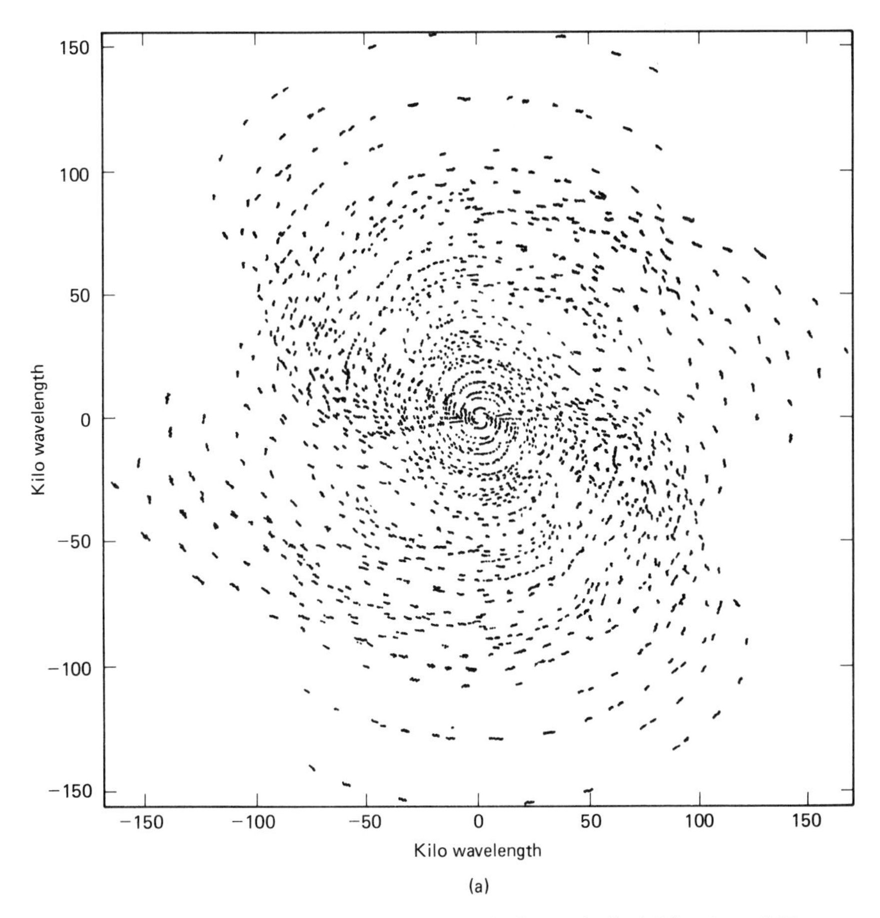

Figure 5.20 (a) Sparsely sampled (u, v)-plane; (b) the synthesized "dirty beam." [By permission of P. P. Kromberg and J. W. Dreher.]

tation requires images of high dynamic range which can only be achieved with beams of low sidelobes. It is therefore necessary to interpolate the visibilities over an entire region of the baseline plane so as to augment the principal solution with invisible image components in order to obtain an astronomically plausible image. One possible way is to interpolate the observed visibility over the entire observed region by some analytic method such as expansion in a series of spheroidal functions [43] before Fourier inversion. However, the computational burden appears to be excessive for images of a large number of pixels. A simpler method to operate in the image plane is to assume a certain *model* for the image with a set of parameters such as position, strength, and shape of components and fit the model to the observed visibility. If the map is to be the total brightness of a source, the positivity constraint can also be used. The most widely used algorithm for image restoration so far is an iterative method known as "CLEAN," introduced by Högbom [17], the

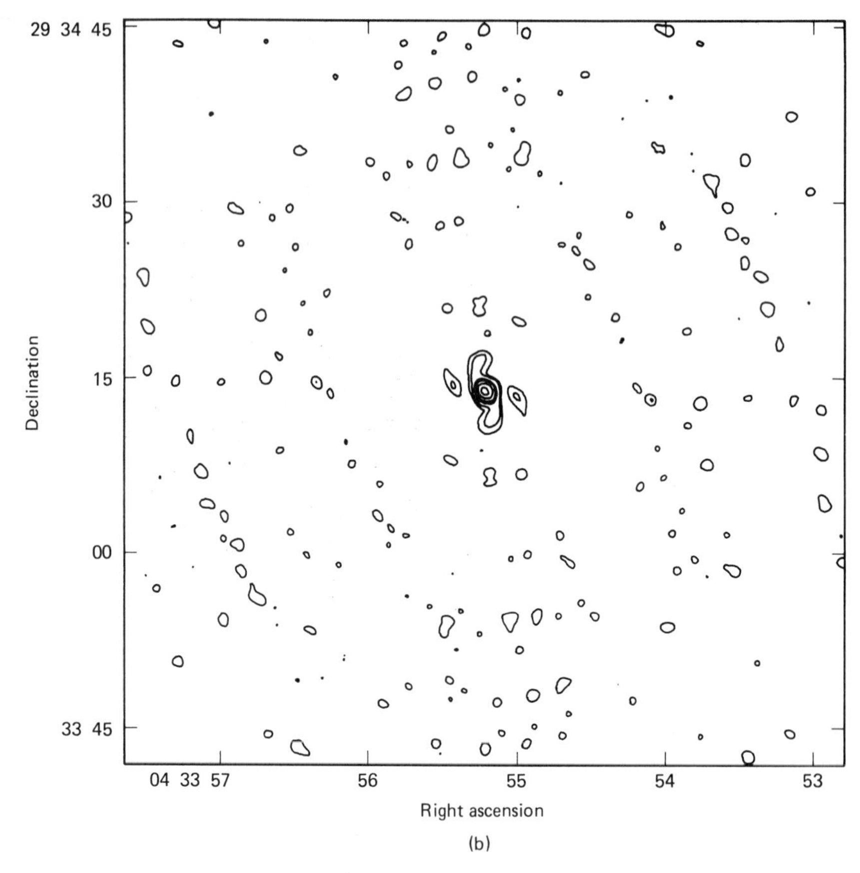

Figure 5.20 *(continued)*

details of which will be given below. Another promising approach based on maximum entropy [18] will be discussed in Section 5.8.

The CLEAN algorithm is based on the a priori assumption that most images of radio sources are essentially blank sky with point-source components distributed in small regions. The positions and strengths of these point-source components are determined by a simple iterative produce. On convolving the point-source components with a "clean" beam, a "clean" map is derived. Thus the CLEAN algorithm is a nonlinear iterative method of deconvolving the synthesized beam $B(l, m)$, also known as the "dirty" beam, from the principal solution $I'(l, m)$, or the "dirty" map, as given in Eq. (5.26). The method is also equivalent to spectral estimation in spatial frequencies in terms of the observed visibilities. Although the method is based on a collection of small sources, it works well even for extended sources. If the image is smooth and much larger than the synthesized beam, the clean map may appear lumpy. Furthermore, the method does not depend on positivity, therefore can be applied to polarization measurements. For these reasons, CLEAN is the most widely used image restoration algorithm in synthesis arrays at the present. It is also the basis of other algorithms to remove the effects of phase and amplitude errors in visibility measurements to be described in Section 5.7.

The iterative procedure operates in the image plane and begins with a dirty map $I'(l, m)$ from which a set of point sources $P(l, m)$ is derived, such that when convolved with the dirty beam $B(l, m)$ the resulting image is almost equal to the dirty map except for a residual $\Delta(l, m)$ defined by

$$\Delta(l, m) = I'(l, m) - \iint P(l', m')B(l - l', m - m')\, dl'\, dm'$$

The procedure consists of the following steps:

1. Locate the maximum in the dirty map and determine its amplitude P.
2. Convolve the dirty beam with a point source (at this location) of amplitude γP, where γ is known as the loop gain.
3. Subtract the result of this convolution from the dirty map.
4. Repeat until the residual $\Delta(l, m)$ is sufficiently small.
5. Convolve the point source collection with an esthetically pleasing clean beam such as a Gaussian function to obtain a smooth, clean map.
6. Add the residual to the clean map.

Figure 5.21 shows various stages of the CLEAN procedure from an observation taken with sample distribution and synthesized beam as shown in Fig. 5.20. The simple image requires only few iterations. Figure 5.22 shows a more complex VLA image derived from several thousand iterations.

The major computational burden of CLEAN lies in the convolution of the point source with the dirty beam. For a large number of iterations, this task is quite considerable. A more efficient FFT-based CLEAN algorithm exploiting the capabilities of array processors has been introduced by Clark [44]. The basic idea here is to separate the operation of peak locating from that of convolution-subtract and perform thc latter on a large number of point sources simultaneously. The algorithm has a minor cycle in approximate point-source location using a truncated beam patch, and a major one in proper subtraction of a set of point sources. Using this algorithm, savings up to 10 times was achieved by Clark.

Steps 1, 2, 4, and 5 are at the control of the map maker. If we happen to know that the image lies within certain a priori known regions or boxes in the field, then we may limit the search to these boxes. If the image has compact components only, the loop gain γ can be as large as unity so as to reduce the number of iterations. This is possible because CLEAN is designed to restore images of this nature. On the other hand, if the image is extended and smooth, a small loop gain will be necessary and more iterations are required. When to stop the algorithm depends on the noise level of the residual in comparison with the clean components. Finally, the choice of the clean beam will affect the resulting clean image; it can be lumpy showing many details, real or unreal, or smooth but of less resolution. By varying these parameters,

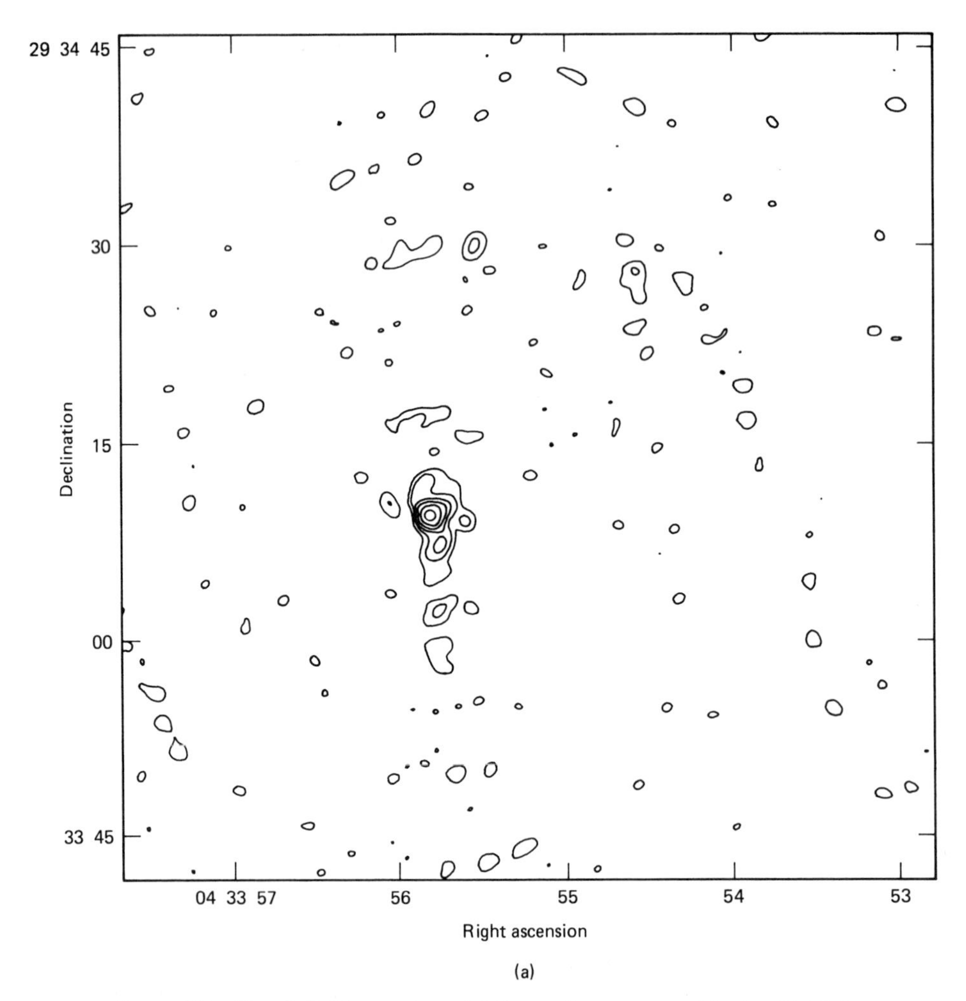

Figure 5.21 The CLEAN procedure: (a) the synthesized map; (b) after CLEAN. [By permission of P. P. Kromberg and J. W. Dreher.]

somewhat different images will result. Since the basis of the method is to interpolate or fill in unobserved visibilities, the final image is a consequence of preconceived astrophysical plausibility. Interpretation of fine details of clean maps should take cognizance of this nonuniqueness nature.

Although CLEAN is extensively used, its mathematical justification is incomplete. Schwarz [45, 46] has considered the mathematical and statistical basis of the method. The convergence toward a necessarily nonunique image can be expressed as the minimization of a norm of the residual, which is defined by

$$Q = \iiiint [I(l, m) - P(l, m)]B(l - l', m - m')[I(l', m') - P(l', m')]\, dl'\, dm'\, dl\, dm$$

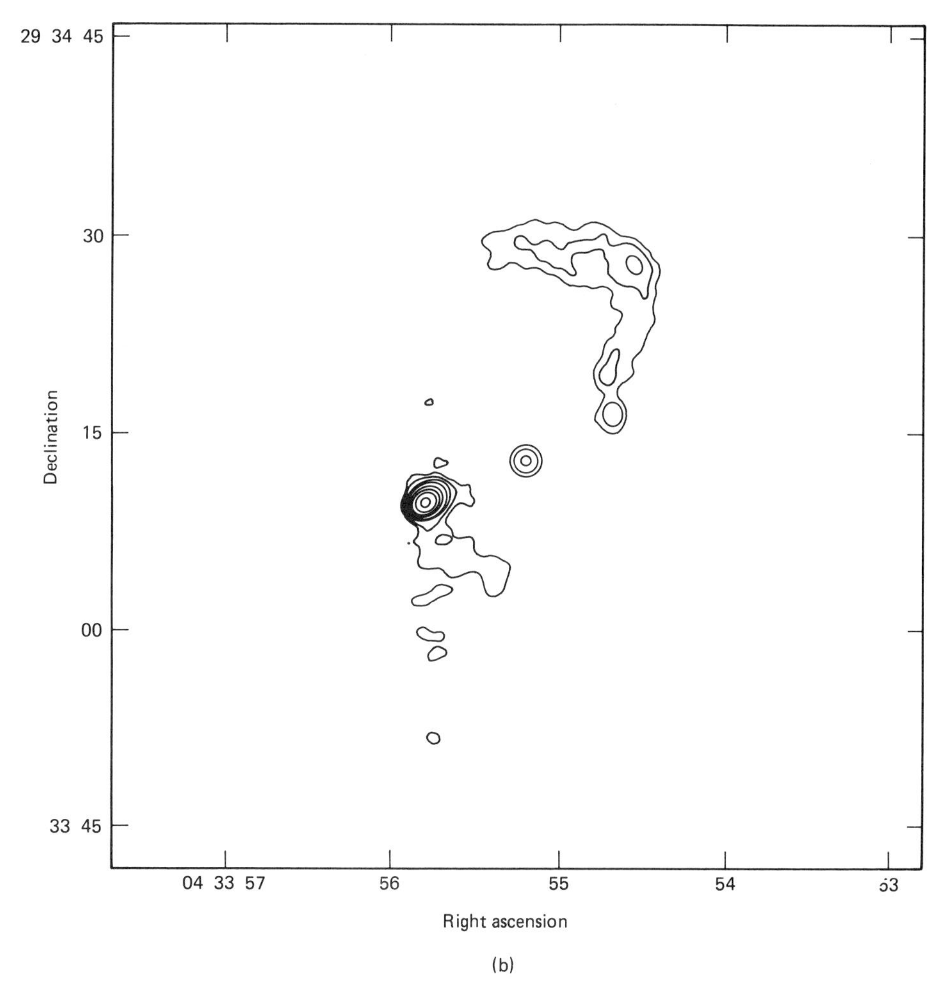

(b)

Figure 5.21 *(continued)*

Let $V_p(u_{hi}, v_{hi})$ be the visibility of the set of point sources $P(l, m)$ at the sampled locations. On using the transform relations of Eqs. (5.24) and (5.25), the norm becomes

$$Q = \sum_h \sum_i W_{hi} \,|\, V(u_{hi}, v_{hi}) - V_p(u_{hi}, v_{hi}) \,|^2$$

Here we have made use of Eq. (5.14), which can be stated as $V(u_{hi}, v_{hi}) = V^*(-u_{hi}, -v_{hi})$ for every element pair. The sum is taken over each element pair, once with real weights W_{hi}. With nonnegative weights $W_{hi} \geq 0$, the norm Q is positive and Schwarz shows that it will converge to zero. Since each point source component of $P(l, m)$ yields a spatial sine-wave component of $V_p(u_{hi}, v_{hi})$, the minimization of Q is equivalent to fitting sine waves to the observed visibilities. The spatial frequency of a

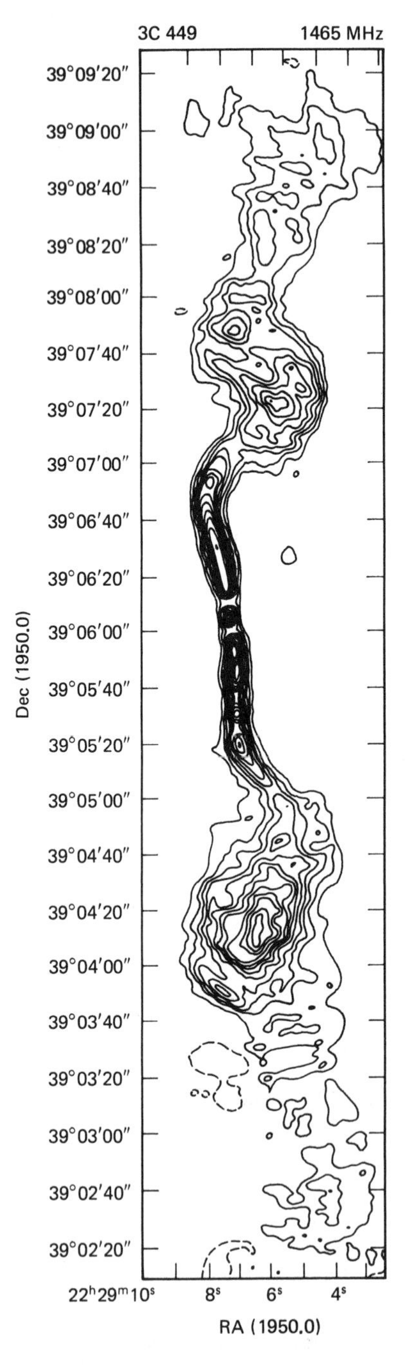

Figure 5.22 VLA map of radio source 3C 449 using CLEAN. [By permission of *Nature* (Vol. 251, p. 437, 1979). From Purley, Willis, and Scott, "Structure of the Radio Jets of 3C 449."]

sine-wave component is determined from the location of the maximum of the Fourier transform of $V(u_{hi}, v_{hi})$, namely, $I'(l, m)$. Thus CLEAN can be interpreted as a spectral estimation method.

Let us now turn to the question of the uniqueness of solution $P(l, m)$ when Q is reduced to zero. It can be shown that if the true image $I(l, m)$ consists of a collection of point sources whose number is less than half the number of the observed visibility samples, CLEAN will converge to the correct solution. Thus, if we have enough a priori knowledge of the true image, we first proceed to locate the image components; then their amplitudes are readily determined. Unfortunately, in the presence of noise, Q cannot be reduced to zero. Thus it is possible only to arrive at some sort of a most probable image. The noise and error performance of CLEAN in the image plane have yet to be properly analyzed.

A synthesis array cannot sample the origin of the baseline space because two array elements cannot be colocated. Furthermore, due to shadowing, some short spacings may also not be available. The result is a synthesized image sitting on a negative depression. Although the use of CLEAN provides interpolated values to these missing samples, incorporation of other measurements into the initial synthesized image can help the convergence considerably. However, since these values are obtained from different instruments [47, 48], they should be merged with care.

When the field of view is large enough so that the baseline sampling can no longer be assumed coplanar and the residual delay to the edge of the field can no longer be neglected, direct Fourier inversion gives a synthesized beam which is position varying. To apply CLEAN with a position-dependent beam would be computationally costly [41]. Means for easing the computation load remain to be developed.

The use of the clean beam in generating a clean map is a necessary final step. Because of the finite beam width of the synthesized beam, it is not possible to resolve image features beyond a limit. It is not likely that the true image $I(l, m)$ is the set of point sources $P(l, m)$ derived from CLEAN. In fact, for an extended image, after the first point source located at the maximum is removed, the maximum of the residual will appear at the first sidelobe because it is negative. The next point source extracted will therefore be located at that position, determined entirely by the synthesized beam. If the extended image has jets or ridges, a linear set of point sources would result. Thus it is necessary to use a clean beam having about the same main beam shape as the synthesized beam but without the offending sidelobes to remove the structure of point-source distribution inherent in CLEAN. Even then, the clean map may appear to be lumpy with knots. A heuristically chosen clean beam will in general have Fourier transform values at the observed sample points different from unity. Thus the visibility of the clean image will no longer agree with the observed visibility. The problem of how to incorporate a finite beam width in CLEAN remains unsolved. One possible approach is to accept the portion of the dirty map above some level and convolve it with the synthesized beam instead of replacing it by a set of point sources [49] and subtract. The process is then repeated until the residual becomes sufficiently low. How well the process applies to poorly sampled observations with large sidelobes in the synthesized beam remains to be seen.

5.7 PHASE AND AMPLITUDE ERRORS; SELF-CALIBRATION OF VISIBILITIES IN IMAGE RESTORATION

In the preceding two sections, image reconstruction from incomplete samples in the baseline space was discussed. The image quality depends on the signal-to-noise ratio and on how accurately the instrument can measure visibility amplitude and phase. Ultimately, all terrestrial astronomical observations are limited by the inhomogeneities of the earth's atmosphere. In principle, a well-designed and well-operated instrument can have both gain and phase instabilities minimized. One exception is in VLBI, where independent local oscillators are used. The best oscillators are stable enough in the short term for coherent integration of visibilities [50]. Their long-term wanderings, however, preclude the measurement of visibility phase, unless satellite phase links [51] can be used. Varying inhomogeneities in the water vapor content of the troposphere cause frequent disturbances in index of refraction. These fluctuations have a scale size ranging from one to tens of kilometers, persisting for minutes to hours [52, 53]. At frequencies below 1 GHz, ionospheric irregularities ranging in size from a few kilometers to 1000 km become important, in particular, the diurnal variations during sunrise and sunset [54]. These disturbances introduce phase errors in the signal arriving at the elements of a synthesis array. At frequencies above 10 to 20 GHz, liquid water in the troposphere (in particular, rain) causes high attenuation [55], introducing large amplitude errors. Since the reconstruction of high-quality images requires amplitude errors of less than a few percent and phase errors below a few degrees, elaborate means must be used to reduce the effects of the inherent instabilities.

There are three principal techniques to handle phase and amplitude errors. The first one is calibration using a nearby reference source. This, of course, needs to be performed frequently, and there must be a reference source close to the object being observed. The second approach is to measure the pertinent quantities and use these values to correct for the uncertainties. For instance, the total precipitable water vapor can be measured using a water vapor radiometer combined with surface temperature, pressure, and humidity, giving an estimation of path length due to the "wet" component of the atmosphere [56]. Using radio emissions from a satellite such as the Navstar, it should be possible to measure the total electron content from which a phase correction can be derived. Finally, by using the fact that the principal errors are caused by a signal received by an individual element while each visibility is associated with two elements, the errors in some of the observed visibilities can be removed. Thus, by appropriate signal processing, an astronomically plausible image can be obtained. We describe this approach below.

To begin with, in the noiseless case it should be possible to reconstruct, in principle, an image from the visibility amplitude only. This is because, in one dimension, the Fourier transform of an image of finite support is an entire function [57]. The squared amplitude of the visibility would have both the zeros of the visibility and their complex conjugates [58]. By invoking the positivity constraint, these zeros of the visibility should be derivable [59]. In fact, an image can be derived from the Fourier transform of a fully sampled visibility amplitude squared $|V(u, v)|^2$ by using an iterative procedure commonly employed in crystallography [60]. Similarly, image reconstruction from phase information only has been investigated, although it is not appropriate for synthesis array [61]. In view of the above, it is not surprising that quality images can be derived from incomplete, imprecise, and noisy samples of visibility.

The majority of phase and gain errors in synthesis array measurements, arising from both instrumental and propagation effects, can be lumped into uncertainties in a complex element gain $G_j(h)$. The observed visibility $V'_{jk}(h)$ for array elements j and k, as can be seen from Eq. (5.16), is related to the true visibility according to

$$V'_{jk}(h) = G_j(h)G_k^*(h)V_{jk}(h) + \epsilon_{jk}(h) \tag{5.33}$$

where $\epsilon_{jk}(h)$ is the contribution of system noise. $V_{jk}(h)$ varies with time h due to baseline $\mathbf{u} = \mathbf{r}_j - \mathbf{r}_k$ scanning by earth's rotation. In addition, both the complex element gain and the noise are also time varying. However, uncertainties in element gain $G_j(h)$ do not render visibility measurements useless because each visibility depends on two elements. Consider three elements, i, j, and k. In the noiseless situation, we see that the visibility phases satisfy the relation

$$\angle V'_{jk}(h) + \angle V'_{ki}(h) + \angle V'_{ij}(h) = \angle V_{jk}(h) + \angle V_{ki}(h) + \angle V_{ij}(h)$$

because the phase errors of the three elements cancel out. Thus, knowing two visibility phases, the other one can be determined, regardless of the phase errors associated with the elements. This relation was first noted by Jennison [62] and is now known as *closure phase*. In the presence of large element phase errors, the visibility measurements do not directly yield true phase values. They do provide a constraint for the true phase values measured with every element triad. A similar closure relation on visibility amplitude exists for the case of four elements [63],

$$\frac{|V'_{ij}(h)|\,|V'_{kl}(h)|}{|V'_{ik}(h)|\,|V'_{jl}(h)|} = \frac{|V_{ij}(h)|\,|V_{kl}(h)|}{|V_{ik}(h)|\,|V_{jl}(h)|}$$

where the gain amplitudes $|G_j(h)|$ cancel out. Each observed *closure amplitude* provides a constraint on true closure amplitude aside from the effect of noise.

The phase and amplitude closures have been employed to reconstruct what is known as *hybrid maps* from phaseless VLB1 data [18, 64, 65] as well as from data with large amplitude errors [19]. Observations with such large errors no longer measure the visibility as a function of baseline space. Instead, closure phase for groups of three different points and closure amplitudes for four distinct positions in the baseline space are measured. These position groups sweep the baseline space as the earth rotates. From the measured closure values an iterative procedure can be used to reconstruct an image as follows:

1. Decide on an initial model.
2. Use the model to derive phase on $N - 1$ baselines and amplitude on N baselines for each observation interval, using Eq. (5.23).
3. From the observed closure relations, derive the phases and amplitudes of the remaining baselines.
4. Form a new model using the derived visibility and CLEAN.
5. Repeat until the map is satisfactory and agrees well with closure values.

The closure phase and amplitude relations are derived assuming that the measured visibilities $V'_{jk}(h)$ are caused by element gain uncertainties in the form of Eq. (5.33). We can therefore bypass the closure relations and seek a map as well as the unknown complex gains G_j, such that when the visibility V_{jk} of the map is substituted into Eq. (5.33) the error from the observed visibility V'_{jk} is minimized in some way. Using such an approach, Schwab [66] has introduced an algorithm that minimizes the sum of the squares of residuals,

$$C = \sum_h \sum_{j<k} W_{jk} \,|\, V'_{jk}(h) - G_j(h) G_k^*(h) V_{jk}(h) \,|^2 \tag{5.34}$$

subject to $\angle G_r(h) = 0$ for some phase reference element. W_{jk} is a weight that should be set proportional to the inverse of the variance of noise $\epsilon_{jk}(h)$. The quantity C is a measure of how well the image satisfies the constraint imposed by the observed imprecise visibilities. The method is known as *self-calibration* because the complex gain $G_j(h)$ is also to be determined. The steps involved are as follows.

1. Begin with an initial model.
2. Find the visibilities at the observed sample points using Eq. (5.23).
3. Solve for the complex gains by minimizing C using a gradient algorithm.
4. Correct observed visibilities according to

$$V_{jk}(h) = G_j^{-1}(h) G_k^{*-1}(h) V'_{jk}(h)$$

5. Form a new model from the above V_{jk} using CLEAN.
6. Repeat, unless the agreement between map and observed data is satisfactory.

The process can be modified to take into account different levels of phase and amplitude errors in each element [67] by including the variance of $|G_j(h)|$ and $\angle G_j(h)$ for the individual elements.

It may appear that the two approaches discussed above are not well founded. However, it must be realized that for large arrays the fraction of visibility phase and amplitude information lost due to element gain uncertainties are small. For an N-element nonredundant array having $N(N-1)/2$ baselines the missing relative element phase information is $N - 1$ and the missing element gain information is N.

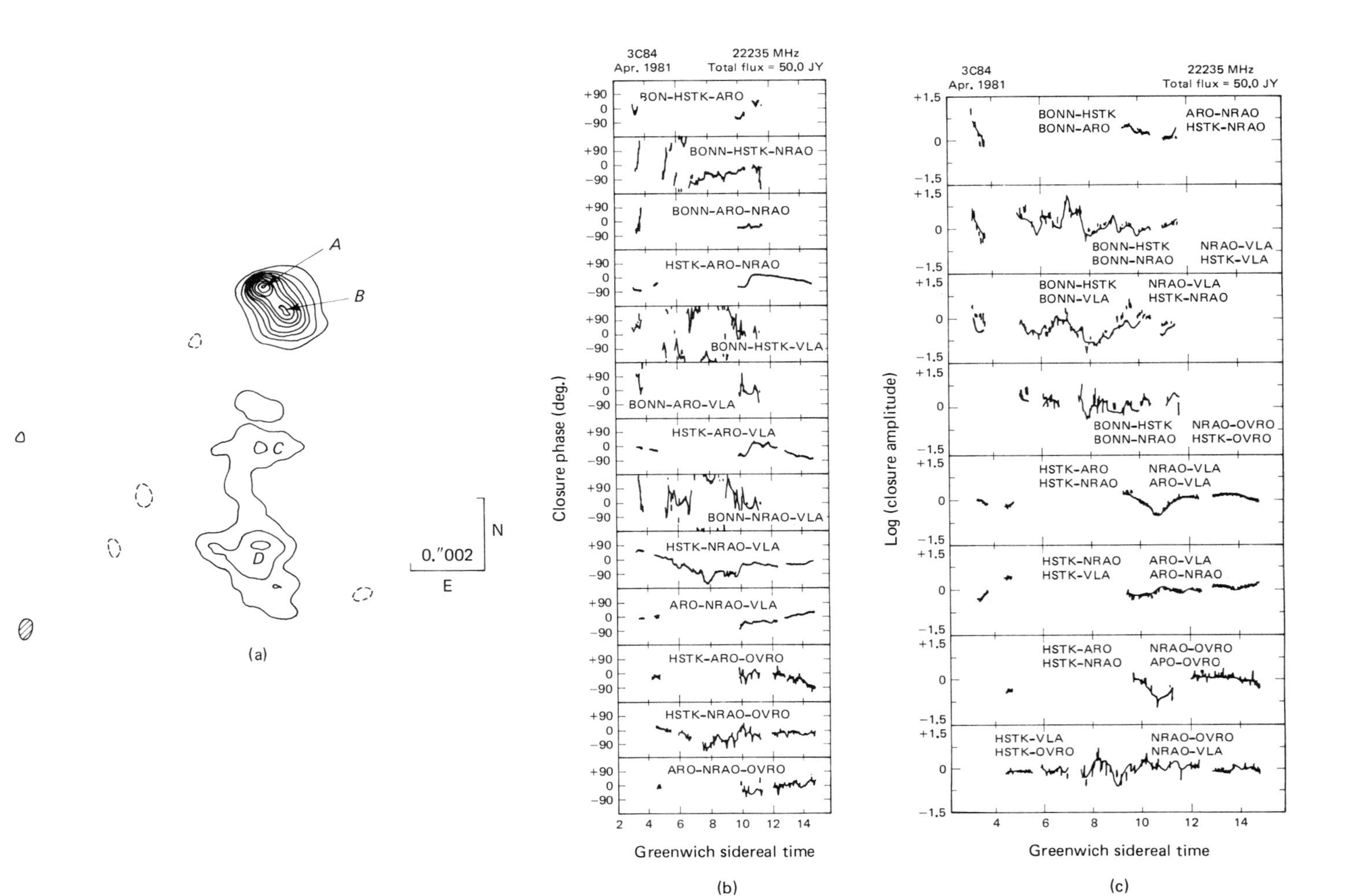

Figure 5.23 (a) VLBI map of 3C 84, (b) based on phase closure and (c) amplitude closure. [By permission of *Astrophysical Journal* (Vol. 265, p. 107, 1983). From paper titled "Asymmetric Structure in the Nuclei of NGC 1275 and 3C 345."]

The fractional loss of phase information and the fractional loss of amplitude information are therefore $2/N$ and $2/(N-1)$, respectively, both of which tend to zero rather rapidly. The only irretrievable pieces of information are the source position and intensity. Thus, with judicious care, images of not too complicated nature can be reconstructed from well-sampled baseline space with reasonable dynamic range when the source is not too weak. Figure 5.23 shows an example constructed from VLBI data without phase and with poor amplitude variations produced by atmospheric effects at wavelengths near 1 cm. Despite the success of self-calibration and closure relations, without which VLBI maps could not be derived, nevertheless, we do not know the circumstances under which the method works. Investigations on the convergence and error are yet to be done.

5.8 MAXIMUM-ENTROPY IMAGE RESTORATION

In the preceding two sections we discussed image restoration from incompletely sampled imprecise visibilities in the presence of noise, based on the algorithm CLEAN. There are, of course, many other alternatives to reconstruct an image agreeing with the observations. We may, for example, impose assumptions on certain statistical properties of the image, such as sharpness or entropy [18, 68]. Among the different possibilities so far, only maximum entropy has been applied to restore large synthesis array images [69, 70]. The method is also widely used in the spectral estimation of time series [71]. In the following, we introduce the *maximum-entropy method* and related image restoration from *Bayes' theorem*, taking explicit account of measurement noise, and the nonlinear optimization techniques required to solve for the image.

According to Jaynes [72, 73], the concept of introducing prior knowledge can be considered as a consequence of Bayes' theorem with a priori knowledge. Let $P[I(l, m) \mid V'_{jk}(h)]$ be the a posteriori probability that image $I(l, m)$ is correct, given the set of observed probabilities $V'_{jk}(h)$. Bayes' theorem states that

$$P[I(l, m) \mid V'_{jk}(h)] = \frac{P[V'_{jk}(h) \mid I(l, m)]P[(I(l, m)]}{P[V'_{jk}(h)]} \tag{5.35}$$

where $P[V'_{jk}(h) \mid I(l, m)]$ is the conditional probability of the set of observed visibilities $V'_{jk}(h)$, given the image $I(l, m)$, and $P[I(l, m)]$ is the a priori probability of $I(l, m)$. Since the image $I(l, m)$ determines the true visibilities $V_{jk}(h)$ uniquely, the conditional probability above can be replaced by that of observed $V'_{jk}(h)$, given the true visibilities $V_{jk}(h)$. Now the relation between the two visibilities is given by Eq. (5.33). Therefore, $P[V'_{jk}(h) \mid V_{jk}(h)]$ can be derived from the error statistics. Let the errors of the measurement $\epsilon_{jk}(h)$ be uncorrelated, zero-mean, complex Gaussian distributed random variables with variance $\sigma_{jk}(h)^2$. The conditional probability of measured visibility $V'_{jk}(h)$, given the true visibilities $V_{jk}(h)$, is therefore

$$P[V'_{jk}(h) \mid V_{jk}(h)] = (2\pi)^{-M} \prod_{\substack{j<k \\ h}} \sigma_{jk}(h)^{-2} \exp\left[-\frac{1}{2} \sum_{\substack{j<k \\ h}} \frac{|\epsilon_{jk}(h)|^2}{\sigma_{jk}(h)^2} \right] \tag{5.36}$$

where M is the total number of baseline space samples. If we can find an appropriate a priori probability $P[I(l, m)]$, then an estimate based on the measurements $V'_{jk}(h)$ that maximizes $P[I(l, m) \mid V'_{jk}(h)]$ can be found because $P[V'_{jk}(h)]$ is a constant and does not effect the maximization. Let the prior probability of various images before having any observational visibilities be expressed as

$$P[I(l, m)] = \exp\left(\frac{S}{\lambda}\right)$$

where S depends on the image $I(l, m)$ and λ is a constant. The maximization of the a posteriori probability of Eq. (5.35) becomes the maximization of the objective function

$$J = S - \lambda C \tag{5.37}$$

where, from Eqs. (5.36) and (5.33), we have

$$\begin{aligned} C &= \frac{\sum_{\substack{j<k \\ h}} |\epsilon_{jk}(h)|^2}{\sigma_{jk}(h)^2} \\ &= \frac{\sum_{\substack{j<k \\ h}} | V'_{jk}(h) - G_j(h)G_k^*(h)V_{jk}(h) |^2}{\sigma_{jk}(h)^2} \end{aligned} \tag{5.38}$$

with $V_{jk}(h)$ derived from $I(l, m)$ according to Eq. (5.22). The result of Bayes' theorem can be interpreted as the maximization of S under the constraint C of a chi-squared statistic with a suitable Lagrange multiplier λ. The quantity C measures the misfit between the observed visibilities and the visibilities obtained from the maximum a posteriori image.

It is important to acknowledge errors in the observational data; otherwise, the reconstructed image may contain features arising from noise only and may result in "super resolution" exceeding the resolution inherent in a set of noisy measurements [10, 69]. The criterion should therefore be the maximization of S under the constraint $C \leq C_{\max}$, where $C_{\max}$ is the largest acceptable misfit in terms of the chi-squared distribution and the degree of freedom M. Note that the self-calibration algorithm described in Section 5.7 is the minimization of C in Eq. (5.34) with weights W_{jk}, which may be different from $1/\sigma_{jk}(h)^2$.

The selection of prior distribution remains a problem. The most probable image can be derived from the Bose–Einstein statistics of photons, or it can be selected from some heuristic quality of image. Even in the maximum-entropy approach, there are two definitions of entropy in use, denoted as H_1 and H_2, where

$$H_1: \quad S = \iint \log I(l, m)\, dl\, dm$$

$$H_2: \quad S = -\iint i(l, m) \log i(l, m)\, dl\, dm,$$

$$i(l, m) = \frac{I(l, m)}{\iint I(l, m)\, dl\, dm}$$

H_2 was considered [73, 74] to best represent the configurational entropy of an image. H_1 was employed by Burg [75] in the estimation of the power spectrum of a time series. To a certain extent, it measures the entropy of the wave field on the ground. Recent works on spectral estimation of time series emphasize the interpretation of Burg's method as fitting an autoregressive process to the time series. However, it is difficult to regard the wave field impinging on a synthesis array as a spatially filtered random noise. Both entropies can be derived from the statistics of photon occupation number in a cell [76]. In the nondegenerate limit when the number of photons in a cell is much less than the degrees of freedom of the photons, then H_2 holds, while in the other limit H_1 is valid. However, it is difficult to reconcile that the emission mechanism directly influences the image of objects in the sky. Although the two entropies are different, their integrands are both real for $I > 0$ and their first derivatives both tend to infinity at $I = 0$. This is equivalent to the positivity constraint, because maximization of S will prevent I from becoming zero or negative. In addition, the two integrands have negative second derivatives, thereby favoring uniform images. This implies that the maximum entropy maps are smooth and of low contrast. As an alternative to maximum entropy, the image contrast has been recently proposed as a priori distribution where

$$S = \pm \iint I^2(l, m)\, dl\, dm$$

It is readily seen that the integral measures the concentration of image near bright spots. When S with the positive sign is maximized [77], an image of maximum sharpness will result. On the other hand, if S with the negative sign is maximized [78], a smooth image will be reconstructed. In view of the different priors that can be adopted, it may be necessary to compare different images obtained from the same set of data for proper astrophysical interpretation.

The solution of the maximization problem is a formidable task. Under the appropriate conditions, the maximization problem of Eq. (5.37) is unique [79]. However, unlike the time-series situation, no explicit solution exists for either of the entropy measures. A numerical solution of the constrained maximization problem is therefore necessary. The problem of nonlinear optimization is well known [80, 81] and many algorithms are available. In view of the large number of pixels required in most images, severe limits are imposed by slow convergence, large storage requirements, and limited numerical precision. An efficient and robust algorithm must be used; otherwise, the computation burden will become excessive. Wernecke and D'Addario [69, 79] replaced the constrained maximization with the unconstrained maximization of J with λ chosen so that the misfit C is acceptable. They used both the *steepest-ascent method* and the *conjugate-gradient method* [82]. In the steepest-ascent algorithm, the *search direction* is defined by ∇J and the $(n + 1)$th iteration is described by

$$I(l, m)^{(n+1)} = I(l, m)^{(n)} + \mu \nabla J^{(n)}$$

with suitable μ. A major disadvantage of the steepest ascent method is that $\nabla J^{(n)}$ and therefore $I(l, m)$ can become negative. In the conjugate-gradient method the search is, instead of along $\nabla J^{(n)}$, directed along the part of $\nabla J^{(n)}$ which is conjugate to one or more previous directions $\mathbf{e}^{(n)}$, where

$$\mathbf{e}^{(n+1)} = \nabla J^{(n+1)} - \frac{\mathbf{e}^{(n)} | \nabla J^{(n+1)} |}{| \nabla J^{(n)} |}$$

Although the method converges faster asymptotically, it remains plagued by negative $I(l, m)$.

An efficient and robust algorithm for maximizing H_2 was developed by Skilling [83]. Since the main computational effort lies in evaluating the gradient $\nabla J^{(n)}$, flexibility is gained by searching not along a line but in a subspace of several dimensions. To facilitate different values of λ, the subspace is constructed from $\nabla S^{(n)}$ and $\nabla C^{(n)}$ separately. The problem of negative $I(l, m)$ is alleviated by using a distance limit $\sum [\delta I(l, m)]^2 / I(l, m) \leq l_0^2$. The algorithm has been applied to images of 10^6 pixels with dynamic range reaching 10^4.

Image restoration from incomplete, imprecise, and noisy data has reached a certain degree of maturity where efficient and robust algorithms based on CLEAN, self-calibration, and maximum entropy are routinely applied to restore images of 10^6 pixels or more. Furthermore, advances will continue in the improvement of convergence, error analysis, computing burden, and in adapting algorithms to problems being investigated.

5.9 SUMMARY

Since their introduction, synthesis radio telescopes have become one of the principal instruments for observational astronomy. With baselines from hundreds of meters to thousands of kilometers, these instruments offer resolutions ranging from tens of minutes to milli-seconds of arc. In addition, with their large collecting areas, they have enormous sensitivities capable of detecting compact objects at great distances as well as nearby diffused clouds. Many important astrophysical phenomena have been discovered and studied in detail. These investigations are made possible by the sophisticated image reconstruction techniques described above to handle the incomplete and noisy data contaminated by propagation and instrumental effects. For instance, distant quasars having compact cores ejecting relativitistic particles along jets which eventually become diffused envelopes have been mapped with scale sizes varying over six orders of magnitude. Closer to the solar system, masering molecules in interstellar clouds have been monitored for their distribution, physical state, kinematics, and chemical composition. New astrophysical demands have led to the planning and development of new major arrays covering whole continents. To further increase baseline length beyond the earth's diameter, satellite-borne array elements have been proposed. The ability to observe through the rapidly varying refraction and attenuation of the atmosphere has prompted the proposal of ground-based synthesis arrays for operation in the low millimeter waves.

The construction of a new instrument inevitably leads to the development and refinement of methods of data acquisition and image reconstruction. As a result, new astrophysical phenomena will be discovered. The new problems generated will require new instruments for their solution. We thus have a symbiotic relation between astrophysics, instruments, and image reconstruction techniques. In the presence of rapidly advancing technology in data acquisition and processing, we expect to see continuing advances in understanding, methodology, and algorithms for reconstructing images with greater details, higher dynamic ranges, and finer temporal frequency resolutions. Such advances will not only lead to new astrophysical discoveries, but also have an impact on related areas.

REFERENCES

1 J. D. Kraus, *Radio Astronomy*, McGraw-Hill, New York, 1966.

2 W. N. Christiansen and J. A. Högbom, *Radiotelescopes*, Cambridge University Press, Cambridge, 1969.

3 K. I. Kellermann and G. L. Verschuur, eds., *Galactic and Extra-Galactic Radio Astronomy*, Springer-Verlag, New York, 1973.

4 M. Ryle, "Radio Telescopes of Large Resolving Power," Science, Vol. 188, pp. 1071–1078, June 1975.

5 M. Ryle, "The 5-km Radio Telescope at Cambridge," Nature, Vol. 239, pp. 435–438, Oct. 1972.

6 J. A. Högbom and W. N. Brouw, "The Synthesis Radio Telescope at Westerbork. Principles of Operation, Performance and Data Reduction," Astron. Astrophys., Vol. 33, pp. 289–301, 1974.

7 A. R. Thompson, B. G. Clark, C. M. Wade, and P. J. Napier, "The Very Large Array," Astrophys. J., Suppl., Vol. 44, pp. 151–167, 1980.

8 J. G. Davies, B. Anderson, and I. Morison, "The Judrell Bank Multi-telescope Radio-Linked Interferometer," Nature, Vol. 288, pp. 64–66, 1980.

9 N. W. Broten, T. H. Legg, J. L. Locke, C. W. McLeish, R. S. Richards, R. M. Chisholm, H. P. Gush, J. L. Yen, and J. A. Galt, "Long Baseline Interferometry: A New Technique," Science, Vol. 156, pp. 1592–1593, June 1967.

10 M. H. Cohen, "Introduction to Very Long Baseline Interferometry," Proc. IEEE, Vol. 61, pp. 1192–1197, Sept. 1973.

11 J. L. Yen, K. I. Kellermann, B. Rayhrer, N. W. Broten, D. N. Fort, S. H. Knowles, W. B. Waltman, and G.. W. Swenson, Jr., "Real-Time Very-Long-Baseline Interferometry Based on the Use of a Communications Satellite," Science, Vol. 198, pp. 289–291, Oct. 1977.

12 M. Born, and E. Wolf, *Principles of Optics*, Pergamon Press, Oxford, 1970.

13 G. W. Swenson, Jr., and N. C. Mathur, "The Interferometer in Radio Astronomy," Proc. IEEE, Vol. 56, pp. 2114–2130, Dec. 1968.

14 E. B. Fomalont, "Earth-Rotation Aperture Synthesis," Proc. IEEE, Vol. 61, pp. 1211–1218, Sept. 1973.

15 A. E. E. Rogers, "Theory of Two-Element Interferometers," in *Methods of Experimental Physics*, Vol. 12, Part C, Academic Press, New York, 1976.

16 W. N. Brouw, "Aperture Synthesis," in *Methods of Computational Physics*, Vol. 14, Academic Press, New York, 1975.

17 J. A. Högbom, "Aperture Synthesis with a Non-regular Distribution of Interferometer Baselines," Astron. Astrophys., Suppl., Vol. 15, pp. 417–426, June 1974.

18 D. N. Fort, and H. K. C. Yee, "A Method of Obtaining Brightness Distribution from Long-Baseline Interometry," Astron. Astrophys., Vol. 50, pp. 19–22, 1976.

19 A. C. S. Readhead, R. C. Walker, T. J. Pearson, and M. H. Cohen, "Mapping Radio Sources with Uncalibrated Visibility Data," Nature, Vol. 285, pp. 137–140, 1980.

20 J. G. Ables, "Maximum Entropy Spectral Analysis," Astron. Astrophys., Suppl., Vol. 15, pp. 383–393, June 1974.

21 J. N. Douglass, F. N. Bash, F. D. Ghigo, G. F. Moseley, and G. W. Torrence, "First Results from the Texas Interferometer: Positions of 605 Discrete Sources," Astrophys. J., Vol. 78, pp. 1–17, Feb. 1973.

22 G. W. Swenson, Jr., and N. C. Mathur, "On the Space-Frequency Equivalence of a Correlator Interferometer," Radio Sci., Vol. 4, pp. 69–71, Jan. 1969.

23 A. E. E. Rogers, "Very Long Baseline Interferometry with Large Effective Bandwidth for Phase-Delay Measurements," Radio Sci., Vol. 5, pp. 1239–1248, Oct. 1970.

24 A. R. Thompson, and L. R. D'Addario, "Frequency Response of a Synthesis Array: Performance Limitations and Design Tolerances," Radio Sci., Vol. 17, pp. 357–369, Mar.–Apr. 1982.

25 A. Bus, E. Raimond, and H. W. van Someran Greve, "A Digital Spectrometer for the Westerbar Synthesis Radio Telescope," Astron. Astrophys., Vol. 98, pp. 251–259, 1981.

26 J. H. Van Vleck and D. Middleton, "The Spectrum of Clipped Noise," Proc. IEEE, Vol. 54, pp. 2–19, Jan. 1966.

27 S. Weinreb, "Digital Radiometer," Proc. IEEE, Vol. 49, p. 1099, June 1961.

28 F. K. Bowers and R. J. Klingler, "Quantization Noise of Correlation Spectrometers," Astron. Astrophys., Suppl., Vol. 15, pp. 373–380, June 1974.

29 J. L. Yen, "The Role of Fast Fourier Transform Computers in Astronomy," Astron. Astrophys., Suppl., Vol. 15, pp. 483–484, June 1974.

30 B. J. New, "The AM 29500 Signal Processing Family," Proc. 1981 Int. Conf. Acoust. Speech Signal Process., pp. 378–381.

31 Y. L. Chow, "On Designing a Supersynthesis Antenna Array," IEEE Trans. Antennas Propag., Vol. AP-20, pp. 30–35, Jan. 1972.

32 C. C. Counselman III, "VLBI Techniques Applied to Problems of Geodesy, Geophysics, Planetary Science, Astronomy and General Relativity," Proc. IEEE, Vol. 61, pp. 1225–1230, Sept. 1973.

33 A. T. Moffet, "Minimum-Redundancy Linear Arrays," IEEE Trans. Antennas Propag., Vol. AP-16, pp. 172–175, Mar. 1968.

34 G. Swarup, and D. S. Bagri, "An Aperture-Synthesis Interferometer at Ooty for Operation at 327 MHz," Proc. IEEE, Vol. 61, pp. 1285–1287, Sept. 1973.

35 J. P. Wild, ed., "The Culgoora Radioheliograph," Proc. IREE (Australian), Vol. 28, pp. 277–384, Sept. 1967.

36 R. N. Bracewell and J. A. Roberts, "Aerial Smoothing in Radio Astronomy," Aust. J. Phys., Vol. 7, pp. 615–640, 1954.

37 R. N. Bracewell, "Strip Integration in Radio Astronomy," Aust. J. Phys., Vol. 9, pp. 198–217, 1956.

38 A. R. Thompson and R. N. Bracewell, "Interpolation and Fourier Transformation of Fringe Visibilities," Astron. J., Vol. 79, pp. 11–24, Jan. 1974.

39 H. S. M. Coxeter, *Introduction to Geometry*, Wiley, New York, 1961.

40 R. M. Mersereau, "The Processing of Hexagonally Sampled Two-Dimensional Signals," Proc. IEEE, Vol. 67, pp. 930–949, June 1979.

41 R. H. Frater and I. S. Docherty, "On the Reduction of Three-Dimensional Interferometer Measurements," Astron. Astrophys., Vol. 84, pp. 75–77, 1980.

42 R. M. Hjellming, ed., "An Introduction to the NRAO Very Large Array," National Radio Astronomy Observatory, New Mexico, 1982.

43 W. R. Burns and S. S. Yao, "A New Approach to Aperture Synthesis Processing," Astron. Astrophys., Vol. 6, pp. 481–485, 1970.

44 B. J. Clark, "An Efficient Implementation of the Algorithm CLEAN," Astron. Astrophys., Vol. 89, pp. 377–378, 1980.

45 V. J. Schwarz, "Mathematical–Statistical Description of the Iterative Beam Removing Technique (Method CLEAN)," Astron. Astrophys., Vol. 65, pp. 345–356, 1978.

46 V. J. Schwarz, "The Method 'CLEAN'—Use, Misuse and Variations," in *Image Formation from Coherence Functions in Astronomy*, International Astronomical Union Colloquium 49, ed. C. van Schounveld, D. Reidel, Boston, 1979, pp. 261–275.

47 R. D. Ekers and A. H. Rots, "Short Spacing Synthesis from a Primary Scanned Interferometer," in *Image Formation from Coherence Functions in Astronomy*, International Astronomical Union Colloquium 49, ed. C. van Schounveld, D. Reidel, Boston, 1979, pp. 61–63.

48 E. Bajaja and G. D. van Albada, "Complementing Aperture Synthesis Radio Data by Short Spacing Components from Single Dish Observations," Astron. Astrophys., Vol. 75, pp. 251–254, 1979.

49 A. E. E. Rogers, "Method of Using Closure-Phases in Radio Aperture Synthesis," Soc. Photo-opt. Instr. Eng., Int. Opt. Comput. Conf., Vol. 231, 1980, pp. 10–17.

50 A. E. E. Rogers and J. Moran, "Coherence Limits for Very Long Baseline Interferometry," IEEE Trans. Instr. Meas., Vol. IM-30, pp. 283–286, 1981.

51 S. H. Knowles, W. B. Waltman, J. L. Yen, J. Galt, D. N. Fort, W. H. Cannon, D. Davidson, W. Potrachenko, and J. Popelar, "A Phase-Coherent Link via Synchronous Satellite Developed for Very Long Baseline Radio Interferometry," Radio Sci., Vol. 17, pp. 1661–1670, 1982.

52 B. Elsmore and M. Ryle, "Further Astrometric Observations with the 5 km Radio Telescope," Mon. Not. R. Astron. Soc., Vol. 174, pp. 411–423, 1976.

53 J. P. Hamaker, "Atmospheric Delay Fluctuations with Scale Size Greater than One Kilometer, Observed with a Radio Interferometer Array," Radio Sci., Vol. 13, pp. 873–891, Sept.–Oct. 1978.

54 T. A. Th. Spoelstran and R. T. Schilizzi, "Phase Errors in Radio Astronomy Interferometers Due to Ionospheric Disturbances," Proc. Symp. Satellite Beacon, Warsaw, May 1980, pp. 315–322.

55 D. C. Hogg and T. C. Chu, "The Role of Rain in Satellite Communications," Proc. IEEE, Vol. 63, pp. 1308–1331, Sept. 1975.

56 D. C. Hogg, F. O. Guirand, and M. T. Decker, "Measurement of Excess Radio Transmission Length on Earth–Space Paths," Astron. Astrophys., Vol. 95, pp. 304–307, 1981.

57 A. A. G. Requicha, "The Zeros of Entire Functions: Theory and Engineering Applications," Proc. IEEE, Vol. 68, pp. 308–328, Mar. 1980.

58 R. E. Burge, M. H. Fiddy, A. H. Greenway, and G. Ross, "The Phase Problem," Proc. R. Soc., Ser. A, Vol. 350, pp. 191–212, 1976.

59 R. H. T. Bates, "Fringe Visibility Intensities May Uniquely Define Brightness Distribution," Astron. Astrophys., Vol. 70, pp. L27–L29, 1978.

60 J. E. Baldwin and P. J. Warner, "Phaseless Aperture Synthesis," Mon. Not. R. Astron. Soc., Vol. 182, pp. 411–422, 1978.

61 M. H. Hayes, J. S. Lim, and A. V. Oppenheim, "Signal Reconstruction from Phase or Magnitude," IEEE Trans. Acoust. Speech Signal Process., Vol. ASSP-28, pp. 672–680, Dec. 1980.

62 R. Jennison, "A Phase Sensitive Interferometer Technique for the Measurement of the Fourier Transforms of Spacial Brightness Distributions," Mon. Not. R. Astron. Soc., Vol. 118, pp. 176–284, 1958.

63 R. Q. Twiss, A. W. L. Carter, and A. G. Little, "Brightness Distribution Over Some Strong Radio Sources at 1427 Mc/s," Observatory, Vol. 80, pp. 153–159, 1960.

64 A. C. S. Readhead and R. W. Wilkinson, "The Mapping of Compact Radio Sources from VLBI Data," Astrophys. J., Vol. 223, pp. 26–36, 1978.

65 W. D. Cotton, "A Method of Mapping Compact Structure of Radio Sources Using VLBI Observations," Astron. J., Vol. 84, pp. 1122–1128, Aug. 1979.

66 F. R. Schwab, "Adaptive Calibration of Radio Interferometer Data," Soc. Photo-opt. Instr. Eng., Int. Optical Comput. Conf., Vol. 231, pp. 18–24, 1980.

67 T. J. Cornwell and P. N. Wilkinson, "A New Method for Making Maps with Unstable Radio Interferometers," Mon. Not. R. Astron. Soc., Vol. 196, pp. 1067–1086, 1981.

68 R. A. Muller and A. Buffinton, "Real-Time Correction of Atmospherically Degraded Telescope Images through Image Sharpening," J. Opt. Soc. Am., Vol. 64, pp. 1200–1210, 1974.

69 S. J. Wernecke and L. R. D'Addario, "Maximum Entropy Image Reconstruction," IEEE Trans. Comput., Vol. C-26, pp. 351–364, Apr. 1977.

70 S. F. Gull and G. J. Daniell, "Image Reconstruction from Incomplete and Noisy Data," Nature, Vol. 272, pp. 686–690, Apr. 1978.

71 S. Haykin, ed., *Nonlinear Methods of Spectral Analysis*, Springer-Verlag, New York, 1979.

72 E. T. Jaynes, "Information Theory and Statistical Mechanics," Phys. Rev., Vol. 106, pp. 620–630, 1957.

73 E. T. Jaynes, "Prior Probabilities," IEEE Trans. Syst. Sci. Cybern., Vol. SSC-4, pp. 227–241, 1968.

74 B. R. Frieden, "Restoring with Maximum Likelihood and Maximum Entropy," J. Opt. Soc. Am., Vol. 62, pp. 511–518, 1972.

75 J. P. Burg, "Maximum Entropy Spectral Analysis," 37th Ann. Int. Mtg., Soc. Explor. Geophysicists, Oklahoma City, Okla., 1967.

76 R. Kikuchi and B. H. Soffer, "Maximum Entropy Image Restoration I. The Entropy Expression," J. Opt. Soc. Am., Vol. 67, pp. 1656–1657, 1977.

77 P. L. Baker, "On the Recovery of Images from Incomplete Interferometric Measurements," Astron. Astrophys., Vol. 94, pp. 85–90, 1981.

78 T. J. Cornwell, "The Use of Bayesian Statistics in Image Estimation from Interferometer Data," *Image Formation from Coherence Functions in Astronomy*, International Astronomical Union Colloquium 49, ed. C. van Schounveld, D. Reidel, Boston, 1974, pp. 227–234.

79 S. Wernecke, "Two-Dimensional Maximum Entropy Reconstruction of Radio Brightness," Radio Sci., Vol. 12, pp. 831–844, Sept.–Oct. 1977.

80 R. L. Fox, *Optimization Methods for Engineering Design*, Addison-Wesley, Reading, Mass., 1971.

81 W. Murray, ed., *Numerical Methods for Unconstrained Optimization*, Academic Press, New York, 1972.

82 R. Fletcher and C. M. Reeves, "Function Minimization by Conjugate Gradients," Comput. J., Vol. 7, pp. 149–154, 1964.

83 J. Skilling, "Maximum Entropy," *Algorithms and Applications*, unpublished.

6

AVI C. KAK
School of Electrical Engineering
Purdue University, West Lafayette, Indiana

Tomographic Imaging with Diffracting and Non-diffracting Sources

6.1 INTRODUCTION

Tomography refers to cross-sectional imaging of objects from either transmission or reflection data. In most cases, the object is illuminated from many different directions either sequentially or simultaneously, and the image is reconstructed from data collected either in transmission or reflection. In this chapter we will only be concerned with reconstructing cross-sectional images from transmission data.

The most spectacular success of tomography is in diagnostic imaging with nondiffracting sources such as x-rays. However, recently there has been an increasing interest in extending this image formation technique to ultrasound, microwaves, and nuclear magnetic resonance (NMR). In the context of medical imaging, this interest has been spurred by the radiation hazards associated with x-rays, which makes them unsuitable for such important applications as the mass screening of female breast for the detection of cancerous tumors. In nonmedical areas, the desire to extend tomographic imaging to ultrasound and microwaves has been generated by the possibility of constructing accurate maps of underground resources for seismic exploration, cross-sectional imaging for nondestructive testing, and other uses.

Extending tomographic imaging to diffracting sources such as ultrasound and microwaves has not been easy and active research is still continuing. Algorithms developed to date are useful only if the object inhomogeneities are very small (less than a few percent of the background value). This limits the usefulness of the current

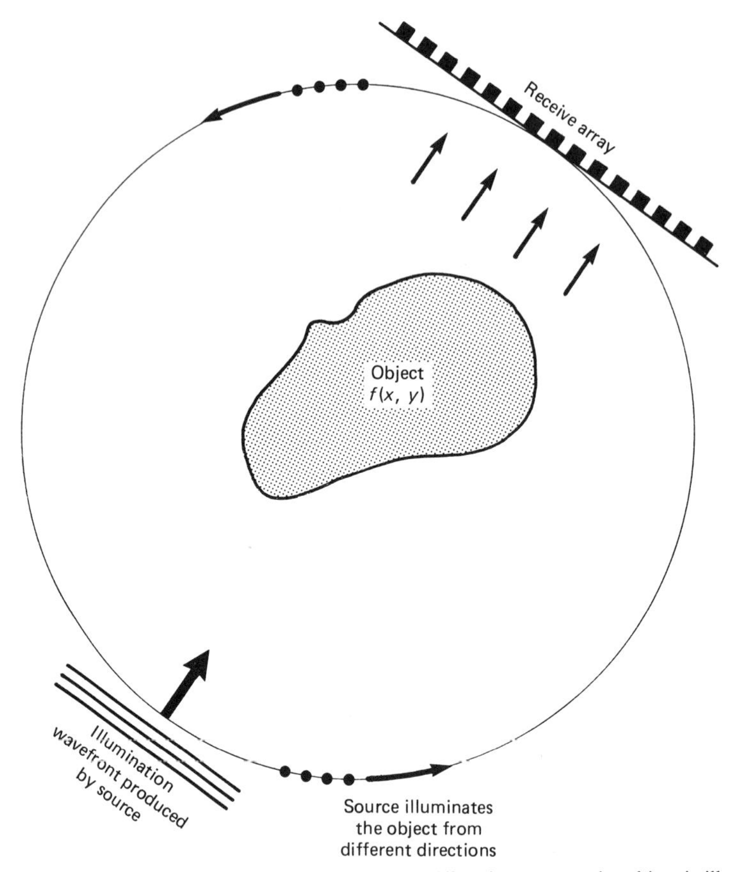

Figure 6.1 For tomographic imaging with a diffracting source, the object is illuminated from various directions and the transmitted fields recorded by a receive array.

algorithms in medical applications; however, they have utility in seismic mapping, sonar imaging, and ocean acoustic tomography. Hopefully, in the not too distant future, we will see algorithms capable of imaging larger inhomogeneities.

Given the very unequal degrees of success that have been achieved with nondiffracting and diffracting sources, the reader will probably be surprised by the organization of this chapter. Departing from the organization of a previous exposition [58], we have not given x-ray tomography its usual preeminence. Instead, this chapter reflects an almost equal balance between diffracting and nondiffracting cases. This was done for three reasons. First, the author expects the readers of a book on array processing to be more interested in tomography with diffracting sources because of its expected applications in sonar and seismic imaging. Second, there is greater research interest at this time in imaging with diffracting sources. Third, theoretically at least, it should be possible to derive all the reconstruction algorithms for nondiffracting sources from those for the diffracting case by taking the zero-wavelength limit.

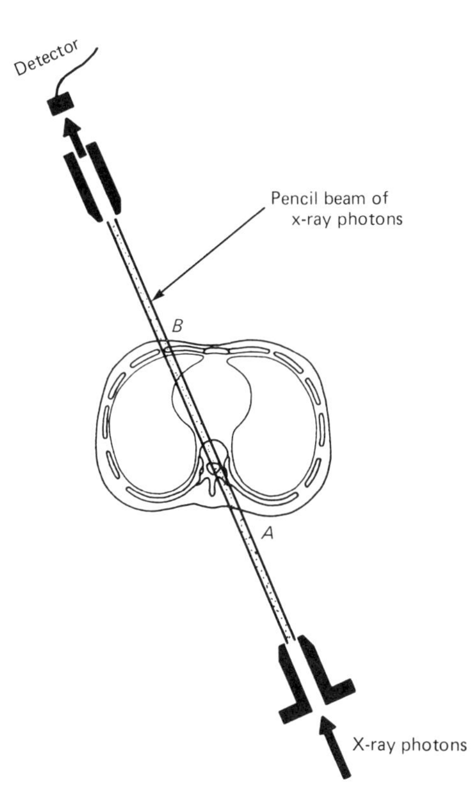

Figure 6.2 Logarithm of the ratio of the beam intensity at B to the incident intensity at A is proportional to the line integral of the x-ray attenuation coefficient from A to B.

From a theoretical standpoint, when diffracting illumination is used, the problem of tomographic imaging may be stated as follows. In Fig. 6.1, $f(x, y)$ represents a two-dimensional distribution of some parameter of the object in a desired cross-sectional plane. This object is illuminated from various directions with diffracting energy such as microwaves or ultrasound, and transmitted fields are measured by receiver arrays as shown. Given the measurements, how does one reconstruct the object distribution? From how many directions must the object be illuminated to ensure a unique solution to the problem of reconstruction?

With diffracting illumination, an exact solution to the problem is made difficult by the complicated nature of the wave–matter interactions. These interactions are described by differential or integral equations which, in general, do not possess closed-form solutions. Two types of solutions, called the Born and the Rytov, are available if we can get away with the assumption of weak interactions. Small inhomogeneities lead to weak interactions. By either theoretical reasoning or computer simulations on simple cross-sectional distributions, it is usually possible to determine the magnitude of the inhomogeneities, which if exceeded would lead to a breakdown in the Born- or Rytov-based solutions.

On the other hand, when we use nondiffracting illumination, such as x-rays, obtaining high-quality solutions becomes a much easier task. In Fig. 6.2 we have illustrated a pencil beam of x-rays penetrating an object, whose transmitted intensity is measured by either a photodetector or an ionization chamber. It can be shown that the logarithm of the ratio of the transmitted beam intensity at B to the incident beam intensity at A is directly proportional to the line integral of the attenuation coefficient of the object along the path AB. The problem of *image reconstruction* then becomes one of inverting such line integral data for recovering a map of the attenuation coefficient. Obtaining a solution is not as difficult now, since different line integrals constitute decoupled pieces of information. In other words, as opposed to the case of diffracting illumination, each unit of measurement may be considered to be a function of only those object elements that are along the line of sight between the focal spot on the x-ray tube and the detector aperture. No doubt, there are always some x-ray photons that are scattered into adjacent detector channels. However, their effect can be substantially minimized by using collimation for each detector channel. Any residual scattering is further gotten rid of by the expedient of deconvolution. (Such residual scattering is usually known as detector crosstalk.)

6.2 SOME APPLICATIONS OF TOMOGRAPHIC IMAGING

We will now discuss some application areas addressed by the techniques of tomographic imaging. Our discussion is intended to be a sampler rather than a complete survey of such areas. However, it is hoped that the examples presented will serve to illustrate the wide diversity of imaging problems where tomographic algorithms find application. The degree of success achieved in all these areas is not the same. Those applications that require diffracting illumination are amenable to successful imaging only if the object inhomogeneities represent small deviations from the background, and if the correlation lengths of these deviations are large compared to a wavelength. Research is continuing in the development of diffraction tomography algorithms for objects that do not satisfy this condition. Objects with weak inhomogeneities may be imaged with either the diffraction tomography algorithms based on the first Born and the first Rytov solutions, or by a combination of digital ray tracing and algebraic reconstruction algorithms. If the inhomogeneity deviations are less than 2 to 3 percent of the background value, it is often possible to ignore the diffraction (and refraction) effects altogether, and treat the problem as in x-ray tomography.

Consider, for example, the problem of earth resources imaging using cross-borehole measurements with either electromagnetic or acoustic sources, an application investigated by Dines and Lytle [32, 68]. As illustrated in Fig. 6.3, high-frequency electromagnetic waves or pulsed seismic (acoustic) waves are transmitted from the transmitter to the receiver. For each position of the transmitter in one borehole, the transmitted fields are measured at a large number of locations in the other borehole. This procedure is repeated for different positions of the transmitter in its borehole. Lytle and Dines have used ray tracing and algebraic methods for

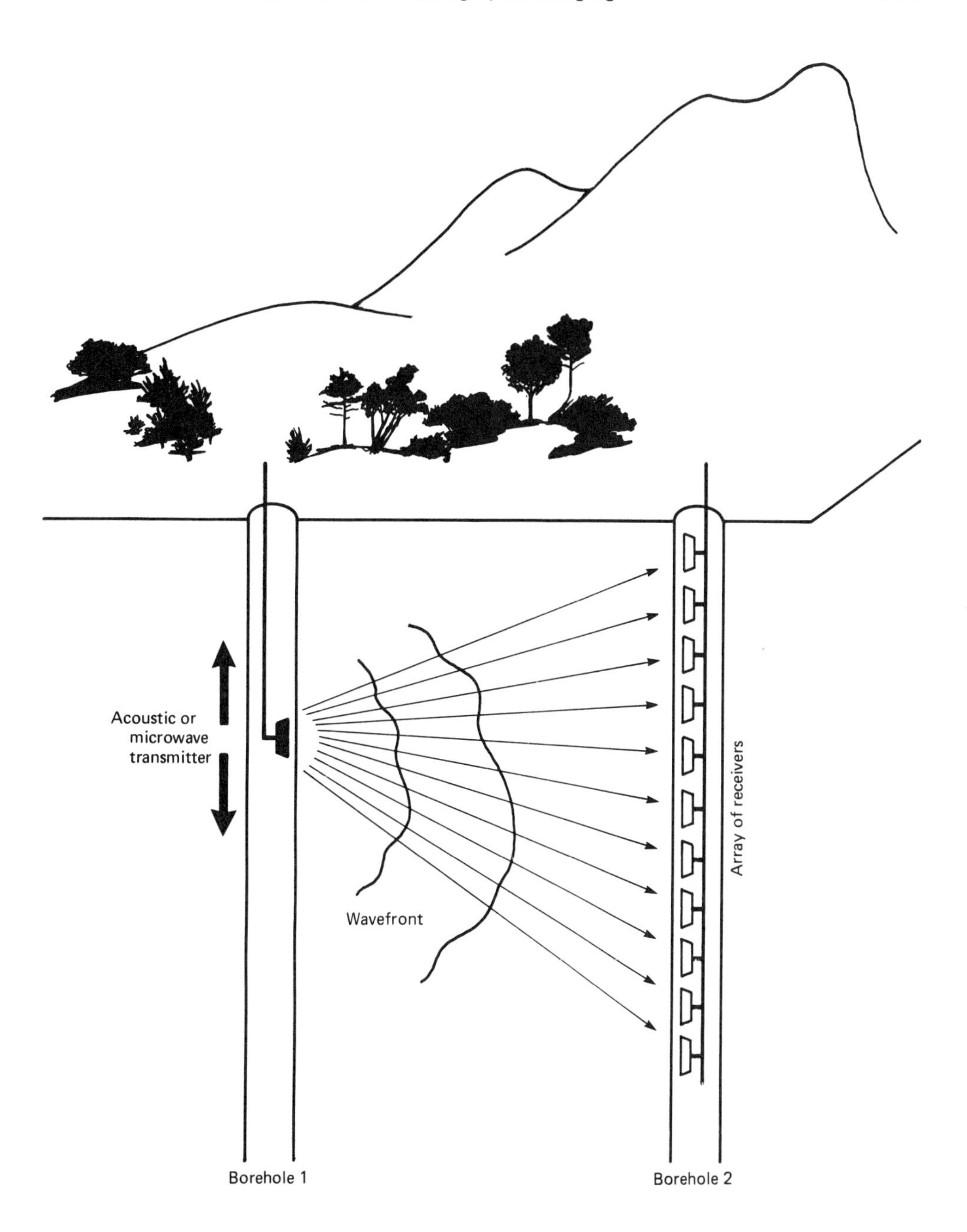

Figure 6.3 Application of tomography to cross-borehole imaging for earth resources mapping. For *each position* of the transmitter in one borehole, the transmitted fields are picked up by the elements of the receive array.

calculating the refractive index and attenuation coefficient distributions between the two boreholes. They have also investigated the use of this technique for determining underground fracturing patterns induced by detonating high explosives in a coal seam, and for locating high contrast underground anamolies such as voids, water pockets, and abandoned tunnels.

One of the first practical applications of tomographic imaging algorithms was due to Bracewell et al. [8, 9] in radio astronomy, where the requirement is to determine the brightness (in terms of the intensity of radio waves) distribution over the whole of the celestial sphere, a small area the size of the sun, or, in the case of "radio stars," over still smaller areas. In these applications, a radio telescope with a long, narrow aperture in one direction was used as a receiver. As a result, each measurement was approximately a strip integral. Earth's rotation was used to make one celestial scan. Scans along different orientations were generated by reorienting the radio telescope. This procedure is also applicable to the calculation of brightness distributions from records of lunar occultations.

Another application of tomographic imaging is in electron microscopy [26]. A transmission electron micrograph is essentially a projection of a specimen onto a plane. Projections taken at various directions are then used for three-dimensional reconstruction of the specimen by the application of tomographic algorithms.

Tomographic imaging with x-rays has had its greatest impact in diagnostic medicine, where it gives physicians three-dimensional views of the inside of the human body. In such systems (see Fig. 6.4, for example), an x-ray source illuminates

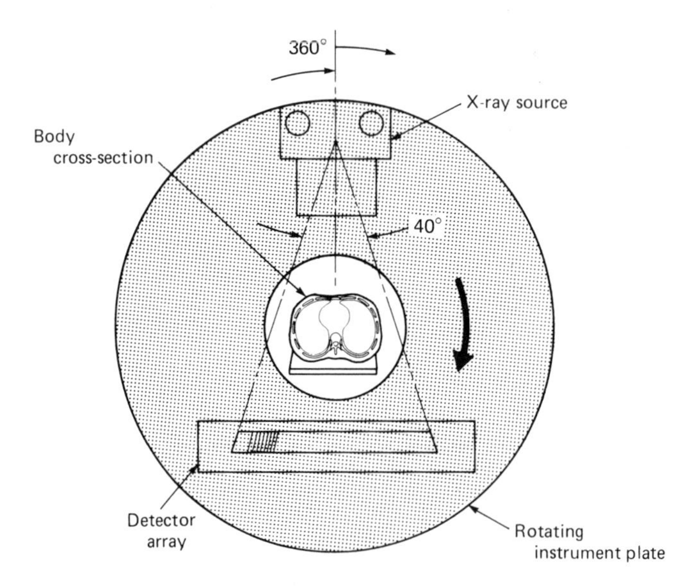

Figure 6.4 Fan-beam scanner. An x-ray source illuminates a cross section of the body, and the transmitted photons are measured by an array of detectors. The measurement system is mounted on a gantry that rotates rapidly around the patient.

a cross section of the body, and the transmitted photons are measured by an array of detectors, usually xenon ionization chambers. This measurement system can be rapidly rotated around the patient to generate a large number of projections, from which the cross-sectional image is reconstructed. For medical diagnostic imaging, tomographic imaging has also been accomplished with radio isotopes injected into the body [13, 63], with ultrasound [25, 31, 38], and with NMR [20, 93]. In the ultrasonic tomography systems that have been built to date, refraction and diffraction are ignored, which consequently produces artifacts in the images even for purely soft tissue imaging. It is hoped that with the expected advances in diffraction tomography it might be possible some day to eliminate these artifacts.

6.3 DEFINITIONS AND THEORETICAL PRELIMINARIES

Imaging with a Diffracting Source

Diffraction tomography algorithms are derived from the following general equation for wave propagation in an inhomogeneous medium:

$$(\nabla^2 + k^2)u(\mathbf{r}) = -G_f(\mathbf{r}) \tag{6.1}$$

where $u(\mathbf{r})$ represents the scalar field and G_f the *forcing function*, which depends on both the object inhomogeneities and the wave field. The constant k is the *complex wavenumber* and is usually calculated from the average properties of the inhomogeneous medium.

For acoustic (or ultrasonic) tomography, $u(\mathbf{r})$ can be the pressure field at position $\mathbf{r}$. For the electromagnetic case, assuming the applicability of a scalar propagation equation, and also assuming that the incident field may approximately be considered to be a plane wave, $u(\mathbf{r})$ may be set equal to the complex amplitude of the electric field along its polarization.

For the acoustic case, the simplest choice of the forcing function G_f is given by

$$G_f(\mathbf{r}) = k^2[n^2(\mathbf{r}) - 1]u(\mathbf{r}) \tag{6.2}$$

where n is the *complex refractive index* at position $\mathbf{r}$, and is equal to

$$n(\mathbf{r}) = \frac{c_0}{c(\mathbf{r})} \tag{6.3}$$

where c_0 is the propagation velocity in the medium in which the object is immersed and $c(\mathbf{r})$ is the propagation velocity at location $\mathbf{r}$ in the object. For the acoustic case where only the compressional waves in a viscous compressible fluid are involved, we have

$$c(\mathbf{r}) = \frac{1}{\sqrt{\rho\kappa}} \tag{6.4}$$

where ρ and κ are the *local density* and the *complex compressibility* at location $\mathbf{r}$.

For the electromagnetic case, a simple choice for the forcing function, G_f, is given by

$$G_f(\mathbf{r}) = 2k^2[n(\mathbf{r}) - 1]u(\mathbf{r}) \tag{6.5}$$

where $n(\mathbf{r})$ is again the refractive index as defined in Eq. (6.3). In that definition, the velocity $c(\mathbf{r})$ is now given by

$$c(\mathbf{r}) = \frac{1}{\sqrt{\mu\epsilon}} \tag{6.6}$$

where μ and ϵ are the *magnetic permeability* and the *dielectric constant* at **r**.

The form for G_f in Eq. (6.2) is valid only provided that we can ignore the first- and higher-order derivatives of the medium parameters. If the inhomogeneous medium can be modeled as a viscous compressible fluid, an exact form for the forcing function is given by

$$G_f = -k^2\gamma_\kappa u + \nabla \cdot (\gamma_\rho \nabla u) \tag{6.7}$$

where

$$\gamma_\kappa = \frac{\kappa - \kappa_0}{\kappa_0} \tag{6.8}$$

$$\gamma_\rho = \frac{\rho - \rho_0}{\rho} \tag{6.9}$$

where κ_0 and ρ_0 are either the compressibility and density of the medium in which the object is immersed, or the average compressibility and the density of the object, depending on how the process of imaging is modeled. On the other hand, if the object is a solid and can be modeled as a linear isotropic viscoelastic medium, the forcing function G_f possesses another more complicated form. Since this form involves tensor notation, it will not be presented here and the interested reader is referred to [21, 75].

Whereas Eq. (6.2) is a simplified form of the exact expression in Eq. (6.7) for the acoustic case, for the electromagnetic case the form of the forcing function in Eq. (6.5) can easily be obtained from the wave propagation equation

$$(\nabla^2 + k^2)u(\mathbf{r}) = 0 \tag{6.10}$$

for a homogeneous medium. If the medium is weakly scattering, we may assume that this equation continues to be valid with the propagation constant k replaced by one that is a function of position, which will be denoted by k_r. This new propagation equation may then be written as

$$(\nabla^2 + k_r^2)u(\mathbf{r}) = 0 \tag{6.11}$$

To make more explicit the deviations in k_r from the average wavenumber k, we write the former as

$$k_r = k[1 + \delta(\mathbf{r})] \tag{6.12}$$

Substituting this expression for k_r in Eq. (6.11), and assuming that $\delta \ll 1$, we can write

$$(\nabla^2 + k^2)u(\mathbf{r}) = -2k^2\,\delta(\mathbf{r})u(\mathbf{r}) \tag{6.13}$$

Since from Eq. (6.12), $\delta(\mathbf{r})$ is equal to

$$\begin{aligned}\delta(\mathbf{r}) &= \frac{k_r}{k} - 1\\ &= \frac{\omega/c(\mathbf{r})}{\omega/c_0} - 1\\ &= \frac{c_0}{c(\mathbf{r})} - 1\\ &= n(\mathbf{r}) - 1\end{aligned} \tag{6.14}$$

which completes the derivation of $G_f(\mathbf{r})$ in Eq. (6.5).

For the electromagnetic case, a final word about Eq. (6.1) with the forcing function given by Eq. (6.5) is in order. Note that Eq. (6.1) is a *scalar* wave propagation equation. Its use implies that there is no depolarization as the electromagnetic wave propagates through the medium. It is known [54] that the depolarization effects can be ignored only if the wavelength is much smaller than the correlation size of the inhomogeneities in the object. If this condition is not satisfied, then strictly speaking we must use the vector wave propagation equation

$$\nabla^2\mathbf{E}(\mathbf{r}) + k^2n^2\mathbf{E}(\mathbf{r}) - 2\nabla\left(\frac{\nabla n}{n}\cdot\mathbf{E}\right) = 0 \tag{6.15}$$

where $\mathbf{E}$ is the electric field vector. A vector theory for diffraction tomography based on this equation has not yet been developed.

In the development of diffraction tomography algorithms here, we will only be using Eq. (6.1) for wave propagation with G_f given by Eq. (6.2) for the acoustic case and Eq. (6.5) for the electromagnetic case. Algorithms for other forms of F_t in the acoustics case are discussed in [73, 75].

When the object is immersed in a medium, the total field at any position can be modeled as a superposition of the incident field, $u_i(\mathbf{r})$, and the scattered field, $u_s(\mathbf{r})$, as given by

$$u(\mathbf{r}) = u_i(\mathbf{r}) + u_s(\mathbf{r}) \tag{6.16}$$

If we assume that the wavenumber for the medium is the same as the average wavenumber for the object, we may say that the incident wave field u_i satisfies the propagation equation

$$(\nabla^2 + k^2)u_i(\mathbf{r}) = 0 \tag{6.17}$$

Substituting Eq. (6.16) in Eq. (6.1), and then using the fact that u_i satisfies Eq. (6.10), we get the following equation for the scattered fields:

$$(\nabla^2 + k^2)u_s(\mathbf{r}) = -G_f(\mathbf{r}) \tag{6.18}$$

We may rewrite this equation as

$$(\nabla^2 + k^2)u_s(\mathbf{r}) = -k^2 f(\mathbf{r})u(\mathbf{r}) \tag{6.19}$$

where $f(\mathbf{r})$ will be called the *object function*. For the acoustic case it is given by

$$f(\mathbf{r}) = n^2(\mathbf{r}) - 1 \tag{6.20}$$

and is obtained by using Eq. (6.2). For the electromagnetic case, the object function is obtained from Eq. (6.5) and is given by

$$f(\mathbf{r}) = 2[n(\mathbf{r}) - 1] \tag{6.21}$$

We define the Fourier transform of the object function by $F(\omega_1, \omega_2)$, with

$$F(\omega_1, \omega_2) = \iint_S f(x, y)e^{-j(\omega_1 x + \omega_2 y)}\, dx\, dy \tag{6.22}$$

where ω_1 and ω_2 are the spatial frequencies along the x- and y-directions, respectively. S is any area in the (x, y)-plane that encloses the object cross section.

Imaging with a Nondiffracting Source

Theoretically, it is possible to derive the reconstruction algorithms for this case from those for diffracting illumination by simply taking the limit as the wavelength approaches zero. However, it is more elegant to present the theory for the nondiffracting case within its own framework. The following definitions are essential to this framework.

Let a two-dimensional function $f(x, y)$ represent the object cross section whose image is desired. (The property distribution of the object that is represented by this function will be left unspecified for the derivation of tomography algorithms.) A line running through the cross section is called a *ray*. The integral of $f(x, y)$ along a ray is called a *line integral*. A measurement made by the source–detector pair shown in Fig. 6.2 yields an approximation to a single line integral. For the case of x-rays, this may be shown as follows. Let the line AB in Fig. 6.2 correspond to a parallel beam of x-rays. Only those photons that are propagating in the direction depicted by the head of the arrow are considered a part of the beam. As the beam propagates, photons are continually lost from the beam either because they are deflected (scattered) or because they are absorbed. At each point, both these losses are accounted for by a constant known as the *linear attenuation coefficient*, which is usually denoted by μ. Let N_{in} be the number of incident photons in the time span of a measurement (usually a few milliseconds). Within the same time instant, let N_d be the number of photons existing on side B. If we assume that all the incident photons have the same energy, the relationship between N_d and N_{in} is given by

$$N_d = N_{in} \exp\left[-\int_{ray} \mu(x, y)\, ds\right] \tag{6.23}$$

where $\mu(x, y)$ is the linear attenuation coefficient at the point (x, y), and ds is an element of length along the ray. From the equation above, we may write

$$\int_{ray} \mu(x, y)\, ds = \ln \frac{N_{in}}{N_d} \tag{6.24}$$

The left-hand side of this equation is simply a ray integral of the function $\mu(x, y)$.

Sets of line integrals are called *projections*. A *parallel projection* is generated if we scan the transmitter–receiver pair in a direction that is perpendicular to the line AB in Fig. 6.2. In Fig. 6.5 we have shown two parallel projections. The data

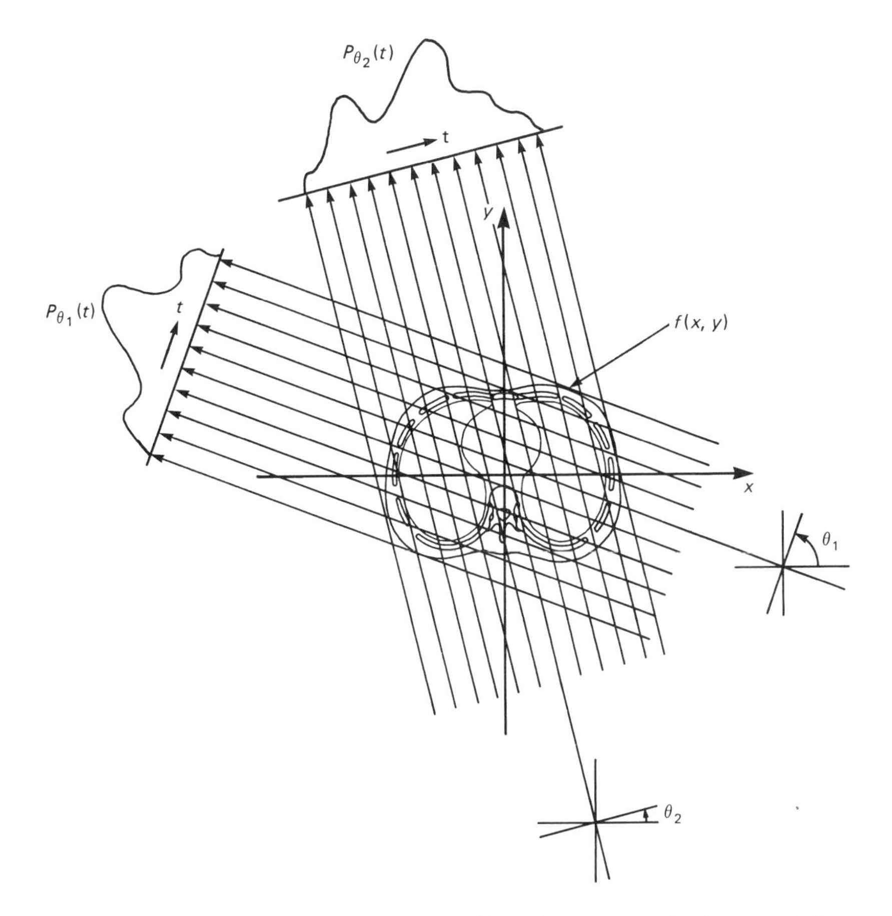

Figure 6.5 Two parallel projections of an object whose cross section is represented mathematically by the function $f(x, y)$.

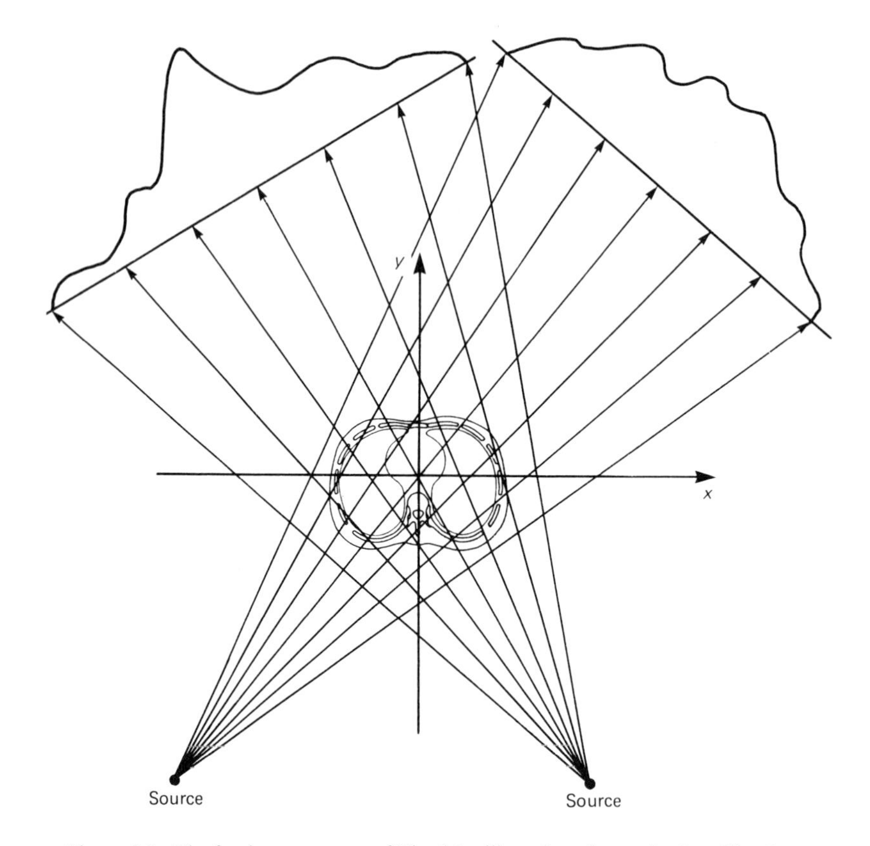

Figure 6.6 The fan-beam scanner of Fig. 6.4 will produce fan projections like those shown here.

generated by the scanner shown in Fig. 6.4 consist of *fan projections*, and in Fig. 6.6 we have shown a couple of those.

A mathematical relationship between a function $f(x, y)$ and its parallel projections may be written down by using the equation of a straight line in a plane. For example, the equation of line AB in Fig. 6.7 is given by

$$x \cos \theta + y \sin \theta = t_1 \tag{6.25}$$

where t_1 is the perpendicular distance of the line from the origin. The integral of the function $f(x, y)$ along this line may be expressed as

$$\begin{aligned} P_\theta(t_1) &= \int_{\text{ray AB}} f(x, y)\, ds \\ &= \int_{-\infty}^{\infty} \int_{-\infty}^{\infty} f(x, y)\, \delta(x \cos \theta + y \sin \theta - t_1)\, dx\, dy \end{aligned} \tag{6.26}$$

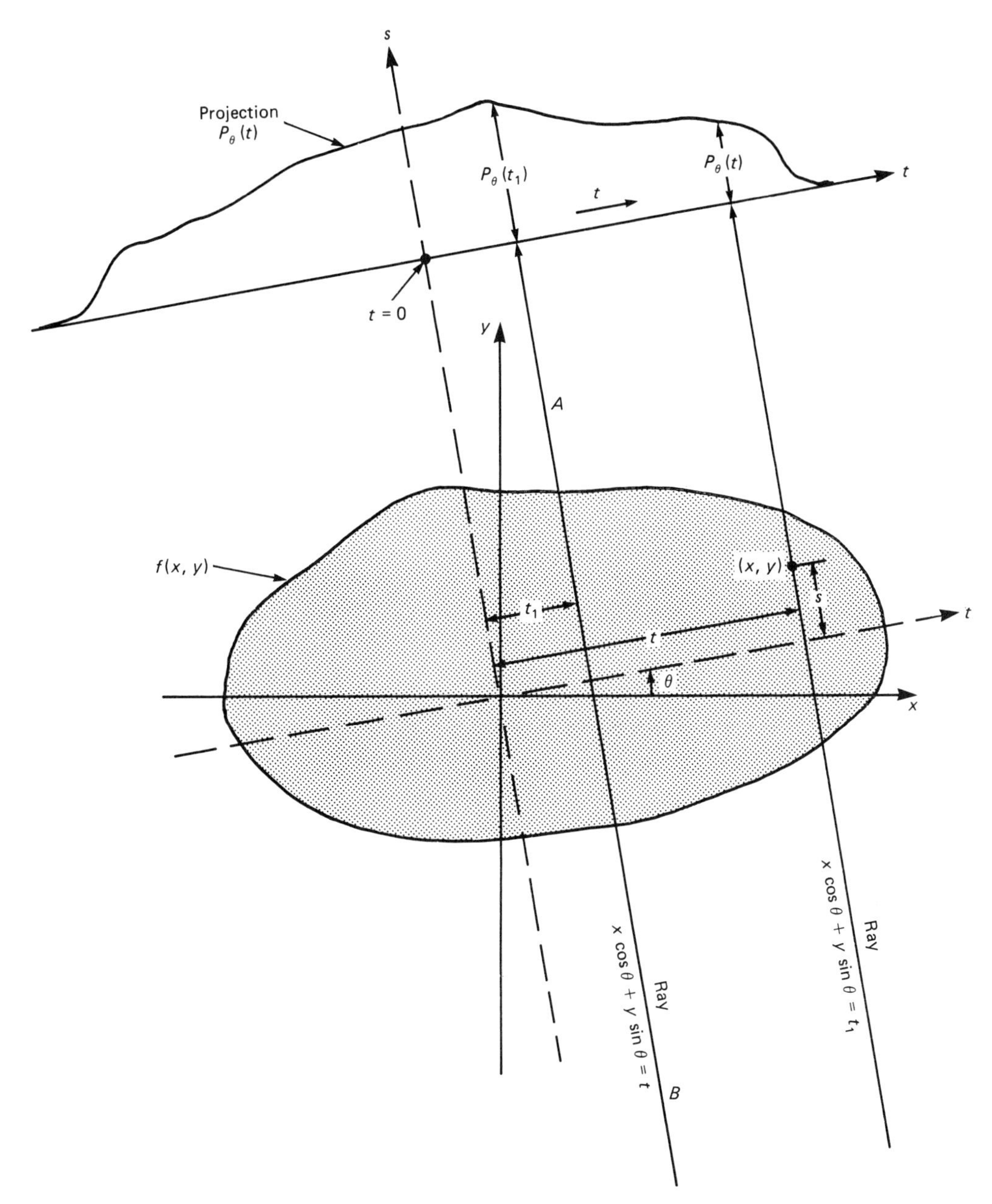

Figure 6.7 The function $P_\theta(t)$ is a parallel projection of $f(x, y)$. Also shown in this figure is the rotated coordinate system represented by s and t axes.

The function $P_\theta(t)$ as a function of t (for a given value of θ) defines the parallel projection of $f(x, y)$ for angle θ. The two-dimensional function $P_\theta(t)$ is also called the *Radon transform* of $f(x, y)$.

Whereas the Fourier transform, $F(\omega_1, \omega_2)$, of the object function will be, as before,

$$F(\omega_1, \omega_2) = \iint_S f(x, y) \exp\left[-j(\omega_1 x + \omega_2 y)\right] dx\, dy \tag{6.27}$$

the Fourier transform of a projection will be denoted by $S_\theta(\omega)$, and is defined as

$$S_\theta(\omega) = \int_{-\infty}^{\infty} P_\theta(t) \exp(-j\omega t)\, dt \tag{6.28}$$

6.4 TWO FUNDAMENTAL THEOREMS FOR TOMOGRAPHIC IMAGING

The Fourier Diffraction Projection Theorem

Fundamental to diffraction tomography is the *Fourier diffraction projection theorem*, which relates the Fourier transform of the measured forward-scattered data with the Fourier transform of the object. *The theorem is valid when the inhomogeneities in the object are only weakly scattering.* The statement of the theorem is:

> When an object, $f(x, y)$ is illuminated with a plane wave as shown in Fig. 6.8, the Fourier transform of the forward-scattered fields measured on line TT gives the values of the two-dimensional transform, $F(\omega_1, \omega_2)$, of the object along a circular arc in the frequency domain, as shown in the right half of the figure.

The importance of the theorem is made obvious by the fact that if an object is illuminated by plane waves from many directions over 360°, the resulting circular arcs in the (ω_1, ω_2)-plane will fill up the frequency domain. The function $f(x, y)$ may then be recovered by Fourier inversion.

Before giving a short proof of the theorem, we would like to say a few words about the dimensionality of the object vis-à-vis that of the wave fields. Although the theorem talks about a two-dimensional object, what is actually meant is an object that does not vary in the z-direction. In other words, the theorem is about a cylindrical object whose cross-sectional distribution is given by the function $f(x, y)$. The forward-scattered fields are measured on a line of detectors along TT in Fig. 6.8. If a truly three-dimensional object was illuminated by the plane wave, the forward-scattered fields would now have to be measured by a planar array of detectors. The Fourier transform of the fields measured by such an array would give the values of the three-dimensional transform of the object over a spherical surface. This was first shown by Wolf [105]. A more recent exposition is in [75], where the authors have also presented a new synthetic aperture procedure for a full three-dimensional reconstruction using only two rotational positions of the object. In this chapter, however, we will continue to work with two-dimensional objects in the sense described here.

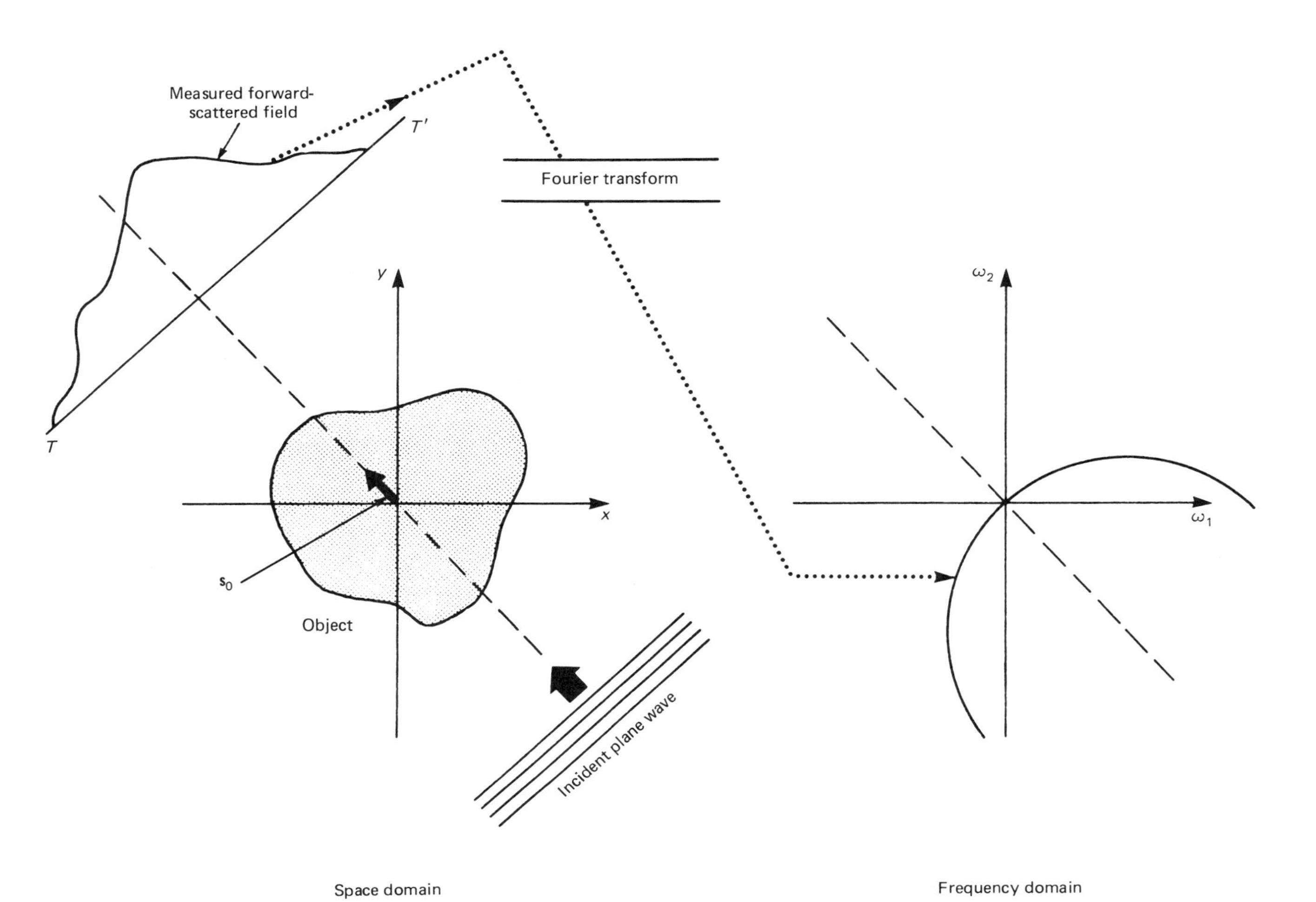

Figure 6.8 The Fourier diffraction projection theorem. In the left half of the figure, a plane wave is illuminating an object. The forward-scattered fields are measured on line TT'. A Fourier transform of this recording gives the values of $F(\omega_1, \omega_2)$ on the circular arc in the frequency domain, as shown on the right.

The theorem is proved by converting the wave propagation equation

$$(\nabla^2 + k^2)u_s(\mathbf{r}) = -k^2 f(\mathbf{r})u(\mathbf{r}) \tag{6.29}$$

which was discussed in Section 6.2, into an integral form by using a *Green's function* appropriate to the boundary conditions. Green's functions are used primarily to solve the wave propagation equation with sources as represented by the terms on the right-hand side [39, 70]. The propagation from the sources is assumed to take place in a homogeneous medium. This formalism can be applied to "solving" the equation above by treating the right-hand-side term as a source function for the scattered field, u_s, in a homogeneous medium of wavenumber k. Although in our case we are dealing basically with a source-free inhomogeneous medium, by writing the Helmholtz equation in the form given above, we have given it the appearance of what would correspond to propagation in a homogeneous medium with sources.

The Green's function is the solution of the differential equation

$$(\nabla^2 + k^2)g(\mathbf{r}_1 \mid \mathbf{r}_0) = -\delta(\mathbf{r}_1 - \mathbf{r}_0) \tag{6.30}$$

and describes the radiated fields from a single point source in a homogeneous medium, taking into account the relevant boundary conditions. For example, the Green's function corresponding to radiation within free space is a spherical wave expanding about the source point:

$$g(\mathbf{r}_1 \mid \mathbf{r}_0) = \frac{\exp(jk|\mathbf{r}_1 - \mathbf{r}_0|)}{|\mathbf{r}_1 - \mathbf{r}_0|} \tag{6.31}$$

where $\mathbf{r}_0$ is the location of the point source and g is the field at location $\mathbf{r}_1$. For our two-dimensional case, a more appropriate Green's function represents the radiated fields in free space (wavenumber: k) from a cylindrical filament at location $\mathbf{r}_0$ in the (x, y)-plane, the filament being perpendicular to the (x, y)-plane. The Green's function in this case is given by

$$g(\mathbf{r}_1 \mid \mathbf{r}_0) = \frac{j}{4} H_0(k|\mathbf{r}_1 - \mathbf{r}_0|) \tag{6.32}$$

where H_0 is the *zero-order Hankel function of the first kind.* Since Eq. (6.32) represents the radiation from a two-dimensional impulse source, the total radiation from all the sources on the right-hand side in Eq. (6.29), must be given by the following superposition:

$$u_s(\mathbf{r}) = \frac{jk^2}{4} \iint_S f(\mathbf{r}_0)u(\mathbf{r}_0)H_0(k|\mathbf{r} - \mathbf{r}_0|)\, d\mathbf{r}_0 \tag{6.33}$$

where S is any area in the (x, y)-plane that encloses the object cross section. The function H_0 has the following plane-wave decomposition:

$$H_0(k|\mathbf{r} - \mathbf{r}_0|) = \frac{1}{\pi} \int_{-\infty}^{\infty} \frac{1}{\beta} \exp\{j[\alpha(x - x_0) + \beta|y - y_0|]\}\, d\alpha \tag{6.34}$$

where

$$\beta = \sqrt{k^2 - \alpha^2} \tag{6.35}$$

Basically, Eq. (6.34) expresses a cylindrical wave as a superposition of plane waves. For $|\alpha| \leq k$, the plane waves are of the ordinary type, propagating along the direction given by $\tan^{-1}(\beta/\alpha)$. However, for $|\alpha| > k$, β becomes imaginary, and the waves decay exponentially, and they are called *evanescent waves*. Evanescent waves are usually of no significance beyond about 10 wavelengths from the source.

In spite of its appearance to the contrary, the integral equation (6.33) is not really a solution for the scattered field in terms of the object distribution. That is because u_s is also a part of the total field $u(\mathbf{r})$ on the right-hand side. In general, it is not possible to solve exactly this integral equation [and therefore, the differential equation (6.29) also] for a closed-form solution for the scattered field. (If such a closed-form solution were possible, then for the purpose of imaging it could perhaps be inverted to yield the unknown object distribution in terms of the measured fields.)

Since in general, it is impossible to solve Eq. (6.33) for the scattered field, approximations must be made. Two types of approximations are available: the Born and the Rytov. The *first-order* Born approximation consists of replacing the total field on the right-hand side in Eq. (6.33) by the incident field. Since the incident field is a plane wave, it can be expressed in the form

$$u_i(\mathbf{r}) = u_0 \exp(jk\mathbf{s}_0 \cdot \mathbf{r}) \tag{6.36}$$

where u_0 represents the complex amplitude of the plane wave which is propagating in the direction of the unit vector $\mathbf{s}_0$ (Fig. 6.8). Replacing $u(\mathbf{r})$ by $u_i(\mathbf{r})$ in Eq. (6.33) and using Eq. (6.36), we get for the scattered field

$$u_s(\mathbf{r}) = \frac{jk^2 u_0}{4\pi} \iint_S f(\mathbf{r}_0) \exp(jk\mathbf{s}_0 \cdot \mathbf{r}_0) \times \int_{-\infty}^{\infty} \frac{1}{\beta} \exp\{j[\alpha(x - x_0) + \beta|y - y_0|]\}\, d\alpha\, d\mathbf{r}_0 \tag{6.37}$$

where we have used the expansion in Eq. (6.34). As pointed out before, the condition $|\alpha| > k$ corresponds to evanescent modes reaching the receive array. Since for all practical purposes, these exponentially decaying modes will make negligible contributions to the recorded data, we can rewrite the equation for the scattered field as follows:

$$u_s(\mathbf{r}) = \frac{jk^2 u_0}{4\pi} \iint_S f(\mathbf{r}_0) \exp(jk\mathbf{s}_0 \cdot \mathbf{r}_0) \times \int_{-k}^{k} \frac{1}{\beta} \exp\{j[\alpha(x - x_0) + \beta|y - y_0|]\}\, d\alpha\, d\mathbf{r}_0 \tag{6.38}$$

In order to show the final steps in the proof of the theorem, we will now assume for notational convenience that the direction of the incident plane wave is along the y-axis. Since in transmission imaging, the scattered fields are measured by a linear array located at a $y = l$ that is greater than any y-coordinate within the object (see Fig. 6.9), $|y - y_0|$ in the expression above may simply be replaced by

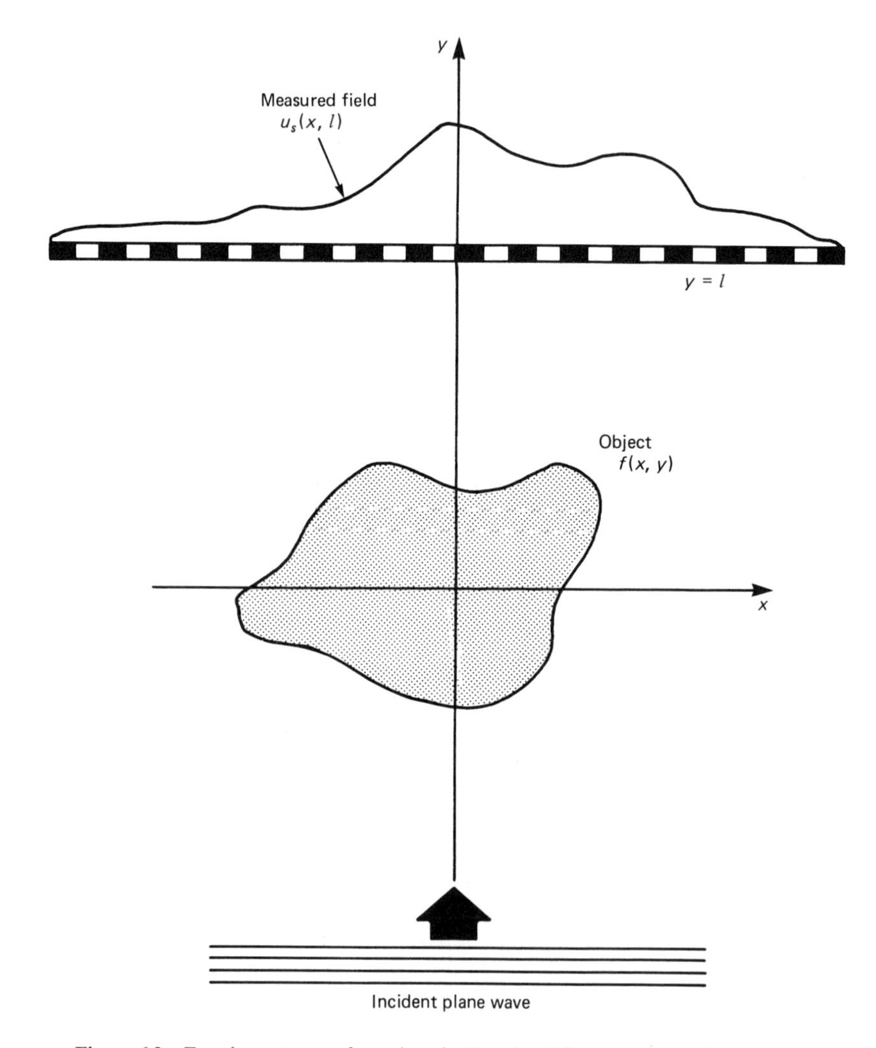

Figure 6.9 For the purpose of proving the Fourier diffraction projection theorem, we assume that the incident plane wave is propagating along positive y-direction and that the fields are measured on the $y = l$ line.

$l - y_0$, and the resulting form rewritten as

$$u_s(x, l) = \frac{jk^2 u_0}{4\pi} \int_{-k}^{k} d\alpha \, \frac{1}{\beta} \exp\,[j(\alpha x + \beta l)] \iint_S dx_0 \, y_0 f(x_0, y_0)$$
$$\times \exp\,\{-j[\alpha x_0 + (\beta - k) y_0]\} \qquad (6.39)$$

Recognizing the inner double integration as the two-dimensional Fourier transform of the object function, $f(x, y)$, this equation can be expressed somewhat more compactly as

$$u_s(x, l) = \frac{jk^2 u_0}{4\pi} \int_{-\infty}^{\infty} d\alpha \, \frac{1}{\beta} \exp\,[j(\alpha x + \beta l)] F(\alpha, \beta - k) \qquad (6.40)$$

where $F(\cdot, \cdot)$ is the Fourier transform of the object function as defined in Eq. (6.22).

Let $U_s(\omega)$ denote the Fourier transform of the one-dimensional function $u_s(x, l)$ with respect to x, that is,

$$U_s(\omega) = \int_{-\infty}^{\infty} u_s(x, l) \exp\,(-j\omega x) \, dx \qquad (6.41)$$

As mentioned before, the physics of wave propagation dictates that the highest angular spatial frequency in the measured scattered field on the line $y = l$ is unlikely to exceed k. Therefore, in almost all practical situations, $U_s(\omega) \simeq 0$ for $|\omega| > k$. This is consistent with neglecting the evanescent modes in Eq. (6.38). Substituting Eq. (6.40) in Eq. (6.41), and using the following property of Fourier integrals,

$$\int_{-\infty}^{\infty} \exp\,[j(\omega - \alpha)x] \, dx = 2\pi\delta(\omega - \alpha) \qquad (6.42)$$

where $\delta(\cdot)$ is the Dirac delta function, we get

$$U_s(\omega, l) = \frac{jk^2 u_0}{2} \frac{1}{\sqrt{k^2 - \omega^2}} \exp\,[j\sqrt{k^2 - \omega^2}\, l] F(\omega, \sqrt{k^2 - \omega^2} - k) \qquad \text{for } |\omega| < k$$
$$(6.43)$$

As ω varies from $-k$ to $+k$, the coordinates $(\omega, \sqrt{k^2 - \omega^2} - k)$ trace out a semicircular arc in the (ω_1, ω_2)-plane as shown in Fig. 6.10, which proves the theorem.

To summarize, if we take the Fourier transform of the forward-scattered data when the incident illumination is propagating along the positive y-axis, the resulting transform will be zero for angular spatial frequencies $|\omega| > k$. For $|\omega| < k$, the transform of the data gives values of the Fourier transform of the object on the semicircular arc AOB shown in Fig. 6.10 in the (ω_1, ω_2)-plane. The endpoints A and B of the semicircle are at a distance of $\sqrt{2}k$ from the origin in the frequency

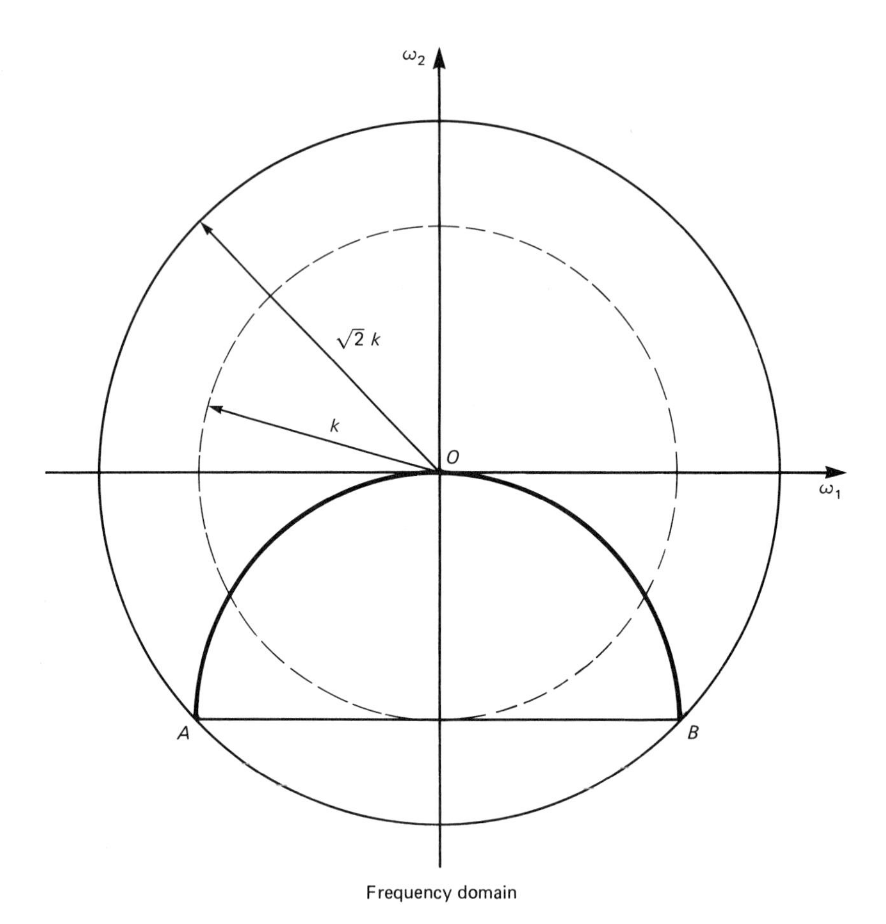

Figure 6.10 Fourier transform of the fields measured in Fig. 6.9 yields values of the two-dimensional transform of the object along the semi-circular arc AOB shown here.

domain. For notational convenience, we rewrite Eq. (6.43) as

$$U_s(\omega) = \frac{jk^2u_0}{2} \frac{1}{\sqrt{k^2-\omega^2}} \exp\,[jl\sqrt{k^2-\omega^2}]Q(\omega) \qquad \text{for } |\omega| < k \tag{6.44}$$

where we have suppressed l in the argument of U_s, and $Q(\omega)$ is equal to the values of the function $F(\omega_1, \omega_2)$ on the arc AOB in Fig. 6.10, as shown by

$$Q(\omega) = F(\omega, \sqrt{k^2-\omega^2} - k) \qquad \text{for } |\omega| < k \tag{6.45}$$

The proof of the theorem given above was based on the first-order Born approximation. The proof can also be established using the Rytov approximation, which for large objects (compared to a wavelength) characterized by very small inhomogeneities (less than 1 percent of the average value) gives results better than the Born approximation.

In the Rytov-based methods, the total field at a point is expressed as

$$u(\mathbf{r}) = \exp\,[\psi(\mathbf{r})] \tag{6.46}$$

The field is expressed entirely as a complex phase denoted by ψ. We may now express the total phase as a sum of complex phase ψ_i that would exist in the absence of scattering inhomogeneities, and deviations ψ_s from this value caused by the process of scattering, as shown by

$$\psi(\mathbf{r}) = \psi_i(\mathbf{r}) + \psi_s(\mathbf{r}) \tag{6.47}$$

In imaging algorithms based on Rytov methods, the measured data consists of the complex phase $\psi_s(\mathbf{r})$ that appears in this expression. The measurement of ψ_s is done by first either recording (after removing the object from the apparatus) or estimating the incident $u_i(\mathbf{r})$ which is equal to $\exp\,[\psi_i(\mathbf{r})]$. We then record the total field $u(\mathbf{r})$, which is equal to $\exp\,[\psi(\mathbf{r})]$. It follows from the expression above that the data ψ_s for imaging would then be given by

$$\psi_s(\mathbf{r}) = \ln u(\mathbf{r}) - \ln u_i(\mathbf{r}) \tag{6.48}$$

To correctly evaluate the logarithm of the nonzero complex numbers $u(\mathbf{r})$ and $u_i(\mathbf{r})$, we need to know the unwrapped phases of these quantities. Since phase unwrapping can be very noise sensitive, we must maintain a high signal-to-noise ratio in the experimental recording of the fields.

Since in the absence of any scattering, the incident plane wave travels unperturbed through the medium, we must have

$$\psi_i(\mathbf{r}) = u_0 \exp\,[jk\mathbf{s}_0 \cdot \mathbf{r}] \tag{6.49}$$

where, as before, $\mathbf{s}_0$ is a unit vector along the direction of the incident plane wave (see Fig. 6.8). Substituting Eq. (6.49) in Eq. (6.46), we may write

$$u(\mathbf{r}) = u_0 \exp\,[jk\mathbf{s}_0 \cdot \mathbf{r} + \psi_s(\mathbf{r})] \tag{6.50}$$

It can be shown that if we substitute this form for $u(\mathbf{r})$ in the propagation equation (6.29), the resultant may be converted into the following integral form [54]:

$$\psi_s(\mathbf{r}) = \frac{1}{u_i(\mathbf{r})} \iint_S g(\mathbf{r} \mid \mathbf{r}_0)[\nabla\psi_s \cdot \nabla\psi_s + k^2 f(\mathbf{r}_0)]u_i(\mathbf{r}_0)\, dx\, dy \tag{6.51}$$

The *first-order* Rytov solution is obtained by setting ψ_s equal to 0 on the right-hand side of Eq. (6.51), which gives

$$\psi_s(\mathbf{r}) = \frac{k^2}{u_i(\mathbf{r})} \iint_S g(\mathbf{r} \mid \mathbf{r}_0) f(\mathbf{r}_0) u_i(\mathbf{r}_0)\, dx\, dy \tag{6.52}$$

Comparing this result with Eq. (6.33), we get the following analytic relationship between the first-order Born and the first-order Rytov approximations:

$$u_s(\mathbf{r}) = u_i(\mathbf{r})\psi_s(\mathbf{r}) \tag{6.53}$$

Because of this simple relationship, any algorithm developed for the first-order Born approximation for the data u_s can be easily extended to the first-order Rytov approximation for the data ψ_s. Therefore, in the rest of the discussion on diffraction tomography, we will only deal with the first-order Born approximation.

The Fourier Slice Theorem

Fundamental to a number of reconstruction techniques for nondiffracting sources such as x-rays is the *Fourier slice theorem.* It relates the one-dimensional Fourier transform of a projection of a function $f(x, y)$ to its two-dimensional Fourier transform. The statement of the theorem is:

> The Fourier transform of a parallel projection of an image $f(x, y)$ taken at an angle θ gives a slice of the two-dimensional transform, $F(\omega_1, \omega_2)$, subtending an angle θ with the ω_1-axis. In other words, the Fourier transform of $P_\theta(t)$ gives the values of $F(\omega_1, \omega_2)$ along line BB shown in Fig. 6.11.

The theorem implies that if an object is illuminated from many different directions, it may be possible to fill up the frequency domain. Theoretically at least, the object could then be reconstructed by direct Fourier inversion. In practice, it is more common to use the filtered-backprojection algorithms for reconstruction, because of their greater accuracy and faster implementations. These algorithms, discussed in Section 6.6, are derived by using the Fourier slice theorem.

To prove the theorem, let us first consider the values of $F(\omega_1, \omega_2)$ on the line $\omega_1 = 0$ in the (ω_1, ω_2)-plane. From Eq. (6.27), we have

$$\begin{aligned} F(\omega_1, 0) &= \int_{-\infty}^{\infty}\int_{-\infty}^{\infty} f(x, y) \exp(-j\omega_1 x)\, dx\, dy \\ &= \int_{-\infty}^{\infty}\left[\int_{-\infty}^{\infty} f(x, y)\, dy\right] \exp(-j\omega_1 x)\, dx \\ &= \int P_0(t) \exp(-j\omega_1 t)\, dt = S_0(\omega_1) \end{aligned} \tag{6.54}$$

since $\int f(x, y)\, dy$ gives the projection of the image for $\theta = 0$. Note also that for this projection, x and t are the same. The Fourier transform $S_\theta(\omega)$ of a projection was defined in Eq. (6.28).

The result above indicates that the values of the Fourier transform $F(\omega_1, \omega_2)$ on the line defined by $\omega_2 = 0$ can be obtained by Fourier transforming the vertical projection of the image. This result can be generalized to show that if $F(\Omega, \theta)$ denotes the values of $F(\omega_1, \omega_2)$ along a line at an angle θ with the ω_1-axis as shown

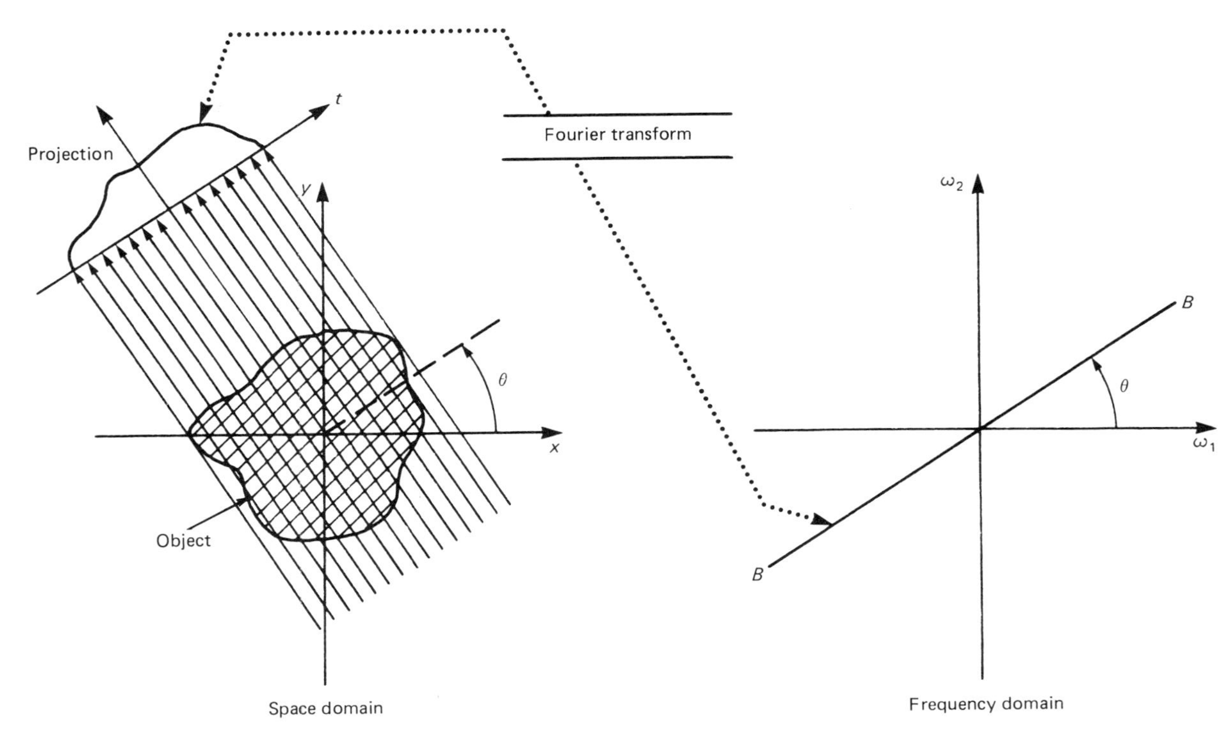

Figure 6.11 The Fourier slice theorem. In the left half of the figure, $P_\theta(t)$ is a parallel projection of $f(x, y)$ at angle θ. Fourier transform of $P_\theta(t)$ gives values of $F(\omega_1, \omega_2)$ on line BB shown on the right.

in Fig. 6.11, and if $S_\theta(\omega)$ is the Fourier transform of the projection $P_\theta(t)$, then

$$F(\Omega, \theta) = S_\theta(\Omega) \tag{6.55}$$

This can be proved as follows. Let $f(t, s)$ be the function $f(x, y)$ in the rotated coordinate system (t, s) in Fig. 6.7. The coordinates (t, s) are related to the (x, y) coordinates by

$$\begin{bmatrix} t \\ s \end{bmatrix} = \begin{bmatrix} \cos\theta & \sin\theta \\ -\sin\theta & \cos\theta \end{bmatrix} \begin{bmatrix} x \\ y \end{bmatrix} \tag{6.56}$$

Clearly, we have

$$P_\theta(t) = \int_{-\infty}^{\infty} f(t, s)\, ds \tag{6.57}$$

Therefore,

$$\begin{aligned} S_\theta(\Omega) &= \int_{-\infty}^{\infty} P_\theta(t) \exp(-j\Omega t)\, dt \\ &= \int_{-\infty}^{\infty} \int_{-\infty}^{\infty} f(t, s)\, ds \exp(-j\Omega t)\, dt \end{aligned} \tag{6.58}$$

Transforming the right-hand side of the equation above into (x, y)-coordinates, we get

$$\begin{aligned} S_\theta(\omega) &= \int_{-\infty}^{\infty} \int_{-\infty}^{\infty} f(x, y) \exp[-j\Omega(x \cos\theta + y \sin\theta)]\, dx\, dy \\ &= F(\omega_1, \omega_2) \qquad \text{for} \quad \begin{cases} \omega_1 = \Omega \cos\theta \\ \omega_2 = \Omega \sin\theta \end{cases} \\ &= F(\Omega, \theta) \end{aligned} \tag{6.59}$$

which proves the theorem. This result is also known as the *projection slice theorem.*

6.5 INTERPOLATION AND A FILTERED-BACKPROPAGATION ALGORITHM FOR DIFFRACTING SOURCES

In our proof of the Fourier diffraction projection theorem, we showed that when an object is illuminated with a plane wave traveling in the positive y-direction, the Fourier transform of the forward-scattered fields gives values of $F(\omega_1, \omega_2)$ on the arc AB shown in Fig. 6.10. Therefore, if an object is illuminated from many different directions, we can, in principle, fill up a disk of diameter $\sqrt{2}\,k$ in the frequency domain with samples of $F(\omega_1, \omega_2)$, which is the Fourier transform of the object, and then reconstruct the object by direct Fourier inversion. Therefore, we can say that diffraction tomography determines the object up to a maximum angular spatial frequency of $\sqrt{2}\,k$. To this extent, the reconstructed object is a low-pass version of the original. In practice, the loss of resolution caused by this bandlimit is negligible, it being more influenced by considerations such as the aperture sizes of the transmitting and receiving elements.

The fact that the frequency-domain samples are available over circular arcs, whereas for fast Fourier inversion it is desired to have samples over a rectangular lattice, is a source of computational difficulty with this direct Fourier inversion approach. To help the reader visualize the distribution of the available frequency-domain information, we have shown in Fig. 6.12 the sampling points on a circular arc grid, each arc in this grid corresponding to the transform of one projection. It should also be clear from this figure that by illuminating the object over 360°, a *double* coverage of the frequency domain is generated; note, however, that this

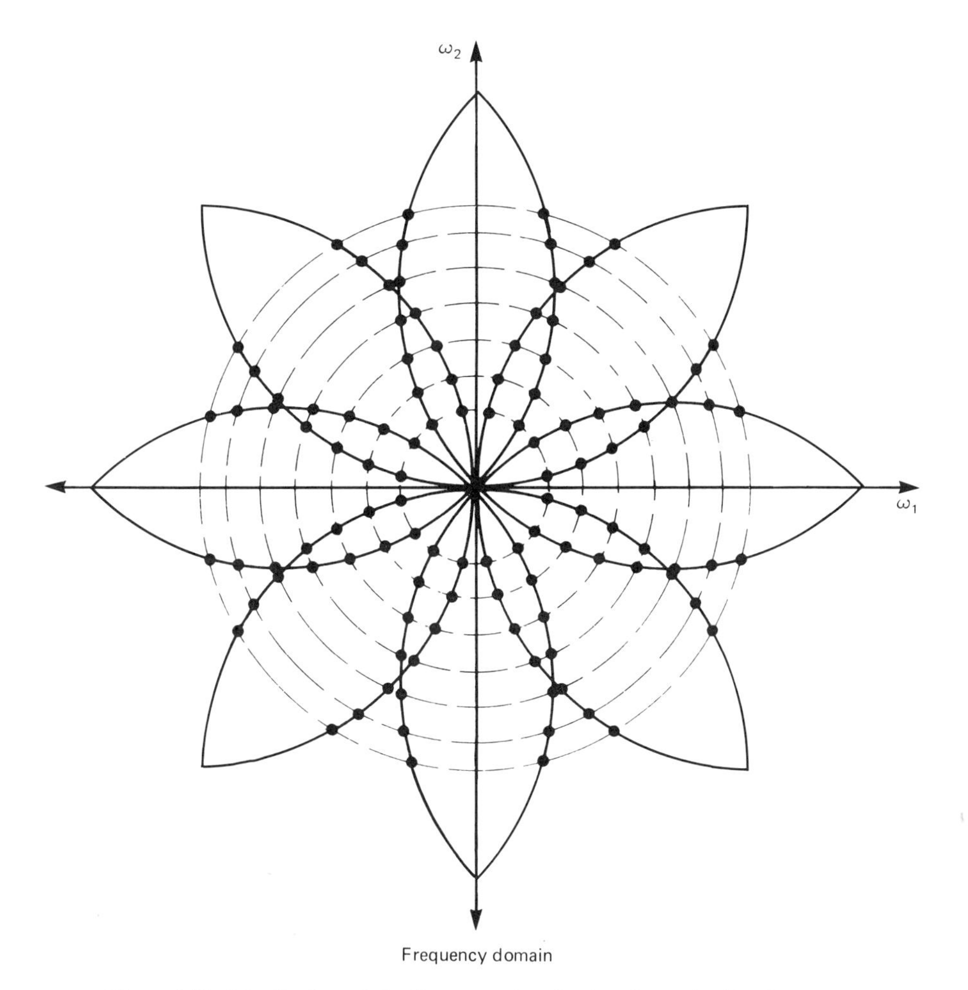

Figure 6.12 Distribution of the frequency-domain data from the Fourier transforms of the projections. Each semicircular arc corresponds to one projection.

double coverage is uniform. We may get a complete coverage of the frequency domain, with illumination restricted to a portion of 360°; however, in that case there would be patches in the (ω_1, ω_2)-plane where we would have a double coverage. In interpolating from circular arc grids to rectangular grids, it is often easier to contend with a uniform double coverage, as opposed to a coverage that is single in most areas and double in patches.

However, for some applications not given to data collection from all possible directions, it is useful to bear in mind that it is not necessary to go all around an object to get a complete coverage of the frequency domain. In principle, it should be possible to get an equal-quality reconstruction when illumination angles are restricted to a 180°-plus interval, the angles in excess of 180° required to complete the coverage of the frequency domain.

In order to discuss frequency-domain interpolation between a circular arc grid on which the data are generated by diffraction tomography, and a rectangular grid suitable for image reconstruction, we must first select parameters for representing each grid and then write down the relationship between the two sets of parameters.

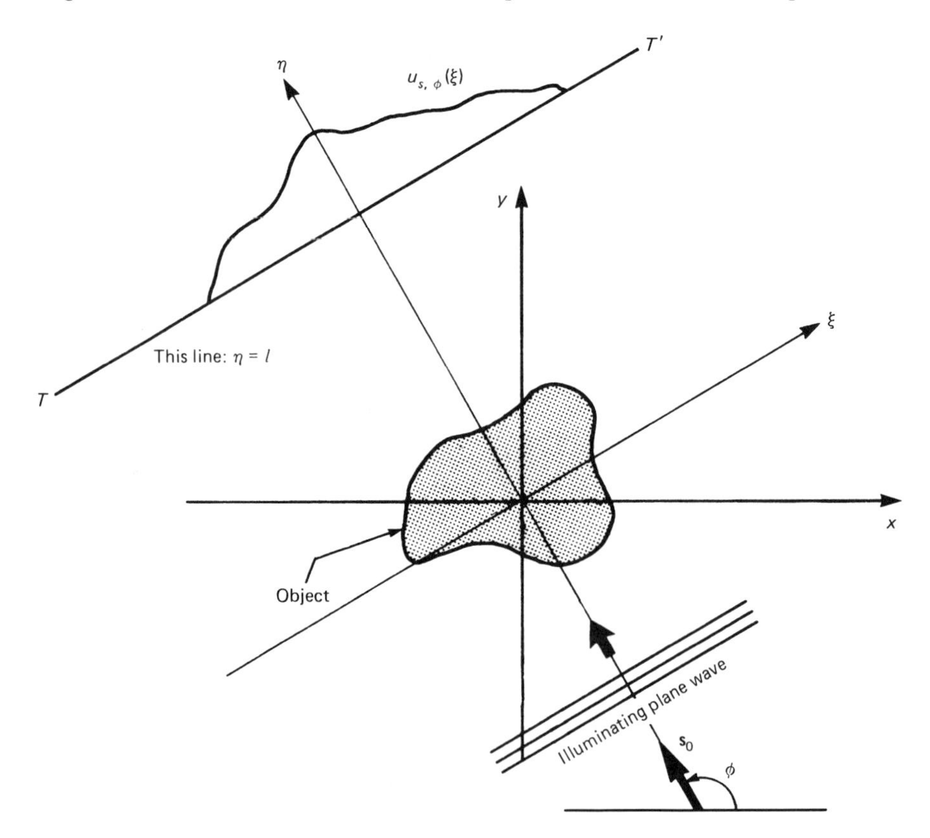

Figure 6.13 The (ξ, η) coordinate system and how the illumination angle ϕ is measured are illustrated here.

In Eq. (6.44), $U_s(\omega)$ was used to denote the Fourier transform of the transmitted data when an object is illuminated with a plane wave traveling along the positive y direction. We now use $U_{s,\phi}(\omega)$ to denote this Fourier transform, where the subscript ϕ indicates the angle of illumination. This angle is measured as shown in Fig. 6.13. Similarly, $Q(\omega)$ used in Eq. (6.44) is now represented by $Q(\omega, \phi)$ to indicate the values of $F(\omega_1, \omega_2)$ along a semicircular arc oriented at an angle ϕ as shown in Fig. 6.14. Therefore, when an illuminating plane wave is incident at angle ϕ, the equality in Eq. (6.44) can be rewritten as

$$U_{s,\phi}(\omega) = \frac{jk^2 u_0}{2} \frac{1}{\sqrt{k^2 - \omega^2}} \exp\left[jl\sqrt{k^2 - \omega^2}\right] Q(\omega, \phi) \qquad \text{for } |\omega| < k \tag{6.60}$$

In most cases the transmitted data will be uniformly sampled in space, and a *discrete* Fourier transform of these data will generate uniformly spaced samples of

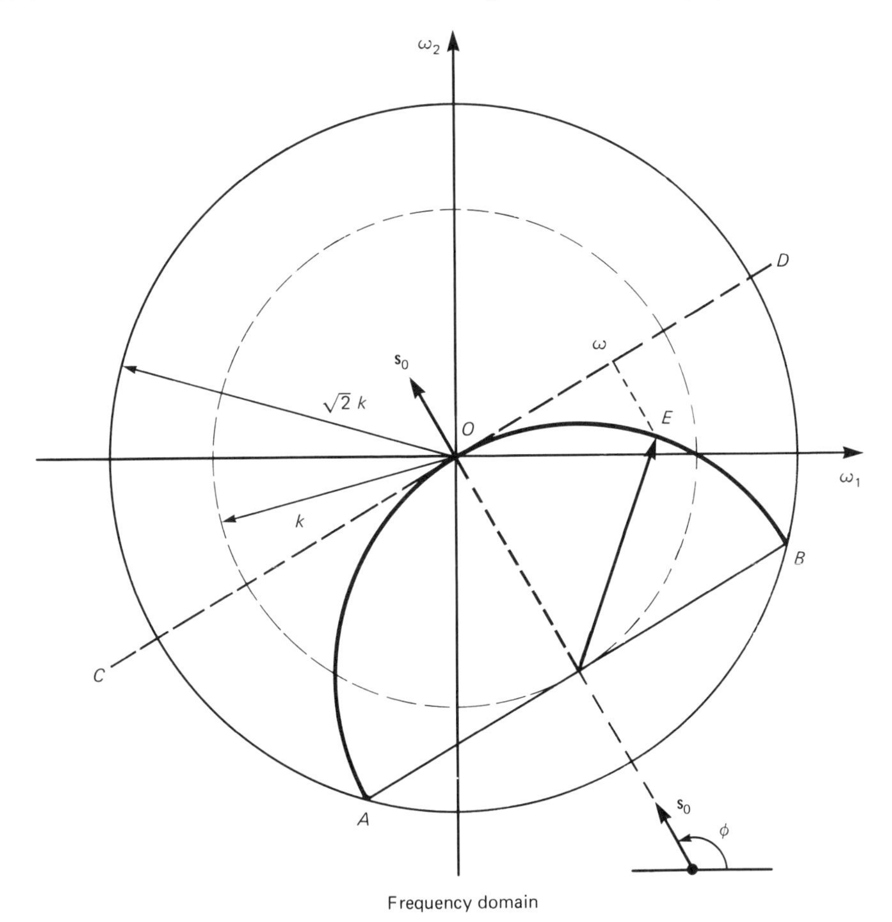

Figure 6.14 When the object is illuminated as shown in Fig. 6.13, the Fourier transform of $u_{s,\phi}(\xi, l)$ gives the values of $F(\omega_1, \omega_2)$ on the arc AOB shown here.

$U_{s,\phi}(\omega)$ in the ω-domain. We will therefore designate each point on the arc AOB by its (ω, ϕ) parameters. [Note that (ω, ϕ) are *not* the polar coordinates of a point on arc AOB in Fig. 6.14. Therefore, ω is *not* the radial distance in the (ω_1, ω_2)-plane. For point E shown, the parameter ω is obtained by projecting E onto line CD.] We continue to denote the rectangular coordinates in the frequency domain by (ω_1, ω_2).

Before we present relationships between (ω, ϕ) and (ω_1, ω_2), it must be mentioned that the points generated by the AO and OB portions of the arc AOB as ϕ is varied from 0 to 2π are to be considered separately. We do this because, as mentioned before, the arc AOB generates a double coverage of the frequency domain as ϕ is varied from 0 to 2π, which is undesirable for discussing a one-to-one transformation between the (ω, ϕ) parameters and the (ω_1, ω_2) coordinates.

We now reserve (ω, ϕ) parameters to denote the arc grid generated by the portion OB as shown in Fig. 6.14. It is important to note that for this arc grid, ω varies from 0 to k and ϕ from 0 to 2π.

We now present the transformation equations between (ω, ϕ) and (ω_1, ω_2). We accomplish this in a slightly roundabout manner by first defining polar coordinates (Ω, θ) in the (ω_1, ω_2)-plane as shown in Fig. 6.15. In order to go from (ω_1, ω_2) to (ω, ϕ), we will first transform from the former coordinates to (Ω, θ) and then from (Ω, θ) to (ω, ϕ). The rectangular coordinates (ω_1, ω_2) are related to the polar coordinates (Ω, θ) by (Fig. 6.15)

$$\Omega = \sqrt{\omega_1^2 + \omega_2^2} \tag{6.61}$$

$$\theta = \tan^{-1} \frac{\omega_2}{\omega_1} \tag{6.62}$$

In order to relate (Ω, θ) to (ω, ϕ), we now introduce a new angle β, which is the angular position of a point (ω_1, ω_2) on arc OB in Fig. 6.15. Note from the figure that the point characterized by angle β is also characterized by parameter ω. The relationship between ω and β is given by

$$\omega = k \sin \beta \tag{6.63}$$

The following relationship exists between the polar coordinates (Ω, θ) on the one hand and the parameters β and ϕ on the other:

$$\beta = 2 \sin^{-1} \frac{\Omega}{2k} \tag{6.64}$$

$$\phi = \theta + \frac{\pi}{2} + \frac{\beta}{2} \tag{6.65}$$

By substituting Eq. (6.64) in (6.63) and then using (6.61), we can express ω in terms of ω_1 and ω_2. This result is shown in Eq. (6.66). Similarly, by substituting Eq. (6.62)

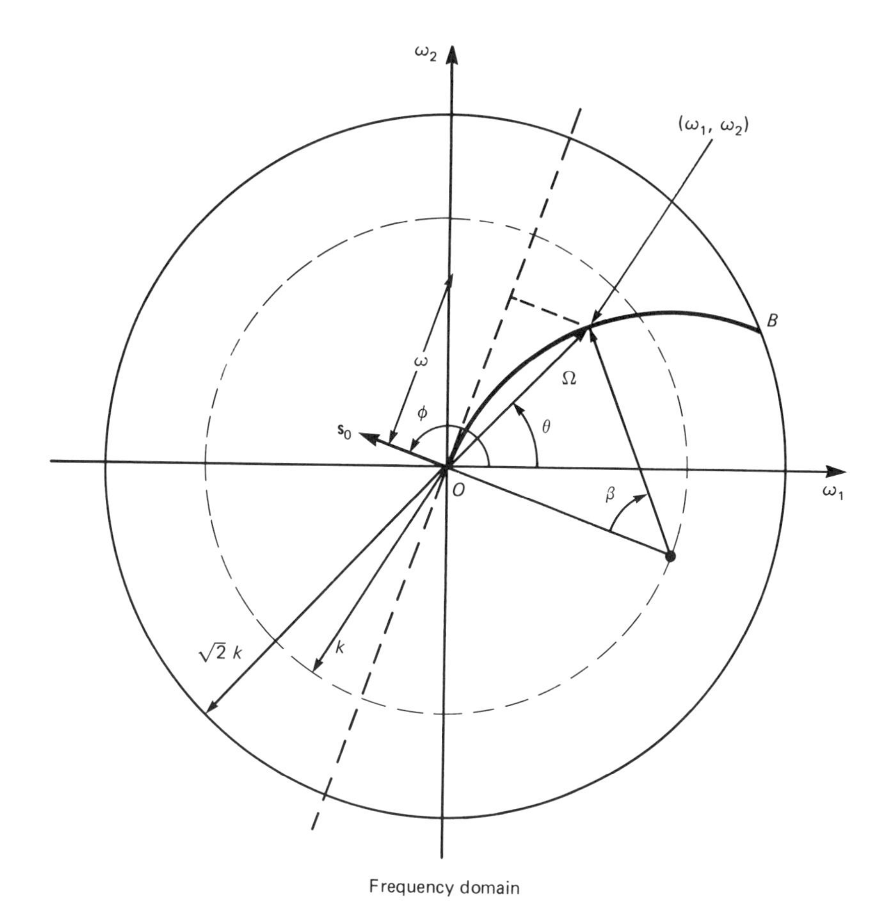

Figure 6.15 Polar coordinates (Ω, θ).

in (6.65), we obtain Eq. (6.67).

$$\omega = \sin \left\{ 2 \sin^{-1} \frac{\sqrt{\omega_1^2 + \omega_2^2}}{2k} \right\} \tag{6.66}$$

$$\phi = \tan^{-1} \frac{\omega_2}{\omega_1} + \sin^{-1} \frac{\sqrt{\omega_1^2 + \omega_2^2}}{2k} + \frac{\pi}{2} \tag{6.67}$$

These are our transformation equations for interpolating from the (ω, ϕ) parameters used for data representation to the (ω_1, ω_2) parameters needed for inverse transformation. To convert a particular rectangular point into (ω, ϕ)-domain, we substitute its ω_1 and ω_2 values in Eqs. (6.66) and (6.67). The resulting values for ω and ϕ may not correspond to any for which $Q(\omega, \phi)$ is known. By virtue of Eq. (6.60), $Q(\omega, \phi)$

will only be known over a uniformly sampled set of values for ω and ϕ. In order to determine Q at the calculated ω and ϕ, we use the following procedure. Given $N_\omega \times N_\phi$ uniformly located samples, $Q(\omega_i, \phi_j)$, we calculate a bilinearly interpolated value of this function at the desired ω and ϕ by using

$$Q(\omega, \phi) = \sum_{i=1}^{N_\omega} \sum_{j=1}^{N_\phi} Q(\omega_i, \phi_j) h_1(\omega - \omega_i) h_2(\phi - \phi_j) \tag{6.68}$$

where

$$h_1(\omega) = \begin{cases} 1 - \dfrac{|\omega|}{\Delta\omega}, & |\omega| \le \Delta\omega \\ 0, & \text{otherwise} \end{cases} \tag{6.69}$$

$$h_2(\phi) = \begin{cases} 1 - \dfrac{|\phi|}{\Delta\phi}, & |\phi| \le \Delta\phi \\ 0, & \text{otherwise} \end{cases} \tag{6.70}$$

$\Delta\phi$ and $\Delta\omega$ are the sampling intervals for ϕ and ω, respectively. When expressed in the manner shown above, bilinear interpolation may be interpreted as the output of a filter whose impulse response is $h_1 h_2$.

The results obtained with bilinear interpolation can be considerably improved if we first increase the sampling density in the (ω, ϕ)-plane by using the computationally efficient method of zero-extending the inverse two-dimensional inverse *fast Fourier transform* (FFT) of the $Q(\omega_i, \phi_j)$ matrix. The technique consists of first taking a two-dimensional inverse FFT of the $N_\omega \times N_\phi$ matrix consisting of the $Q(\omega_i, \phi_j)$ values, zero-extending the resulting $N_\omega \times N_\phi$ array of numbers to, let us say, $mN_\omega \times nM_\phi$, and then taking the FFT of this new array. The result is an mn-fold increase in the density of samples in the (ω, ϕ)-plane.

After computing $Q(\omega, \phi)$ at each point of a rectangular grid by the procedure outlined above, the object $f(x, y)$ is obtained by a simple two-dimensional inverse FFT. To illustrate the accuracy of the interpolation-based algorithms, we will use the image in Fig. 6.16(a) as a test "object" for showing some computer simulation results. Figure 6.16(a), with its gray levels as shown in Fig. 6.16(b), is a modification of the Shepp and Logan "phantom" described in Section 6.6 to the case of diffraction imaging. The gray levels shown in Fig. 6.16(b) represent the refractive index values. This test image is a superposition of ellipses, with each ellipse being assigned a refractive index value as shown in Table 6.1. A major advantage of using an image like Fig. 6.16(a) for computer simulation is that we can obtain easily the transforms of the diffracted projections of each ellipse therein. As explained in [80], the transforms of diffracted projections of an ellipse are obtained by simply sampling the two-dimensional transform of the ellipse on a circular arc grid. When weak scattering is assumed, the transform of a diffracted projection for the entire object is the sum of the such transforms for individual ellipses. *We must mention that by generating the diffracted projection data for computer simulation by this procedure, we are testing only the accuracy of the reconstruction algorithm, without checking whether or*

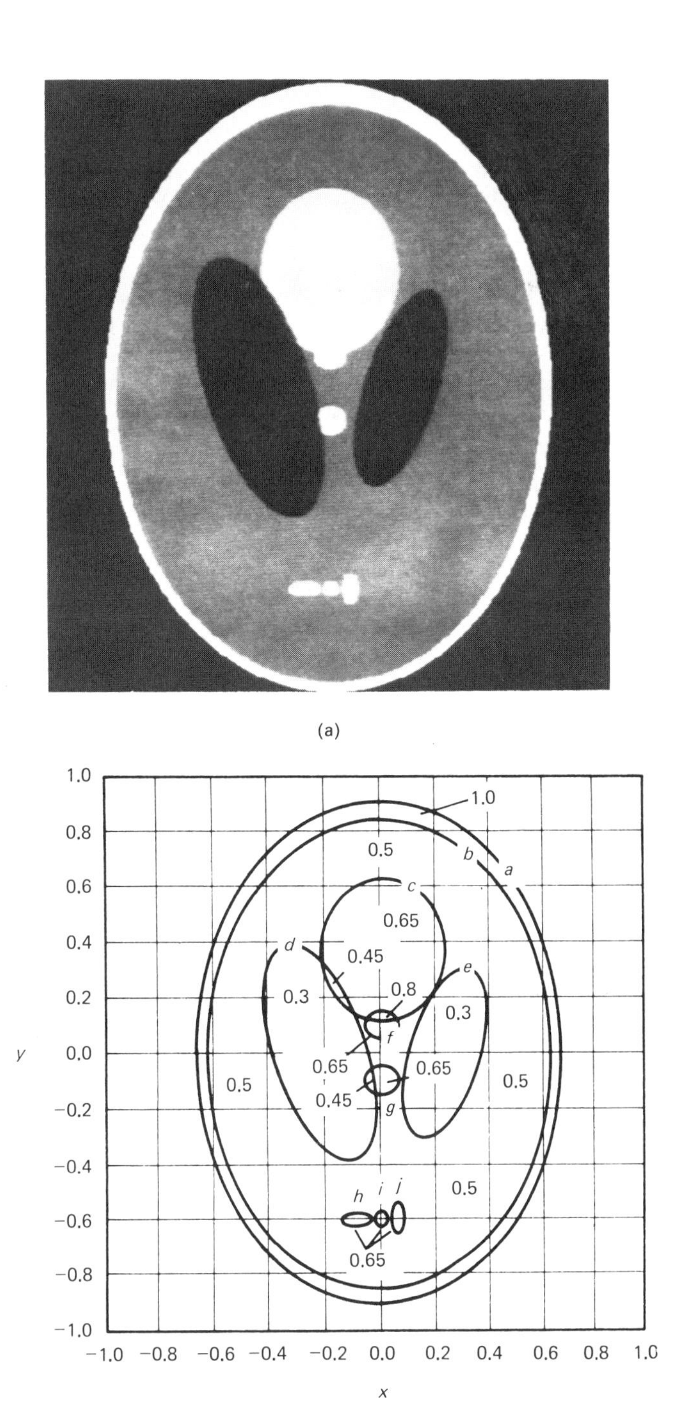

Figure 6.16 (a) Test "object" used for testing by computer sumulation the accuracy of reconstruction algorithms for diffraction tomography. (b) The gray levels in (a) represent the refractive index distribution in the test image. The numerical values of this distribution are shown here.

TABLE 6.1 THE PARAMETERS OF THE COMPONENT ELLIPSES OF THE TEST IMAGE USED FOR RECONSTRUCTION WITH DIFFRACTION TOMOGRAPHY ALGORITHMS

Ellipse	Coordinates of the center	A major axis	B minor axis	α rotation angle	ρ gray level
a	(0, 0)	0.92	0.69	90°	1.0
b	(0, −0.184)	0.874	0.6624	90°	−0.5
c	0.22, 0)	0.31	0.11	72°	−0.2
d	(−0.22, 0)	0.41	0.16	108°	−0.2
e	(0, 0.35)	0.25	0.21	90°	0.1
f	(0, 0.1)	0.046	0.046	0	0.15
g	(0, −0.1)	0.046	0.046	0	0.15
h	(−0.08, −0.605)	0.046	0.023	0	0.15
i	(0, −0.605)	0.023	0.023	0	0.15
j	(0.06, −0.605)	0.046	0.023	90°	0.15

not the "test object" satisfies the underlying assumption of weak scattering. In order to test this crucial assumption, we must generate exactly on a computer the forward-scattered data of the object. For multicomponent objects, such as the one shown in Fig. 6.16(a), it is very difficult to do so due to the interactions between the components. (However, it is possible to generate the exact forward-scattered fields for simple objects like a cylinder, and then reconstruct from these fields by using the algorithms presented here. Using this procedure, a comparison of the Born- and Rytov-approximations-based reconstructions has been presented by Slaney and Kak [94].)

Computer simulation results are shown in Figs. 6.17 and 6.18. For Fig. 6.17, prior to bilinear interpolation, the space-domain zero-padding technique in the preceding section was employed to increase the (ω, ϕ)-plane sampling density eightfold, from the original 90×64 sized array to a 360×128 array. When instead of using 90 sampling points per projection, we use 128, and then by zero-padding technique increase the size of the (ω, ϕ) array from 128×64 to 512×128 (again an eightfold increase in sampling density) prior to applying bilinear interpolation, the resulting reconstruction is shown in Fig. 6.18. Comparing Figs. 6.17(a) and 6.18(a), we see the rings and other interference artifacts when only 90 sampling points are used for each projection. For the reconstructions shown in Figs. 6.17(a) and 6.18(a), the VAX-11/780 processing time was 2.2 min.

It has recently been shown by Devaney [29] that there is an alternative method for reconstructing images from the diffracted projection data. This procedure, called the *filtered-backpropagation method,* is similar in spirit to the filtered-backprojection techniques which (due to their superior numerical accuracy) have been one factor in the enormous success of x-ray tomography. (The filtered-backprojection algorithms for nondiffracting sources are discussed in the next section.) Unfortunately, whereas the filtered-backprojection algorithms also possess efficient implementations, the same cannot be said for the filtered-backpropagation

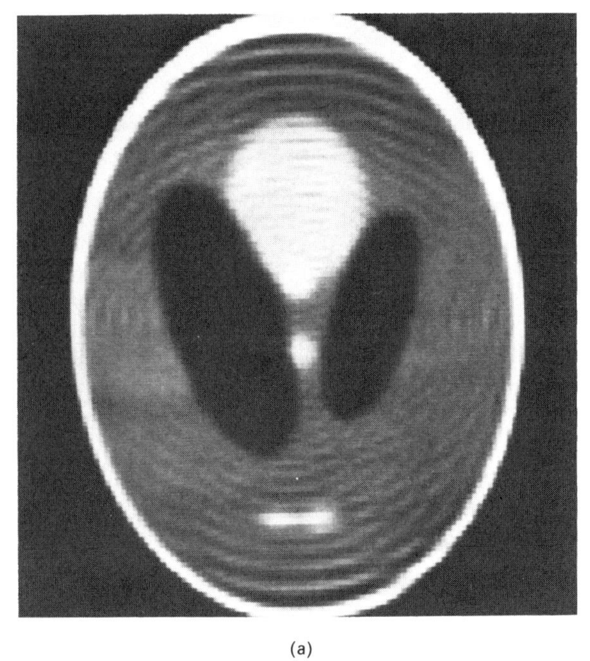

(a)

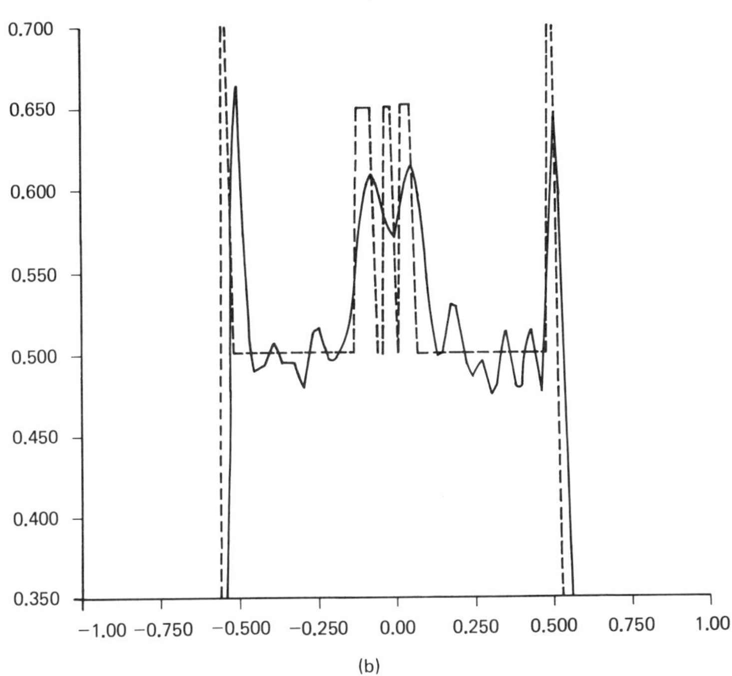

(b)

Figure 6.17 (a) A 128 × 128 reconstruction obtained with bilinear interpolation from 64 projections and 90 samples per projection. Prior to bilinear interpolation, the frequency-domain sampling density was increased eightfold by the zero-padding technique. (b) Numerical comparison of the true and reconstructed values on line $y = -0.605$.

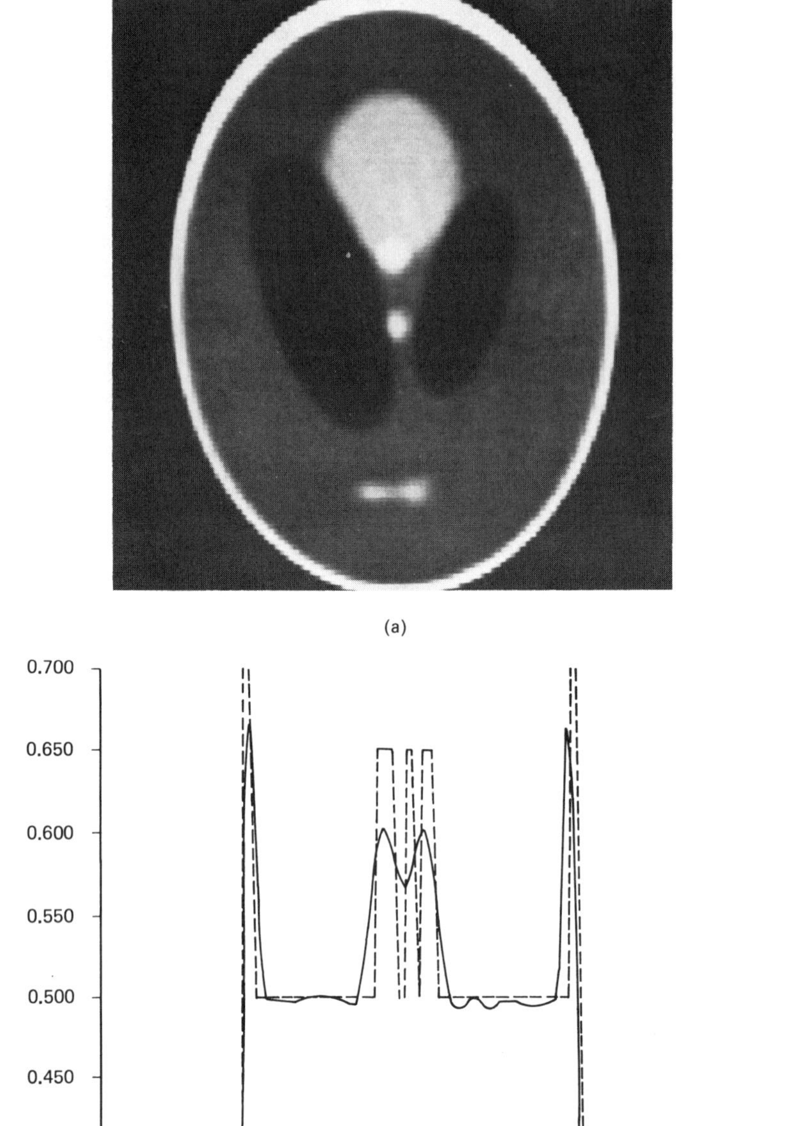

Figure 6.18 (a) A 128 × 128 reconstruction obtained with bilinear interpolation from 64 projections and 128 samples per projection. Prior to applying bilinear interpolation, the sampling density in the (ω, ϕ) domain was increased eightfold by using the zero-padding technique. (b) Numerical comparison of the true and reconstructed values on the line $y = -0.605$ through the test image.

algorithms. The latter class of algorithms is highly demanding of computer time, much more so than the interpolation procedure discussed above. With regard to accuracy, they do not seem to possess any advantage, especially if interpolation is carried out after increasing the sampling density by appropriate zero-padding as discussed above. The point to be made here is that with a given amount of computer processing effort, whereas for the nondiffracting case the filtered-backprojection algorithms are more accurate than algorithms based on frequency-domain interpolation, for the case of imaging with a diffracting source the opposite seems to be true [80].

The filtered-backpropagation algorithm may be derived by first writing the inverse transform of Eq. (6.27) in the polar coordinates (Ω, θ) as follows:

$$f(x, y) = \frac{1}{4\pi^2} \int_0^{2\pi} \int_0^{\sqrt{2}k} F_p(\Omega, \theta) \exp [j\Omega(x \cos \theta + y \sin \theta)]\Omega \, d\Omega \, d\theta \tag{6.71}$$

where the upper limit on Ω follows from the discussion in Section 6.4. The function $F_p(\Omega, \theta)$ is the object transform $F(\omega_1, \omega_2)$ represented in polar coordinate. That is,

$$F_p(\Omega, \theta) = F(\omega_1, \omega_2) \tag{6.72}$$

at $\omega_1 = \Omega \cos \theta$ and $\omega_2 = \Omega \sin \theta$. Devaney has shown that by using the transformation equations given at the beginning of this section, the integral in Eq. (6.71) can be expressed as

$$f(x, y) = \frac{1}{4\pi^2} \int_0^{2\pi} d\phi \int_{-k}^{k} \frac{j}{ku_0} \exp (-jkl) U_{s,\phi}(\omega) |\omega| \times \exp [j(\sqrt{k^2 - \omega^2} - k)(\eta - l)] \exp (j\omega\xi) \, d\omega \tag{6.73}$$

where

$$\xi = x \sin \phi - y \cos \phi \tag{6.74}$$

and

$$\eta = x \cos \phi + y \sin \phi \tag{6.75}$$

The coordinates (ξ, η) are illustrated in Fig. 6.13. To bring out the filtered-backpropagation implementation, we write here separately the inner integration:

$$\Pi_\phi(\xi, \eta) = \frac{1}{2\pi} \int_{-\infty}^{\infty} \Gamma_\phi(\omega) H(\omega) G_\eta(\omega) \exp (j\omega\xi) \, d\omega \tag{6.76}$$

where

$$H(\omega) = \begin{cases} |\omega|, & |\omega| \le k_0 \\ 0, & |\omega| > k_0 \end{cases} \tag{6.77}$$

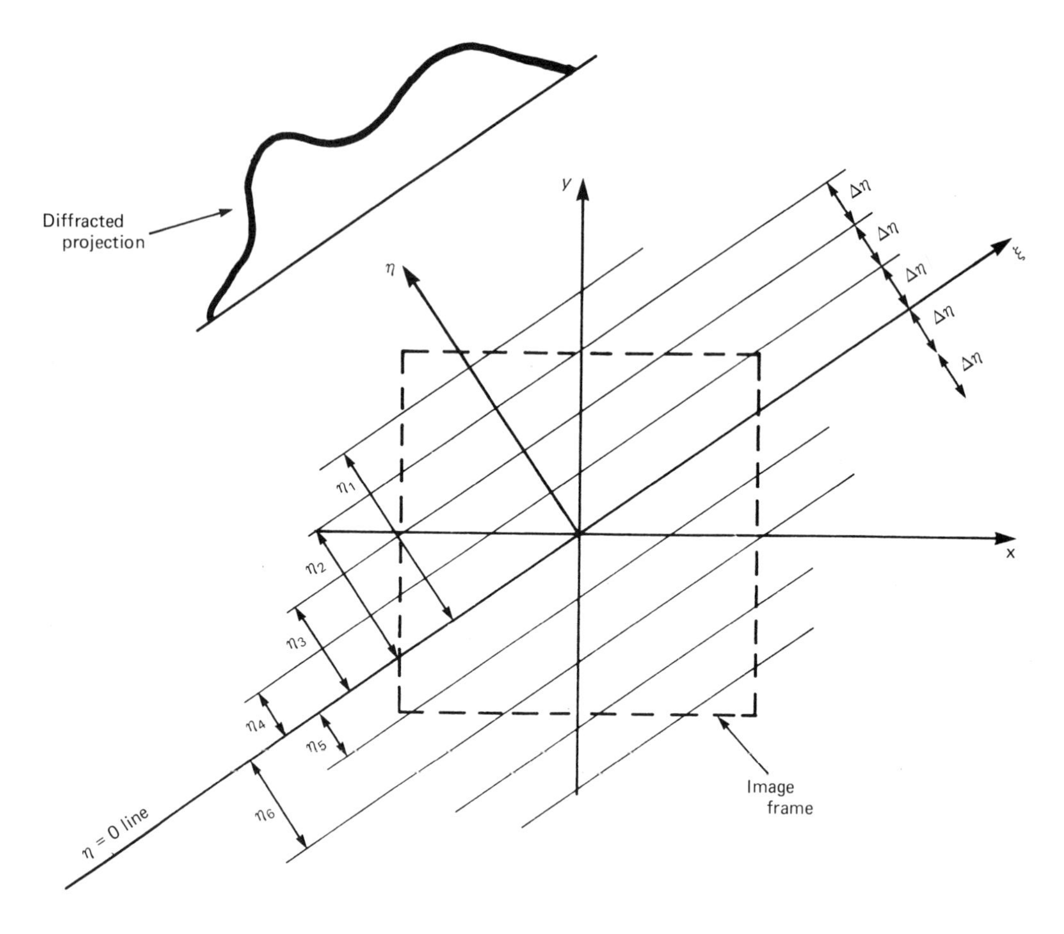

Figure 6.19 For each η = constant line shown in this figure, the diffracted projection must be filtered with a different transfer function.

and

$$G_\eta(\omega) = \begin{cases} \exp\,[j(\sqrt{k^2-\omega^2}-k)(\eta-l)], & |\omega| \le k \\ 0, & |\omega| > k \end{cases} \tag{6.78}$$

where, except for a scalar multiplicative constant, $\Gamma_\phi(\omega)$ is the same as $U_{s,\phi}(\omega)$. Without the extra filter function $G_\eta(\omega)$, the rest of Eq. (6.76) corresponds to the filtering operation of the projection data in x-ray tomography (see the next section). The filtering as called for by the transfer function $G_\eta(\omega)$ is depth dependent due the parameter η, which is equal to $x \cos \phi + y \sin \phi$.

In terms of the filtered projections $\Pi_\phi(\xi, \eta)$ in Eq. (6.76), the reconstruction

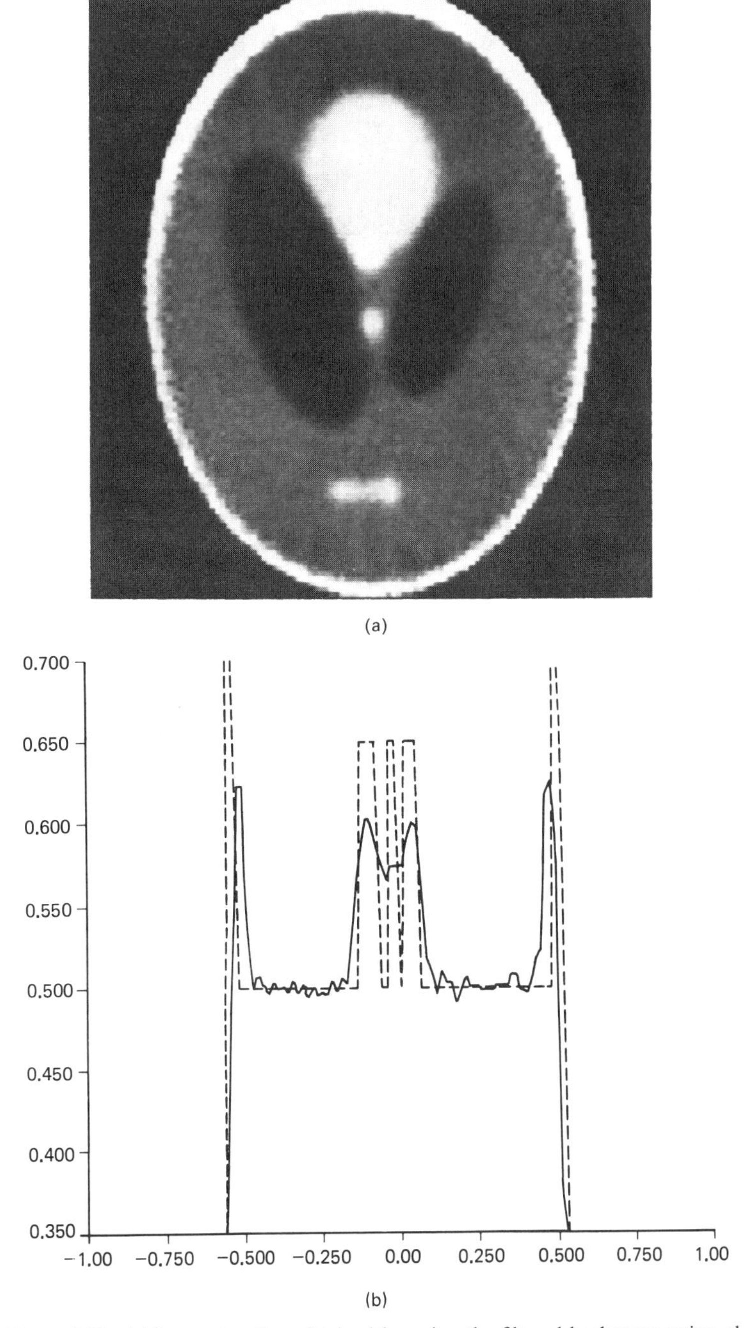

Figure 6.20 (a) Reconstruction obtained by using the filtered-backpropagation algorithm on 64 projections and 128 samples in each projection. $N_\eta = 128$. (b) Numerical comparison of the true and the reconstructed values on the line $y = -0.605$.

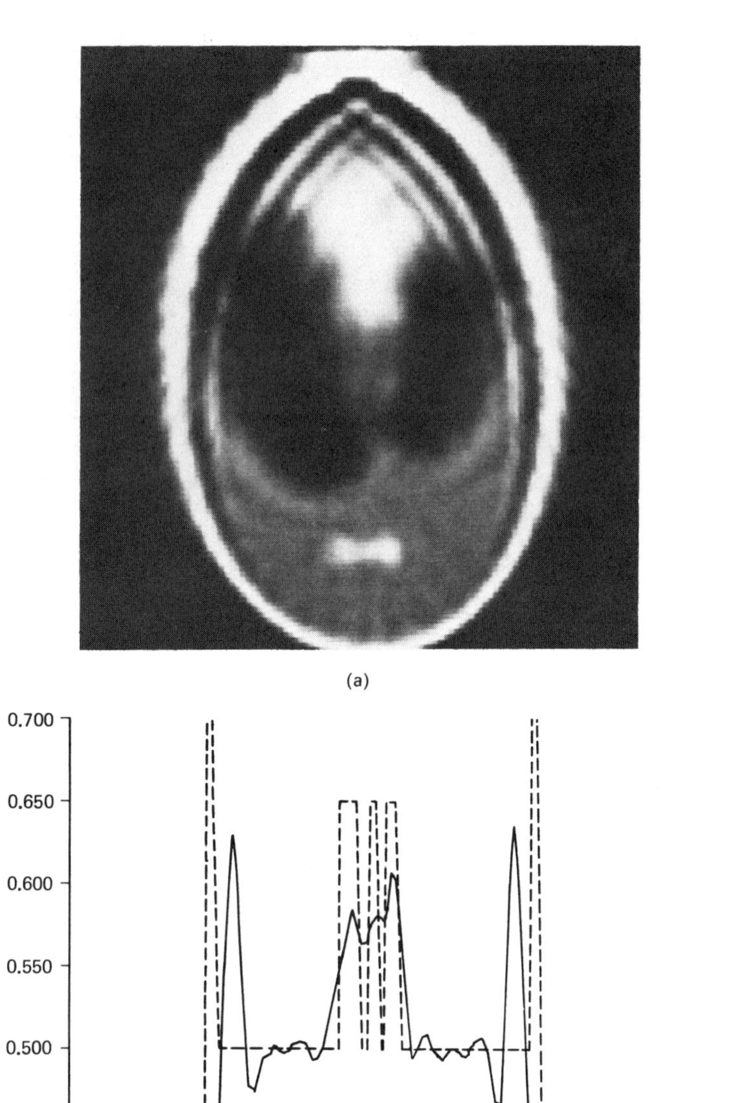

Figure 6.21 (a) Reconstruction obtained by using the modified filtered-backpropagation algorithm on 64 projections and 128 samples per projection. The filter function corresponds to what would yield local accuracy at the site of the three small ellipses. (b) Numerical comparison of the true and the reconstructed values on the line $y = -0.605$.

integral of Eq. (6.73) may be expressed as

$$f(x, y) = \frac{1}{2\pi} \int_0^{2\pi} d\phi \; \Pi_\phi(x \sin \phi - y \cos \phi, x \cos \phi + y \sin \phi) \tag{6.79}$$

The computational procedure for reconstructing an image on the basis of Eqs. (6.76) and (6.79) may be presented in the form of the following steps:

Step 1. In accordance with Eq. (6.76), filter each projection with a separate filter for each depth in the image frame. For example, if we chose only nine depths, as shown in Fig. 6.19, we would need to apply nine different filters to the diffracted projection shown there. (In most cases for 128 × 128 reconstructive, the number of discrete depths chosen for filtering the projection will also be around 128. If they are much less than 128, spatial resolution will be lost.)

Step 2. To each pixel (x, y) in the image frame, in accordance with Eq. (6.79), allocate a value of the filtered projection that corresponds to the nearest depth line.

Step 3. Repeat the preceding two steps for all projections. As a new projection is taken up, add its contribution to the current sum at the pixel located at (x, y).

The depth-dependent filtering in Step 1 makes this algorithm computationally very demanding. For example, if we choose N_η depth values, the processing of each projection will take $(N_\eta + 1)$ fast Fourier transformations (FFTs). If the total number of projections is N_ϕ, this translates into $(N_\eta + 1)N_\phi$ FFTs. For most $N \times N$ reconstructions, both N_η and N_ϕ will be approximately equal to N. Therefore, Devaney's filtered-backpropagation algorithm requires approximately N^2 FFTs compared to $4N$ FFTs for bilinear interpolation. (For precise comparisons, we must mention that the FFTs for the case of bilinear interpolation are longer due to zero-padding.)

Figure 6.20 shows the reconstruction obtained by using the filtered-backpropagation algorithm with $N_\eta = 128$ on the same original data set as was used for Fig. 6.18. The VAX-11/780 CPU processing time for the image in Fig. 6.20 was 50 min.

Devaney [29] has also proposed a modified filtered-backpropagation algorithm in which $G_\eta(\omega)$ is simply replaced by a single $G_{\eta_0}(\omega)$, where $\eta_0 = x_0 \cos \phi + y_0 \sin \phi$, (x_0, y_0), being the coordinates of the point where local accuracy in reconstruction is desired. (Elimination of depth-dependent filtering reduces the number of FFTs to $2N_\phi$.) When this algorithm was implemented to yield accurate reconstruction in the vicinity of the three small ellipses in the test image, the overall reconstructed image is shown in Fig. 6.21. The VAX-11/780 processing time was 8 min.

6.6 FILTERED-BACKPROJECTION ALGORITHMS FOR NONDIFFRACTING SOURCES

Reconstruction Algorithm for Parallel Projection Data

The algorithm that is currently being used in almost all applications using x-rays (or γ-rays) is the filtered-backprojection algorithm. It has been shown to be extremely accurate and amenable to fast implementation. We present it in this section for the case of parallel projection data. Fan projections are considered in the next section.

The algorithm will be derived by using the Fourier slice theorem. This theorem is brought into play by rewriting the inverse transform of Eq. (6.27) in polar coordinates and rearranging the limits of the integrations therein. *The derivation of the algorithm is perhaps one of the most illustrative examples of how we can obtain a radically different computer implementation by simply rewriting the fundamental expressions of the underlying theory.*

This section also includes a discussion on different aspects of the computer implementation of this algorithm. We have shown how fast Fourier transform (FFT) can be used to speed up the filtering part of the algorithm and further how the backprojection can be implemented without virtually any multiplications.

If, as before, (Ω, θ) are the polar coordinates in the (ω_1, ω_2)-plane, the integral in Eq. (6.27) can be expressed as follows:

$$\begin{aligned} f(x, y) &= \int_0^{2\pi} \int_0^{\infty} F(\Omega, \theta) \exp[j\Omega(x \cos\theta + y \sin\theta)]\Omega \, d\Omega \, d\theta \\ &= \int_0^{\pi} \int_0^{\infty} F(\Omega, \theta) \exp[j\Omega(x \cos\theta + y \sin\theta)]\Omega \, d\Omega \, d\theta \\ &\quad + \int_0^{\pi} \int_0^{\infty} F(\Omega, \theta + 180^\circ) \exp\{j\Omega[x \cos(\theta + 180^\circ) \\ &\quad + y \sin(\theta + 180^\circ)]\}\Omega \, d\Omega \, d\theta \end{aligned} \tag{6.80}$$

Using the property

$$F(\Omega, \theta + 180^\circ) = F(-\Omega, \theta) \tag{6.81}$$

the expression above for $f(x, y)$ may be written as

$$\begin{aligned} f(x, y) &= \int_0^{\pi} \left[\int_{-\infty}^{\infty} F(\Omega, \theta) |\Omega| \exp(j\Omega t) \, d\Omega \right] d\theta \\ &= \int_0^{\pi} \left[\int_{-\infty}^{\infty} S_\theta(\Omega) |\Omega| \exp(j\Omega t) \, d\Omega \right] d\theta \\ &= \int_0^{\pi} \left[\int_{-\infty}^{\infty} S_\theta(\omega) |\omega| \exp(j\omega t) \, d\omega \right] d\theta \end{aligned} \tag{6.82}$$

where, as noted before,

$$t = x \cos \theta + y \sin \theta \tag{6.83}$$

To obtain Eq. (6.82) we have used the Fourier slice theorem [see Eq. (6.55)] to rewrite $F(\Omega, \theta)$ as $S_\theta(\Omega)$. For the purpose of computer implementation, it is useful to break Eq. (6.82) into the two operations shown below.

$$f(x, y) = \int_0^\pi Q_\theta(x \cos \theta + y \sin \theta)\, d\theta \tag{6.84}$$

where

$$Q_\theta(t) = \int_{-\infty}^{\infty} S_\theta(\omega) |\omega| \exp(j\omega t)\, d\omega \tag{6.85}$$

These formulas for reconstruction say that from each projection $P_\theta(t)$ we first calculate a "filtered projection" $Q_\theta(t)$ by using Eq. (6.85), and then use Eq. (6.84) to reconstruct the function $f(x, y)$.

To discuss the filtering part described by Eq. (6.85), we note that in practice projections are bandlimited. Let W be any angular frequency (in radians per centimeter) greater than the smallest beyond which the spectral energy in all the projections may be ignored. If the projection data are sampled with a sampling interval of τ centimeters, and if we assume that the sampled data does not suffer from aliasing errors, W is given by

$$W = \frac{2\pi}{2\tau} \tag{6.86}$$

With bandlimited projections, Eq. (6.85) may be expressed as

$$Q_\theta(t) = \int_{-\infty}^{\infty} S_\theta(\omega) H(\omega) \exp(j\omega t)\, d\omega \tag{6.87}$$

where

$$H(\omega) = |\omega|\, b_W(\omega) \tag{6.88}$$

and

$$b_W(\omega) = \begin{cases} 1, & |\omega| < W \\ 0, & \text{otherwise} \end{cases} \tag{6.89}$$

The function $H(\omega)$, shown in Fig. 6.22, represents the transfer function of a filter with which the projections must be processed. The impulse response, $h(t)$, of this filter is given by the inverse Fourier transform of $H(\omega)$, as shown by

$$h(t) = \int_{-\infty}^{\infty} H(\omega) \exp(j\omega t)\, d\omega \tag{6.90}$$

$$= \frac{1}{2\tau^2} \frac{\sin(2\pi t/2\tau)}{(2\pi t/2\tau)} - \frac{1}{4\tau^2} \left[\frac{\sin(\pi t/2\tau)}{(\pi t/2\tau)} \right]^2 \tag{6.91}$$

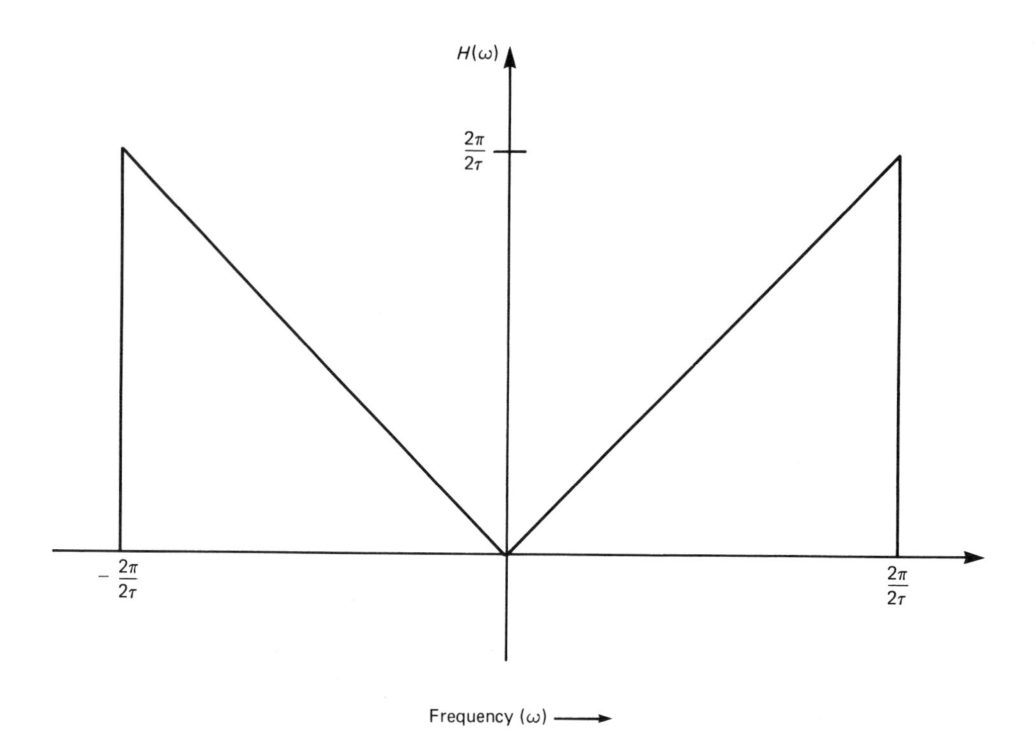

Figure 6.22 Transfer function of the filter with which the projection must be processed prior to backprojection.

where we have used Eq. (6.86). Since the projection data are measured with a sampling interval of τ, for digital processing the impulse response need only be known with the same sampling interval. The samples, $h(n\tau)$, of $h(t)$ are given by

$$h(n\tau)=\begin{cases}\dfrac{1}{4\tau^2}, & n=0\\ 0, & n \text{ even}\\ -\dfrac{1}{n^2\pi^2\tau^2}, & n \text{ odd}\end{cases} \tag{6.92}$$

This function is shown in Fig. 6.23.

Since both $P_\theta(t)$ and $h(t)$ are now bandlimited functions, they may be expressed as

$$P_\theta(t)=\sum_{k=-\infty}^{\infty} P_\theta(k\tau)\frac{\sin\left[W(t-k\tau)\right]}{W(t-k\tau)} \tag{6.93}$$

$$h(t)=\sum_{k=-\infty}^{\infty} h(k\tau)\frac{\sin\left[W(t-k\tau)\right]}{W(t-k\tau)} \tag{6.94}$$

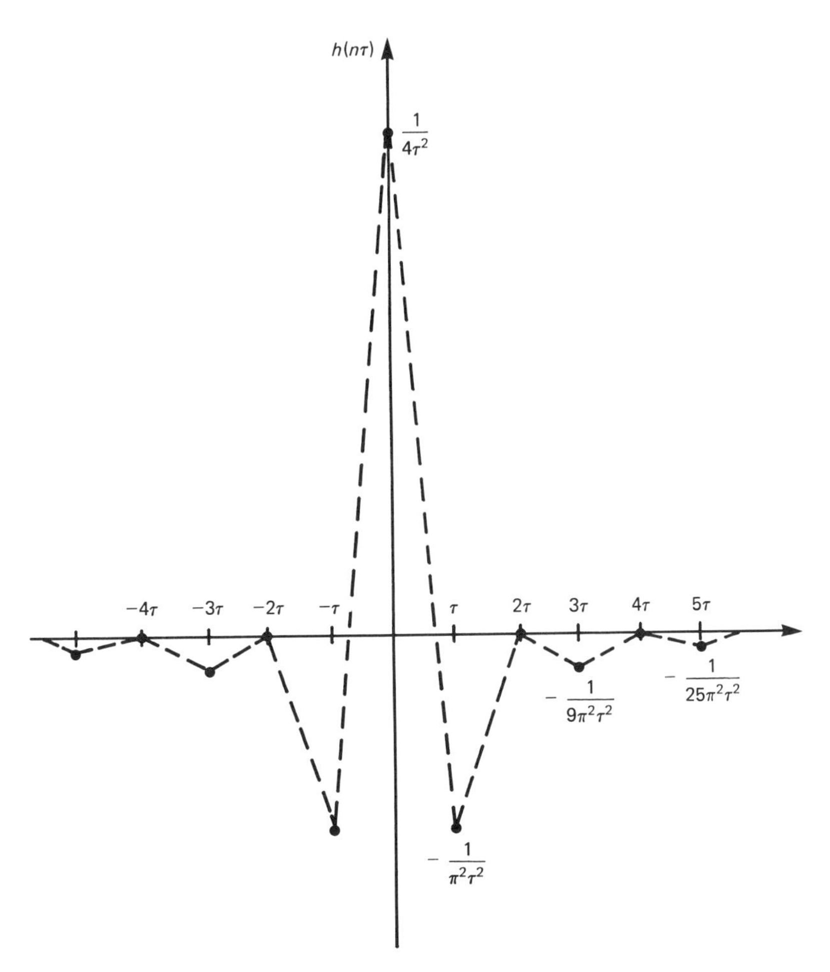

Figure 6.23 Unit sample response corresponding to the transfer function of Fig. 6.22.

By using the convolution theorem, which says that the inverse transform of a product of two functions in the ω-domain is equivalent to their convolutions in the t-domain, we can write Eq. (6.87) as

$$Q_\theta(t) = \int_{-\infty}^{\infty} P_\theta(t')h(t - t')\, dt' \tag{6.95}$$

Substituting Eqs. (6.93) and (6.94) in Eq. (6.95), we get the following result for the values of the filtered projection at the sampling points of $P_\theta(k\tau)$:

$$Q_\theta(n\tau) = \tau \sum_{k=-\infty}^{\infty} h(n\tau - k\tau)P_\theta(k\tau) \tag{6.96}$$

In practice each projection is of only finite extent. Suppose that each $P_\theta(k\tau)$ is zero outside the index range $k = 0, \ldots, N-1$. We may now write the following two equivalent forms of Eq. (6.96):

$$Q_\theta(n\tau) = \tau \sum_{k=0}^{N-1} h(n\tau - k\tau) P_\theta(k\tau) \tag{6.97a}$$

or

$$Q_\theta(n\tau) = \tau \sum_{k=-(N-1)}^{N-1} h(k\tau) P_\theta(n\tau - k\tau), \qquad n = 0, 1, 2, \ldots, N-1 \tag{6.97b}$$

Equation (6.97) implies that in order to determine $Q_\theta(n\tau)$ for $n = 0, 1, \ldots, N-1$, the length of the sequence $h(n\tau)$ used should be from $n = -(N-1)$ to $n = (N-1)$.

The discrete convolution in Eq. (6.97) may be implemented directly on a general-purpose computer. However, it is much faster to implement it in the frequency domain using FFT algorithms. (By using specially designed hardware, direct implementation of Eq. (6.97) can be made as fast or faster than the frequency-domain implementation.) For the frequency-domain implementation we have to keep in mind the fact that we can now only perform periodic (or circular) convolutions. The convolution required in Eq. (6.97) is aperiodic. To eliminate the inter-period interference artifacts inherent to periodic convolution, we pad the projection data with a sufficient number of zeros. It can easily be shown that if we pad $P_\theta(k\tau)$ with zeros so that it is $(2N-1)$ elements long, we avoid interperiod interference over the N samples of $Q_\theta(k\tau)$. Of course, if we want to use the base-2 FFT algorithm, which is often the case, the sequences $P_\theta(k\tau)$ and $h(k\tau)$ have to be zero-padded so that each is $(2N-1)_2$ elements long, where $(2N-1)_2$ is the smallest integer that is a power of 2 and that is greater than $2N-1$. Therefore, the frequency-domain implementation may be expressed as

$$Q_\theta(n\tau) = \tau \times \text{IFFT}\{\text{FFT}[P_\theta(n\tau) \text{ with ZP}] \times \text{FFT}[h(n\tau) \text{ with ZP}]\} \tag{6.98a}$$

where FFT and IFFT denote, respectively, the fast Fourier transform and the inverse fast Fourier transform; ZP stands for zero padding. We usually obtain superior reconstructions when some smoothing is also incorporated in Eqs. (6.97) and (6.98a). For example, in Eq. (6.98a) smoothing may be implemented by multiplying the product of the two FFTs by a Hamming window [44]. When such a window is incorporated, Eq. (6.98a) may be rewritten as

$$\begin{aligned} Q_\theta(n\tau) = \tau \times \text{IFFT}\{&\text{FFT}[P_\theta(n\tau) \text{ with ZP}] \\ &\times \text{FFT}[h(n\tau) \text{ with ZP}] \times [\text{smoothing window}]\} \end{aligned} \tag{6.98b}$$

After the filtering of each projection is accomplished, the reconstructed image $f(x, y)$ may then be obtained by a discrete approximation to the integral in Eq. (6.84), that is,

$$\hat{f}(x, y) = \frac{\pi}{K} \sum_{i=1}^{K} Q_{\theta_i}(x \cos\theta_i + y \sin\theta_i) \tag{6.99}$$

where the K angles θ_i are those for which the projections $P_\theta(t)$ are known. $\hat{f}(x, y)$ is a discrete version of $f(x, y)$ as reconstructed by using the implementation as presented here.

Equation (6.99) calls for each filtered projection Q_{θ_i} to be "backprojected." This can be explained as follows. To every point (x, y) in the image plane there corresponds a value of t $(= x \cos \theta + y \sin \theta)$ for a given value of θ. The contribution that Q_{θ_i} makes to the reconstruction at (x, y) is its value for the corresponding value of t. This is illustrated further in Fig. 6.24. It is easily shown that for the indicated angle θ_i, the value of $t = (x \cos \theta_i + y \sin \theta_i)$ is the same for all (x, y) on the line LM. Therefore, the filtered projection Q_{θ_i} will make the same contribution to the reconstruction at all these points. Accordingly, in reconstruction each function $Q_{\theta_i}(t)$ is smeared back over the image plane. The sum (multiplied by π/K) of all such smearings results in the reconstruction image.

Note that the value of $x \cos \theta_i + y \sin \theta_i$ in Eq. (6.99) may not correspond to one of the values of t for which Q_{θ_i} is determined in Eq. (6.97) or in (6.98). However, Q_{θ_i} for such t may be approximated by suitable interpolation; often, linear interpolation is adequate. Sometimes, in order to eliminate the computations required for interpolation, preinterpolation of the functions $Q_\theta(t)$ is also used. In this technique,

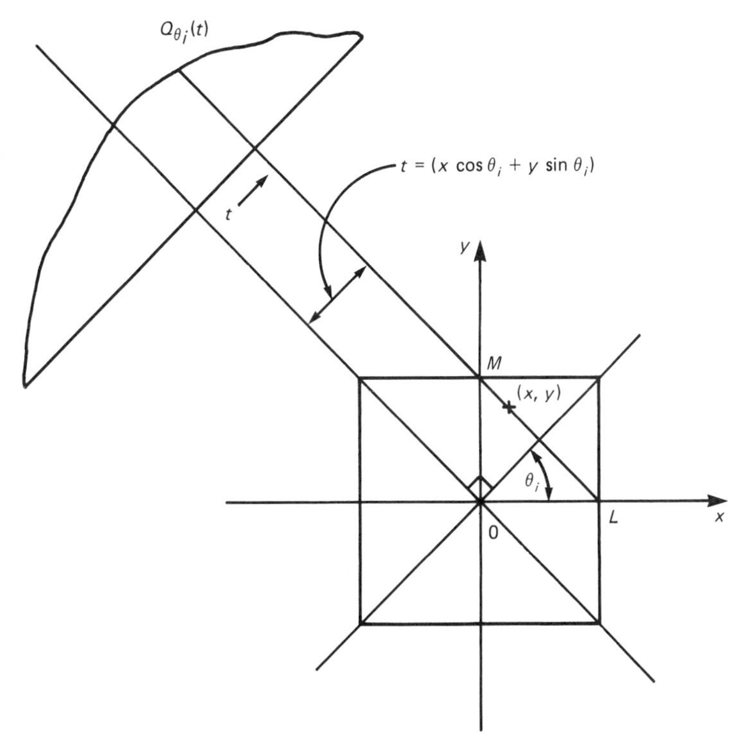

Figure 6.24 When we backproject, the filtered projection $Q_\theta(t)$ makes the same contribution to all the pixels on line LM in the image frame.

which can be combined with the computation in Eqs. (6.98), prior to backprojection the function $Q_\theta(t)$ is preinterpolated onto 10 to 1000 times the number of points in the projection data. From this dense set of points we simply retain the nearest neighbor to obtain the value of Q_{θ_i} at $x \cos \theta_i + y \sin \theta_i$. A variety of techniques are available for preinterpolation. In one method that is particularly attractive because of the speed with which it can be implemented, prior to performing the IFFT in Eq. (6.98), the *frequency-domain function* is padded with a large number of zeros. The inverse transform then yields the preinterpolated Q_θ. Note that with preinterpolation and with appropriate programming, the backprojection for parallel projection data can be accomplished with virtually no multiplications.

To illustrate the accuracy of this algorithm, we now show some computer simulation results. These results will be presented for the test image in Fig. 6.25(a), whose gray levels are shown in Fig. 6.25(b). (Note that the gray levels in Fig. 6.25(a) are not the same as those in Fig. 6.16(a), although the two images look the same.) This image, first proposed by Shepp and Logan [92], is often used for testing algorithms for nondiffracting sources because it approximately represents a cross section of the human head, which is known to place greatest demands on a tomographic system with regards to its accuracy. Most of the useful information in the image is contained in the 1 percent variations in gray levels within the "skull." This test image is a superposition of 10 ellipses, each characterized by a gray level in its interior as shown in Fig. 6.25b and Table 6.2. Ellipses are particularly convenient to use because it is possible to write simple analytical expressions for their projections. These expressions are given in [87, Chap. 8] and in [60].

Using the implementation in Eq. (6.98a), Fig. 6.26 shows the reconstructed values on the line $y = -0.605$ for the image in Fig. 6.25. Figure 6.26(a) shows the complete reconstruction. The number of rays used in each projection was 127 and the number of projections was 100. To make convolutions aperiodic the projection data were padded with zeros to make each projection 256 elements long. Figure 6.26(b) illustrates the remarkable accuracy of this method.

Reconstruction Algorithms for Fan-Beam Projections

The theory in the preceding subsections dealt with reconstructing images from their parallel projections. For generating parallel data a source–detector combination has to scan linearly over the length of a projection, rotate through a certain angular interval, and then scan linearly over the length of the next projection. This usually results in scan times that in medical applications are as long as a few minutes. A much faster way to generate the line integrals is by using the system depicted in Fig. 6.4, which generates fan projections shown in Fig. 6.6. In such scanners the projections are recorded on the fly as the gantry rotates around the patient. This reduces the data collection time to a couple of seconds.

It is possible to rearrange the fan data into parallel projection data. If we want to transform a given number of fan projections into the same number of parallel

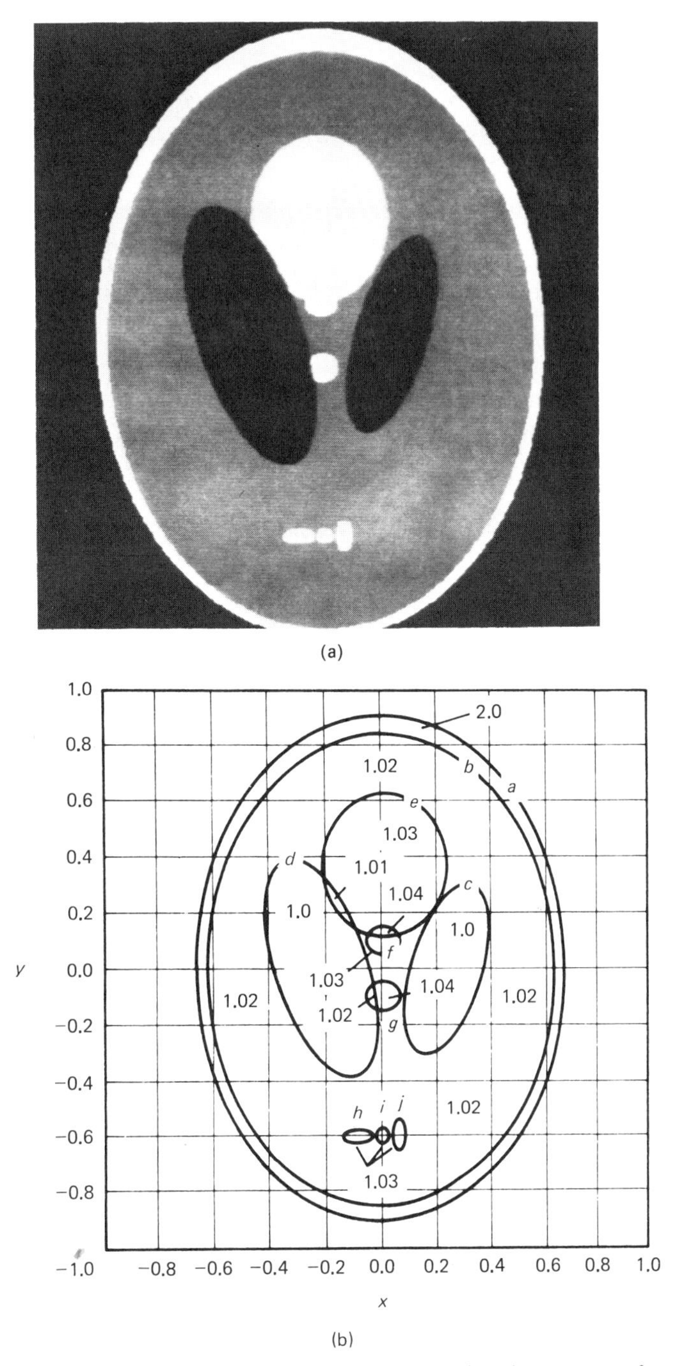

Figure 6.25 (a) The image shown here is used for testing the accuracy of reconstruction algorithms for tomographic imaging with nondiffracting sources. Although it is similar in appearance to the image of Fig. 6.16(a), its gray-level distribution is different. (b) The test image of (a) is a superposition of 10 ellipses shown here. On a scale of 0 to 2, the gray-scale distribution of the test image is also shown.

TABLE 6.2 THE PARAMETERS OF THE COMPONENT ELLIPSES OF THE TEST IMAGE USED FOR RECONSTRUCTION WITH ALGORITHMS DESIGNED FOR THE NONDIFFRACTING CASE

Ellipse	Coordinates of the center	A major axis	B minor axis	α rotation angle	ρ gray level
a	(0, 0)	0.92	0.69	90°	2
b	(0, −0.0184)	0.874	0.662	90°	−0.98
c	(0.22, 0)	0.31	0.11	72°	−0.02
d	(−0.22, 0)	0.41	0.16	108°	−0.02
e	(0, 0.35)	0.25	0.21	0	0.01
f	(0, 0.1)	0.046	0.046	0	0.01
g	(0, −0.1)	0.046	0.046	0	0.01
h	(−0.08, −0.605)	0.046	0.046	0	0.01
i	(0, −0.605)	0.023	0.023	0	0.01
j	(0.06, −0.605)	0.046	0.023	90°	0.01

projections, much intraprojection interpolative computation would be required before the algorithm described in the preceding section could be used. If we are willing to record a very large number of fan projections, it is possible to avoid this difficulty if we constrain the angles at which the fan projections are measured, and also the angular interval between the rays within each fan projection. It is then possible to convert the fan data into equivalent parallel form merely by *rebinning,* and this rebinning can be accomplished on the fly, by which is meant that as each fan projection is recorded, its rays can be immediately reindexed to correspond to the parallel form. The resulting parallel projection data are still not uniformly spaced, and some interpolation may be required to rectify that. The reader is referred to [31, 60, 87] for further discussion on this rebinning algorithm. According to this author's experience, the most accurate procedure to reconstruct an image from fan data is by direct filtered backprojection, which is discussed in this section. The derivation of a filtered-backprojection algorithm for the fan case is analytically more complex and computationally more time consuming, but it possesses the same elegance in its implementation as the algorithm discussed in the preceding subsection.

There are two types of fan projections, depending on whether a projection is sampled at equiangular or equispaced intervals. This difference is illustrated in Fig. 6.27. In (a) we have shown an equiangular set of rays. If the detectors for the measurement of line integrals are arranged on the straight line D_1D_2, this implies unequal spacing between them. The second type of fan projection is generated when the rays are arranged such that the detector spacing on a straight line is now equal [Fig. 6.27(b)]. The algorithms that reconstruct images from these two different types of fan projections are different. In this section, to illustrate how we obtain a direct reconstruction algorithm for fan projections, we show its derivation for the case corresponding to Fig. 6.27(b). A similar derivation for the case of equiangular fan data may be found in [60, 87].

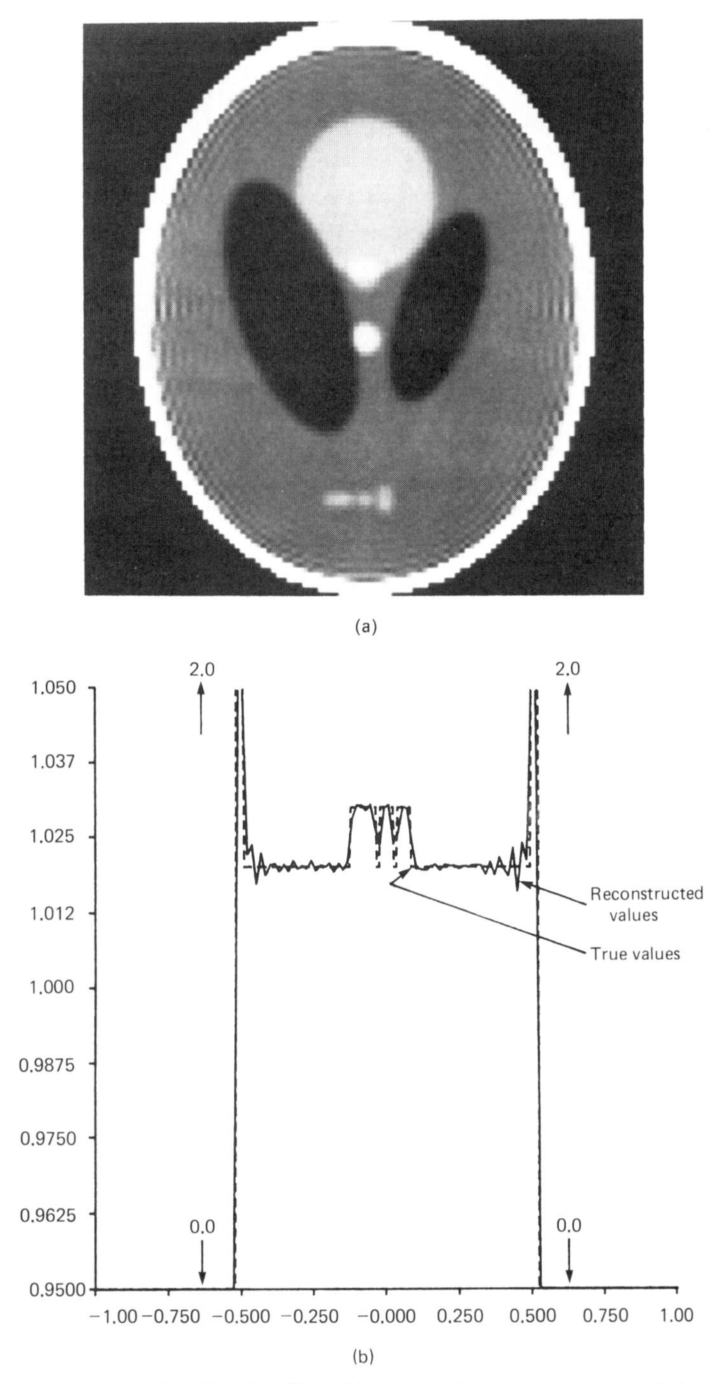

Figure 6.26 (a) A 128 × 128 filtered-backprojection reconstruction of the test image in Fig. 25 from 100 parallel projections over 180° and 127 rays in each projection. (b) Numerical comparison of the true and reconstructed values on line $y = -0.605$.

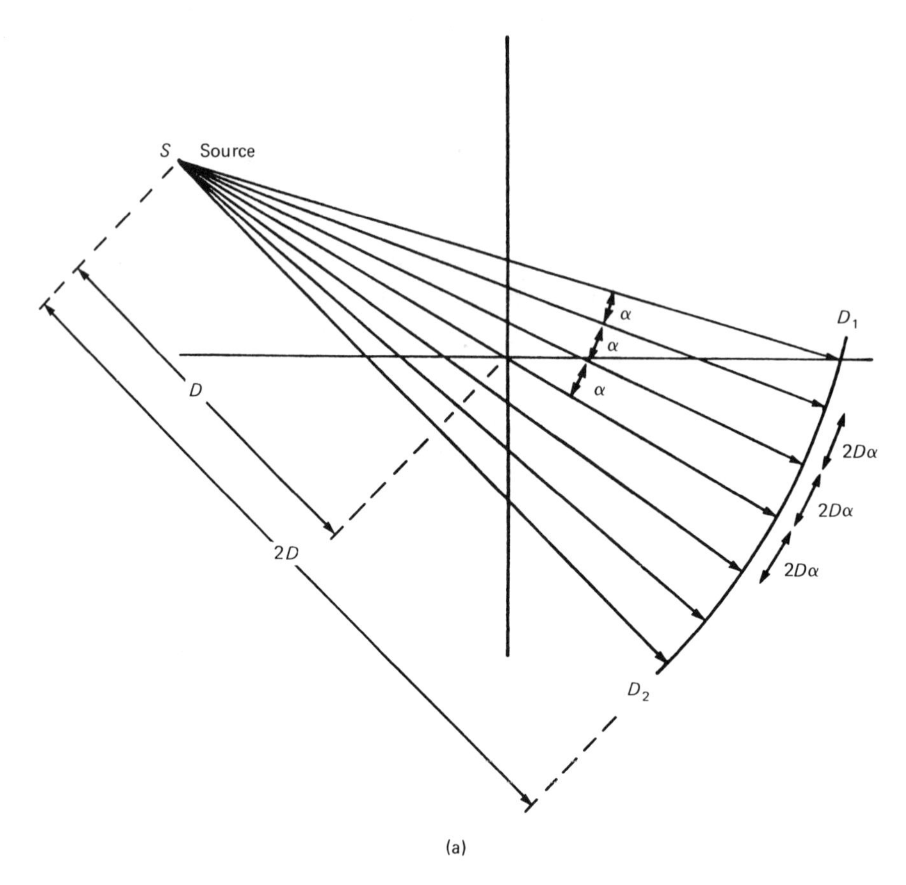

Figure 6.27 A fan-beam projection may be taken with either equiangular rays, as shown in (a); or with rays that result in equispaced detectors on a line, as shown in (b).

Our approach consists of analytically transforming the filtered-backprojection formulas presented in the preceding section and reexpressing them for the case of fan projections. We then show how the resulting integrands can be written in a form so as to permit a filtered-backprojection type of computer implementation, although, of course, the filtering transfer function is different from that discussed previously, and the backprojection now involves some weighting coefficients.

Let $R_\beta(s)$ denote a fan projection as shown in Fig. 6.28(a), where s is the distance along the straight line corresponding to the detector bank. Although the projections are measured on a line such as D_1D_2 in Fig. 6.28(a), for theoretical purposes it is more efficient to assume the existence of an imaginary detector line $D_1'D_2'$ passing through the origin. We now associate the ray integral along SB with point A on $D_1'D_2'$, as opposed to point B on D_1D_2. Thus in Fig. 6.28(b) we associate a fan projection $R_\beta(s)$ with the imaginary detector line $D_1'D_2'$. Now consider a ray

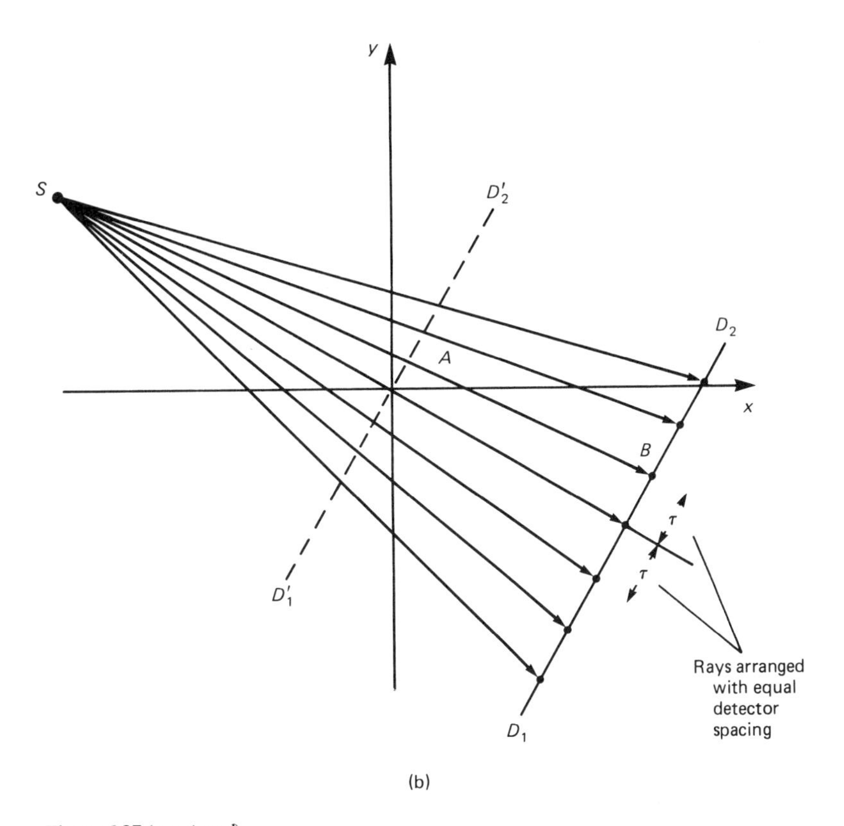

Figure 6.27 *(continued)*

SB in the figure; the value of s for this ray is the length of OA. If parallel projection data were generated for the object under consideration, the ray SB would belong to a parallel projection $P_\theta(t)$ with θ and t as shown in Fig. 6.28(b). The relationships between β and s for the fan case and θ and t for the parallel case are given by

$$\begin{aligned} t &= s \cos \gamma & \theta &= \beta + \gamma \\ &= \frac{sD}{\sqrt{D^2 + s^2}}, & &= \beta + \tan^{-1} \frac{s}{D} \end{aligned} \tag{6.100}$$

where use has been made of the fact that angle AOC is equal to angle OSC, and where D is the distance of the source point S from the origin O.

By combining Eqs. (6.84) and (6.95), we can write the following expression for the reconstructed image in terms of parallel projections:

$$f(x, y) = \int_0^\pi \int_{-t_m}^{t_m} P_\theta(t) h(x \cos \theta + y \sin \theta - t) \, dt \, d\theta \tag{6.101}$$

where t_m is the value of t for which $P_\theta(t) = 0$ with $t > t_m$ in all projections. This

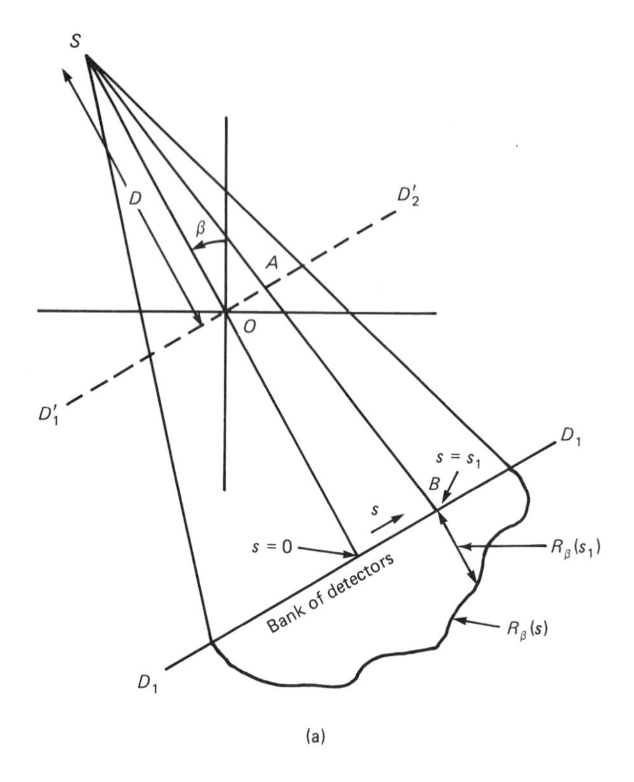

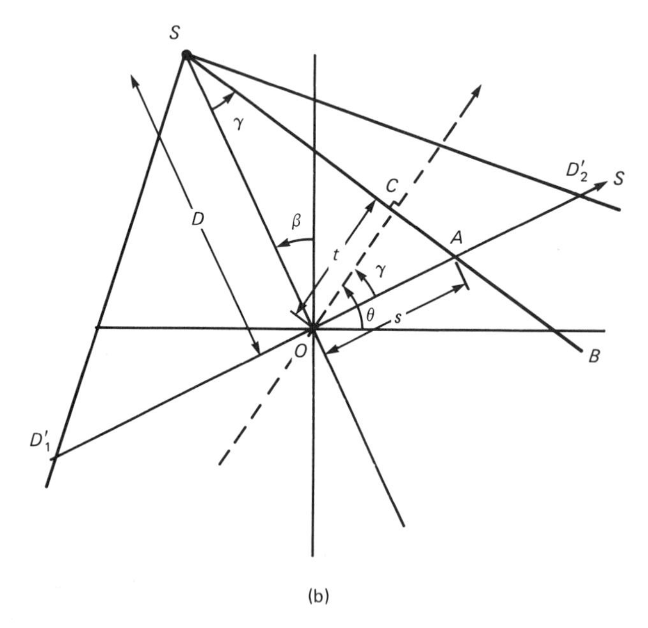

Figure 6.28 (a) For the case of a fan beam with equispaced detectors, each projection is denoted by the function $R_\beta(s)$. (b) Shown here are the various parameters used in the derivation of the fan-beam reconstruction algorithm.

equation only requires the parallel projections to be collected over 180°. For fan data, algorithms are more easily derived if the projections are measured over all of 360°. So we start out by rewritting this expression as

$$f(x, y) = \frac{1}{2} \int_0^{2\pi} \int_{-t_m}^{t_m} P_\theta(t) h(x \cos \theta + y \sin \theta - t) \, dt \, d\theta \tag{6.102}$$

Derivation of the algorithm becomes easier when the coordinates (x, y) are expressed in the polar form (r, ϕ), as shown by

$$x = r \cos \phi, \qquad y = r \sin \phi \tag{6.103}$$

The expression in Eq. (102) can now be written as

$$f(r, \phi) = \frac{1}{2} \int_0^{2\pi} \int_{-t_m}^{t_m} P_\theta(t) h(r \cos (\theta - \phi) - t) \, dt \, d\theta \tag{6.104}$$

Using the transformation relationships in Eq. (6.100), this double integration may be expressed as

$$f(r, \phi) = \frac{1}{2} \int_{-\beta_1}^{\beta_2} \int_{-s_m}^{s_m} P_{\beta+\gamma}\left(\frac{sD}{\sqrt{D^2 + s^2}}\right) h\left[r \cos\left(\beta + \tan^{-1}\frac{s}{D} - \phi\right) - \frac{Ds}{\sqrt{D^2 - s^2}}\right] \times \frac{D^3}{(D^2 + s^2)^{3/2}} \, ds \, d\beta \tag{6.105}$$

where

$$\beta_1 = -\tan^{-1}\frac{s_m}{D}$$

$$\beta_2 = 2\pi - \tan^{-1}\frac{s_m}{D} \tag{6.106}$$

and where we have used

$$dt \, d\theta = \frac{D^3}{(D^2 + s^2)^{3/2}} \, ds \, d\beta \tag{6.107}$$

In Eq. (6.106) s_m is the largest value of s in each projection and corresponds to t_m for the parallel projection data. As β varies from the lower limit at $-\tan^{-1}(s_m/D)$ to its upper limit at $2\pi - \tan^{-1}(s_m/D)$, it covers an angular interval of 360°. Since all functions of β in Eqs. (6.105) and (6.106) are periodic with period 2π, these limits may be replaced by 0 and 2π, respectively. Also, the expression

$$P_{\beta+\gamma}\left(\frac{sD}{\sqrt{D^2 + s^2}}\right)$$

corresponds to the ray integral along SB in the parallel projection data $P_\theta(t)$. The identity of this ray integral in the fan-projection data is simply $R_\beta(s)$. Introducing these changes in Eq. (6.105), we get

$$f(r, \phi) = \frac{1}{2} \int_0^{2\pi} \int_{-s_m}^{s_m} R_\beta(s) h \left[r \cos\left(\beta + \tan^{-1}\frac{s}{D} - \phi\right) - \frac{Ds}{\sqrt{D^2 - s^2}} \right] \times \frac{D^3}{(D^2 + s^2)^{3/2}} \, ds \, d\beta \tag{6.108}$$

In order to express this formula in a filtered-backprojection form, we first analyze the argument of h. The argument may be written as

$$r \cos\left(\beta + \tan^{-1}\frac{s}{D} - \phi\right) - \frac{Ds}{\sqrt{D^2 - s^2}}$$

$$= r \cos(\beta - \phi) \frac{D}{\sqrt{D^2 + s^2}} - [D + r \sin(\beta - \phi)] \frac{s}{\sqrt{D^2 + s^2}} \tag{6.109}$$

We now introduce two new variables that are easily calculated in a computer implementation. The first of these, denoted by U, is for each pixel (x, y) the ratio of SP (Fig. 6.29) to the source-to-origin distance. Note that SP is the projection of the source-to-pixel distance SE on the central ray. Thus

$$U(r, \phi, \beta) = \frac{\overline{SO} + \overline{OP}}{D}$$

$$= \frac{D + r \sin(\beta - \phi)}{D} \tag{6.110}$$

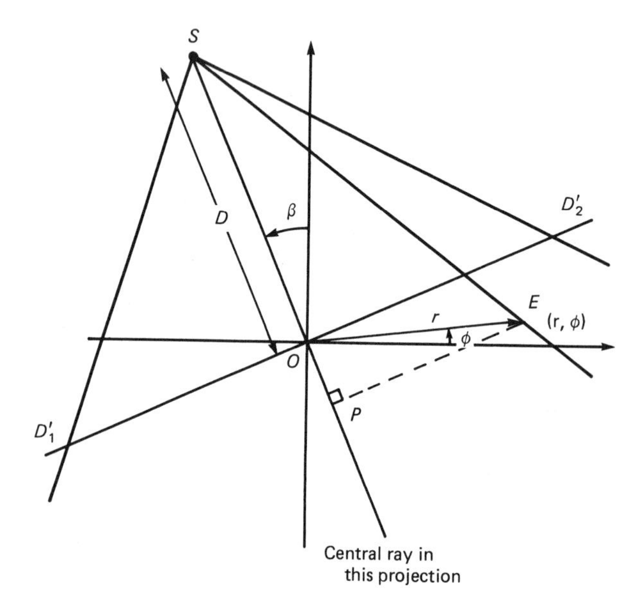

Figure 6.29 For a pixel at polar coordinates (r, ϕ), the variable U_r is the ratio of SP to the source-to-origin distance.

The other parameter we want to define is the value of s for the ray that passes through the pixel (r, ϕ) under consideration. Let s' denote this value of s. Since s is measured along the imaginary detector line $D'_1D'_2$, it is given by the distance OF. Since

$$\frac{s'}{\overline{SO}} = \frac{\overline{EP}}{\overline{SP}}$$

we have

$$s' = D \frac{r \cos(\beta - \phi)}{D + r \sin(\beta - \phi)} \tag{6.111}$$

Equations (6.110) and (6.111) can be utilized to expess (6.109) in terms of U and s':

$$r \cos\left(\beta + \tan^{-1}\frac{s}{D} - \phi\right) - \frac{Ds}{\sqrt{D^2 + s^2}} = \frac{s'UD}{\sqrt{D^2 + s^2}} - \frac{sUD}{\sqrt{D^2 + s^2}} \tag{6.112}$$

Substituting Eq. (6.112) in (6.108), we get

$$f(r, \phi) = \frac{1}{2}\int_0^{2\pi}\int_{-s_m}^{s_m} R_\beta(s) h\left((s' - s)\frac{UD}{\sqrt{D^2 + s^2}}\right)\frac{D^3}{(D^2 + s^2)^{3/2}}\, ds\, d\beta \tag{6.113}$$

To show that this integral indeed implies a filtered-backprojection implementation, we will examine the argument of $h(\cdot)$. In order to do so, we recall from the preceding subsection the fact that this function is nominally the inverse Fourier transform of $|\omega|$ in the frequency domain, which follows from Eq. (6.90) if we ignore for a moment the bandlimit W on the projections. Therefore, we may write, at least symbolically,

$$h(t) = \int_{-\infty}^{\infty} |\omega| \exp(j\omega t)\, d\omega \tag{6.114}$$

(Strictly speaking, this integration exists only as a distribution function.) Therefore,

$$h\left((s' - s)\frac{UD}{\sqrt{D^2 + s^2}}\right) = \int_{-\infty}^{\infty} |\omega| \exp\left\{j\omega(s' - s)\frac{UD}{\sqrt{D^2 + s^2}}\right\} d\omega \tag{6.115}$$

Using the transformation

$$\omega' = \omega \frac{UD}{\sqrt{D^2 + s^2}} \tag{6.116}$$

we can rewrite Eq. (6.115) as follows:

$$h\left((s' - s)\frac{UD}{\sqrt{D^2 + s^2}}\right) = \frac{D^2 + s^2}{U^2D^2}\int_{-\infty}^{\infty} |\omega'| \exp[j2\pi(s' - s)\omega']\, d\omega'$$

$$= \frac{D^2 + s^2}{U^2D^2} h(s' - s) \tag{6.117}$$

Substituting this in Eq. (6.114), we get

$$f(r, \phi) = \int_0^{2\pi} \frac{1}{U^2} \int_{-\infty}^{\infty} R_\beta(s) g(s' - s) \frac{D}{\sqrt{D^2 + s^2}} \, ds \, d\beta \tag{6.118}$$

where

$$g(s) = \tfrac{1}{2} h(s) \tag{6.119}$$

For the purpose of computer implementation, Eq. (6.118) may be interpreted as a weighted filtered-backprojection algorithm. To show this, we rewrite Eq. (6.118) as follows:

$$f(r, \phi) = \int_0^{2\pi} \frac{1}{U^2} Q_\beta(s') \, d\beta \tag{6.120}$$

where

$$Q_\beta(s) = R'_\beta(s) * g(s) \tag{6.121}$$

and

$$R'_\beta(s) = R_\beta(s) \frac{D}{\sqrt{D^2 + s^2}} \tag{6.122}$$

The symbol * denotes the operation of convolution. Equations (6.120) to (6.122) suggest the following steps for computer implementation:

Step 1. Assume that each projection $R_\beta(s)$ is sampled with a sampling interval of a. The known data then are $R_{\beta_i}(na)$, where n takes integer values with $n = 0$ corresponding to the central ray passing through the origin; β_i are the angles for which fan projections are known. The first step is to generate for each fan projection $R_{\beta_i}(na)$ the corresponding modified projection $R'_{\beta_i}(na)$ given by

$$R'_{\beta_i}(na) = R_{\beta_i}(na) \frac{D}{\sqrt{D^2 + n^2 a^2}} \tag{6.123}$$

Step 2. Convolve each modified projection $R'_{\beta_i}(na)$ with $g(na)$ to generate the corresponding filtered projection:

$$Q_{\beta_i}(na) = R'_{\beta_i}(na) * g(na) \tag{6.124}$$

where the sequence $g(na)$ is given by the samples of Eq. (6.119):

$$g(na) = \tfrac{1}{2} h(na) \tag{6.125}$$

Substituting in this the values of $h(na)$ given in Eq. (6.92), we get for the impulse

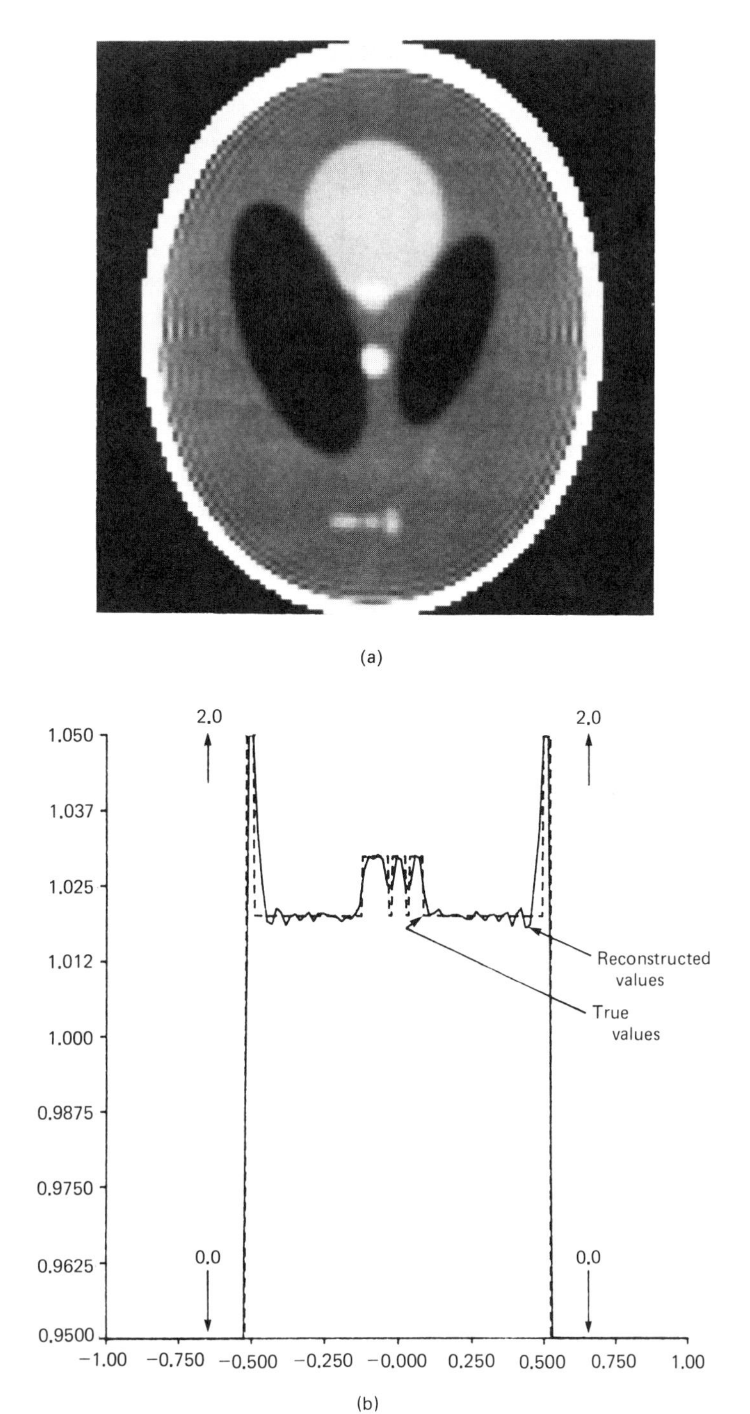

Figure 6.30 (a) Fan-beam reconstruction of the image of Fig. 6.25(a). The fan angle is 45°, the number of projections 200 over 360°, and the number of rays in each projection 100. (b) Numerical comparison of the true and reconstructed values on the $y = -0.605$ line.

response of the convolving filter,

$$g(na) = \begin{cases} \dfrac{1}{8a^2}, & n = 0 \\ 0, & n \text{ even} \\ -\dfrac{1}{2n^2\pi^2a^2}, & n \text{ odd} \end{cases} \tag{6.126}$$

When the convolution of Eq. (6.124) is implemented in the frequency domain using an FFT algorithm, the projection data must be padded with a sufficient number of zeros to avoid distortion due to interperiod interference.

In practice, superior reconstructions are obtained if a certain amount of smoothing is included with the convolution in Eq. (6.124). If $k(na)$ is the impulse response of the smoothing filter, we can write

$$Q_{\beta_i}(na) = R_{\beta_i}(na) * g(na) * k(na) \tag{6.127}$$

In a frequency-domain implementation this smoothing may be achieved by a simple multiplicative window such as a Hamming window.

Step 3. Perform a *weighted* backprojection of each filtered projection along the corresponding fan. The sum of all the backprojections is the reconstructed image

$$f(x, y) = \Delta\beta \sum_{i=1}^{M} \frac{1}{U^2(x, y, \beta_i)} Q_{\beta_i}(s') \tag{6.128}$$

where s' identifies the ray that passes through (x, y) in the fan for the source located at angle β_i. Of course, this value of s' may not correspond to one of the values of na at which Q_{β_i} is known. In that case interpolation is necessary.

Figure 6.30(a) shows a reconstruction obtained by using the steps outlined here. The number of projections used was 100, and the number of rays in each projection 127. The angle of the fan beam was 45°. The original image is the same as in Fig. 6.25(a). Figure 6.30(b) shows the reconstructed values on the line $y = -0.605$.

6.7 ALGEBRAIC RECONSTRUCTION ALGORITHMS

An entirely different approach for tomographic imaging consists of assuming that the cross section consists of an array of unknowns, and then setting up algebraic equations for the unknowns in terms of the measured projection data. Although conceptually this approach is much simpler than the transform-based methods discussed in previous sections, for medical applications it lacks the accuracy and the speed of implementation. However, there are situations where it is not possible to measure a large number of projections, or the projections are not uniformly distributed over 180° or 360°. An example of this situation is earth resources imaging using cross-borehole measurements discussed in Section 6.2. Problems of this type are perhaps more amenable to solution by algebraic techniques.

Many imaging problems that suffer from inadequate number of projections (or an incomplete coverage by them), also suffer from refraction and diffraction effects. As will be obvious from the discussion to follow in this section, in algebraic methods it is essential to know ray paths that connect transmitter and receiver positions. When refraction and diffraction effects are substantial (medium inhomogeneities exceed 10% of the average background value and the correlation length of these inhomogeneities is comparable to a wavelength), it becomes impossible to predict these ray paths. If algebraic techniques are applied under these conditions, we often obtain meaningless results.

If the refraction and diffraction effects are small (medium inhomogeneities less than 2 to 3 percent of the average background value and the correlation width of these inhomogeneities much greater than a wavelength), in some cases it is possible to combine algebraic techniques with digital ray-tracing techniques [1], and devise an iterative procedure in which we first construct an image ignoring refraction, and then trace rays connecting transmitter and receiver locations through this distribution, and finally these rays are used to construct a more accurate set of algebraic equations. By computer simulation, it is possible to show improvements obtained by such an iterative approach; however, experimental verifications have not yet been obtained.

Space limitations prevent the author from discussing here the combined ray tracing and algebraic reconstruction algorithms. Our aim in this section is merely to introduce the reader to the algebraic approach for image reconstruction. We first show how we may construct a set of linear equations whose unknowns are elements of the object cross section. The "method of projections" for solving these equations is then presented. This is followed by the various approximations that are used in this method to speed up its computer implementation.

In Fig. 6.31 we have superimposed a square grid on the image $f(x, y)$, and we assume that in each cell $f(x, y)$ is constant. Let f_m denote this constant value in the mth cell, and let N be the total number of cells. For algebraic techniques a ray is defined somewhat differently. A ray is now a "fat" line running through the (x, y)-plane. To illustrate this, we have shaded the ith ray in Fig. 6.31, where each ray is of width τ. In most cases the ray width is approximately equal to the image cell width. A line integral will now be called a *ray sum.*

Like the image, the projections will also be given a one-index representation. Let p_m be the ray sum measured with the mth ray as shown in Fig. 6.31. The relationship between the f_i's and p_i's may be expressed as

$$\sum_{j=1}^{N} w_{ij} f_j = p_i, \qquad i = 1, 2, \ldots, M \tag{6.129}$$

where M is the total number of rays (in all the projections) and w_{ij} is the weighting factor that represents the contribution of the jth cell to the ith ray integral. The

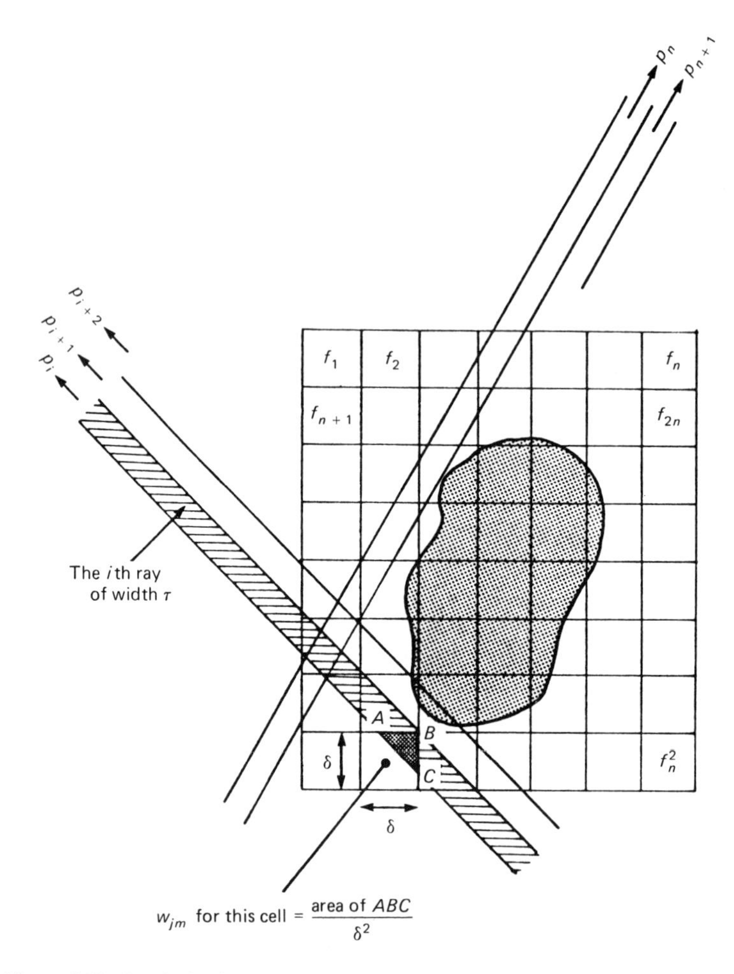

Figure 6.31 In algebraic reconstruction methods, a square grid is superimposed over the unknown image. Image values are assumed to be constant within each cell of the grid.

factor w_{ij} is equal to the fractional area of the jth image cell intercepted by the ith ray as shown for one of the cells in Fig. 6.31. Note that most w_{ij}'s are zero since only a small number of cells contribute to any given ray sum.

If M and N were small, we could use conventional matrix theory methods to invert the system of equations (6.129). However, in practice N may be as large as 65,000 (for 256 × 256 images), and in most cases for images of this size, M will also have the same magnitude. For these values of M and N the size of the matrix $[w_{ij}]$ in Eq. (6.129) is 65,000 × 65,000 which precludes any possibility of direct matrix inversion. Of course, when noise is present in the measurement data and when $M < N$, even for small N it is not possible to use direct matrix inversion, and some least-squares methods may have to be used. When both M and N are large, such methods are also computationally impractical

For large values of M and N there exist very attractive iterative methods for solving Eq. (6.129), which are based on the "method of projections" as first proposed by Kaczmarz [56], and later elucidated further by Tanabe [98]. To explain the computational steps involved in these methods, we first write Eq. (6.129) in an expanded form:

$$\begin{aligned} w_{11}f_1 + w_{12}f_2 + w_{13}f_3 + \cdots + w_{1N}f_N &= p_1 \\ w_{21}f_1 + w_{22}f_2 + \qquad + \cdots + w_{2N}f_N &= p_2 \\ &\vdots \\ w_{M1}f_1 + w_{M2}f_2 + \qquad + \cdots + w_{MN}f_N &= p_M \end{aligned} \tag{6.130}$$

A grid representation with N cells gives an image N degrees of freedom. Therefore, an image, represented by $(f_1, f_2, \ldots, f_N)$, may be considered to be a single point in an N-dimensional space. In this space each of the equations above represents a hyperplane. When a unique solution to these equations exists, the intersection of all these hyperplanes is a single point giving that solution. This concept is illustrated further in Fig. 6.32, where for the purpose of display, we have considered the case of only two variables f_1 and f_2 satisfying the following equations:

$$\begin{aligned} w_{11}f_1 + w_{12}f_2 &= p_1 \\ w_{21}f_1 + w_{22}f_2 &= p_2 \end{aligned} \tag{6.131}$$

The computational procedure for locating the solution in Fig. 6.32 consists of first starting with an initial guess, projecting this initial guess on the first line, reprojecting the resulting point on the second line, then projecting back onto the first line, and so on. If a unique solution exists, the iterations will always converge to that point. A computer implementation of this procedure follows.

For the computer implementation of this method, we first make an initial guess at the solution. This guess, denoted by $f_1^{(0)}, f_2^{(0)}, \ldots, f_N^{(0)}$, is represented vectorially by $\mathbf{f}^{(0)}$ in the N-dimensional space. In most cases, we simply assign a value of zero to all the f_j's. This initial guess is projected on the hyperplane represented by the first of the equations in (6.130) giving $\mathbf{f}^{(1)}$, as illustrated in Fig. 6.32 for the two-dimensional case. $\mathbf{f}^{(1)}$ is projected on the hyperplane represented by the second equation in (6.130) to yield $\mathbf{f}^{(2)}$, and so on. When $\mathbf{f}^{(j-1)}$ is projected on the hyperplane represented by the jth equation to yield $\mathbf{f}^{(j)}$, the process can be mathematically described by

$$\mathbf{f}^{(j)} = \mathbf{f}^{(j-1)} - \frac{(\mathbf{f}^{(j-1)} \cdot \mathbf{w}_j - p_j)}{\mathbf{w}_j \cdot \mathbf{w}_j} \mathbf{w}_j \tag{6.132}$$

where $\mathbf{w}_j = (w_{j1}, w_{j2}, \ldots, w_{jN})$, and $\mathbf{w}_j \cdot \mathbf{w}_j$ is the dot product of $\mathbf{w}_j$ with itself.

In applications requiring a large number of views and where reconstructions are made on large matrices, the difficulty with using Eq. (6.132) can be in the

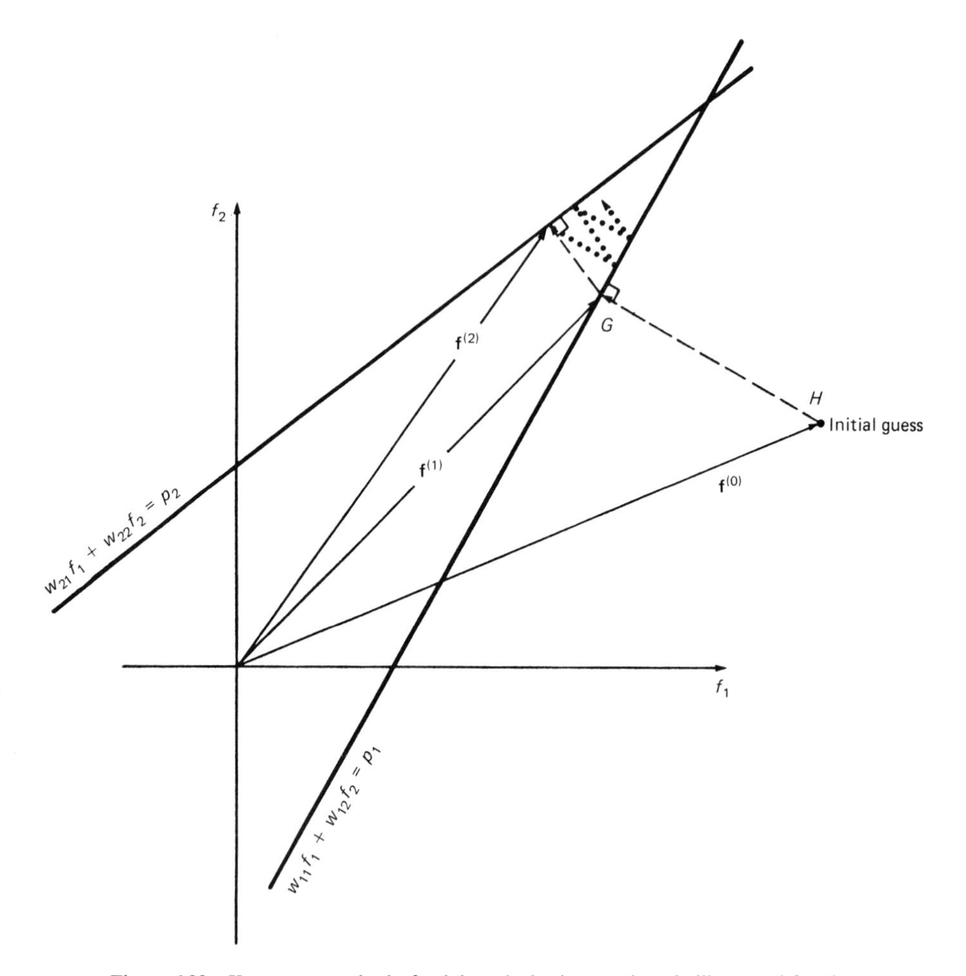

Figure 6.32 Kaczmarz method of solving algebraic equations is illustrated for the case of two unknowns. One starts with some arbitrary initial guess and then projects onto the line corresponding to the first equation. The resulting point is now projected onto the line representing the second equation. If there are only two equations, this process is continued back and forth, as illustrated by the dots until convergence is achieved.

calculation, storage, and fast retrieval of the weight coefficients w_{ij}. Consider the case where we wish to reconstruct an image on a 100×100 grid from 100 projections with 150 rays in each projection. The total number of weights, w_{ij}, needed in this case is 10^8, which is an enormous number and can pose problems in fast storage and retrieval in applications where reconstruction speed is important. This problem is somewhat eased by making approximations such as considering w_{ij} to be only a function of the perpendicular distance between the center of ith ray and the center of the jth cell. This perpendicular distance can be computed at run time.

To get around the implementation difficulties caused by the weight coefficients, a myriad of other algebraic approaches have also been suggested, many of which are approximations to Eq. (6.132). To discuss these more implementable approximations we first recast Eq. (6.132) in a slightly different form:

$$f_m^{(j)} = f_m^{(j-1)} + \frac{p_j - q_j}{\sum_{k=1}^{N} w_{jk}^2} w_{jm} \tag{6.133}$$

where

$$\begin{aligned} q_j &= \mathbf{f}^{(j-1)} \cdot \mathbf{w}_j \\ &= \sum_{k=1}^{N} f_k^{(j-1)} w_{jk} \end{aligned} \tag{6.134}$$

These equations say that when we project the $(j-1)$th solution onto the jth hyperplane [jth equation in Eq. (6.130)], the gray level of the mth element, whose current value is $f_m^{(j-1)}$, is obtained by correcting this current value by $\Delta f_m^{(j)}$, where

$$\Delta f_m^{(j)} = f_m^{(j)} - f_m^{(j-1)} = \frac{p_j - q_j}{\sum_{k=1}^{N} w_{jk}^2} w_{jm} \tag{6.135}$$

Note that while p_j is the measured ray sum along the jth ray, q_j may be considered to be the computed ray sum for the same ray based on the $(j-1)$th solution for the image gray levels. The correction Δf_m to the mth cell is obtained by first calculating the difference between the measured ray sum and the computed ray sum, normalizing this difference by $\sum_{k=1}^{N} w_{jk}^2$, and then assigning this value to all the image cells in the jth ray, each assignment being weighted by the corresponding w_{jm}.

With the preliminaries presented above, we now discuss three different computer implementations of the algebraic algorithms. These are represented by the acronyms ART, SIRT, and SART. In the author's opinion at this time, for reconstructions on large matrices from a large number of projections, the SART algorithm is probably the best. It seems to give decent results in only one iteration, compared to many required by SIRT, and at the same time does not suffer from the "salt and pepper" noise that characterizes ART reconstructions.

Algebraic Reconstruction Techniques

In many algebraic reconstruction technique (ART) implementations w_{jk}'s in Eq. (6.135) are simply replaced by 1's and 0's, depending on whether the center of the kth image cell is within the jth ray. This makes the implementation easier because such a decision can easily be made at computer run time. In this case the denominator in Eq. (6.135) is given by $\sum_{k=1}^{N} w_{jk}^2 = N_j$, which is the number of image cells whose centers are within the jth ray. The correction to the mth image cell from the

jth equation in Eq. (6.135) may now be written as

$$\Delta f_m^{(j)} = \frac{p_j - q_j}{N_j} \tag{6.136}$$

for all the cells whose centers are within the jth ray. We are essentially smearing back the difference $(p_j - q_j)/N_j$ over these image cells. In Eq. (6.136), q_j's are calculated using the expression in Eq. (6.134), except that one now uses the binary approximation for w_{jk}'s.

The approximation in Eq. (6.136), although easy to implement, often leads to artifacts in the reconstructed images, especially if N_j is not a good approximation to the denominator. Superior reconstructions may be obtained if Eq. (6.136) is replaced by

$$\Delta f_m^{(j)} = \frac{p_j}{L_j} - \frac{q_j}{N_j} \tag{6.137}$$

where L_j is the length (normalized by δ, see Fig. 6.31) of the jth ray through the reconstruction region.

The pixel corrections as computed by Eqs. (6.136) or (6.137) are applied on a ray-by-ray basis in each projection. In other words, we take one equation at a time from the set in Eq. (6.130), for each equation we calculate the computed ray sum by using Eq. (6.134) with the appropriate approximation for w_{jk}'s, we subtract this computed ray sum from the measurement p_j, and assign this difference to pixels according to Eq. (6.136). After we go through all the equations in Eq. (6.130), we have a solution that is denoted by $\mathbf{f}^{(M)}$. We now go back to the first equation in Eq. (6.130) and iterate the whole procedure, which yields $\mathbf{f}^{(2M)}$, and so on. Iterations are stopped when the computed changes of the image cell values are negligible compared to their current values. Tanabe [98] has shown that assuming that w_{jk}'s are known precisely, and further assuming that a unique solution $\mathbf{f}_s$ to the system of equations in Eq. (6.130) exists, then

$$\lim_{k \to \infty} \mathbf{f}^{(kM)} = \mathbf{f}_s \tag{6.138}$$

A few comments about the convergence of the algorithm are in order. If in Fig. 6.32 the two hyperplanes had been perpendicular to each other, the reader may easily show that given for an initial guess any point in the (f_1, f_2)-plane, it is possible to arrive at the correct solution in only two steps like Eq. (6.132). On the other hand, if the two hyperplanes have only a very small angle between them, the value of k in Eq. (6.138) may acquire a large value (depending on the initial guess) before the correct solution is reached. Therefore, the sequence in which the equations in (6.130) are taken up has a considerable influence on the rate of convergence to the solution. The reader is also referred to [60] for further comments about the convergence and also about what happens when we have an overdetermined or underdetermined system of equations.

ART reconstructions usually suffer from *salt and pepper noise*, which is caused by the inconsistencies introduced in the set of equations by the approximations commonly used for w_{jk}'s. The result is that the computed ray sums in Eq. (6.134) are usually a poor approximation to the corresponding measured ray sums. The effect caused by such inconsistencies is exacerbated by the fact that as each equation corresponding to a ray in a projection is taken up, it changes some of the pixels just altered by the preceding equation in the same projection. The SIRT algorithm described briefly below also suffers from these inconsistencies in the forward process [appearing in the computation of q_i's in Eq. (6.134)], but by eliminating the continual and competing pixel update as each new equation is taken up, it results in smoother reconstructions.

It is possible to reduce the effects of this noise in ART reconstructions by relaxation, in which we update a pixel by $\alpha \cdot \Delta f_m^{(j)}$, where α is less than 1. In some cases, the relaxation parameter α is made a function of the iteration number; that is, it becomes progressively smaller with increase in the number of iterations. The resulting improvements in the quality of reconstruction are usually at the expense of convergence.

Simultaneous Iterative Reconstruction Technique

In the simultaneous iterative reconstruction technique (SIRT) approach, which at the expense of slower convergence usually leads to better-looking images than those produced by ART, we again use Eq. (6.136) or (6.137) to compute the change $\Delta f_m^{(j)}$ in the mth pixel caused by the jth equation in (6.130). However, the value of the mth cell is not changed at this time. Before making any changes, we go through all the equations, and then only at the end of each iteration are the cell values changed, the change for each cell being the average value of all the changes computed for that cell. This constitutes one iteration of the algorithm. In the second iteration, we go back to the first equation in (6.130) and the process is repeated.

Simultaneous Algebraic Reconstruction Technique

We have recently developed a variation on the algebraic approaches discussed above, which we call the simultaneous algebraic reconstruction technique (SART), that seems to combine the best of ART and SIRT. In only one iteration, we can obtain reconstructions of good quality and numerical accuracy. In SART, discussed in detail in [2], we have improved the quality of the forward process by writing equations (6.129) as

$$p_j = \sum_{m=1}^{P_j} f(s_{jm})\, \Delta s \tag{6.139}$$

where s_{jm}'s are an equispaced set of points along the jth ray in the (x, y)-plane as shown in Fig. 6.33, and P_j is the total number of such points along the ray. [Note that $f(x, y)$ denotes the image values as a function of the continuous Cartesian coordinates in the (x, y)-plane.] In our implementation, we have assumed that the

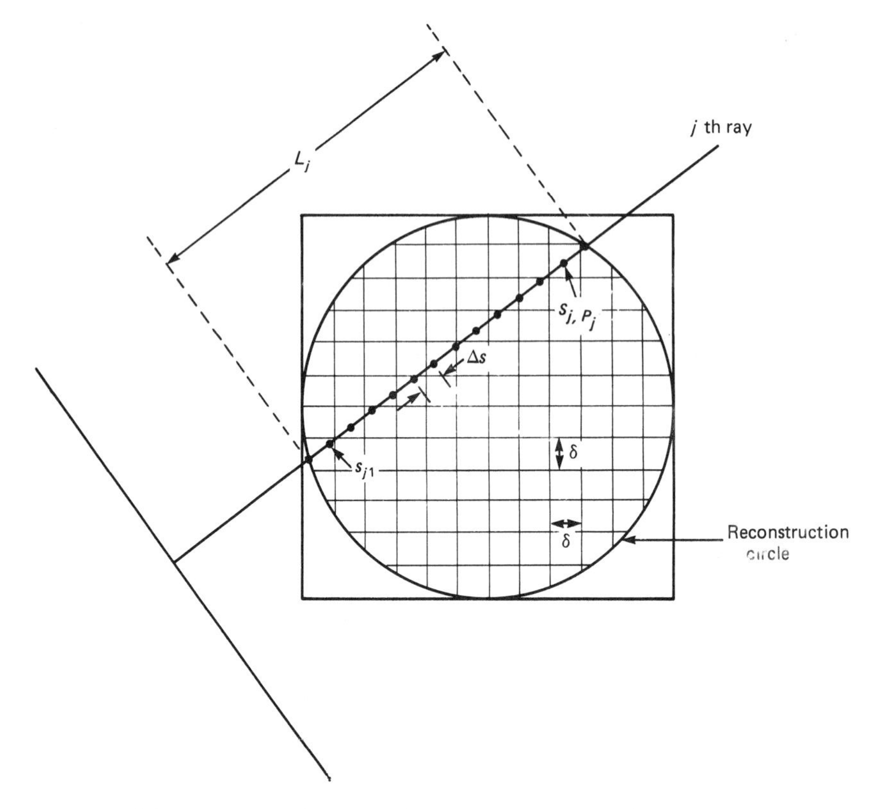

Figure 6.33 A ray sum is expressed as a sum of image values along a set of equispaced points on a straight line through the reconstruction circle. The physical length, L_j, of the jth ray is given by the intercept of the straight line and the circle. Only the pixels within the circle are used for formulating the algebraic equations.

object is enclosed by a circle, as shown in Fig. 6.33. We call this circle the *reconstruction circle.* The value $f(s_{jm})$ is now represented in terms of the four nearest f_n's by bilinear interpolation. In general, we may write

$$f(s_{jm}) = \sum_{i=1}^{N} d_{ijm} f_i \tag{6.140}$$

In bilinear interpolation, for each s_{jm} only the four nearest d_{ijm}'s are nonzero and their sum equals unity. Combining the two equations above, we get

$$p_j = \sum_{i=1}^{N} w_{ij} f_i \tag{6.141}$$

where

$$w_{ij} = \sum_{m=1}^{P_j} d_{ijm} \Delta s \tag{6.142}$$

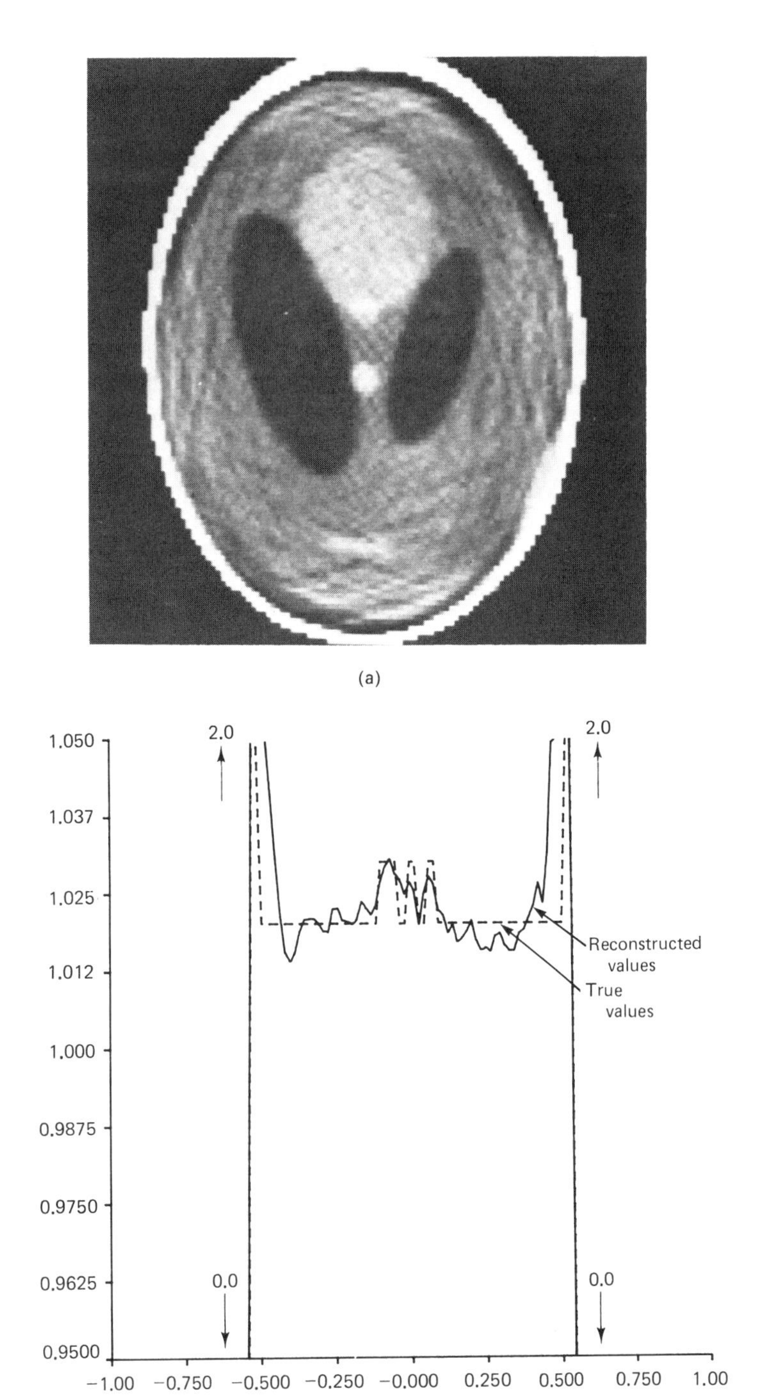

Figure 6.34 Single iteration algebraic reconstruction of the image of Fig. 6.25 with 127 projections and 100 rays in each projection. The reconstruction is on a 128 × 128 matrix. (b) Numerical comparison of the true and reconstructed values on line $y = -0.605$.

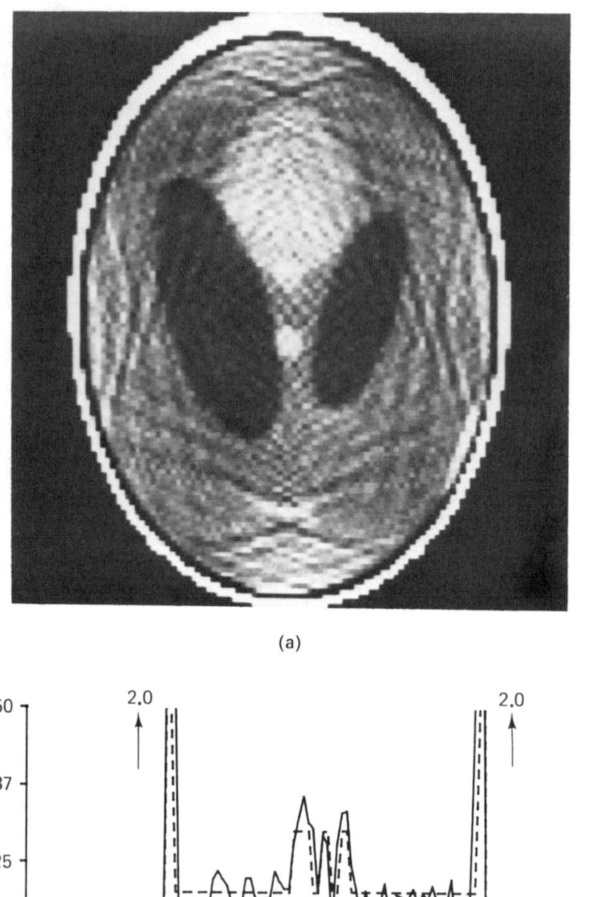

(a)

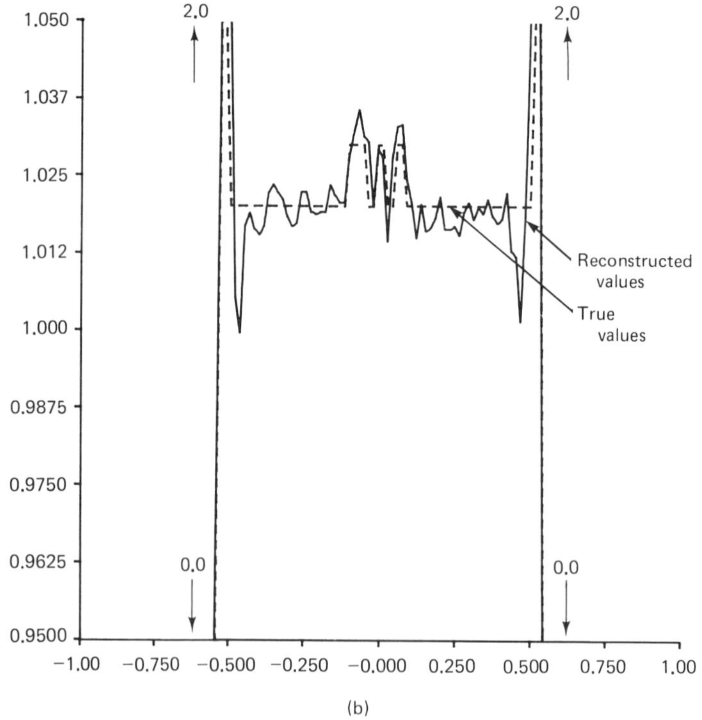

(b)

Figure 6.35 (a) Two-iteration version of the reconstruction shown in Fig. 6.34(a); (b) a numerical comparison of the true and reconstructed values on line $y = -0.605$.

For the overall accuracy of this procedure, it is important that the weights corresponding to the pixels at the beginning and the end of each ray be such that $\sum_{i=1}^{N} w_{ij}$ is equal to the actual physical length, L_j. As far as the choice of Δs (and, therefore, P_j) is concerned, we have found that setting it equal to half of the spacing δ of the sampling lattice provides a good trade-off between the accuracy of representation and computational cost.

In addition to the procedure above for computing w_{ij}'s, we deviate from the traditional ART and SIRT algorithms in two more ways. All the pixel corrections, Δf_m's, are computed for *one* complete projection, and only then are the pixels updated. Also, the back-distribution of the pixel corrections is longitudinally weighted with a Hamming window, the maximum of the window being at the center of the reconstruction circle.

Using the SART implementation, we have shown in Fig. 6.34(a) a *single* iteration algebraic 128 × 128 reconstruction of the test image of Fig. 6.25(a). The data used for the reconstruction consisted of 100 parallel projections over 180° with 127 rays in each projection. Figure 6.34(b) shows the reconstructed values on the line $y = -0.605$. Figure 6.35 shows the corresponding results for two iterations. Note that the improvement over a single iteration is marginal.

6.8 BIBLIOGRAPHICAL NOTES

The current excitement in tomographic imaging originated with Hounsfield's invention [52, 53] of the computed tomography (CT) scanner in 1972, which was indeed a major breakthrough. His invention showed that it is possible to get high-quality cross-sectional images with an accuracy now reaching 1 part in 1000, in spite of the fact that the projection data does not strictly satisfy theoretical models underlying the efficiently implementable reconstruction algorithms. (In x-ray tomography the mismatch with the assumed theoretical models is caused primarily by the polychromaticity of the radiation used.) His invention also showed that it is possible to process a very large number of measurements (now approaching 1 million) with fairly complex mathematical operations, and still get an image that is incredibly accurate. The success of x-ray CT has naturally led to research aimed at extending this mode of image formation to ultrasound and microwave sources.

The paper by Mueller et al. [71] was responsible for focusing the interest of many researchers on the area of diffraction tomography, although from a purely scientific standpoint the technique can be traced back to the now classic paper by Wolf [105], and a subsequent article by Iwata and Nagata [55]. The small perturbation approximations that are used for developing the diffraction tomography algorithms have been discussed by Ishimaru [54] and Morse and Ingard [70]. Diffraction tomography falls under the general subject of inverse scattering. The issues relating to the uniqueness and stability of inverse scattering solutions are addressed in [4, 28, 76, 89]. The filtered-backpropagation algorithm for diffraction tomography was first advanced by Devaney [29, 30]. More recently, Pan and Kak

[80] showed that by using bilinear interpolation followed by direct Fourier inversion, we can obtain reconstructions of quality comparable to that produced by the filtered-backpropagation algorithm. For an $N \times N$ image reconstructed from N diffracted projections, the former approach requires approximately $4N$ FFTs, whereas the latter technique requires approximately N^2 FFTs. This difference translates into great savings of computation time with the algorithm based on bilinear interpolation. More details on bilinear interpolation used by Pan and Kak can be found in [83]. Interpolation-based algorithms were first studied by Carter [14] and Mueller et al. [72]. The reader is also referred to [36] about how in some cases it may be possible to avoid the interpolation and still be able to reconstruct an object with direct two-dimensional Fourier inversion. A diffraction-tomography approach that requires only two rotational positions of the object has been advanced by Nahamoo and Kak [75], and its computer implementation has been studied by Pan and Kak [81]. Diffraction tomography based on the reflected data has been studied in great detail by Norton and Linzer [77]. A comparison of Born and Rytov approximations is presented in [61] and [94].

Because of the absence of any refraction or diffraction, with x-rays the problem of tomographic imaging reduces to reconstructing an image from its line integrals. The first mathematical solution to the problem of reconstructing a function from its projections was given by Radon [84] in 1917. More recently, some of the first investigators to examine this problem either theoretically or experimentally (and often independently) include (in a roughly chronological order): Bracewell [8], Olendorf [78], Cormack [22, 23], Kuhl and Edwards [63], DeRosier and Klug [27], Tretiak et al. [100], Rowley [88], Berry and Gibbs [7], Ramachandran and Lakshiminarayanan [85], Bender et al. [6], and Bates and Peters [5]. A detailed survey of the work done in computed tomographic imaging up to 1979 appears in [58].

The idea of filtered-backprojection was first advanced by Bracewell [9] and later independently by Ramachandran and Lakshminarayanan [85]. The superiority of the filtered-backprojection algorithms over the algebraic techniques was first demonstrated by Shepp and Logan [92]. Its development for the fan-beam data was first made by Lakshminarayanan [65] for the equispaced collinear-detectors case and later extended by Herman and Naparstek [49] to the case of equiangular rays. The fan-beam algorithm derivation presented here was first developed by Scudder [91]. Many authors [3, 62, 64, 67, 99] have proposed variations on the filter functions of the filtered-backprojection algorithms discussed in this chapter. The reader is referred particularly to [62, 67] for ways to speed up the filtering of the projection data by using binary approximations and/or inserting zeros in the unit sample response of the filter function. Images may also be reconstructed from fan-beam data by first sorting them into parallel projection data. Fast algorithms for ray sorting of fan-beam data have been developed by Wang [103], Dreike and Boyd [34], Peters and Lewitt [82], and Dines and Kak [31]. The reader is referred to [74] for a filtered-backprojection algorithm for reconstructing from data generated by using very narrow-angle fan beams that rotate *and* traverse *continuously* around the object. The reader is also referred to [50, 51] for algorithms for nonuniformly sampled projections data, and to [10, 66, 79, 90, 97] for reconstructions from incomplete and limited projections. Space limitations have prevented us from discussing algorithms for doing full three-dimensional reconstructions [18, 19]. We have also not gone through the circular harmonic transform method of image reconstruction as proposed by Hansen [45, 46].

As with the case with diffracting sources, tomographic imaging with nondiffracting sources may also be accomplished, although less accurately, by direct Fourier inversion instead of the filtered-backprojection method presented in this chapter. This was first shown by Bracewell [8] for radioastronomy, and later independently by DeRosier and Klug [27] in electron microscopy and Rowley [88] in optical holography. Several workers who applied this method to radiography include Tretiak et al. [100], Bates and Peters [5], and Mersereau and Oppenheim [69]. To utilize two-dimensional FFT algorithms for image formation, the direct Fourier approach calls for frequency-domain interpolation from a polar grid to a rectangular grid. For some recent methods to minimize the resulting interpolation error, the reader is referred to [95]. More recently, Wernecke and D'Addario [104] have proposed a maximum-entropy approach to direct Fourier inversion (see Section 5.8). Their procedure is especially applicable if for some reason the projection data are insufficient.

Aliasing artifacts in tomographic imaging with nondiffracting sources have been studied by Brooks et al. [11, 12] and Crawford and Kak [24]. With regard to the properties of noise in images reconstructed with filtered backprojection, Shepp and Logan [92] first showed that when filtered backprojection algorithms are used, the variance of the noise is directly proportional to the area under the square of the filter function. This derivation was based on the assumption that the variance of the measurement noise is the same for all the rays in the projection data, a condition that is usually not satisfied. A more general expression (not using this assumption) for the noise variance was derived by Kak [58], who has also introduced the concept of "the relative-uncertainty image." For tomographic imaging with x-rays, Tretiak [101] has derived an algorithm-independent lower bound on the noise variance in a reconstructed image. Noise properties of x-ray tomograms have also been discussed by Chesler et al. [17].

The algebraic approaches to image reconstruction have been studied in great detail by Gordon et al. [40–43], Herman [47, 48], Budinger and Gullberg [13], Gilbert [37], Oppenheim [79], and Andersen and Kak [2]. Article [35] is a recent discussion on the convergence rates of the algebraic techniques. Ramakrishnan et al. [86] have shown how by orthogonalization of the equations, we can increase the speed of convergence of a reconstruction algorithm. It appears that for objects that are only lightly refracting, a combination of algebraic reconstruction and digital ray tracing may be the best approach to obtaining high-quality reconstructions [15, 16]. A survey of digital ray-tracing and ray-linking algorithms for this purpose was presented by Andersen and Kak in [1]. If a refracting object has special symmetries, then as shown by Vest [102], it may be possible to reconstruct the object without ray tracing. In some cases it is possible to ignore refraction entirely and treat the problem as one of imaging with a nondiffracting source. Some early ultrasonic reconstructions using this approach were reported by Glover and Sharp [38]. As shown by Dines and Kak [32], Kak [58], and Crawford and Kak [25], in such ultrasonic reconstructions it is possible to maintain the basic x-ray type of algorithm, yet introduce artifact-removing features such as corrections for multipath effects, frequency-dependent attenuation, and so on. For procedures on how we may contend with frequency-dependent attenuation in tomographic reconstructions, the reader is referred to [57].

The reader is also referred to [58, 59] for a survey of medical tomographic imaging. For applications in radio astronomy, where the aim is to reconstruct the "brightness" distribution of a celestial source of radio waves from its strip integral measurements taken with special antenna beams, the reader is referred to [8, 9]. For electron-microscopy applications, where one attempts to reconstruct the molecular structure of complex biomolecules from transmission micrograms, the reader should consult [26, 43]. The applications of this technique in optical interferometry, where the aim is to determine the refractive-index field of an optically transparent medium, are discussed in [7, 88, 96]. The applications of tomography in earth resources imaging are presented in [33, 68].

6.9 SUMMARY

This chapter dealt with cross-sectional imaging (tomography) with diffracting and nondiffracting sources. We listed a large number of areas where tomography concepts can be used for image formation. These areas included medical imaging, earth resources imaging, celestial imaging, etc. We then presented two key theorems which form the foundations of tomographic reconstruction algorithms. These were called the Fourier Projection Theorem for the case of nondiffracting sources, and the Fourier Diffraction Projection Theorem for diffracting sources. Subsequently, we presented the reconstruction algorithms. We also presented the algebraic approaches to reconstruction, since these methods are expected to play an important role in the imaging of refracting media. We concluded with bibliographical notes on tomography to give the reader a chronological perspective on the area.

ACKNOWLEDGMENTS

The authors's expertise in tomographic imaging is a result of active collaboration with a remarkable collection of people in the School of Electrical Engineering at Purdue University. Included among these are his former students David Nahamoo and Carl Crawford, and his present students Anders Andersen, Malcolm Slaney, and Mani Azimi. Without the exchanges of ideas with them, the overview presented here would not have been possible. The author would also like to express his appreciation to S. X. Pan, currently a visiting scholar at Purdue, for many stimulating discussions about diffraction tomography. The reconstructions shown in this chapter were produced by Carl Crawford, Anders Andersen, Malcolm Slaney, and S. X. Pan. The author's work in tomographic imaging was funded initially by numerous research grants from the National Institutes of Health. More recently, his work in this area has also been supported by the Walter Reed Army Institute of Medical Research. He has also benefited from his many interactions with industrial organizations, often on a consulting basis.

BIBLIOGRAPHY AND REFERENCES

1 A. H. Andersen and A. C. Kak, 'Digital Ray Tracing in Two-Dimensional Refractive Fields," J. Acoust. Soc. Am., Vol. 72, pp. 1593–1606, 1982.

2 A. H. Andersen and A. C. Kak, "Simultaneous Algebraic Reconstruction Technique: A New Implementation of the ART Algorithm," Ultrasonic Imaging, Vol. 6, Jan. 1984.

3 N. Baba and K. Murata, "Filtering for Image Reconstruction from Projections," J. Opt. Soc. Am., Vol. 67, pp. 662–668, 1977.

4 H. P. Baltes, ed., *Inverse Source Problems in Optics*, Springer-Verlag, New York, 1978.

5 R. H. T. Bates and T. M. Peters, "Towards Improvements in Tomography," N. Z. J. Sci., Vol. 14, pp. 883–896, 1971.

6 R. Bender, S. H. Bellman, and R. Gordon, "ART and the Ribosome: A Preliminary Report on the Three-Dimensional Structure of Individual Ribosomes Determined by an Algebraic Reconstruction Technique," J. Theor. Biol., Vol. 29, pp. 483–487, 1970.

7 M. V. Berry and D. F. Gibbs, "The Interpretation of Optical Projections," Proc. R. Soc. Lond., Vol. A314, pp. 143–152, 1970.

8 R. N. Bracewell, "Strip Integration in Radio Astronomy," Aust. J. Phys., Vol. 9, pp. 198–217, 1956.

9 R. N. Bracewell and A. C. Riddle, "Inversion of Fan-Beam Scans in Radio Astronomy," Astrophys. J., Vol. 150, pp. 427–434, 1967.

10 R. N. Bracewell and S. J. Wernecke, "Image Reconstruction over a Finite Field of View," J. Opt. Soc. Am., Vol. 65, pp. 1342–1346, 1975.

11 R. A. Brooks and G. Dichiro, "Statistical Limitations in X-ray Reconstruction Tomography," Med. Phys., Vol. 3, pp. 237–240, 1976.

12 R. A. Brooks, G. H. Weiss, and A. J. Talbert, "A New Approach to Interpolation in Computed Tomography," J. Comput. Assist. Tomogr., Vol. 2, pp. 577–585, 1978.

13 T. F. Budinger and G. T. Gullberg, "Three-Dimensional Reconstruction in Nuclear Medicine Emission Imaging," IEEE Trans. Nucl. Sci., Vol. NS-21, pp. 2–21, 1974.

14 W. H. Carter, "Computational Reconstruction of Scattering Objects from Holograms," J. Opt. Soc. Am., Vol. 60, pp. 306–314, 1970.

15 S. Cha and C. M. Vest, "Interferometry and Reconstruction of Strongly Refracting Asymmetric-Refractive-Index Fields," Opt. Soc. Am., Vol. 4, pp. 311–313, 1979.

16 S. Cha and C. M. Vest, "Tomographic Reconstruction of Strongly Refracting Fields and Its Application to Interferometric Measurement of Boundary Layers," Appl. Opt., Vol. 20, pp. 2787–2794, 1981.

17 D. A. Chesler, S. J. Riederer, and N. J. Pele, "Noise Due to Photon Counting Statistics in Computer X-ray Tomography," J. Comput. Assist. Tomogr., Vol. 1, pp. 64–74, 1977.

18 M. Y. Chiu, H. H. Barrett and R. G. Simpson, "Three-Dimensional Reconstruction from Planar Projections," J. Opt. Soc. Am., Vol. 70, pp. 755–762, July 1980.

19 M. Y. Chiu, H. H. Barrett, R. G. Simpson, C. Chou, J. W. Arendt, and G. R. Gindi, "Three-Dimensional Radiographic Imaging with a Restricted View Angle," J. Opt. Soc. Am., Vol. 69, pp. 1323–1330, Oct. 1979.

20 Z. H. Cho, H. S. Kim, H. B. Song, and J. Cumming, "Fourier Transform Nuclear Magnetic Resonance Tomographic Imaging," Proc. IEEE, Vol. 70, pp. 1152–1173, 1982.

21 B. D. Coleman, "Foundations of Linear Viscoelasticity," Rev. Mod. Phys., Vol. 9, pp. 433–450, 1981.

22 A. M. Cormack, "Representation of a Function by Its Line Integrals with Some Radiological Applications," J. Appl. Phys., Vol. 34, pp. 2722–2727, 1963.

23 A. M. Cormack, "Representation of a Function by Its Line Integrals with Some Radiological Applications," J. Appl. Phys., Vol. 35, pp. 2908–2913, 1964.

24 C. R. Crawford and A. C. Kak, "Aliasing Artifacts in Computerized Tomography," Appl. Opt., Vol. 18, pp. 3704–3711, 1979.

25 C. R. Crawford and A. C. Kak, "Multipath Artifact Corrections in Ultrasonic Transmission Tomography," Ultrason. Imaging, Vol. 4, pp. 234–266, 1982.

26 R. A. Crowther, D. J. DeRosier, and A. Klug, "The Reconstruction of a Three-Dimensional Structure from Projections and Its Applications to Electron Microscopy," Proc. R. Soc. Lond., Vol. A317, pp. 319–340, 1970.

27 D. J. DeRosier and A. Klug, "Reconstruction of Three-Dimensional Structures from Electron Micrographs," Nature, Vol. 217, pp. 130–134, 1968.

28 A. J. Devaney, "Nonuniqueness in the Inverse Scattering Problem," J. Math. Phys., Vol. 19, pp. 1525–1531, 1978.

29 A. J. Devaney, "A Filtered Backpropagation Algorithm for Diffraction Tomography," Ultrason. Imaging, Vol. 4, pp. 336–350, 1982.

30 A. J. Devaney, "A Computer Simulation Study of Diffraction Tomography" IEEE Trans. Biomed. Engrg., Vol. BME-30, pp. 377–386, 1983.

31 K. A. Dines and A. C. Kak, "Measurement and Reconstruction of Ultrasonic Parameters for Diagnostic Imaging," Technical Report TR-EE-77-4, School Electr. Eng., Purdue University, West Lafayette, Ind., 1976.

32 K. A. Dines and A. C. Kak, "Ultrasonic Attenuation Tomography of Soft Biological Tissues," Ultrason. Imaging, Vol. 1, pp. 16–33, 1979.

33 K. A. Dines and R. J. Lytle, "Computerized Geophysical Tomography," Proc. IEEE, Vol. 67, pp. 1065–1073, 1979.

34 P. Dreike and D. P. Boyd, "Convolution Reconstruction of Fan-Beam Reconstructions," Comput. Graph. Image Proc., Vol. 5, pp. 459–469, 1977.

35 P. P. B. Eggermont, G. T. Herman, and A. Lent, "Iterative Algorithms for Large Partitioned Linear Systems, with Application to Image Reconstruction," Linear Algebra Appl., Vol. 40, pp. 37–67, 1981.

36 A. F. Fercher, H. Bartelt, H. Becker, and E. Wiltschko, "Image Formation by Inversion of Scattered Data: Experiments and Computational Simulation," Appl. Opt., Vol. 18, pp. 2427–2439, 1979.

37 P. Gilbert, "Iterative Methods for the Reconstruction of Three-Dimensional Objects from Their Projections, J. Theor. Biol., Vol. 36, pp. 105–117, 1972.

38 G. H. Glover and J. L. Sharp, "Reconstruction of Ultrasound Propagation Speed Distribution in Soft Tissue: Time-of-Flight Tomography," IEEE Trans. Sonics Ultrason., Vol. SU-24, pp. 229–234, 1977.

39 J. W. Goodman, *Introduction to Fourier Optics*, McGraw-Hill, New York, 1968, Chap. 3.

40 R. Gordon, R. Bender, and G. T. Herman, "Algebraic Reconstruction Techniques (ART) for Three-Dimensional Electron Microscopy and X-ray Photography," J. Theor. Biol., Vol. 29, pp. 470–481, 1971.

41 R. Gordon, R. Bender, and G. T. Herman, "Algebraic Reconstruction Techniques (ART) for Three-Dimensional Electron Microscopy and X-ray Photography," J. Theor. Biol., Vol. 29, pp. 471–481, 1971.

42 R. Gordon, "A Tutorial on ART (Algebraic Reconstruction Techniques)," IEEE Trans. Nucl. Sci., Vol. NS-21, pp. 78–93, 1974.

43 R. Gordon and G. T. Herman, "Reconstruction of Pictures from Their Projections," Commun. ACM, Vol. 14, pp. 759–768, 1971.

44 R. W. Hamming, *Digital Filters*, Prentice-Hall, Englewood Cliffs, N. J., 1977.

45 E. W. Hansen, "Theory of Circular Image Reconstruction," J. Opt. Soc. Am., Vol. 71, pp. 304–308, Mar. 1981.

46 E. W. Hansen, "Circular Harmonic Image Reconstruction: Experiments," Appl. Opt., Vol. 20, pp. 2266–2274, July 1981.

47 G. T. Herman and S. Rowland, "Resolution in ART: An Experimental Investigation of the Resolving Power of an Algebraic Picture Reconstruction," J. Theor. Biol., Vol. 33, pp. 213–223, 1971.

48 G. T. Herman, A. Lent, and S. Rowland, "ART: Mathematics and Applications: a Report on the Mathematical Foundations and on Applicability to Real Data of the Algebraic Reconstruction Techniques," J. Theor. Biol., Vol. 43, pp. 1–32, 1973.

49 G. T. Herman and A. Naparstek, "Fast Image Reconstruction Based on a Radon Inversion Formula Appropriate for Rapidly Collected Data," SIAM J. Appl. Math., Vol. 33, pp. 511–533, 1977.

50 B. K. P. Horn, "Density Reconstructions Using Arbitrary Ray Sampling Schemes," Proc. IEEE, Vol. 66, pp. 551–562, 1978.

51 B. K. P. Horn, "Fan-Beam Reconstruction Methods," Proc. IEEE, Vol. 67, pp. 1616–1623, 1979.

52 G. N. Hounsfield, "A Method of Apparatus for Examination of a Body by Radiation Such As X-ray or Gamma Radiation," Patent Specification 1,283,915, The Patent Office, London, 1972.

53 G. N. Hounsfield, "Computerized Transverse Axial Scanning (Tomography): Part 1. Description of System," Br. J. Radiol., Vol. 46, pp. 1016–1022, 1973.

54 A. Ishimaru, *Wave Propagation and Scattering in Random Media*, Vol. 2, Academic Press, New York, 1978.

55 K. Iwata and R. Nagata, "Calculation of Refractive Index Distribution from Interferograms Using Born and Rytov's Approximation," Jap. J. Appl. Phys., Vol. 14, Suppl. 14-1, p. 383, 1975.

56 S. Kaczmarz, "Angenaherte Auflosung von Systemen linearer Gleichungen," Bull. Acad. Pol. Sci. Lett., A, pp. 355–357, 1937.

57 A. C. Kak and K. A. Dines, "Signal Processing of Broadband Pulse Ultrasound: Measurement of Attenuation of Soft Biological Tissues," IEEE Trans. Biomed. Eng., Vol. BME-25, pp. 321–344, 1978.

58 A. C. Kak, "Computerized Tomography with X-ray, Emission and Ultrasound Sources," Proc. IEEE, Vol. 67, pp. 1245–1272, 1979.

59 A. C. Kak, Guest Ed., "Computerized Medical Imaging," special issue of IEEE Trans. Biomed. Eng., Feb. 1981.

60 A. C. Kak, "Image Reconstruction from Projections," in *Digital Image Processing Techniques*, ed., M. P. Ekstrom, Academic Press, New York, 1983 (to appear).

61 M. Kaveh, M. Soumekh, and R. K. Mueller, "A Comparison of Born and Rytov Approximation in Acoustic Tomography," Acoust. Imaging, Vol. 10, 1981.

62 S. K. Kenue and J. F. Greenleaf, "Efficient Convolution Kernels for Computerized Tomography," Ultrason. Imaging, Vol. 1, pp. 232–244, 1979.

63 D. E. Kuhl and R. Q. Edwards, "Image Separation Radio-Isotope Scanning," Radiology, Vol. 80, pp. 653–661, 1963.

64 Y. S. Kwoh, I. S. Reed, and T. K. Truong, "A Generalized $|w|$-Filter for 3-D Reconstruction," IEEE Trans. Nucl. Sci., Vol. NS-24, pp. 1990–1998, 1977.

65 A. V. Lakshminarayanan, "Reconstruction from Divergent Ray Data," Technical Report 92, Dept. of Computer Science, State University of New York at Buffalo, 1975.

66 R. M. Lewitt and R. H. T. Bates, "Image Reconstruction from Projections," Optik, Vol. 50, Part I: pp. 19–33; Part II: pp. 85–109; Part III: pp. 189–204; Part IV: pp. 269–278, 1978.

67 R. M. Lewitt, "Ultra-fast Convolution Approximation for Computerized Tomography," IEEE Trans. Nucl. Sci., Vol. NS-26, pp. 2678–2681, 1979.

68 R. J. Lytle and K. A. Dines, "Iterative Ray Tracing between Boreholes for Underground Image Reconstruction," IEEE Trans. Geosci. Remote Sensing, Vol. GE-18, pp. 234–240, 1980.

69 R. M. Mersereau and A. V. Oppenheim, "Digital Reconstruction of Multidimensional Signals from Their Projections," Proc. IEEE, Vol. 62, pp. 1319–1338, 1974.

70 P. M. Morse and K. V. Ingard, *Theoretical Acoustics*, McGraw-Hill, New York, 1968.

71 R. K. Mueller, M. Kaveh, and G. Wade, "Reconstructive Tomography and Applications to Ultrasonics," Proc. IEEE, Vol. 67, pp. 567–587, 1979.

72 R. K. Mueller, M. Kaveh, and R. D. Iverson, "A New Approach to Acoustic Tomography Using Diffraction Techniques," Acoustical Holography, Vol. 8, ed. A. F. Metherell, Plenum Press, New York, 1980, pp. 615–628.

73 R. K. Mueller, "Diffraction Tomography I: The Wave Equation," Ultrason. Imaging, Vol. 2, pp. 213–222, 1980.

74 D. Nahamoo, C. R. Crawford, and A. C. Kak, "Design Constraints and Reconstruction Algorithms for Transverse-Continuous-Rotate CT Scanners," IEEE Trans. Biomed. Eng., Vol. BME-28, pp. 79–97, 1981.

75 D. Nahamoo and A. C. Kak, "Ultrasonic Diffraction Imaging," Technical Report TR-EE-82-20, School Electr. Eng., Purdue University, West Lafayette, Ind., 1982.

76 M. Z. Nashed, "Operato-theoretic and Computational Approaches to Ill-Posed Problems with Application to Antenna Theory," IEEE Trans. Antennas Propag., Vol. AP-29, 1981.

77 S. J. Norton and M. Linzer, "Ultrasonic Reflectivity Imaging in Three Dimensions: Exact Inverse Scattering Solutions for Plane, Cylindrical, and Spherical Apertures," IEEE Trans. Biomed. Eng., Vol. BME-28, pp. 202–220, 1981.

78 W. H. Oldendorf, "Isolated Flying Spot Detection of Radiodensity Discontinuities Displaying the Internal Structural Pattern of a Complex Object," IRE Trans. Biomed. Eng., Vol. BME-8, pp. 68–72, 1961.

79 B. E. Oppenheim, "Reconstruction Tomography from Incomplete Projections," in *Reconstruction Tomography in Diagnostic Radiology and Nuclear Medicine*, ed. M. M. TerPogossian et al., University Park Press, Baltimore, 1975.

80 S. X. Pan and A. C. Kak, "A Computational Study of Reconstruction Algorithms for Diffraction Tomography: Interpolation vs. Filtered-Backpropagation", IEEE Trans. Acoust. Speech Signal Process, Vol. ASSP-31, pp. 1262–1275, 1983.

81 S. X. Pan and A. C. Kak, "A Reconstruction Algorithm for Synthetic Aperture Diffraction Tomography and Its Implementation" (submitted for publication).

82 T. M. Peters and R. M. Lewitt, "Computed Tomography with Fan-Beam Geometry," J. Comput. Assist. Tomogr., Vol. 1, pp. 429–436, 1977.

83 P. M. Prenter, *Splines and Variational Methods*, Wiley, New York, Chap. 5.

84 J. Radon, "Uber die Bestimmung von Funktionen durch ihre Intergralwerte langs gewisser Mannigfaltigkeiten," ("On the Determination of Functions from Their Integrals along Certain Manifolds,"), Ber. Saechs. Akad. Wiss., Vol. 29, pp. 262–279, 1917.

85 G. N. Ramachandran and A. V. Lakshminarayanan, "Three-Dimensional Reconstructions from Radiographs and Electron Micrographs: Application of Convolution Instead of Fourier Transforms," Proc. Natl. Acad. Sci., USA, Vol. 68, pp. 2236–2240, 1971.

86 R. S. Ramakrishnan, S. K. Mullick, R. K. S. Rathore, and R. Subramanian, "Orthogonalization, Bernstein Polynomials, and Image Restoration," Appl. Opt., Vol. 18, pp. 464–468, 1979.

87 A. Rosenfeld and A. C. Kak, *Digital Picture Processing*, 2nd ed., Vol. 1, Academic Press, New York, 1982.

88 P. D. Rowley, "Quantitative Interpretation of Three-Dimensional Weakly Refractive Phase Objects Using Holographic Interferometry," J. Opt. Soc. Am., Vol. 59, pp. 1496–1498, 1969.

89 T. K. Sarkar, "Some Mathematical Considerations in Dealing with the Inverse Problem," IEEE Trans. Antennas Propag., Vol. AP-29, pp. 373–379, 1981.

90 T. Sato, S. J. Norton, M. Linzer, O. Ikeda, and M. Hirama, "Tomographic Image Reconstruction from Limited Projections Using Iterative Revisions in Image and Transform Spaces," App. Opt., Vol. 20, pp. 395–399, Feb. 1980.

91 H. J. Scudder, "Introduction to Computer Aided Tomography," Proc. IEEE, Vol. 66, pp. 628–637, June 1978.

92 L. A. Shepp and B. F. Logan, "The Fourier Reconstruction of a Head Section," IEEE Trans. Nucl. Sci., Vol. NS-21, pp. 21–43, 1974.

93 L. Shepp, "Computerized Tomography and Nuclear Magnetic Resonance," J. Comput. Assist. Tomogr., Vol. 4, p. 94, 1980.

94 M. Slaney, A. C. Kak and L. E. Larsen, "Limitations of Imaging with First Order Diffraction Tomography," IEEE Trans. Microwave Theory and Tech., July 1984 (to appear).

95 H. Stark, J. W. Woods, I. Paul, and R. Hingorani, "Direct Fourier Reconstruction in Com-

puter Tomography," IEEE Trans. Acoust. Speech Signal Process., Vol. ASSP-29, pp. 237–244, 1981.

96 D. W. Sweeney and C. M. Vest, "Reconstruction of Three-Dimensional Refractive Index Fields from Multi-directional Interferometric Data," Appl. Opt., Vol. 12, pp. 1649–1664, 1973.

97 K. C. Tam and V. Perez-Mendez, "Tomographical Imaging with Limited Angle Input," J. Opt. Soc. Am., Vol. 71, pp. 582–592, May 1981.

98 K. Tanabe, "Projection Method for Solving a Singular System," Numer. Math., Vol. 17, pp. 203–214, 1971.

99 E. Tanaka and T. A. Iinuma, "Correction Functions for Optimizing the Reconstructed Image in Transverse Section Scan," Phy. Med. Biol., Vol. 20, pp. 789–798, 1975.

100 0. Tretiak, M. Eden, and M. Simon, "Internal Structures for Three-Dimensional Images," Proc. 8th Int. Conf. Med. Biol. Eng., Chicago, 1969.

101 O. J. Tretiak, "Noise Limitations in X-ray Computed Tomography," J. Comput. Assist. Tomogr., Vol. 2, pp. 477–480, 1978.

102 C. M. Vest, "Interferometry of Strongly Refracting Axisymmetric Phase Objects," Appl. Opt., Vol. 14, pp. 1601–1606, 1975.

103 L. Wang, "Cross-section Reconstruction with a Fan-Beam Scanning Geometry," IEEE Trans. Comput., Vol. C-26, pp. 264–268, 1977.

104 S. J. Wernecke and L. R. D'Addario, "Maximum Entropy Image Reconstruction," IEEE Trans. Comput., Vol. C-26, pp. 351–364, 1977.

105 E. Wolf, "Three-Dimensional Structure Determination of Semi-transparent Objects from Holographic Data," Opt. Commun., Vol. 1, No. 4, pp. 153–156, 1969.

Index

A

Adaptive antenna for angle of arrival estimation, 234
Air wave, 29
Algebraic image reconstruction algorithms, 408
  algebraic reconstruction technique (ART), 413
  simultaneous algebraic reconstruction technique (SART), 415
  simultaneous iterative reconstruction technique (SIRT), 415
Array gain, 120
Array gain improvement ratio, 120, 122
Array processing, 1
Autoregressive/maximum entropy (AR/ME) spectral estimation. *See* Maximum entropy method
Autoregressive-moving average (ARMA) method, 288

B

Bearing time recorder (BTR), 157
Born approximation, 367
Burg formula, 258
Burg technique, 252

C

CLEAN, 331
Complex coherence, 140
Conjugate-gradient method, 229, 345
Cramér-Rao bound, 218
Cross-spectral density matrix (CSDM), 125

D

Deconvolution, 44, 73
  predictive, 84
Deghosting filter, 75
Delay-and-sum (DS) beamformer, 119
Digital correlator, 305
Dilational wave equation, 16
Dispersion analysis, 68

E

Eigenvalue/eigenvector decomposition, 269
Elastic wave equation, 15
Enhanced spectrum estimator, 175
Estimator-subtractor noise canceller, 138

F

F-K plot, 33, 46
Fast Fourier transform (FFT) cross-spectrum analyzer, 307
Fermat's principle, 30
Filtered-back propagation algorithm
  for diffracting sources, 374
  for nondiffracting sources, 390
Forward-backward linear prediction (FBLP) method, 262
  limitations, 267
  modified, 275
  noiseless data, 270
  sensitivity to parameter variations, 280
Fourier diffraction projection theorem, 364
Fourier slice theorem, 322, 372

G

Geophone, 10
Golden section search, 213
Gram-Schmidt orthonormalization, 226
Green's function, 9, 366

H

Head wave, 29
High resolution array processing. *See* Parametric spectral estimation methods
High wavenumber noise, 161

I

Image reconstruction in radio astronomy, 2, 293
  phase-amplitude errors, 338

Image reconstruction in radio astronomy (*Cont.*)
using Fourier inversion, 320
using maximum entropy, 342
using the method "CLEAN", 330
Image reconstruction in tomography, 354
Imaging,
with a diffracting source, 357
with a nondiffracting source, 360

K

Kumaresan-Prony method, 277

L

Lamé constants, 15
Large aperture seismic array (LASA), 41
Lattice filter, 255
Least squares method. *See* Forward-backward linear prediction method
Levinson recursion, 52, 254
multichannel generalization, 61
Linear prediction, 243

M

Maximum entropy method, 51, 116, 165, 170, 249, 342
Maximum likelihood (parameter) estimation, 208
Maximum likelihood receiver
nonsymmetric multipath, 225
symmetric multipath, 208
Maximum likelihood (spectral estimation) method, 55, 116, 287
Minimum complexity implementation of optimum sonar beamformer, 132
Minimum norm solution, 269
Minimum-variance distortionless signal response (MVDR) beamformer, 117
constrained, 123
spectral analysis, 169
Multichannel filtering, 44, 59
Multidimensional filtering, 44, 59, 185
Multipath model
diffuse, 203, 286
specular, 200
Mutual coherence, 300
Mutual spectral density, 300

N

Navier's equation, 16
Noise estimator-subtractor, 117
Normal equations, 51, 247
Null-steering system, 130

P

p wave, 17
Parametric spectral estimation methods, 51, 164, 207, 242
Passive sonar equation, 120
Periodogram, 48, 205
Polarization, 4, 68
Prediction-error filter, 51, 244
minimum phase property, 251
whitening property, 251
Principle of maximum entropy, 47
Projection slice theorem. *See* Fourier slice theorem
Prony's method, 270
Pseudo-inverse of a matrix, 271

R

Radar array processing, 2, 194
Radiation pattern, 206
Radio sources,
wave fields, 296
Rotational sampling of baseline space, 308
Rytov approximation, 370

S

Salt and pepper noise, 414
Seismic data acquisition, 31

Seismic experiment, 8
  models, 20
  objectives, 25
Seismic imaging problem, 9
Seismic signal processing, 2, 6
Seismic wavelet, 20
Shear wave, 17
Signal-nulling matrix filter, 135
Sonar array processing, 2, 115
Sonar array processor realizations, 145
  closed-loop, 151
  direct inverse-estimate-plug, 147
  estimate-invert-plug, 146
  orthonormalization, 147
Sonar optimum beamforming
  basic considerations, 119
Spatial correlation matrix, 199
Spectral estimation, 46
  model-based methods, 51, 164, 242
  window-based methods, 48, 207
Stacking filters, 80
  CDP stack, 81
Standard beamwidth (BW), 207

T

$\tau - p$ methods, 91
Tomographic imaging, 2, 351
  applications, 354
Travelling waves, 3
Two-dimensional, space-time Fourier transform, 185

V

Van Cittert-Zernike theorem, 299
Velocity analysis, 77
  normal moveout (NMO) correction, 78
  semblance measure, 79
Velocity filtering, 63
Vertical seismic profile (VSP), 94
  cumulative attenuation, 96
Very long baseline interferometer (VLBI), 294

W

Wave equation, 15, 358
Wave propagation, 3
Wavenumber, 3, 33
Wavenumber vector, 3
Weight-and-sum methods, 37
Wiener-Khinchin theorem, 297

Y

Yule-Walker equations. *See* Normal equations

Z

Zoeppritz equations, 18